BIOFERTILIZERS & ORGANIC FARMING

THE EDITORS

Himadri Panda, M.Sc. FIC, FICS is a Industrial and Technical Consultant of GIDC. He is a Fellow of Essential Oil Association of India, Indian Pulp & Paper Technical Association, Indian National Science Congress, Oil & Colour Chemist's Association (U.K.). He is also a Member of Chinese Academy of Forestry. He was also Chief Chemist (R&D) of I.T.R. Co. Ltd., Bareilly (U.P.) and Former Chief Chemist (Q.C. & R&D) Tarpina Private Ltd., Ramnagar, Uttaranchal.

Dr. Dharamvir Hota is presently the emeritus professor in the department of agriculture science and economics, an appointment he has enjoyed since 1998. He was a senior researchers at the International Institute of Agricultural Science, where he worked on the use of genetically modified crops for structure determination of molecular modeling.

He has been teaching and doing research on agricultural systems in and around the country. He teaches soil science, organic agriculture and land use, and conducts research on the effects of organic and conventional farming systems on soil, crop productivity and quality, with publications appearing in various internationally reputed journals.

BIOFERTILIZERS & ORGANIC FARMING

Edited by
Himadri Panda
&
Dharamvir Hota

2023
Gene-Tech Books
New Delhi - 110 002

ISBN: 978-81-89729-20-2

Published by : **GENE-TECH BOOKS**
4762-63/23, Ansari Road, Darya Ganj,
NEW DELHI - 110 002
Phone: 011-43003222
e-mail: genetechbooks@yahoo.co.in

Laser Typesetting : **Classic Computer Services**
Delhi - 110 035

Printed at : **Replika Press Pvt Ltd**

PRINTED IN INDIA

Preface

Supporting increasing population levels on a near stabilized agricultural land places a heavy burden on the soil source–particularly its nutrient supplying power. India is one of the several such countries where the stress on the soils is increasing. During the past 50 years, the number of persons supported by each hectare of land has doubled from three to six. With severe droughts, unavailability of water resources and traditional agriculture cause lower production of food grains, India experienced severe food crisis and was compelled to depend on concessional food imports. Our inability to survive the crisis through domestic surplus of food and consequent humiliating dependence on international aid to made the Indian planners, look desperately for strategies which may help in rapid progress on agricultural front. The problem is further compounded in several areas due to the excessive use of chemical fertilizers which resulted into considerable deterioration in the quality of indigenous soil. Indiscriminate use of inorganic fertilizers and pesticides obstructed healthy growth of millions of bacteria which otherwise thrive in soils and thereby completely disturbed the natural mechanism of providing several micronutrients required for healthy growth of plants and protection against pests and diseases. Infact, a chase for high productivity started a vicious circle of high doses of fertilizers and pesticides leading deterioration in the quality of soil thereby making increasingly higher use of fertilizers and pesticides inevitable over the period.

Intensive agriculture with the use of agrochemicals in large amount has, no doubt, resulted in manifold increase in the productivity of farm commodities but the adverse effect of these chemicals are clearly visible on soil structure, soil microflora, quality of water, food, fodder and food materials. Pesticides and nitrates from fertilizers have been detected in groundwater in many agricultural region. High concentration of nitrates in drinking water affects human health particularly infants and can prove fatal in some cases. Traces of banned DDT and HCH isomers have been found in soil, cereals, pulses, vegetable oil, human milk, butter, fat, fish, meat, eggs, vegetables, fruits, animal feed and drinking water. Thus organic farming is certainly an answer for making available safe food and clean environment.

Pollution free environment is essential for sustaining life of all living beings on the earth. Ecofriendly farming or organic farming or natural farming has emerged as the only answer to bring sustainability to agriculture. Ecofriendly farming is a farming of integration of biological, cultural and natural inputs including integrated diseases and pest management practices. It not only advocates for stopping or restricting the use of pesticides but also emphasizes the need of farming which should create an ecological balance and a micro-environment suitable for health and growth for soil microflora, plants and animals.

Nature is always curious, novel and ever changing. It has created such living beings, biofertilizers agents, on this earth. Out of 20,000 cultivated plant species, man could exploit only 100 species for their food requirement. Out of 10,000 cereals only eight, *viz.*, wheat, rice, maize, sorghum, barley, oats, bajra and triticale are being used for food. Among 3000 tropical fruits only mango, papaya, banana and pineapple are cultivated worldwide. Very little attention was paid on microorganism although they are highest in number, more diversified and very simple in their lifestyle, among them man can utilize very few of them for his benefit. Some microbes forecast natural calamities.

Since biofertilizers are introduced very recently there is inadequate awareness among the workers. Biofertilizers could effectively supplement the nutrient requirement of crops through chemical fertilizers, to meet the soaring demand for food, fibre and fuel.

Worldwide, there is an increasing awareness about sustainable agriculture practices in view of energy shortage, food safety and environmental concerns, arising out of chemical farming. Sustainable agriculture is one in which the goal is permanence, achieved through the utilization of renewable resources. This leads to development of concept of organic natural farming.

Integrated plant nutrition can be best if it is practised on scientific facts, local conditions and microeconomics. We hope this publication will create a balanced, objective and science based appreciation for meeting the nutrient needs of agriculture. This book has been written for agricultural planners, soil scientists, biologists, microbiologists, students, teachers, fertilizer industry, personnel, research and development units, organisations engaged in biofertilizer production, training centres, all those interested in the efficient use and recycling of wastes, resource management and sustainable farming.

Himadri Panda

Dharamvir Hota

Contents

Chapter 1
Integrated Plant Nutrition Systems

The greatest challenge facing mankind in the 21st century is to produce the basic necessities of food, feed, fibre, fuel and raw materials from 0.14 ha or less land per caput. While the use of mineral fertilizers is the quickest and surest way of boosting crop production, their cost and other constraints frequently deter farmers from using them in recommended quantities and in balanced proportions. As a consequence of this and other constraints there seems to be no option but to fully exploit potential alternative sources of plant nutrients. Complementary use of available renewable sources of plant nutrients (organic/biological) along with mineral fertilizers is of great importance for the maintenance of soil productivity, *i.e.*, soil structure, soil bioactivity, soil exchange capacity and water holding capacity. Results from various cropping systems and ecologies illustrate that positive interactions result from the integrated use of mineral fertilizers and organic/biological sources of plant nutrients within the framework of Integrated Plant Nutrition System (IPNS).

In response to this need, FAO programmes on IPNS are receiving priority—an area in which FAO has been active for over a decade. This chapter provides a brief account of the concept and components of IPNS, present state of the art on the subject, initiatives and programmes of the FAO and some IPNS models which can be transferred across locations. As several of the results obtained in India have been discussed in other chapters in this volume, results included here are largely from other countries. Experimental results cited are more in the nature of examples and a detailed technical review of all available results is neither intended nor is it within the scope of this chapter.

IPNS Concept and Material

The basic concept underlying IPNS is the maintenance or adjustment of soil fertility and of plant nutrient supply to an optimum level for sustaining the desired crop productivity through optimization of the benefits from all possible sources of plant nutrients in an integrated manner. The appropriate combination of mineral fertilizers, organic manures, crop residues, compost or N-mixing crops varies according to the system of land use and ecological, social and economic conditions. Contrary of the Low External Input (LEI) and Organic Farming approaches, the IPNS involve a low to medium external input approach, taking into account a holistic view of soil fertility and plant nutrition management for a targeted yield based not only on cropping and farming systems but also on distinct geographical areas or villages as a dynamic system. The IPNS approach can be modulated; a factor of targeted yields in any area according to land, water and climate potentials.

The cropping system rather than an individual crop, and the farming system rather than an individual field, are the focus of attention in this approach for developing IPNS practices for major agro-ecological zones and for various categories of farms. The best associations of various types of plant nutrients in different fields are identified for a balanced plant nutrition and high yield, at the same time sustaining soil fertility and controlling nutrient losses. It is envisaged that locally-available materials of plant or animal origin as by-products of agricultural activities be used or, where such materials are not abundantly available, *in situ* production of organics be attempted. In areas and farms where non-cultivated areas are available (forests, pastures and fallows), biomass production is attempted for nutrient transfer to cultivated areas in a sustainable system. In areas and farms where land is scarce and limited to the fields, increasing biomass production in fallows and on the borders of the plots can be attempted. Introduction of green manuring, of living or dead cover in the cropping system can also contribute to nutrient supplies.

Alternative sources of organic material which have considerable potential are: quick-growing leguminous shrubs grown as a part of the cropping system and incorporated into the soil at an appropriate stage as green manure; leguminous trees grown in hedgerows and their lopping used as mulch materials or incorporated into the soil of the cropped alleys between them; forage or food legumes properly inoculated with *Rhiozobium* grown in the cropping sequence; and the use of Azolla of blue-green algae with wetland crops.

The following phenomena must be understood in the context of organic recycling and IPNS:

1. *In situ* recycling of crop residues rather than their non-agricultural use and wastage will bring back a certain amount of nutrients to the same field. However, they are not new sources and are deficient in the same nutrients (like P, S, K, Ca, Mg) as those of the soil.
2. *In situ* growing of leguminous trees and their biomass incorporation will bring atmospheric N into the system, and occasionally, through their deep roots, nutrients leached beyond root zone and sub-soil mineralized nutrients are brought to the surface layer for the use of annual crops.
3. When biomass is brought from outside the plot or farm and/or cattle graze on uncultivated land, it is basically a transfer of nutrients from one place to another. So, except for N where biofixation can introduce nutrients into the soil-crop system, IPNS is a process which limits the losses of the system and not a process which supplies plant nutrients for the whole system.

State of the Art of IPNS

Attempts made in several countries to complement the use of mineral with organic sources of plant nutrients have generated useful, though limited information on the complimentary and synergistic effects of these materials on the yield of crops. Some examples of results are described in the following sections.

Farmyard Manures (FYM)

In many countries such as India and China, cow dung is abundantly available for use as an organic source of plant nutrients. In Indian trials, the rice yield was increased by 450 kg/ha (47 per cent) by the application of 5.6 t/ha of FYM alone and by 600 kg/ha (63 per cent) when an equivalent amount of N (67 kg/ha of N) was applied as urea (Table 1.1). Applied together, the yield increase was 1,120 kg/ha (118 per cent) over no N or no FYM (control plot). In China, under low soil fertility conditions, the FYM (15 t/ha) effect on wheat yield exceeded the effect of 70 kg/ha N. When these two sources of nutrients were applied together, the yield increase over no treatment control was 460 kg/ha (35 per cent), while the yield increase with 140 kg/ha N alone was only 300 kg/ha (23 per cent).

Table 1.1: Effect of Complimentary Application of FYM and Mineral Fertilizers on the Grain Yield of Rice and Wheat

Country	*Crop*	*FYM (t/ha)*	*Fertilizer N (kg/ha)*	*Yield (kg/ha)*	*% Yield Increase Over Control*
India	Rice	0	0	950	–
		5.6	0	1400	47
		0	67	1550	63
		5.6	67	2070	118
China	Wheat	0	0	1330	–
		15	0	1570	18
		0	70	1530	15
		15	70	1790	35
		0	140	1630	23
		15	140	1990	50

In another set of long-term experiments conducted at different locations in India representing varying agro-ecological situations, positive effects of FYM were observed. Thus, integrated use of organic manures with mineral fertilizers improved the soil and crop productivity and increased the yield response.

Green Manures and Crop Residues

Green manuring with legumes has long been known to be beneficial for sustainable crop productivity. With intensification of agriculture and the advent of mineral fertilizers, farmers' enthusiasm for this practice declined. The lack of interest arises from farmers' perception of inadequate economic return from the labour and from the time the green manure crop occupies the land. To overcome this objection, an innovative technique is being studied in some countries, namely to plant a food legume, harvest the pods just prior to full maturity and turn the biomass into the soil when still green.

One such attempt was made at the International Rice Research Institute (IRRI), Philippines, for both upland and lowland rice conditions. The pre-rice grain legume was cowpea (*Vigna unguiculata*), of which the whole biomass was turned in at 75 per cent flowering stage and compared with the treatment where the grain was harvested before turning in the residues. Virtually no difference was found in the two years, either for upland or lowland situations, in the rice yield between the two processes. The gain from either method, averaged over the two years, was 800 kg/ha under lowland and 500 kg/ha under upland conditions, compared with either a pre-rice fallow or a pre-rice cowpea from which all the above ground parts had been removed (Table 1.2).

Table 1.2: Grain Yield of Rice as Affected by Pre-rice Land Use

Sl.No.	Pre-rice Treatment	Lowland (kg/ha)		Upland (kg/ha)	
		1986	1987	1986	1987
1.	Fallow	4,800	4,500	1,200	1,300
2.	Cowpea incorporated into soil at flowering	5,700	5,400	1,600	2,000
3.	Cowpea grown, grain harvested and straw incorporated into soil	5,500	5,200	1,700	2,000
4.	Cowpea grown, both grain and straw removed	4,900	4,500	1,400	1,500

Under lowland conditions, incorporating the whole biomass of cowpea at flowering, and growing rice without any fertilizer produced yields which were equivalent to applying 36 kg/ha N in 1986 and 58 kg/ha N in 1987 to the rice crop grown after a previous fallow or a pre-rice cowpea from which all the above-ground parts had been removed. Under upland conditions, these advantages were equivalent to applying 70 kg/ha N in 1986 and 65 kg/ha N in 1987 after a pre-rice fallow. The root residues of cowpea were equivlent to 30 kg/ha N in 1986 and 15 kg/ha N in 1987 under upland situations. Farmers can thus choose to increase yields or save on fertilizer Nor a combination, depending on the economic and agronomic conditions.

Similarly, results of joint research by the Centre for Soil Research, Indonesia and IRRI during 1986/87 showed that on lowland rice incorporation of *Sesbania rostrata* at 45 days, along with an application of 60 kg/ha N enhanced rice yield by 500 kg/ha over the application of 120 kg/ha N alone (Table 1.3).

Table 1.3: Effect of N Fertilization and *Sesbania rostrata* Green Manuring (SR) on the Yield of Lowland Rice (IR-36) in East Java, Indonesia

Fertilizers kg/ha					Grain Yield kg/ha
N			P_2O_5	K_2O	
0			0	0	2000
60			45	30	3700
0	+	SR at 45 days	45	30	3100
60	+	SR at 45 days	45	30	4600
120	+	SR at 45 days	45	30	4100

While beneficial effects of mulching are widely recognized, procuring mulch material in sufficient quantity is a serious practical problem in many areas. Management of crop residue as a source of mulch is, therefore, closely linked with cropping systems, tillage methods, agro-ecological conditions, farming systems and particularly, cattle population in the farm. A range of cultural practices is available, mainly in humid tropics, to provide adequate amounts of residue mulch for soil protection and fertility enhancement, *e.g.*, cover crops, conservation tillage, alley cropping, use of industrial by products, etc. The sustainability of each practice depends on the local biophysical and socio-economic environment.

Legumes in Cropping Systems and Fertilizer Use

Forage legumes are well recognized as capable of improving soil fertility through symbiotic N fixation to the extent that the non-legumes subsequently grown benefit from the residual fertility.

Medicago or annual *Trifolium* species are estimated to contribute 50–70 kg/ha N for the subsequent crop.

There is evidence that grain legumes also benefit the subsequent crop through enrichment of soil fertility. In an experiment conducted over a three-year period at Milingano (Tanzania), P applied to mungbean (*Vigna radiata*) not only increased the grain yield of the legume by 70 kg/ha (25 per cent) but also of the subsequent maize by 1000 kg/ha (59 per cent). Phosphorus application to maize via the previous mungbean crop was as effective as direct P application to maize, and the additional yield of mungbean made the system economically more attractive.

Biological Sources or Nitrogen

Biological fixation of atmospheric N by heterotrophic organisms like cyanobacteria (blue-green algae) individually or in association with the aquatic fern *Azolla* has been in use in rice production systems in many countries. Extensive trials have been made with blue-green algae and *Azolla* in several countries. In an experiment conducted in India, when mineral fertilizers were used with blue-green algae on rice, the yield was increased by 90–220 kg/ha (3 to 5 per cent over a base yield of about 4000 kg rice/ha) by their combined application.

From work done with *Azolla* in the Philippines it was observed that top-dressing *Azolla* with a basal application of 30 kg/ha N increased rice yield by 800 kg/ha (31 per cent) over mineral N alone or 400 kg/ha (13 per cent) over *Azolla* alone (Table 1.4). The *Azolla* was equivalent to about 30 kg/ha N.

Table 1.4: Grain Yield of Lowland Rice as Affected by Mineral N and Incorporation of *Azolla* into the Soil

Treatment	*Yield (kg/ha)*
T1 30 kg/ha N	2,600
T2 *Azolla* top-dressed	3,000
T3 30 kg/ha N + *Azolla* top-dressed	3,400
% increase T2 over T1	15.0
% increase T3 over T1	31.0
% increase T3 over T2	13.0

The importance of management of *Azolla* in the process of its incorporation into the soil and its interaction with added fertilizer phosphorus has been shown. In an experiment conducted at IRRI, Philippines, it was noted that *Azolla* growth without P was poor. Its inoculation in the paddy field increased rice grain yield only in the plots which were not puddled. Mid-season puddling increased rice growth over shadowing the effectiveness of inoculation. *Azolla* growth was profuse in plots receiving P and the rice grain yield increased substantially with *Azolla* incorporation. An increase in rice yield of 1050 kg/ha (71 per cent) was obtained by a combination of *Azolla* inoculation, its mid-season incorporation (39 days after inoculation) and P addition over control. With only *Azolla* inoculation, but without its incorporation or P addition the yield advantage did not exceed 25 per cent. It thus becomes evident that proper management of *Azolla* and its incorporation in the soil are crucial factors determining its effectiveness (Table 1.5).

Table 1.5: Grain Yield (kg/ha) of Rice as Affected by *Azolla* Inoculation, Method of its Incorporation and Phosphorus Addition

Azolla Treatment	*No P_2O_5*		*32 kg/ha P_2Q_5*	
	No Pudding	*Mid-season Pudding*	*No Pudding*	*Mid-season Pudding*
No inoculation	1,480	2,360	1,860	2,250
Azolla inoculation	1,850	2,160	2,020	2,530
% difference due to *Azolla* inoculation	+ 25	– 8	+ 9	+ 12

From this brief review presented and a wide analysis of documentation, the following conclusions can be drawn on the present state of understanding with regard to IPNS:

1. Organic/biological sources of plant nutrients complement mineral fertilizers in meeting nutrient requirements of crops. The magnitude of the contribution will vary according to sources and agro-ecological conditions.
2. In many situations, synergistic effects due to such combined application could be expected, thus increasing the Fertilizer Use Efficiency (FUE).
3. Residual effects of added organic sources in the cropping system could also be expected together with an improvement in soil physical condition.
4. Under high input production systems, where productivity cannot be further increased with incremental use of mineral fertilizers alone, addition of organic sources could again increase yields through increased soil productivity and higher FUE.

FAO Programmes in the Development of IPNS

FAO has taken a pioneering initiative in developing the IPNS approach within the following conceptual programmes:

Availability or Organic Sources and Potential of BNF

When planning programmes and activities in organic recycling, it is worthwhile, if not indispensable, to consider all the possible uses of a particular material and to examine if recycling it as a manure is really the best choice. Once this is clear, one can start looking into how the actually available organic materials can be recycled and used as manure/mulch/soil conditioner in an optimal way. Only then can sound ideas be developed on how to integrate the available nutrient carriers (organic, biological and mineral) in an environment-friendly way which is also socially and. economically acceptable to the farmer.

To have a first hand appreciation of the potential availability or various sources, status reports for selected countries in the Asia and the Pacific Region and the Africa Region, covering various aspects of soil fertility and crop nutrition, were prepared. These documents show that the prospects of various organic and biological sources differ from country to country. It would, however, be necessary to quantify their actual availability for field application and, also, such assessments need to be extended to other countries. The following arc the main nutrient sources, in order of priority:

Table 1.6: Main Nutrient Sources

Sl.No.	Country	Main Organic and Biological Sources
1.	Bangladesh	Animal wastes; BNF (*Rhizobium*); green manuring
2.	Indonesia	BNF (*Rhizobium*); recycling of legume crop residues and rice straw; animal wastes
3.	Nepal	Hill areas animal wastes; BNF (*Rhizobium*) Tarai areas BNF (*Rhizobium*); green manuring
4.	Pakistan	Animal wastes; BNF (*Rhizobium*); green manuring
5.	Sri Lanka	Recycling of rice straw and legume crop residues; BNF (*Rhizobium*)
6.	Thailand	BNF (*Rhizobium*); crop residues; agro-industrial wastes
7.	Burkina Faso	Animal wastes; crop residues; BNF (*Rhizobium*)
8.	Guinea Bissau	Crop residues, BNF (*Rhizobium* and *Azolla*)
9.	Madagascar	Animal wastes; crop residues, particularly rice straw; BNF (*Rhizobium* and *Azolla*)
10.	Rwanda	Animal wastes (in Butare and Gitarama regions); BNF (*Rhizobium*); crop residues
11.	Sudan	Animal wastes; crop residues; BNF (*Rhizobium*)
12.	Tanzania	BNF (*Rhizobium*); crop residues
13.	Zaire	Crop residues; forest leaves; BNF (*Rhizobium*)
14.	Zambia	Animal wastes (certain areas in southern, western and central provinces); crop residues; BNF (*Rhizobium*)

IPNS Models

After the identification of potential organic/biological resources, some simple models of nutrient scheduling underlying the concepts and application of IPNS for some selected cropping systems have been developed. Their details have been presented elsewhere. Attempts are being made to develop computer models for an IPNS calculation programme of plant nutrition recommendations, based on general agronomic assumptions. Farming systems should be observed and understood before proposing a new approach. Involvement of farmers in the whole innovation process is probably the most efficient way to build up sustainable and socially acceptable models adapted to different local situations.

Field Trial Network

To generate further information on combined applications of organic/biological sources with mineral fertilizers in different agro-ecological situations and dominant cropping systems, some research institutions have been sub-contracted to initiate long-term experiments. To bypass the limitations on the number of treatments and trials that could be carried out by local research institutions in any country, and to benefit from the results of others having similar agro-ecological conditions, this field activity is being expanded into a collaborative network programme within the framework of Technical Cooperation between Devcloping Countries. These trials will complement the existing FAO sponsored Fertilizer Programme in most of the countries. To start with, trials at fixed sites representing major agro-ecological conditions are being undertaken in India, Indonesia, Laos, Nepal, Pakistan, Tanzania and Thailand. In time and depending on financial resources, the network programme is proposed to be extended to other countries.

Results obtained from field experimentation can be validated against predictive models. Field experimentation can bring about a step-by-step improvement by identifying and dealing with each major constraint to plant nutrition and crop production. Moderate improvements in traditional farming may be the initial step in a long-term strategy aimed at transforming low-input subsistence farming into science-based profitable agriculture with high or moderate inputs. Thus, the above network activity is only catalytic and a beginning. Much more remains to be done to determine integrated plant nutrition and adapted practices for sustainable agriculture, according to the goals of target groups of farmers.

Promotion and Technology Transfer

Promotion of Symbiotic BNF Activities

Parallel to the field trials, the BNF technology is being developed through a number of technical cooperation projects. Major emphasis is on assisting national research institutions in the selection and production of the best *Rhizobium* strains, and in setting guidelines for maintaining high standards of inoculant quality. An expert consultation on *Rhizobium* inoculant quality was held in March 1991.

Technology Transfer

To facilitate the transfer of IPNS technology, a data bank on available results is being established.

Promising treatment combinations from the field trial network will be included in the national demonstration programme/block demonstrations within the framework of the FAO Fertilizer Programme field projects, currently operational in 21 countries.

Transfer of this technology for the benefit of small farmers.

To foster applications of technical results, FAO continues to publish publications and training materials and on appropriate methods of production, conservation and efficient use of organic/biological sources along with efficient use of mineral fertilizers.

Future Research and Development Needs

Research Needs

It is apparent that there is a need for:

1. More information on integrated nutrient recommendations for cropping systems as a whole, taking into account balanced plant nutrition through the complementary synergistic effect of combined use of both mineral and organic/biological sources for sustained crop production and enhancing labour productivity.
2. Although general principles may be the same, technological packages for sustainable management of soil and plant nutrient resources are site-specific and depend on farming systems, farm size, availability of inputs, and socio-economic factors. Area specific and on-farm synthesis of packages is needed on the basis of components and sub-systems such as those described above. Agronomic productivity, economic profitability, and ecological compatibility of such packages needs to be assessed through appropriate research on well-defined, representative sites.
3. Recommendations for different agro-ecological situations, taking into account available nutrient resources, farming systems and the choice of improved agro techniques lo alleviate soil constraints for plant nutrition.
4. The application of IPNS on a large scale requires some adaptive research, demonstration programmes backed by training in good management practices, and a policy improvement and infrastructure which stimulates the efficient use of both internal and off-farm inputs.

Development or National Strategy

To provide a framework for national strategies for the conservation and regeneration of soil productivity, the following activities are being initiated. These complement the ongoing activities on IPNS:

1. Selected case studies of national fertilizer and fertility situations and strategies with the following components:
   (*a*) "National accounts" for plant nutrients, assessing supplies and losses and preparing balance sheets and projections by crop and homogenous agro-ecological zone, together with identification of critical areas and total future requirements at country level.
   (*b*) Identification and appraisal of all currently and potentially available sources of plant nutrients–organic and mineral–at country level. Review of technical and economic potentials. Assessment of options for use of these additional/alternative sources of nutrient supplies.
   (*c*) Formulation of alternative strategies to meet future demand for soil fertility improvement and maintenance, including policy options with regard to prices, incentives, subsidies and taxes.
   (*d*) Preparation of national action plans for the sustainable development and maintenance of soil productivity.

   A series of national programmes and projects may then follow, expanding upon experience gained through more than 25 years of activities of the FAO Fertilizer Programme.,
2. Preparation of a Covenant on Good Fertilizer Use Practices.
3. Preparation of a block demonstration programme on sustainable agricultural practices, incorporating elements of sustainable development of soil productivity and crop production such as Integrated Nutrient Management, Integrated Pest Management, Improved Soil and Water Conservation and Integrated Crop Husbandry Management, and its implementation within the framework of the ongoing block demonstration activities of the FAO Fertilizer Programme and other related field programmes.

Conclusions

Soil productivity can be improved and maintained by integrated plant nutrition systems, which address two important problems: the low and decreasing crop yields caused by plant nutrient depletion and inefficient fertilizer use in the developing world, as well as the less extensive but similarly serious problem of pollution by excessive and inappropriate plant nutrient application in developed and locally in developing countries.

The reality is that there are limitations to both low input and high input systems. Solutions will most probably have to be tailored to individual agro-ecological and socio-economic situations and are likely to involve elements of both low-input and high input systems. In most cases, minimizing external inputs through optimal combinations of plant nutrient sources available within the farm will be one of the objectives. Increased efficiency, as well as the reduction of off-farm inputs, will contribute to environmental conservation and decrease or eliminate pollution of soils and groundwater.

Each country will need to use balance sheet approaches to match the needs of its present and future populations with the potentials of its land and water resources under different input levels, and build a long-term plan of action on the results. Depending on the situation in each country, this may imply changes in policy or legislation, prices, incentives or taxes, as well as institutions.

A strategy or action plan for the sustainable development of soil productivity through IPNS would involve the mobilization of advisory services as well as of land users' associations. It may also require technical assistance, aid-in-kind and other supply options, and revolving funds and national production of mineral fertilizers.

Chapter 2

Organic Manures: Their Nature and Characteristics

Organic materials are valuable by-products of farming and allied industries, derived from plant and animal sources. Organic manures which are bulky in nature but supply the plant nutrients in small quantities are termed bulky organic manures, *e.g.*, farmyard manure, rural and town compost, night-soil, green manure, etc. whereas those containing higher percentage of major plant nutrients like nitrogen, phosphorus and potash are concentrated organic manures, *e.g.*, oil-cakes, blood and meat-meals, fish-meal, guano, shoddy and poultry manure, etc.

Farmyard Manure

This is the traditional organic manure and is most readily available to the farmers. In Western countries, it is the product of decomposition of the liquid and solid excreta of the livestock, stored in the farm along with varying amounts of straws or other litter used as bedding. Indian litter is rarely used as bedding because the straw is utilized as fodder. A portion of cattle-dung is used as fuel in rural homes. Cattle-urine is absorbed in the soil spread over the floor of the shed but no extra soil is use-d for effective absorption of this fraction.

On an average, well-rotted farmyard manure (FYM) contains 0.5 per cent N, 0.2 per cent P_2O_5 and 0.5 per cent K_2O. Based on this analysis, an average dressing of 25 tonnes per hectare of farmyard manure supplies 112 kg of N, 56 kg of P_2O_5 and 112 kg of K_2O. These quantities are not fully available to the crops in the year of application. Nitrogen is very slow-acting and less than 30 per cent of it is generally available to the first crop. About 60 to 70 per cent of the phosphate and about 75 per cent of the potash become available to the immediate crop. The rest of the plant nutrients become available to the subsequent crops. The phenomenon of availability of plant nutrients to the subsequent crop is known as residual effect.

Under the tropical climatic conditions of this country, the organic matter is quickly lost and fresh applications are necessary to obtain increased yields and maintain soil fertility.

High doses of farmyard manure can be applied under intensive irrigated cropping conditions, *e.g.*, about 25 tonnes per hectare for sugarcane, vegetables, potatoes, rice, etc. 12.5 tonnes for irrigated or rain-fed crops where the rainfall is medium to heavy (about 125 cm) and from 5 to 7 tonnes in dry areas where the rainfall is low (about 50 cm). In dry-farming areas (rainfall below 50 cm), application of 2.5 tonnes of farmyard manure per hectare gives significant increase in crop yield.

The method of application of farmyard manure generally adopted in our country is defective. Most of the cultivators unload farmyard manure in small piles in the fields and leave it as such for a month or so before it is spread and subsequently ploughed in or disced in the field. Plant nutrients are lost considerably during the exposure of the manure to sun and rains. In summer, it results in rapid drying and considerable loss of nitrogen, whereas in the rainy season the available nitrogen and a good portion of soil humus are washed away. To derive maximum benefit, the farmyard manure immediately on being carted to fields should be spread and mixed into the soil. The manure can also be applied in furrows.

Compost

Compost manures are the decayed refuse like leaves, twigs, roots, stubble, *bhusa,* crop residue and hedge clippings, street, refuse collected in towns and villages, water hyacinth, saw-dust and bagasse. The process of decomposition is hastened by adding nitrogenous material like cowdung, night-soil, urine or fertilizers. A large number of soil micro-organisms feed on these wastes and convert it into well-rotted manure. The final product is known as compost.

Farmyard manure and compost possess the same characteristics. The method of application of compost is the same as that of farmyard manure. The chemical composition of compost prepared from different sources is indicated in Table 2.1.

Table 2.1

Sources of Manure	*Nitrogen (N) %*	*Phosphorus (P_2O_5) %*	*Potassium (K_2O) %*
Farm litter	0.5	0.2	0.5
Water hyacinth	2.0	1.0	2.3
Town compost (street refuse, night-soil, urine, etc.)	1.5	1.0	1.5

Sheep and Goat Manure

The droppings of sheep and goats make a very good manure. Panning is, therefore, a common practice of ensuring the use of sheep and goat-droppings in the fields. Sheep and goat manure contains 3 per cent N, 1 per cent P_2O_5 and 2 per cent K_2O.

Poultry Manure

This is a rich organic manure, since liquid and solid excreta are excreted together resulting in no urine loss. Poultry manure ferments very quickly. If left exposed, it may lose up to 50 per cent of its nitrogen within 30 days.

Poultry manure can be applied to the soil directly as soon as possible. After application, it should be worked into the surface of the soil. If the droppings come from the cages or dropping pits, superphosphate may be added to these at the rate of 1 kg per day. per hundred birds. This improves the fertilizing quality and helps the control of flies and odour.

The average chemical composition of the poultry manure is shown in Table 2.2.

Table 2.2

Type	Moisture %	Nitrogen %	Phosphorus %	Potassium %
Fresh	75	1.47	1.15	0.48
Floor litter	24	3.03	2.63	1.40

Oil-cakes

Oilseeds are generally rich in manurial ingredients. After oil extraction, the oil-cakes are rich in nitrogen and also contain phosphorus and potash.

Cultivators apply both edible and non-edible oil-cakes to the soil as manure. Edible oil-cakes are more profitable as cattle feeds. As such, non-edible cakes should be used as manures.

The percentage of nitrogen ranges from 2.5 in mahua to 7.9 in decorticated safflower cakes. The P_2O_5 contents in oil-cakes vary from 0.8 to 3 per cent and K_2O from 1.2 to 2.2 per cent.

Oil-cakes though insoluble in water are quick-acting organic manures, their nitrogen becoming quickly available to the plants in about a week for ten days after application. Mahua oil-cake, however, takes about two months to nitrify. The solvent-extracted oil-cakes are somewhat more quick-acting than the *ghani*-hydraulic or expeller-pressed oil-cakes. The quantity of organic matter that gets added in normal application of oil-cakes is too small to cause improvement in physical properties of soil.

Oil-cakes need to be well-powdered before application so that they can be spread evenly and are easily decomposed by microorganisms. They can be applied a few days before sowing or as top-dressing. Mahua-cakes should, however, be applied quite in advance of sowing time. Oil-cakes are more effective in moist soil and in wet weather than in dry soil and in dry weather. In fresh condition, oil-cakes should not be put in contact with germinating seeds or young plants as they become permeated with fungi and molds in the soil.

The use of oil-cakes on foodgrain crops like wheat and rice is not recommended now on economic grounds. Cakes, specially ground-nut and coconut, are extensively applied for top-dressing of sugar-cane crop. Farmers growing betel leaves also use oil-cakes.

Meal Group of Manures

These are all quick-acting manures suitable for all types of soil and for all crops. In this group come blood-meal (generally used in grape cultivation), meat-meal, fish-meal, horn- and hoof-meal and bone-meal. Meat-meal and blood-meal are applied like oil-cakes whereas fish-meal should preferably be powdered. Horns and hooves of slaughtered or dead animals are converted into horn and hoof-meal by cooking in the bone-digester, and then drying and powdering them. Their average chemical composition is given in Table 2.3.

Bone-meal

Sterilized bone-meal is an important mineral supplement in livestock feed; yet it is used chiefly as phosphatic fertilizer. Small quantities of nitrogen are also applied to the soil through bone-meal. The availability of phosphorus from bone-meal depends on the particle size; the finer the particles, the greater the phosphorus availability. It is available to farmers in two forms: (*a*) raw bone-meal; and (*b*) steamed bone-meal.

Table 2.3

Type	*Nitrogen* %	*Phosphorus* %	*Potassium* %
Blood-meal	10.12	1.2	1
Meat-meal	10.5	2.5	0.5
Fish-meal	4–10	3.9	0.3–1.5
Horn and hoof-meal	13	–	–
Raw bone-meal	3–4	20–25	–
Steamed bone-meal	1–2	25–30	–

Raw bone-meal consists of crushed bones. The percentage of phosphoric acid and nitrogen varies with the quality of bones and the age of the animals from which these are obtained. Normally, the bones of grown-up animals contain more of phosphoric acid and less nitrogen than those of the young ones. According to the standards laid down by Indian Standards Institution, the raw bone-meal must pass wholly through 2.3 mm I.S. sieve of which not more than 30 per cent shall be retained on 850 micron I.S. sieve.

Steamed bone-meal is obtained by treating the bones with steam under pressure and is generally preferred to raw bone-meal. Steaming increases the percentage of phosphoric acid and reduces the nitrogen content of the bone-meal. It also removes the fat from the bones which make bone-meal very porous and easy to grind. Steamed bone-meal also decomposes more rapidly in the soil than raw bone-meal According to I.S.I. standards, not less than 90 per cent of the material should pass through 1.18 mm I.S. sieve. Other specifications laid down by I.S.I. for the chemical composition of raw and steamed bone-meal are described in Table 2.4.

Table 2.4

Sl.No.		*Raw Bone-meal*	*Steamed Bone-meal*
1.	Moisture percentage by weight (maximum)	8	7
2.	Total phosphates percentage (as P_2O_5) by weight (minimum)	20	22
3.	Available phosphates (as P_2O_5) soluble in 2 per cent citric acid solution, percentage by weight (minimum)	8	16
4.	Nitrogen percentage by weight (minimum)	3	–

Bone-meal is considered useful for all soils. Best results are, however, obtained on acidic soils and soils having good drainage. It is less effective on heavy clay and calcareous soils. It has particularly notable effect on soils which are well-supplied with organic matter. Paddy, wheat and other cereals respond very well to bone-meal particularly in acidic soils. Sugarcane, vegetables, fruits, leguminous crops, pastures and grasses are all benefited by this manure.

Bone-meal is applied to the soil at sowing time or just before it. Its use as top-dressing is not recommended. It is preferably drilled in the soil. A dose of 112–224 kg per hectare is sufficient for most cereal crops. For vegetables and fruits, about 500 to 600 kg of bone-meal per hectare is applied.

Sewage, Sludge and Sullage

In big cities provided with underground sewage and flush system a good deal of night-soil and urine dissolved in water used for flushing is available. This is known as sewage. In smaller towns, the bathroom and kitchen washings, urine, etc. are carried in open gutters. This is known as sullage. Both these are rich in nitrogen and other plant nutrients.

Sewage has two components: (*a*) Solid portion or sludge; and (*b*) liquid portion or sewage water. The sewage should first be settled to remove the major portion of the suspended solids and effluent, which is 99 per cent constituent of sewage. The sludge should be separately collected and dried and used as manure on lands not treated with sewage. The sludges may be classified as (*a*) settled sludge, produced by plain sedimentation, (*b*) digested sludge, resulting from anaerobic decomposition of sedimented sludge, (*c*) activated sludge, produced by a special rapid aerobic treatment of sewage that results in coagulation and settling of suspended material, (*d*) digested activated sludge, and (*e*) chemically precipitated sludge.

Characteristics

The characteristics of sewage show considerable variations, obtained from one source to another. The quality of sewage also changes as a result of discharge of industrial wastes into the sewer system. The characteristics of sewage waste waters obtained from different cities, are presented in Table 2.5. On an average, the sewage of Indian cities contains 50 ppm of nitrogen, 15 ppm of phosphorus and 30 ppm of potassium. Sludge on an average contains 1.5 to 3.5 per cent N, 0.75 to 4 per cent P_2O_5 and 0.3 to 0.6 per cent K_2O.

Table 2.5: Chemical Characteristics of Sewage from Indian Cities

City	*pH*	*Total Solids (ppm)*	*Dissolved Solids (ppm)*	*Suspended Solids*	*Total N (ppm)*	*P_2O_5 (ppm)*	*K_2O (ppm)*	*Cl' (ppm)*	*BOD 9 (5 days) 20°C (mg/l)*
Aligarh	9.0	1470	55	515	55	16	2	205	260
Kanpur	7.0	1500	900	600	50	15	40	85	250
Hissar	7.8	1985	1910	75	58	14	72	500	370
Nagpur	7.2	1200	1000	200	60	20	45	30	350
Mysore	6.9	890	570	315	29	13	41	3	–
Chennai	7.3	1700	1200	500	60	22	55	254	350
Tuticorin	6.8	1480	1240	240	56	10	–	392	420

Water being a scarce commodity there is a need to treat the waste waters for reusing it and preventing the cause of pollution. There are several methods available for the treatment of the domestic waste waters.

Activated sludge process consists of aerating the sewage either by diffusion or by mechanical means. The activated sludge (biological floc) harbours microorganisms, which decompose organic substances present in the sewage aerobically. Sludge is separated from the effluents in a settling tank after a suitable interval. An overall reduction of 85–95 per cent of suspended solids and BOD and 90 to 98 per cent reduction in bacteria could be obtained through the use of this method combined with primary treatment. The effluent thus obtained usually has suspended solids of 20 ppm, which is well within safe limits.

Trickling filteration process involves distribution of sewage on a bed of inert material, like bricks, coke and sand. The organisms growing on the surface of these solids as slimy films act as purifying agents. This process is more effective in treatment of industrial effluents.

After the removal of the solid portion or sludge, sewage water is known as treated effluent and is used for purposes of irrigation and fertilization. Since available sewage and sullage water are in quite a concentrated form, its dilution with the raw water in the proportion 1 : 1 has been found to be beneficial. The raw water may either be from well, tube-well or river. Sewage and sullage waters have been estimated to give 30 to 50 per cent increase in yield of food crops and 100 per cent increase in fodder yield as compared with ordinary irrigation water. But as sewage may contain harmful bacteria, the growing of food crops that are eaten rather raw or low-lying vegetables such as tomatoes, radish, onion, garlic, carrot, etc. is not recommended on sewage farms. Fodder crops like oats, berseem and lucerne, cereals like rice, wheat, *bajra, jowar* etc., fruit trees like papaya, banana, etc. and sugar-cane, used for *gur* or sugar-making but not for chewing, and vegetables eaten in a cooked form could be grown with considerable profit. The vegetable crops generally grown with sewage irrigation are cabbage, cauliflower, turnip, potatoes, brinjal, lady's fingers, beans and leafy vegetables.

Point of Caution

The municipal sewage may also contain high concentrations of heavy metals which may prove toxic to plants. The elements of general concern are B, Cd, Co, Cr, Cu, Hg, Ni, Pb, Se and Zn. The presence of large quantity of heavy metals in sullage such as Zn, Cu, Ni, Cd, Pb and further pollution with industrial organic wastes makes them unsuitable for land application. Cadmium is one of the elements that is taken up by most crops and accumulates in plants and thus endangers animal and human health. The excessive amounts of these metals are added into sewage through discharge of industrial wastes without pre-treatment. These elements can be toxic to plants grown on sewage amended soils or to the animals which consume these plants. In view of this, it is important to monitor the build-up of heavy and toxic substances which commonly occur in sewage/sullage and pose serious hazard to plants.

Chapter 3
Livestock and Human Wastes: Characteristics and Value

Knowledge of livestock waste characteristics is fundamental to the development of feasible waste management and their efficient utilization. Basic information on the frequency of animal manure excretion, quantity of manure and their characteristics permits specific recovery of waste components, by-product development, fertilizer value and reuse of manure as animal feed and as soil conditioner. The term livestock waste means (*i*) fresh excrement including both solid and liquid portions, (*ii*) total excrement, including the bedding material, litter to absorb the liquid component, (*iii*) the material after liquid run-off, evaporation of water and other volatile components and leaching of soluble nutrients, and (*iv*) material obtained following aerobic or anaerobic storage of livestock manure.

The feasible approach to characterize livestock waste to obtain random samples of the solid and liquid waste and analyse for their ingredients. To determine the relationship between animal feed intake and waste characteristics, nutritional trials are conducted and the quantity and quality of the wastes are estimated. The results obtained from nutritional trials are accurate but this is a time-consuming and costly process. The quantity and quality of the wastes are characterized after finding out the digestibility coefficients for the feed components, *viz.*, organic matter, crude fibre, nitrogen free extract, ether extract, crude protein, and after working out the mineral balance in the animals, *viz.*, nitrogen, phosphorus, potassium, calcium and magnesium. Many nutritional trials have been conducted with different livestock like cattle, buffalo, goat, poultry, pig, etc. in national laboratories, veterinary and animal science colleges of agricultural universities in India which can form the basis for assessing the livestock waste characteristics.

The characteristics of livestock wastes are functions of the digestibility, composition of the feed ration and the species of animals and their physiology. The wastes from ruminants such as cattle, buffalo, goat and sheep have a different composition than the wastes obtained from pigs and poultry which are highly digestible. The faeces of livestock consist chiefly of undigested food which has escaped bacterial and digestive enzyme action. Faeces also contain residue from digestive fluids, waste mineral matter, worn-out cells from the intestinal linings, mucus, bacteria and foreign matter such as dirt consumed alongwith food. Undigested protein is excreted in the faeces and the excess nitrogen from the digested protein is excreted in the urine as uric acid or urea. Potassium is absorbed during digestion but eventually most of it is excreted through urine. Calcium, magnesium, iron and phosphorus are excreted mostly in the faeces.

A number of methods are used to describe the characteristics of livestock wastes. They can be described on the basis of pollutional nature in terms of B.O.D. (Biological Oxygen Demand), C.O.D. (Chemical Oxygen Demand), solids per cent, volatile matter content, nutrient and fertilizer value. The data normally available pertains to the quantity of solid or liquid or combined manure in terms of kg or litre per animal per day. This method of quantification is most realistic in estimating the gross wastes generated at a particular livestock production unit. Besides, other parameters like available nutrient content in terms of fertilizer value, *viz.*, nitrogen, phosphoric acid and potassium oxide can also be determined depending upon the utility of the waste. Livestock wastes are generated as a semi-solid and has to be handled and utilized in this condition. However, liquid or slurry waste system has been considered where the waste can be handled as liquid and transported by pumps and spreaders.

The available information on the quantitative and qualitative nature of livestock excreta should be used to assess and to develop order of magnitude information concerning the potential livestock waste availability. Although it is difficult to apply average livestock waste production values to a specific location, knowledge of average values is very useful for assessing the potential of these wastes and their effective utilization. Hence, the quantity of manure excretion, their characteristics for different livestock from the available literature of metabolism and nutritional trials will be considered.

Bovine Manure

For accurate quantitative assessment of cattle and buffalo dung and urine excretion and their characteristics, the results of nutritional trials conducted at National Dairy Research Institute, Karnal and Indian Veterinary Research Institute, Izatnagar on Sahiwal and Tharparkar cattle and Murrah buffaloes were taken into consideration. The dry matter intake, digestibility, quantity of dung and urine excreted and the nitrogen, phosphorus and potassium balance were obtained to assess the quantitative and qualitative nature of the cattle manure.

The quantity of wet dung and urine excreted by Sahiwal and Tharparkar cattle and Murrah buffaloes is presented in Tables 3.1 to 3.3. These Tables provide the information on dry matter intake, dry matter excreted, wet dung and urine produced per day depending on the body weight in respect of different groups of cattle and buffaloes, *viz.*, male calf, heifer, dry and lactating cows, and buffaloes. The range of the above parameters depending on body weight, etc. can be seen in Tables 3.1–3.3.

The lactating cattle and buffaloes, in general, had the highest dry matter intake as also more excretion of dung and urine. This was followed by dry animals, heifers and malo calf. The excretion coefficient of the feed will depend on the nutritive value and digestibility of the dry matter. Even though, the digestibility of the feed consumed by the animal depends on the nature of the feed, it was observed that the Tharparkar breed had the higher excretion coefficient. In the case of Murrah buffalo, the male calves and the non-lactating buffaloes have higher excretion coefficient. The moisture content of the dry dung is a function of type of feed, environmental temperature and humidity. So, the moisture content of the feed, water intake and the season influence the quantity of wet dung and urine excreted.

The composition of the organic fractions in the dung as well as the C : N ratio are presented in Table 3.4. The dung consisted of about 75 to 85 per cent moisture, 15 to 25 per cent organic matter and 2 to 5 per cent mineral matter. The organic matter of dung mainly comprised of 78 to 90 per cent of total carbohydrates (crude fibre + nitrogen free extract), 9 to 18 per cent of crude protein and 2 to 5 per cent

Table 3.1: Quantity of Wet Dung and Urine Excreted per day by Sahiwal Cattle

Group of Cattle	*Body Weight (kg)*	*Dry Matter Intake (kg)*	*Dry Matter Excreted (kg)*	*Wet Dung (kg)*	*Urine (l)*
Male calf	140	3.226	1.302	6.960	3.725
	170	3.906	1.632	8.342	5.675
	190	3.928	1.491	7.144	6.595
	210	4.399	1.615	8.372	5.803
	240	4.429	1.824	9.583	5.426
	250	5.129	1.886	9.430	6.25
	260	5.488	1.808	9.040	8.757
	275	5.550	2.037	10.190	10.914
Heifer	170	4.235	1.560	6.425	5.620
	200	5.505	1.866	10.610	9.950
	225	6.125	1.958	10.435	9.565
	250	6.250	2.180	10.525	11.590
Dry cow	325	5.120	1.920	7.760	10.350
	350	5.325	1.995	8.335	12.960
	375	5.576	2.191	8.760	11.155
	420	6.025	2.250	9.090	12.550
Lactating cow	350	9.560	3.850	22.140	14.120
	375	10.401	4.537	26.650	17.580
	400	11.525	4.755	31.860	21.875

Table 3.2: Quantity of Wet Dung and Urine Excreted per day by Tharparker Cattle

Group of Cattle	*Body Weight (kg)*	*Dry Matter Intake (kg)*	*Dry Matter Excreted (kg)*	*Wet Dung (kg)*	*Urine (l)*
Male calf	120	3.896	1.780	8.825	7.560
	170	3.418	1.520	7.608	6.920
	200	4.227	1.680	8.420	7.455
	225	4.264	1.863	9.325	7.925
Heifer	170	4.125	1.925	9.650	9.820
	190	4.879	2.050	9.325	8.525
	220	5.129	2.105	10.320	11.910
	240	5.425	2.125	10.825	14.105
Dry cow	380	4.250	1.950	6.525	8.740
	400	4.511	2.066	7.170	10.230
	420	4.820	2.250	7.435	12.265
Lactating cow	270	9.875	4.525	22.495	13.805
	385	8.957	4.425	21.530	11.640
	400	10.546	4.561	22.250	13.145
	420	11.250	4.625	25.730	14.570

ether extract. In the case of Sahiwal cattle, the excretion coefficient of crude fibre varied from 20.8 to 44.8 per cent, whereas that of nitrogen free extract varied from 30.4 to 48.2 per cent. In the case of Murrah buffaloes, the excretion coefficient of crude fibre varied from 20.2 to 37.3 per cent, whereas that of nitrogen free extract varied from 35.3 to 56.8 per cent. In the case of lactating Sahiwal cattle, the excretion coefficient of crude protein was only 17.3 per cent, whereas in Murrah buffalo it was 30.8 per cent. The C : N ratio of the dung for cattle and buffalo varied from 19.57 to 49.83 depending upon the feed material. With berseem feeding the C : N ratio of dung excreted by Sahiwal lactating cattle was 28.85, whereas with lucerne hay feeding the buffalo heifer excreted dung with a C : N ratio of 19.57. With wheat *bhusa* feeding, the buffalo male calves excreted dung with a C : N ratio of 40.77, whereas with wheat straw and *jowar* feeding, the dry buffalo excreted dung with a C : N ratio of 49.83. Hence, a higher C : N ratio was observed with feeding of *bhusa* or straw rather than with green fodder.

Table 3.3: Quantity of Wet Dung and Urine Excreted per day by Murrah Buffalo

Group of Cattle	*Body Weight (kg)*	*Dry Matter Intake (kg)*	*Dry Matter Excreted (kg)*	*Wet Dung (kg)*	*Urine (l)*
Male calf	100	2.856	1.420	5.465	3.690
	125	3.250	1.580	6.794	5.205
	150	3.950	2.050	9.510	5.710
	200	4.476	2.105	10.255	8.910
	250	4.911	2.280	11.560	9.515
Heifer	100	4.665	1.450	6.880	4.515
	150	5.445	2.022	7.425	6.183
	250	5.525	2.125	9.125	7.500
	300	5.975	2.314	11.570	9.025
Dry buffalo	350	4.805	2.050	10.380	6.810
	400	6.025	2.994	16.025	7.120
	500	6.525	3.225	17.120	7.605
	580	7.387	3.406	21.216	10.260
Lactating buffalo	400	7.282	3.420	17.705	10.250
	450	8.932	3.572	20.250	11.527
	500	9.520	3.625,	24.705	13.510
	550	9.250	3.420	20.305	13.120
	575	13.560	4.977	27.985	14.825
	600	14.905	5.349	28.745	14.250
	625	15.350	5.477	30.250	15.650

The nitrogen and phosphorus balance of Sahiwal and Tharparkar cattle and Murrah buffaloes is presented in Table 3.5. The excretion of nitrogen through dung varied from 38 to 53 per cent of the total nitrogen excretion. In the lactating cattle, where the nitrogen-outgo through milk was of the order or 18.3 per cent, the N-outgo through dung was 19.7 per cent. In the case of lactating Murrah buffalo with 20.3 per cent N-outgo through milk, the N-outgo through dung was 37.4 per cent and 42.3 per cent through urine. The phosphorus excretion through dung varied from 89 to 98 per cent of the total P-excreted. In the case of lactating animals with 23 to 36 per cent P-outgo through milk, the dung P-outgo ranged from 63 to 79 per cent only (Table 3.5). The excretion of nitrogen and phosphorus not only depended on their intake by animals but also on the age-group of animal, on the season and the metabolic body size of the animal.

Table 3.4: Composition of Dung Excreted by Different Age Groups of Sahiwal Cattle and Murrah Buffaloes

Sl.No.	Group of Animal	Body wt. (kg)	Feed Used	Dry Matter Excreted (kg)	Organic Matter Excreted (kg)	Per cent Composition: Ether Extract	Crude Fibre	Nitrogen Free Extract	Crude Protein	C : N Ratio
(A)	**Cattle**									
1.	Male calf	200	Oats silage + wheat straw	1.530	1.334	4.35	23.01	58.54	14.10	25.67
2.	Heifer	220	Hybrid napier	2.023	1.316	3.30	15.00	64.98	16.72	21.70
3.	Milch cow	370	Berseem	4.479	3.200	4.53	30.18	52.73	12.56	28.85
4.	Dry cow	350	Berseem + oat straw	2.128	1.702	3.32	29.52	53.50	13.66	26.54
(B)	**Buffalo**									
1.	Male calf	240	Wheal *bhusa*	2.128	1.822	2.35	19.81	68.95	8.89	40.77
2.	Heifer	260	Lucerne hay	2.134	1.895	3.01	24.17	54.30	18.52	19.57
3.	Milch buffalo	580	Maize + lucerne hay	4.602	4.309	3.25	9.70	73.36	13.69	26.48
4.	Dry buffalo	434	Wheat straw + *jowar*	3.018	2.577	1.59	22.89	68.33	7.18	49.83

Table 3.5: Nitrogen and Phosphorus Excretion in Different Age Groups of Sahiwal Milch Breed of Cattle and Murrah Buffaloes

Sl.No.	Group of Animal	Nitrogen (g)					Phosphorus (g)				
		Intake	Total Excretion	In Dung	In Urine	In Milk	Intake	Total Excretion	In Dung	In Urine	In Milk
(A)	**Cattle**										
1.	Male calf	88.7	76.8	29.9 (38.9%)	46.9 (61.1%)	–	12.8	8.2	7.7 (93.9%)	0.5 (6.1%)	–
2.	Heifer	137.0	131.0	50.0 (38.1%)	81.0 (61.9%)	–	5.5	2.57	2.5 (97.3%)	0.07 (2.7%)	
3.	Milch cow	376.5	372.0	73.8 (19.7%)	231.2 (62.0%)	67.9 (18.3%)	45.5	38.33	24.09 (63.22%)	0.37 (0.9%)	13.77 (35.9"/.)
4.	Dry cow	96.0	95.0	39.0 (41.0%)	56.0 (59.0%)	–	91.60	841	8.11 (96.8%)	0.3 (3.2%)	–
(B)	**Buffalo**										
1.	Male calf	84.4	64.0	26.0 (40.6%)	38.0 (59.4%)	–	5.85	5.03	4.9 (97.4%)	0.13 (2.6%)	–
2.	Heifer	146.2	134.0	56.0 (41.8%)	78.0 (58.2%)	–	6.94	2.57	2.5 (97.3%)	0.07 (2.7%)	–
3.	Milch buffalo	305.0	264.6	99.0 (37.4%)	112.0 (42.3%)	53.6 (20.3%)	53.6	44.10	33.5 (75.9%)	0.3 (0.8%)	10.3 (22.3%)
4.	Dry buffalo	72.2	56.4	29.8 (52.8%)	26.6 (47.2%)	–	13.9	6.2	5.5 (88.7%)	0.7 (11.3%)	–

Figures within parenthesis indicate the per cent excretion of the mineral through dung, urine and milk.

The potassium oxide excretion is normally not reported in metabolism trials but the available data pertains to male calves and bullocks only (Table 3.6). The outgo of K_2O through dung varies from 8 to 17 per cent only of the total K_2O excretion, whereas the urine contains most of the potash excreted.

Table 3.6: Excretion of Potash in Cattle

Sl.No.	Group of Animal	Feed Used	Body Weight (kg)	K_2O in g/day			
				Intake	Total Outgo	Dung	Urine
1.	Male calf	Guinea grass	100	29.35	25.13	2.76	22.37
			150	37.73	31.89	3.57	28.32
			200	46.59	45.27	7.88	37.39
			250	47.83	41.82	9.81 (15.8%)	32.01 (84.2%)
2.	Bullock	Ragi straw	–	83.05	74.00	6.23 (8.4%)	67.77 (91.6%)
		Bolaram hay	–	25.25	22.00	3.70 (16.8%)	18.30 (83.2%)

Figures within parenthesis indicate the percentage excretion of K_2O through dung and urine.

Goat and Sheep Excreta

The goat with a body weight of 20 and 40 kg excreta 0.320 to 0.625 kg dung and 0.374 to 0.498 litres urine, whereas the sheep with a body weight of 25–40 and 50–60 kg excreted 0.370 to 1.430 kg dung and 0.350 to 0.950 litres urine per head per day. The chemical composition of their excreta showed that dung had a dry matter content of 42 to 48 per cent which constituted 46–51 per cent of the dry matter intake. The organic fraction of the dung comprised 5.2 to 9.3 per cent crude protein, 1.4 to 1.9 per cent ether extract, 27.8 to 36.4 per cent crude fibre, 40 to 47 per cent nitrogen free extract and 0.35 to 0.77 per cent ash.

The, excretion of nitrogen and phosphorus in goat dung and urine is presented in Table 3.7 and that of nitrogen, phosphorus and potash of sheep excreta is presented in Table 3.8. It was observed that the dung comprised 30 to 50 per cent total nitrogen excretion, all phosphoric and 90.95 per cent of potassium. In castrated ram, the urination was observed less than once per hour and yielded 150 ml per urination. The total output or urine ranged, from 1700 to 2000 ml/day/head. The urine comprised mainly 68.85 per cent of urea-N and 11.16 per cent ammonia-N. The average composition of N, P_2O_5 and K_2O in goat and sheep dung comprised 0.65 per cent, 0.5 per cent and 0.03 per cent and that in urine 1.70 per cent, 0.02 per cent and 0.25 per cent, respectively.

Table 3.7: The Excretion of Nitrogen and Phosphorus in Goat Dung and Urine/Head/Day

Expt.No.	Body wt. (kg)	Nitrogen (g)			Phosphorus (g)		
		Intake	Dung	Urine	Intake	Dung	Urine
1.	29–34	6.00–9.40	2.31–3.88	3.01–6.00	1.06–1.81	1.01–1.90	0.01
2.	35–40	8.50–16.80	2.70–5.90	4.80–6.20	1.20–3.50	1.18–2.46	0.02–0.04

Table 3.8: The Excretion of Nitrogen and Phosphorus and Potash in Sheep Dung and Urine (g)/Head/Day

Expt. No.	Body wt. (kg)	D.M. Intake (g)	N			P			K		
			Intake	Dung	Urine	Intake	Dung	Urine	Intake	Dung	Urine
1.	38–45	500	21.8–23.6	3.8–4.8	12.8–17.1	2.20–2.83	1.97–2.00	0.03–0.12	15.7–24.8	0.1–0.9	14.4–20.4
2.	36–50	800	34.9–37.8	5.8–6.7	20.6–27.2	3.52–4.48	2.73–3.30	0.03–0.05	25.2–39.6	0.8–2.1	21.8–28.4
3.	36–40	–	4.4–16.25	2.88–6.41	2.63–7.73	0.51–11.61	0.96–10.15	0.02–0.13	–	–	–
4.	Adult	–	41.3–47.8	7.1–12.0	20.6–24.6	2.55–2.75	1.5–2.3	0.03–0.07	–	–	–

Pig and Hog Excreta

The quantity of dung and urine excreted by hogs of 10 to 50 kg body weight ranged from 0.5 to 2.5 kg and 1.3 to 4.0 litres/head/day respectively, whereas the excretion of dung and urine by hogs of 51 to 90 kg body weight ranged from 1.6 to 4.5 kg and 2.9 to 4.0 litres/head/day respectively. The quantity of excretion of dung and urine by pigs of a body weight of 33 to 45 kg ranged from 0.57 to 1.57 kg and from 2.1 to 3.6 litres /head/day respectively.

The moisture content of hog and pig manure ranged from 69.5 to 76.7 per cent. The average excretion of hog manure was about 6.8 per cent of the body weight/hog/day. The dung excreted by pigs and hogs contained 0.60 per cent N, 0.50 per cent P_2O_5 and 0.20 per cent K_2O, whereas urine comprised 0.40 per cent N, 0.10 per cent P_2O_5 and 0.50 per cent K_2O. One tonne of hog manure or 265 gallons contained 14 lb N, 8 lb P_2O_5 and 8 lb K_2O.

Poultry Excreta

This is a rich organic manure, since liquid and solid excreta are excreted together. The excreta from hen 1 to 6 months old ranged from 229 to 888 g/head/day depending on the age, etc. The poultry excreted about 5 per cent of its body weight/day. The chicken up to 13-week age on an average excreted about 274 g/head/day. The chemical composition of the fresh poultry excreta of adult layers is given by Jai Kishan and Hussain. The dry matter content of the fresh poultry excreta ranged from 20–30 per cent and it comprised 20.2 per cent ash. The organic fraction of the excreta contained 21.5 per cent crude protein, 1.9 per cent ether extract, 13.4 per cent crude fibre and 42.9 per cent nitrogen free extract.

The manurial constituents of the poultry excreta are given by Garner. The fresh poultry excreta comprised 80 per cent moisture, 0.76 per cent N, 0.63 per cent P_2O_5 and 0.22 per cent K_2O, whereas the partly dried and kiln dried excreta had only 30 and 10 per cent moisture respectively and 4 to 5 times of the above nutrients. The deep litter with sawdust and straw comprised 1.8 per cent N, 2.3 percent P_2O_5 and 1.4 per cent K_2O.

Nitrogen excretion by adult layers ranged between 0.65 and 0.75 9 against nitrogen intake of 2.65 and 2.90 g/head/day whereas the phosphorus excretion showed a narrow range of 0.36 to 0.39 g against an intake of 0.62 to 0.70 g/head/day.

Human Excreta

The daily per capita production of suspended solids in human excreta ranged from 0.135 to 0.270 kg/head/day. The composition of night-soil collected from Central Prison, Nagpur analysed by NEERI, 1976 is as follows:

Moisture 86.7 per cent, total solids 13.3 per cent, volatile solids 11.6 per cent, nitrogen 4.0 per cent, phosphorus, 1.53 per cent and potassium 1.08 per cent. The urine excretion of human beings ranged from 0.5 to 3.0 litres/head/day and the average excretion of wet faeces was 0.234 kg/head/day. An estimation of the manurial value of faeces and urine of human being showed that 2.10 g N, 1.64 g P_2O_5 and 0.73 g K_2O/person/day was excreted and the corresponding values for urine were 12.10, 1.80 and 2.22 g/person/day.

Chapter 4
Potential of Organic Materials and Plant Nutrients

The importance of agricultural wastes in general and agro-industrial products in particular has been recognised during the recent years and literature on Indian local organic resources and their possible utilisation has been compiled.

Organic Resources and Potential

India has vast potential of manurial resources and major resources are listed below:

Livestock and Human Wastes

1. Cattle-shed wastes such as cattle and buffalo dung, and urine.
2. Other livestock and human excreta.
3. Byproducts of slaughter-houses and animal carcases: Blood and meat wastes, bones, horns and hooves, leather and hair wastes.

Crop Residues, Tree Wastes and Aquatic Weeds

1. Crop wastes of cereals, pulses and oilseeds (wheat, paddy, *bajra, jowar*, gram, *moong, urad*, cowpea, *arhar, masoor*, ground-nut, linseed, etc.).
2. Stalks of corn, cotton, tobacco, sugar-cane trash, leaves of cotton, jute, tapioca, arecanut, tree leaves, water hyacinth, forest litter, etc.

Green Manure

Sunnhemp (*Crotalaria juncea*), *dhaincha* (*Sesbania aculeata*), cluster beans (*Cyumopsis tetragonoloha*), *senji* (*Melilotus parviflora*), cowpea, (*Vigna catjang*), horse-gram (*Dilichos biflorus*), pillipeasara (*Phaseolus trilobus*), berseem (*Trifolium alexandrinum*), etc.

Urban and Rural Wastes

1. Rural and urban-solid wastes.
2. Urban liquid wastes-sewage and sullage.

Agro-industries Byproducts

1. Oil-cakes.
2. Paddy husk and bran
3. Bagasse and pressmud
4. Sawdust
5. Fruit and vegetable wastes
6. Cotton, wool and silk wastes, and
7. Tea and tobacco wastes.

Marine Wastes

Fish meal and seaweeds.

Tank silts.

Livestock Wastes

Cattle and Buffalo Dung

The estimate of annual production or bovine dung in India on the basis of Livestock Census, 1966 was estimated to be 344.5 million tonnes and 1335 million tonnes. Garg reported that the bovine urine production was approximately 370 million tonnes per annum. But no systematic survey was conducted to estimate the dung and urine excretion in India. Hence, an attempt has been made to formulate a relationship between the dung excretion and feed intake as well as urine excretion on the basis of feed availability for cattle and buffaloes. The quality of wet dung and urine excreted by 178.865 million cattle and 57.941 million buffaloes was assessed on the basis of the feed availability data reported from the pilot surveys conducted by IARS, Delhi.

The digestibility coefficients of different feeding materials in metabolism trials were worked out for the green and dry fodder and concentrate, taking into consideration their dry matter content. The excretion coefficient was calculated on the basis of 100 per cent digestibility coefficient and the moisture content of the dung has been taken as 80 per cent. The dung excretion formula was calculated by taking the digestibility coefficient of 65 per cent for green fodder, 55 per cent for dry fodder and 70 per cent for concentrate feeding from the available literature on nutritional trials conducted at NDRI, Karnal and IVRI, Izatnagar and other trials.

The relationship of feed intake to wet dung excretion is given below:

$$\text{Wet dung (kg)} = 0.525 \text{ green fodder} + 2.25 \text{ dry fodder} + 1.5 \text{ concentrate (kg)} \quad \text{.................I}$$

Further the quantity of urine excreted was studied and the relationship between urine and dung excretion and dry matter intake was worked out for cattle and buffalo.

$$\text{Urine (l)} = 0.677 \times \text{wet dung (kg)} \quad \text{.................II}$$

$$\text{Urine (l)} = \frac{1.52 \times \text{dry matter intake (kg)}}{1.03} \quad \text{.................III}$$

By using the above formulae for urine and dung excretion and the feed availability in India, the excretion pattern for cattle and buffalo in different states of India was calculated.

From the total annual excretion of dung and urine from both cattle and buffaloes, their potentiality of the plant nutrients was worked out taking an average N, P_2O_5 and K_2O content of dung and urine as follows:

Wet dung: N = 0.15 per cent, P_2O_5 = 0.1 per cent, K_2O = 0.05 per cent

Urine: N = 0.20 per cent, P_2O_5 = 0.01 per cent, K_2O = 0.20 per cent

The total organic matter of the bovine excreta was calculated taking the dry matter content of dung as 20 per cent and the organic matter content as 94 per cent of the dry matter.

The daily feed availability for cattle and buffalo in different states of India was reported. The average feed availability for Indian cattle was worked out to be 3.311 kg green fodder, 4.134 kg dry fodder and 0.186 kg concentrate. The daily feed availability for buffaloes in India was worked out to be 4.10 kg green fodder, 4.08 kg dry fodder and 0.177 kg concentrate. It was observed that in the northern stales like Punjab, Haryana and Union Territories of Delhi and Chandigarh, the adult females are fed with 9.10 to 9.25 kg green fodder for cattle and 11.05 to 13.0 kg for buffaloes. But, moderate green fodder supply to adult female cattle and buffaloes was observed in Rajasthan, Bihar, Assam, Jammu and Kashmir and Uttar Pradesh. In other states, where less green fodder is available, more of dry fodder (*bhusa* and hay) was given. In the case of Kerala the lowest feeding of both green and dry fodder was observed.

The average daily out turn of wet dung from the different groups of cattle and buffaloes in different states of India is presented in Table 4.1. The highest excretion rate of more than 15 kg dung per animal was noticed in IHaryana, Punjab, Himachal Pradesh, Maharashtra, Delhi and Chandigarh. The lowest excretion rate of 3.62 to 4.96 kg dung per animal was recorded from Kerala. The all-India average daily dung excretion pattern in kg for cattle and buffalo group is given below:

Cattle: Young stock = 4.90; adult female = 11.26 and adult male = 15.59; Buffaloes: Young stock = 5.67, adult female = 14.78 and adult male = 14.86. But excretion rate of 2.9 to 7.0 kg in different states in India was reported by Goel from the IARS survey on the estimate of dung evacuated only in the households, and in this study, the quantity of dung evacuated out in grazing fields and roadside was excluded.

Table 4.2 presents the daily outturn of urine from different groups of cattle and buffaloes in India. The average excretion of urine/day/cattle and buffalo was 7.36 and 8.02 litres, respectively. The urine excretion/day/cattle was 4.24, 7.09 and 9.10 litres for young stock, adult female and adult male respectively. But the corresponding urine excretion value/day/buffalo was 4.96, 10.01 and 8.63 litres.

Table 4.3 presents the state-wise annual out turn of wet dung and urine from different groups of cattle and buffaloes in India. The annual excretion of wet dung from cattle amounted to 744.565 million tonnes and that from buffaloes amounted to 258.022 million tonnes. Among cattle the percentage production of dung by different age-groups is as follows: young stock up to 3 years = 11.85, adult females = 30.89 and adult male = 57.26. In the case of buffaloes, the percentage production of dung by different age-groups is as follows: young stock up to 3 years = 16.51, adult females = 64.43 and adult males = 19.06. Among cattle 60 per cent or the dung was produced by cattle from Uttar Pradesh, Maharashtra, Rajasthan, Madhya Pradesh and Tamil Nadu, whereas from the total buffalo dung nearly 73 per cent of the dung was produced by buffaloes from Uttar Pradesh, Andhra Pradesh, Rajasthan, Punjab, Bihar, Maharashtra and Madhya Pradesh. Out of the total annual dung production of 1002.587 million tonnes, Uttar Pradesh alone contributed to 18.9 per cent and 11.3 per cent by Maharashtra and 8.0 per cent by Andhra Pradesh. According to the IARS survey the contribution of Uttar Pradesh to the total dung production was 18.7 per cent. The NCAER survey estimated the total annual bovine dung production at 1,335 million tonnes taking an average outturn of roughly 18 kg dung excretion per adult cattle, 23 kg dung per adult buffalo and 9 kg dung per young stock of cattle and buffalo.

Table 4.1: Daily Outturn of Wet Dung per Animal for Different Age-groups of Cattle and Buffaloes in India (kg)

Sl.No.	State	Cattle				Buffalo			
		Y.S.	A.F.	A.M.	Average per Cattle	Y.S.	A.F.	A.M.	Average per Buffalo
1.	Andhra Pradesh	3.97	7.60	18.23	11.41	3.36	9.84	20.00	10.20
2.	Assam	7.03	12.60	11.70	10.52	6.23	13.05	17.38	13.32
3.	Bihar	4.44	8.85	11.37	8.50	8.03	20.49	7.13	13.91
4.	Gujarat	3.95	8.21	15.90	10.77	4.82	13.46	18.16	9.59
5.	Haryana	8.19	21.04	25.25	17.54	7.76	24.88	26.33	16.95
6.	Himachal Pradesh	8.19	21.04	25.25	18.56	7.76	24.88	26.33	19.44
7.	Jammu and Kashmir	7.03	12.60	11.70	10.51	6.23	13.05	17.38	9.54
8.	Kerala	1.00	5.15	5.63	3.62	0.92	7.50	4.93	4.96
9.	Karnataka	5.83	11.86	15.02	11.59	4.88	12.51	12.61	10.09
10.	Madhya Pradesh	3.77	7.26	8.02	6.40	4.62	12.17	11.10	8.96
11.	Maharashtra; Goa, Daman, Diu	8.38	16.30	21.65	16.74	9.69	19.54	21.07	16.68
12.	Nagaland	4.52	10.32	12.31	8.94	6.38	18.50	10.73	12.32
13.	Orissa	4.62	11.85	13.41	10.61	5.13	15.94	14.31	12.73
14.	Punjab	8.19	21.04	25.25	18.58	7.76	24.88	26.33	14.66
15.	Rajasthan	5.76	14.09	16.63	12.30	6.20	18.28	18.22	12.71
16.	Tamil Nadu	4.01	16.64	17.36	14.37	4.70	16.70	20.42	13.62
17.	Uttar Pradesh	4.59	9.19	19.22	13.20	4.97	14.12	14.88	10.99
18.	West Bengal	4.52	10.32	12.31	9.53	6.38	18.50	10.73	12.15
19.	Andaman and other Islands	1.00	5.15	5.63	3.83	0.92	7.50	4.93	4.82
20.	Chandigarh	6.27	19.64	24.24	15.32	6.18	28.43	16.09	20.55
21.	Delhi	6.27	19.64	24.24	17.38	6.18	28.43	16.09	20.24
22.	Manipur	4.52	10.32	12.31	9.27	6.38	18.50	10.73	12.13
23.	Pondicherry	4.01	16.64	17.36	13.09	4.78	16.70	20.42	13.11
24.	Tripura, Meghalaya, Mizoram	4.52	10.32	12.31	8.69	6.38	18.50	10.73	12.92
	All-India Average	4.90	11.26	15.59	11.33	5.67	14.78	14.86	11.60

Y.S.: Young stock; A.F.: Adult female; A.M.: Adult male.

Table 4.2: Daily Outturn of Urine from Different Age-groups of Cattle and Buffaloes in India (in litres)

Sl.No.	State	Cattle			Average per Cattle	Buffalo			Average per Buffalo
		Y.S.	A.F.	A.M.		Y.S.	A.F.	A.M.	
1.	Andhra Pradesh	3.30	4.70	10.38	6.88	3.40	6.33	11.34	6.49
2.	Assam	6.09	8.12	6.94	7.04	5.69	8.54	10.44	8.70
3.	Bihar	2.97	5.94	6.60	5.43	6.94	13.01	4.04	9.40
4.	Gujarat	3.35	5.11	9.14	6.57	4.10	8.11	10.65	6.50
5.	Haryana	6.04	13.82	20.50	11.55	6.89	16.01	16.06	11.82
6.	Himachal Pradesh	7.28	13.52	15.17	12.13	7.01	15.92	16.16	13.16
7.	Jammu and Kashmir	6.15	8.08	6.98	7.11	6.64	8.67	10.59	8.42
8.	Karnataka	4.97	7.47	8.61	7.27	4.19	7.87	7.30	6.63
9.	Kerala	0.89	3.34	3.30	2.45	0.67	5.23	3.12	3.30
10.	Madhya Pradesh	3.19	4.50	4.72	4.14	3.87	7.70	5.94	5.93
11.	Maharashtra; Goa, Daman, Diu	7.04	10.03	12.35	10.32	8.28	12.32	11.92	10.99
12.	Nagaland	3.90	6.51	7.17	5.82	5.51	11.68	6.25	7.95
13.	Orissa	4.12	7.64	8.23	6.95	4.03	10.16	8.48	8.26
14.	Punjab	7.26	13.58	15.14	12.11	6.87	16.28	16.21	12.76
15.	Rajasthan	5.04	8.98	10.03	8.04	5.34	11.74	10.77	9.05
16.	Tamil Nadu	6.21	10.29	9.94	8.73	4.00	10.56	11.46	8.62
17.	Uttar Pradesh	3.99	5.81	11.26	8.15	4.32	8.93	8.91	7.31
18.	West Bengal	4.02	6.34	7.35	6.09	5.67	12.21	6.21	7.78
19.	Andaman and Nicobar Islands (Dadra, Nagar Haveli; Laccadive, Minicoy; Amindive, Islands)	0.90	2.47	3.38	2.04	0.65	9.37	4.11	4.65
20.	Chandigarh	9.22	11.82	14.19	10.45	6.25	19.05	9.80	14.85
21.	Delhi	8.65	13.11	14.56	11.16	5.75	18.32	9.71	13.53
22.	Manipur	4.09	6.09	7.15	5.92	4.08	12.01	7.32	8.40
23.	Pondicherry	3.40	9.96	9.71	8.13	4.85	11.35	14.02	9.81
24.	Tripura, Meghalaya, Mizoram	4.06	6.65	7.37	5.63	6.15	9.51	7.61	8.85
	All-India Average	4.24	7.09	9.10	7.36	4.96	10.01	8.63	8.02

Y.S.: Young stock; A.F.: Adult female; A.M.: Adult male.

The statewise annual out turn of urine from different groups of cattle and buffaloes in India is given in Table 4.3. The annual out turn of urine from cattle amounted to 480.148 million tonnes, whereas from buffaloes it amounted to 178.753 million tonnes. Among cattle the percentage distribution of urine by different age-groups is as follows: young stock = 16.23, adult female = 30.80 and adult male = 52.97. In the case of buffaloes, the percentage production of urine for different age-groups of animal is as follows: young stock = 22.42, adult female = 60.24 and adult male = 17.32. The total annual out turn of urine from cattle and buffaloes in India was 658.90 million tonnes and the overall dung excretion worked out to be about 1.552 times of urine excretion.

Table 4.3: Annual Outturn of Wet Dung and Urine from Cattle and Buffaloes in India

Sl.No.	State	Annual Wet Dung (Million tonnes)			Annual Urine Excretion (Million tonnes)		
		Cattle	Buffalo	Total	Cattle	Buffalo	Total
1.	Andhra Pradesh	52.309	25.080	77.389	32.350	17.219	49.569
2.	Assam	22.368	3.008	25.376	15.131	2.087	17.218
3.	Bihar	49.355	19.202	68.557	30.440	13.001	43.441
4.	Gujarat	25.496	12.760	38.262	15.949	8.475	24.424
5.	Haryana	16.644	15.569	32.213	10.647	11.189	21.836
6.	Himachal Pradesh	15.283	3.895	19.181	9.923	2.692	12.615
7.	Jammu and Kashmir	8.010	1.715	9.725	5.498	1.561	7.059
8.	Karnataka	42.926	11.874	54.800	27.698	8.198	35.896
9.	Kerala	3.674	0.864	4.538	2.631	0.585	3.216
10.	Madhya Pradesh	61.658	19.074	80.732	41.185	12.919	54.104
11.	Maharashtra	88.298	19.688	107.986	57.052	13.639	70.691
12.	Nagaland	0.301	0.047	0.348	0.204	0.030	0.234
13.	Orissa	45.027	6.471	51.498	30.037	4.344	34.381
14.	Punjab	24.755	26.435	51.190	16.417	19.510	35.927
15.	Rajasthan	55.876	22.020	77.896	37.692	15.623	53.315
16.	Tamil Nadur	56.152	13.875	70.027	35.364	9.330	44.694
17.	Uttar Pradesh	128.252	50.780	179.032	80.108	34.575	114.683
18.	West Bengal	41.969	3.718	45.687	27.905	2.454	30.359
19.	Chandigarh	0.028	0.015	0.043	0.020	0.061	0.081
20.	Andaman and Nicobar Islands	0.034	0.093	0.127	0.015	0.014	0.029
21.	Dadra, Nagar Haveli, Laccadive, Minicoy, Amindive Islands	0.063	0.007	0.070	0.031	0.005	0.036
22.	Delhi	0.404	0.968	1.372	0.285	0.651	0.936
23.	Goa, Daman, Diu	0.766	0.241	0.007	0.481	0.157	0.638
24.	Manipur	1.045	0.240	1.285	0.654	0.164	0.818
25.	Lakshadweep	0.006	–	0.006	0.001	–	0.001
26.	Meghalaya	1.568	0.211	1.779	0.991	0.153	1.144
27.	Mizoram	0.079	0.011	0.090	0.053	0.007	0.060
28.	Pondicherry	0.419	0.057	0.476	0.275	0.044	0.319
29.	Tripura	1.800	0.095	1.895	1.111	0.066	1.177
	Total	**744.565**	**258.022**	**1002.587**	**480.148**	**178.753**	**658.901**

If the entire wet dung and urine excreted by the bovines is conserved for manurial purposes, its potentiality for soil nutrients has been worked out as under: 188.380 million tonnes organic matter, 2,822 million tonnes nitrogen, 1.069 million tonnes of phosphoric acid and 1,819 million tonnes of potassium oxide (Table 4.4). According to Garg the annual production of dung and urine was estimated to be 1300 million tonnes and 370 million tonnes respectively and the potential total soil nutrients had been worked out as 4.89 million tonnes nitrogen, 1.37 million tonnes phosphoric acid and 3.85 million tonnes potassium oxide. According to IARS survey the available bovine dung that could be collected from households was 344.5 million tonnes and its manurial potential had been worked out as 1.206 million tonnes nitrogen, 0.517 million tonnes phosphoric acid and 0.689 million tonnes potash. They reported that nearly 29 per cent of the dung collected in households are burnt as fuel cakes, 69 per cent used for making manure and 2 per cent used for other purposes. However, the present report gives a realistic estimate as to the possible excretion of total dung and urine from the bovines in India.

Table 4.4: Potentiality of Total Annual Dung and Urine Excretion of Bovines for Soil Nutrients

Type of Animal and Manure	*Annual Excretion (Million tonnes)*	*N*	*P_2O_5*	*K_2O*	*Organic Matter*
		(Million tonnes)			
Cattle					
Dung	744.565	1.117	0.745	0.372	139.872
Urine	480.148	0.960	0.048	0.960	–
Buffalo					
Dung	258.022	0.387	0.258	0.129	48.508
Urine	178.753	0.358	0.018	0.358	–
Total	–	2.822	1.069	1.819	188.380

Table 4.5: Annual Excretion of Dung and Urine of Livestock and Human Beings in India

Sl.No.	*Animal Type*	*Population (Millions)*	*Daily Excretion*		*Annual Excretion (Million tonnes)*		*Total Excretion (Million tonnes)*	*Per Cent*
			Dung (kg)	*Urine (l)*	*Dung (kg)*	*Urine (l)*		
1.	Cattle and buffalo	236.806	11.597	7.623	1002.587	658.901	1661.488	82.71
2.	Sheep and goats	108.419	0.300	0.200	12.228	7.918	20.146	1.00
3.	Pigs	6.456	2.000	2.000	4.596	3.990	5.586	0.43
4.	Poultry	136.768	0.068	–	3.395	–	3.395	0.20
5.	Other livestock	3.301	5.000	3.300	6.024	4.095	10.119	0.50
6.	Human beings	625.800	0.133	1.200	30.380	274.100	304.480	15.16
	Total				**1059.210**	**949.004**	**2008.214**	**100**

Other Livestock and Human Excreta

According to Livestock Census 1972, India has a population of 108.419 million sheep and goats, 6.456 million pigs, 136.768 million poultry and 3.301 million other livestock including, 0.966 million horses and ponies, 1.126 million camels, and 1.209 million other livestock. The present human population is 625.8 million. The annual excretion. of dung and urine by different livestock and human beings is given in Table 4.5. The annual excretion of bovine dung and urine comprises 82.71 per cent of the total excretion by all other livestock and human beings. The sheep and goat excreta comprised

12.228 million tonnes dung and 7.918 million tonnes urine/year. The pig excreta comprised 4.596 and 3.990 million tonnes of dung and urine respectively. The poultry excreta per annum was 3.395 million tonnes. The excreta from other livestock comprised 6.024 and 4.095 million tonnes dung and urine per year respectively. The human excreta comprised 15.16 per cent of the total livestock and human excreta. The human beings excreted annually 30.380 million tonnes faeces and 274.100 million tonnes urine.

The annual manurial potential of bovine excreta alone is 2.822, 1.069 and 1.819 million tonnes N, P_2O_5 and K_2O, respectively (Table 4.4). The manurial potential of human excreta is 3.228.0.776 and 0.715 million tonnes N, P_2O_5, and K_2O, respectively (Table 4.6). The total annual manurial potential of all livestock and human excreta is 6.414, 1.973 and 2.662 million tonnes, N, P_2O_5 and K_2O, respectively.

Table 4.6: Manurial Potential of Livestock and Human Excreta

Animal	Annual Excretion (million tonnes)		Chemical Composition						N	P_2O_5 (million tonnes)	K_2O
	Dung	Urine	N		P_2O_5		K_2O				
			Dung	Urine	Dung	Urine	Dung	Urine			
Cattle	744.565	480.148	0.15	0.20	0.1	0.01	0.05	0.20	2.077	0.793	1.332
Buffalo	258.022	178.753	0.15	0.20	0.1	0.01	0.05	0.20	0.745	0.276	0.487
Goat & sheep	12.228	7.918	0.65	1.70	0.50	0.02	0.03	0.25	0.214	0.063	0.020
Pigs	4.596	3.990	0.60	0.40	0.50	0.10	0.20	0.50	0.044	0.027	0.029
Poultry	3.395	–	0.80	–	0.60	–	0.30	–	0.027	0.020	0.010
Other livestock	6.024	4.095	0.50	1.20	0.30	–	0.30	1.00	0.079	0.018	0.069
Human beings	30.380	274.100	1.60	1.00	1.20	0.15	0.55	0.20	3.228	0.776	0.715
Total									**6.414**	**1.973**	**2.662**

Slaughter-house Wastes

There are about 3,000 slaughter-houses in the country handling annually nearly 40 million sheep and goats and 1.5 million buffaloes. About 12 million dead large animals are available annually. No systematic steps are taken towards organised collection of bones from dead animals and their utilisation.

Bonemeal

According to the report of the Directorate of Marketing and Inspection, it is estimated that about 4.5 lakh tonnes of bones are available every year in the country, out of which 1.36 lakh tonnes are collected and utilised by bone crushing mills.

Bonemeal is obtained as powder by crushing of bones, and it is used as fertilizer. Steamed bonemeal is obtained by treating the bones with steam under pressure and is used chiefly as phosphatic fertilizer. It also contains about 1 to 2 per cent nitrogen besides P_2O_5 (25 to 30 per cent). Enormous amount of nitrogen and phosphate can be supplied by proper utilisation of bonemeal potential.

Bonemeal is considered useful for most of the soils and the best results are obtained in acidic soils. It is less effective in heavy clay and calcareous soils. It is more useful in soils well supplied with organic matter. Paddy, wheat and other cereals responded very well to application of bonemeal, particularly in acidic soils it can also be used for sugarcane, vegetables, fruits and legume crops.

Bonemcal is likely to be contaminated with *Salmonella,* spores of *Bacillus anthracis* causing an anthrax disease in cattle and other pathogenic organisms. It can be made safe by sterilization.

Blood and Meat-meal

Blood-meal is used as nitrogenous fertilizer or as animal feed. The method for collection is fault and therefore a large amount of blood is wasted. Availability of blood-meal is estimated about 55,000 tonnes and of this only one-third is utilized. It contains 10–12 per cent N, 1–2 per cent P_2O_5 and 1 per cent K_2O and its C/N ratio ranges between 3 and 4. It decomposes readily in soils. It can be used at any time during, growth of crops. Solid slaughter-house wastes consist of waste meat, intestines, offal, etc. and has good manurial value. It is practically a waste at present. It is estimated that 0.12 lakh tonnes of meat-meal can be produced from dead animal wastes. If dried and ground, it will make a good fertilizer containing 8–10 per cent N and 3 per cent phosphoric acid with a C/N ratio between 2 and 3.

Hoof and Horn-meal

This is obtained by grinding hoofs and horns of animals after drying. It contains 10–15 per cent N, 1 per cent P_2O_5 and 2.5 per cent lime.

Leather Wastes

It is estimated that nearly 5,000 tonnes of leather wastes can be collected from slaughtered and fallen animals of total estimated to be 52 million heads for use of organic fertilizer. This remains unutilised at present.

Crop Residues and Aquatic Weeds

Crop Wastes

The potential of crop residues/straw of some of major cereal crops and pulses is given in Table 4.7. The straw yields have been worked out on the basis of average grain-straw ratio of different crops as indicated in the Table 4.7. It is clear from the table that there is huge quantity of renewable crop residues produced every year in the country. The five major crops alone yield approximately 141.2 million tonnes of straw and approximately 10 million tonnes of legumes residues. On average, cereal straw and residues on maturity contain about 0. 5 per cent nitrogen, 0.6 per cent P_2O_5 and 1.5 per cent K_2O. The quantity of nutrients in legume residues is much higher than in cereal straw. The nutrient potential of cereal straw/residues is 0.7 million tonnes of nitrogen, 0.84 million tonnes of P_2O_5 and 2.1 million tonnes of K_2O. Even if 50 per cent of these crop residues are utilised as animal feed, the rest should be mobilised for recycling for their plant nutrient potential and other beneficial effect on soils and plants. The role of crop wastes in maintenance of soil organic matter under tropical and sub-tropical conditions needs no emphasis. Crop residues can be recycled either by composting, or by way of mulch or direct incorporation in the soil. Farmers should be advised by the Extension workers to conserve these manurial resources and suggest proper methods for utilisation of crop wastes.

Water Hyacinth

Water hyacinth (*Eichhornia crassipes*) is a free floating weed plant which grows luxuriantly in ponds, lakes and water reservoirs. It is estimated that total acreage under this weed is about 292,000 ha in Bengal, Bihar, Assam, Eastern U.P., Andhra Pradesh, Tamil Nadu, Orissa and Kerala. The adverse effects of such uncontrolled growth on agriculture, fisheries, transport and human health are obvious and the necessity for its collection and consequent disposal are needed.

Table 4.7: Potential of Crop Residues/Wastes for Recycling as Fertilizer/Annual Feed

Crop	Average Grain Straw Ratio	Area (000 Hectares)	Average Yield/ha (kg)	Production in (000Tonnes)	Dry Matter Yield (000 Tonnes)
Paddy	(1 : 1.02)	39688	1246	74474.2	89,369.04
Jowar	(1 : 1.5)	16101	592.9	525.0	14,287.5
Wheat	(1 : 0.65)	20112	1409	28336.0	18,418.40
Bajra	(1 : 1.5)	11583	494	5726.0	8589.0
Maize	(1 : 1.5)	5996	1173	7036.0	19,554.0
Total Cereals	–	103,522.0	1040	107,698.0	–
Total Pulses	(1 : 0.75)	24,665.0	533	13135.0	9851.25

Paddy production figures have been estimated from rice production figures with the standard ratio of 2 : 3.

Source of Mulch and Manure

Water hyacinth can be used as soil mulch, green manure and compost. Recently attempts have been made to use water hyacinth for biogas production without loss of plant nutrients. Water hyacinth is used as mulch in tea gardens during dry season for conservation of soil moisture and regulation of temperature. The plants can directly be ploughed in the soil and allowed to decompose for a month or so before sowing of a crop. However, the problem is that as plants are bulky, it is difficult to handle and transport to long distances.

Composting

The fresh plant contains 95.5 per cent moisture, 3.5 per cent organic matter, 1 per cent ash, 0.04 per cent nitrogen; 0.06 per cent phosphorus (P_2O_5) and 0.2 per cent potash (K_2O). This is a good source of potassium. It can be converted into compost without additional source of nitrogen. Addition of small amount of soil will accelerate the process or composting of water hyacinth. Since yields of water hyacinth are of the order of 250 tonnes per hectare per year, there is a potential for producing 3 million tonnes of compost annually in the country which will provide on dry basis about 20.5 kg N, 11.0 kg P_2O_5 and 25.0 kg K_2O per tonne. Water hyacinth compost is good for crops like rice, potato, maize, jute and vegetables. The recommended doses of compost for rice, maize and jute is 20, 5 and 7.5 tonnes per hectare, respectively.

Forest-litter Manure

It is estimated that about 15 million tonnes of compost can be obtained from forest-litter annually without in any way adversely affecting the natural regeneration or the forests. If a portion or the surface litter is removed in a regular manner, the manurial value of forest-litter is as good as from compost. Fifteen million tonnes of forest litter manure may contain 0.075, 0.03 and 0.075 million tonnes of N, P_2O_5 and K_2O respectively. At present, however, considerable amount of leaf litter is burnt and huge quantities of plant nutrients allowed to go waste.

Green Manuring

Legume plants are grown for fixing atmospheric nitrogen through *Rhizobium* symbiosis and plants after 8 weeks or growth are incorporated in soil to improve its fertility for raising another crop. Sunnhemp, *dhaincha,* clusterbeans, *senji,* cowpea, *moong, urid,* fodder legumes, etc. are used as green

Table 4.8: Available Potential and Production of Organic Manures through Local Manurial Resources

Sl.No.	Manurial Resources	Potential Qty	Achievements During Different Years (Million tonnes)									
			1950-51	1955–56	1960–61	1965–66	1973–74	1974–75	1975–76	1976–77	1977–78	1978–79
1.	Rural Compost (million tonnes)	600.00	1.40	–	66.00	119.60	150.81	167.70	184.78	195.52	310.00	350.00
2.	Green Manure (million hectares)	30.00	–	–	4.25	8.04	3.71	4.48	5.69	6.00	6.00	6.00
3.	Urban Compost (million tonnes)	15.00	1.40	2.12	2.85	3.50	4.27	4.42	4.80	5.30	5.80	6.50
4.	Sewage/sullage	800 m.g.d.*	–	–	–	12,267	24.000	24.000	26,208	30,000	35,000	40,000
						(Area under irrigation in hectares)						

* m.g.d.: Million gallons/day.

manuring crops. *Dhaincha* and sunnhemp are more popular. Green manuring is confined to certain areas and in extensive agriculture. This practice has not extended in recent years. There are certain other practical difficulties such as lack of water supply for growth or green manure crops which are ascribed to its non-acceptability by the farmers.

Joffe indicated that the following single crop was only benefited due to green manuring and the favourable effect was not due to its contribution towards the improvement or organic matter and nitrogen content of soil. It was reported that the decomposition of legume residues was much faster as compared to farmyard manure and cereal residues and did not improve organic matter status or soils. Singh also indicated that 'legume effect' was more important as his experiments showed that increase in yields was not due to organic matter or nitrogen additions by green manure. His experiments also showed that berseem, *senji* and pea left through their root and stubble 89.4, 53.1 and 20.7 kg nitrogen per hectare. Although the practice of green manuring cannot be followed in intensive agriculture on a large scale, but certainly a fodder or grain legume can be included in multiple cropping sequence. Crop rotation involving sugarcane, cotton and *arhar* (pigeonpea) is ideal. Moreover, growing of legumes can have a certain amount of expensive nitrogenous fertilizers by improving the nitrogen status of soils. However, green manuring practices which do not interfere with the production of main crops should be popularised.

Green-leaf Manuring

Green leaves from trees, *viz.*, *Thespesia populnea. Cassia auriculata, Pongamia glabra, Melia azadirachta, Calotropis gigantea, Adhatoda vasica*, etc., are collected and used for green manuring in southern parts of India. Weeds, *e.g.*, *Croton sparsifolorus, Lucas aspera, Stachytarpheta indica* are also utilised for green-leaf manuring.

Some wild legume plants, *viz.*, *Giliricidia meculata, Pongatnia glabra, Calotropis gigantea, Tephrosia purpurea, Ipomoea carnea, Cassia tora, Sesbania* spp., *Indigofera teysmann, Tephrosia candida* can be grown on bunds and wastelands for utilising their vegetative parts for green-leaf manuring. A programme for raising wild 1egumes on bunds and wastelands may be developed for increasing its scope for its adoption.

Rural and Urban Wastes

Rural and Urban Solid Wastes

The potential availability of rural and town compost is estimated to be 600 million and 15 million tonnes respectively (Table 4.8). However, with the present efforts only 310–350 million tonnes of compost is prepared in villages by additional and improved methods of composting. The present level of production of town compost is only of order of 6.5 million tonnes. However, with the set up of mechanised plants, the potential estimate of city waste is of the order of 50 million tonnes per year.

Sewage and Sullage

At present about 36 million urban population is served by drainage system producing about 292,000 million gallons sewage per annum (Table 4.8). Out of this potential, about 91,250 million gallons are utilised on organised sewage farms. It is estimated that during 1978–79 about 4,000 ha of land would be under sewage irrigation. Sewage has important components—water, plant nutrients and organic matter which are badly required in Indian agriculture. There is a scope for further expansion of sewage farming programmes now, and in future more cities will be provided with drainage systems.

Besides the necessity of sewage farming from point of view of utilization of the resources, it is also an effective method to avoid pollution. Unrestricted discharge of city liquid wastes in rivers and streams and on land results in pollution of environment and is a public health hazard.

There are two hundred and twenty sewage farms located in different parts of the country, about 102 such farms are located in six states, *viz.*, Punjab. Uttar Pradesh, Tamil Nadu, Haryana, Gujarat, and Madhya Pradesh. The present utilization of wastewaters is only about 31 per cent of the total potential. The average NPK content of an Indian city sewage is 50 ppm N, 15 ppm P_2O_5 and 30 ppm K_2O. Sewage sludge could form an important component of composting.

Waste stabilisation ponds have been recognised as effective and economical units for treatment of domestic sewage as well as industrial wastes. It is essentially a microbiological process involving simultaneous activity of bacteria and algae in presence of light, atmospheric oxygen and nutrients in waste waters. Effluents from stabilisation ponds contain algal cells and other nutrients. In addition to oxygenation due to photosynthetic activity, cellular algae is a good source of feed for the growth of edible variety of fish. The effluents from such treatment contain appreciable quantities of organic substances, nitrogen, phosphorus, potassium, etc. In oxidation ponds a part of these nutrients is removed from solution and concentrated in algal cells. The treated sewage effluent is safe and will not cause environmental pollution. The raw sewage contains 60–70 ppm N, 20–25 ppm total P_2O_5 and 40–45 ppm total K_2O and the secondary treated sewage contains 15–20 ppm N, 15–20 ppm total P_2O_5 and 35–40 ppm total K_2O.

Substantial volume of treated or partly treated sewage is usually being led into natural water streams. Thus the nutrients in waste water are not being effectively utilised. The total nutrients from urban and rural communities is substantial which should be recycled.

Agro-Industrial Wastes

Agro-industries are based not only on crops, such as rice, sugarcane, jute, tea, coffee, fruits and vegetables but also based on forest products (non-edible oilseeds, wood, lac, etc.), marine products (prawns, fish, frogs and seaweeds) and slaughterhouse wastes and dead animals. Agro-industries wastes are available in substantial quantities at processing sites whereas animal wastes and crop residues are available at farms and in a scattered way. The characteristics and possible use as organic manures of some potential wastes are discussed below.

Oilseed Industry

Major oilseeds occupy an important position in the agricultural economy and are grown in area of about 16 million hectares. Groundnut is the most important crop followed by rape-mustard, *sesamum*, linseed and castor. Oilseed crops are essential part of human diet as well as provide important industrial raw material. Oilcakes obtained as byproducts are mostly used as cattle feed and manure. Groundnut hulls, obtained during the shelling of groundnut can be used as manure.

Non-edible Oilcakes

Oilcakes are the residues left after extraction of oil from oilseeds. About 0.3 million tonnes of non-edible cake is produced annually. Non-edible cakes such as *neem, karanj, mahua,* castor, etc., are used as organic fertilizer.

The manurial value of these cakes lies mainly in its nitrogen content although it contains small quantities of P_2O_5 and K_2O. The nitrogen content varies from 3 to 9 per cent, depending on the nature of oilcake. The C/N ratio is low ranging between 3 and 15 for different types of oilcakes. Due to low C/N ratio its decomposition rate is faster than cereal and legume residues and other bulky organic manures. This nitrifies very quickly and about 60 to 80 per cent of its nitrogen is converted in available form within 2 to 3 months time. Castor, groundnut, cottonseed, *mahua*, rape-seed, *neem*, *karanj* oilcakes are used as organic manures. *Mahua* cake is poor in nitrogen and takes longer time to nitrify. It is better to apply *mahua* cakes about 2 months in advance to soil before sowing of the crop. Oilcakes on economic grounds are not recommended for cereals like wheat or rice. Cakes are extensively applied for sugarcane crop and betel leaves. Castor cake is supposed to be good vermicide against white ants (termites). Recent investigations at IARI have shown that due to application of oilcakes to soil the population of plant parasitic nematodes was decreased.

Rice Milling Industry

Rice Husk and Rice Bran

Rice husk is the largest product of the rice milling industry comprising 20 to 25 per cent of paddy. Paddy yields about 5 to 7 per cent bran. The availability of rice husk is about 15 million tonnes annually. A typical paddy husk sample contains 42.6 per cent cellulose, 20.1 per cent lignin, 18.6 per cent pentosans and 18.7 per cent ash. The physical and chemical nature is unlike other crop residue wastes. It is a poor source of manure and its N content varies from 0.3 to 0.4 per cent, P_2O_5 0.2 to 0.3 per cent and potash 0.3 to 0.5 per cent. There is a problem of its disposal in certain areas. It is used as fuel and for improving physical conditions of saline and alkali soils. It can be used as bedding material for animals and in composting.

Rice bran yield is about 2.5 million tonnes annually. It has limited scope as fertilizer since this is exploited for production of rice bran oil.

Sugar Milling Industry

Bagasse

It is one of the most important byproducts of sugar industry. The fibre content of Indian sugarcane is 12 to 17 per cent and 33 per cent is bagasse. About 5.3 million tonnes of dry bagasse is annually produced. The bagasse produced in the country is almost entirely used as fuel in boilers of sugar factories. Recent investigations have shown that bagasse is a valuable material for production of pulp, paper, boards, etc. However, a portion of bagasse could be utilised as both for fuel and manure if it is processed through biogas plants. The nitrogen and P_2O_5 per cent of bagasse is approximately 0.25 pet cent and 0.12 per cent and compost produced out of it will have a nitrogen of 1.4 per cent and 0.4 per cent of P_2O_5. It is estimated that about 14 million tonnes of organic manure per year can be produced from this byproduct.

A pilot plant is in operation at the National Sugar Institute, Kanpur which utilises bagasse and sugarcane trash for production of biogas. A composted fertilizer/manure is obtained within 40 to 45 days by this anaerobic process against longer period, normally required in compost pits. The approximate composition of each charge is 80 per cent cellulosic materials (bagasse), 12 per cent animal dung, 5 per cent bonemeal or super-phosphate and 3 per cent calcium carbonate with moisture of about 70 per cent. The aerobic decomposition is allowed for 5 days by blowing compressed air resulting in rise of temperature to 60–80°C followed by anaerobic digestion for methane production. On dry basis, spent slurry contains 1.5–1.8 per cent N, 1.0–1.3 per cent P_2O_5 and 0.6–0.8 per cent K_2O. About 200 cu metre of biogas are obtained per tonne of organic matter contained in agricultural wastes. The biogas has a calorific value of 5,200 K cals per cu metre. The composition of the biogas produced is methane 55–60 per cent, carbon dioxide 30–35 per cent and hydrogen 5–10 per cent.

Press Mud

About 2 million tonnes of press mud is produced annually from the sugar factories. The bulk of the filter/cake is now used as manure in the fields, it contains about 1.25 per cent nitrogen and 2 per cent of P_2O_5 and 20 to 25 per cent organic matter. Compost prepared from press mud contains 1.4 per cent nitrogen and 1 to 1.5 per cent P_2O_5. Since it is very high in lime (up to 45 per cent), its application is useful in acidic soils.

Forest Mill Wastes

Sawdust

The total of sawdust waste in the country from sawmill and plywood manufacture is estimated about 2.2 million tonnes. Sawmills alone account for 2.0 million tonnes of sawdust. It is a wide C/N ratio material (500 : 1) low in nitrogen (0.11 per cent) and also low in phosphate (0.20 per cent). It has limited scope to use as such as fertilizers, although some investigations have shown that it can be used as organic manure in tropical soils. Dried sawdust has good liquid absorbing capacity and can absorb 2–4 times more moisture than cereal straw/residues. Thus it can be used as a good absorbant for soaking urine in cattlesheds and as bedding materials for cattle which can be composted and converted in valuable organic manure. Its value as mulching material and for control of parasitic nematodes has also been recently reported.

Vegetable and Fruit Processing Industry

Vegetable and Fruit Wastes

A large quantity of wastes are obtained from peals, cores, pits, wine, steua and other materials from vegetable and fruit processing plants. There is a wide scope for utilisation of these wastes as fertilizer. The total quantity available from this industry is more than 25,000 tonnes from mango, pineapple, citrus fruits, apples, green peas, tomato, etc. annually. It is estimated that about 10,000 tonnes of compost could be produced out of these wastes.

Cotton Mill

Cotton Wastes/Byproducts

Cotton is an important commercial fibre crop of India. The present area under the crop is about 8 million hectares and production is about 6.5 million bales, valued at about rupees 800 crores. The main products of cottonseed are oilcake or meal, linters and hulls. Out of these, cotton oilcakes can be used as feed and fertilizer and the main wastes/byproducts arising from cotton are (1) cotton stalks, (2) cotton linters and hulls, (3) cotton leaves and other plant parts, (4) cotton dust. Rainfed cotton crop yields two-and-a-half tonnes of cotton-stalks per hectare and under irrigated crop the yield of cotton stalks is about 5 tonnes/ha. Total production of cotton stalks on all-India basis is 12.0 million tonnes which are generally used for fuel purposes or burnt in the field for disposal. The necessary technology for converting it into manure should be developed.

Cotton Dust

Cotton textile mills mainly in their blowing rooms produce a large quantity of this waste, textile mills in India are expected to produce 30,000 to 33,000 tonnes of this waste per year.

The chemical analysis of the cotton dust in per cent is given below: Moisture 8.0, organic matter 70.0, carbon 41.0, nitrogen 1.4, P_2O_5 0.6, K_2O 1.2 and pH 6.2. This material is ideally suited for composting and application to soils. The waste has plant nutrients (NPK) in greater proportion than routine city refuse. Its C/N ratio is ideal for composting. Studies carried out at CPHERE, Nagpur have shown that moisture content of 50–60 per cent with a few turnings, a good quality compost can be prepared within 20 days. One tonne of raw material is estimated to give 0.6 to 0.7 tonne of finished compost. The cost of production works out to be only Rs 6 per tonne.

Tea Industry

During the course of tea production, processing and storage, about 10 million kg of tea waste becomes available in the form of fluffs, stalks and sweepings. It is an important raw material for extraction of caffein. The decaffeinated tea wastes can be used as a manure or an animal feed. Spent tea waste has 0:28 to 3.5 per cent N, 0.4 per cent P_2O_5 and 1.5 per cent K_2O with C/N ratio of 9 to 11.

Tobacco Wastes and Tobacco Seed-cake

It is estimated that out of total of about 62,000 tonnes of such waste available annually, nearly 41,000 tonnes are used for manurial purposes. The tobacco wastes used for manuring contain 0.5 to 1.0 per cent N, 0.8 per cent P_2O_5 and 0.8 per cent K_2O. About 10,000 tonnes of tobacco seed-cake is also available for manurial purposes and it contains 4 to 4.5 per cent N, 7 to 15 per cent P_2O_5 and 5 to 5.5 per cent K_2O.

Jute Sticks

The total quantity of jute sticks produced in India is about 2.5 million tonnes. It is generally used for thatching, hedging and for fuel purposes. Cellulose content of jute stick is quite high as compared to bagasse, rice and wheat straw. It may be utilised in biogas after pretreatment or by chopping it to smaller particle size.

Fisheries and Marine Industry

Seafood canning industry is an important industry in almost all the maritime states of the country. The wastes arise mainly during processing due to low market value of non-edible varieties, etc. Prawn shell and head, trash fish and frog legs are the main byproducts available in this country. A small quantity of prawn waste is used either as manure or for supplementing cattle and poultry feeds. Trash fish which constitute 15 to 30 per cent of the total marine catches are generally not preferred for table purpose because of their small size, bony nature and poor taste. These are converted into fish-meal and fish-manure.

In frog industry only the hind legs of certain frogs are processed while the head and the rest of the body which constitute about 65 per cent of the total weight are generally discarded.

Fish-meal processed by drying non-edible fish or waste from fish industry is a well-recognised balanced organic manure containing significant amount of nitrogen and phosphorus. Fish-meal is also produced by processing in steam digesters and then drying. The nitrogen content varies from 4 to 10 per cent, phosphoric acid from 3 to 9 per cent and K_2O about 1 per cent. It has a C/N ratio of about 4 to 5. The use of fish-manure is restricted to coastal areas.

Marine Algae and Seaweeds

Marine algae and seaweeds form a good source of organic manures amounting to about 10,000 to 15,000 tonnes annually. Seaweed contains 1 to 2 per cent P_2O_5 and 2 to 7 per cent K_2O and a number of trace elements.

Tank Silt

It consists of a large proportion of finer soil particles of silt and clay and organic matter carried by run-off water from the surrounding soil to the tanks during heavy rains. Pond and harbour muds have long been in use as soil ameliorant. The amount of fertilizing constituents present in lank-silt is not high, 0.3 per cent N, 0.3 per cent P_2O5 and 0.3 per cent K_2O. It is considered as active culture of microorganisms, particularly the nitrogen-fixing ones. The residual effect of tank and river-silt is well established. It has been estimated that millions of tonnes of tank and river silt are available for application on land but remain unexploited every year.

Chapter 5

Preparation, Processing and Preservation of Organic Manures

Preparation of Organic Manures

Raw Materials

Compost and farmyard manure are processed from agricultural organic wastes such as straw, leaves, paddy husk, groundnut husk, sugarcane trash, bagasse, cattle dung and urine and habitation wastes like city garbage, nightsoil, sewage and sullage and vegetable matter. The materials undergo intensive decomposition under thermophilic and mesophilic conditions in heaps or pits with adequate moisture and finally yield a brown to dark coloured humified material in four to six months which is more stable in form, valuable for replenishments of plant nutrients, maintenance of soil organic matter and in improving the physical and microbiological conditions of the soil.

Principles of Composting

Action of Microorganisms

The biodegradation process is carried out by different groups of heterotrophic microorganisms, bacteria, fungi actinomycetes and protozoa. The role of cellulolytic and lignolytic microorganisms in decomposition of crop wastes and residues is of prime importance. Microorganisms involved in the process derive their energy and carbon requirements from the decomposition of carbonaceous materials and for every 10 parts of carbon, 1 part of nitrogen is required for building up of their cell protoplasm. Fungi are more efficient in carbon assimilation than bacteria and actinomycetes. Thus carbon dioxide evolution is comparatively less when fungi are more active in biodegradation than bacterial.

When organic materials are broken down in presence of oxygen, the process is called as aerobic decomposition. Under aerobic conditions, living organisms which utilise oxygen, decompose organic matter and assimilate some of the carbon, nitrogen, phosphorus, sulphur and other nutrients for synthesis of their cell protoplasm. Heterotrophs derive energy from the decomposition of organic matter, resulting on production of carbon dioxide, humic substances and release of available plant nutrients. Carbon serves both as energy source and is also required for cell protoplasm, greater amount of carbon is assimilated than nitrogen. Generally about two-thirds of the carbon is respired/evolved as CO_2 and the remaining one-third is combined with nitrogen in the living cells.

Aerobic decomposition of organic materials is most common in nature and generally occurs in arable soils and in forest soil surfaces where animal droppings and organic residues are stabilised into humus, with involvement of different groups of microflora. In the aerobic process, there are no nuisance problems such as foul odour associated with it as is produced under anaerobic conditions due to intermediate compounds.

A great deal of exothermic energy is released during the oxidation of carbon to carbon dioxide. Organic materials in compost heaps or piles under proper insulation generate substantial amount of heat which increases the temperature up to 65–70°C. However, if the temperature exceeds 65–70°C, the microbial activity is decreased due to thermal kill of the microorganisms and the stabilization of organic matter is slowed down. Thermophilic organisms develop when the temperature exceeds above 45°C, and they thrive but in the temperature range of 45–65°C. The major reactions likely to occur under aerobic decomposition system are as follows:

Sugars $(CH_2O) \times + XO_2 \longrightarrow \times CO_2 + \times H_2O$ + Energy
Celluloses
Hemicelluloses
Lignins

Proteins (Organic N) $\longrightarrow NH_3 \longrightarrow NO_2^- \longrightarrow NO_3^-$

Organic sulphur $S + \times O$ □ ⇒ SO_4

Organic phosphate $\longrightarrow H_3PO_4 \longrightarrow Ca(HPO_4)_2$
(Phytin, Lecithin)

Anaerobic micro-organisms break down organic materials by a process of reduction in absence of oxygen. First a special group of acid producing bacteria, facultative heterotrophs degrade organic matter into fatty acids, aldehydes and alcohol, etc. Then a group of bacteria convert the intermediate products to methane, ammonia, carbon dioxide and hydrogen. Oxygen is also required for the anaerobic process but its source is chemical compounds and not free dissolved oxygen. Like aerobic process, the organisms use nitrogen, phosphorus and other nutrients in developing cell protoplasm. During this process the decomposition is not complete and there is less production of carbon dioxide and intermediates, like organic acid will occur in greater amounts. So is the case of nitrogen containing substances such as ammonia. Due to lesser microbial biomass production and carbon assimilation, there greater production of methane. This type of fermentation takes place in gober-gas plants, waste stabilisation ponds in marshy soils, in buried organic materials devoid of oxygen or with low oxidation-reduction potential. Intensive reduction of organic matter is also known as a putrefactive process accompanied by foul odours of hydrogen sulphide and of reduced sulphur containing organic compounds such as mercaptans.

As compared to aerobic process where the release of energy is much greater (484–674 K Cal/glucose molecule) only about 26 K Cal of energy per gram molecule of glucose are released. The energy of carbon is in the methane gas and the resultant energy from gober-gas plants in India is utilised for cooking purpose and can also be used in running engines. The biochemical reactions that occur in anaerobic decomposition of wastes are as follows:

$$(CH_2O)\times \longrightarrow \times CH_3COOH$$

$$CH_3COOH \longrightarrow CH_4 + CO_2$$

$$\text{Organic N} \longrightarrow NH_3$$

$$2H_2S + CO_2 + \text{Light} \longrightarrow (CH_2O)\times + S_2 + H_2O$$

$$\text{Organic P} \longrightarrow \text{reduced P}$$

Factors in Process or Composting

The most important factors in decomposition of organic wastes are as follows:

(*i*) carbon-nitrogen ratio, (*ii*) blending or proportioning of wastes, (*iii*) moisture and aeration, (*iv*) temperature (*v*) reaction, (*vi*) microorganims involved, (*vii*) use of inoculants, (*viii*) calcium phosphate, and (*ix*) destruction of pathogenic organisms.

C : N Ratio

The carbon-nitrogen ratio of organic materials is the most important aspect of composting. The process of conversion of organic materials into manure is chiefly microbiological and is, therefore, influenced by the proportions of carbonaceous and nitrogenous materials that are present in organic wastes to start with. Microorganisms need carbon for growth and nitrogen for protein synthesis. It was found that a C : N ratio of 30 of raw materials could be most desirable for efficient composting. The C : N ratio between 26 to 40 as reported by many workers, provide for rapid land efficient composting. If the organic material is poor in nitrogen or in other words, carbon nitrogen ratio is wide, biological activity diminishes and several successions/cycles of organisms may be required to degrade carbonaceous materials. Immobilised nitrogen is recycled on the death of some of the organisms. Thus, the limited nitrogen is recycled by reducing the carbon content of the organic wastes. At low C : N ratio, ammonia is formed which under favourable conditions can be further oxidised to nitrite) and nitrate. Other nutrients, like phosphorus, potassium, sulphur and micronutrients are also essential for microbial growth.

Lower C : N ratios under unfavourable conditions cause increasing loss of ammonia and higher values may prolong the period of composting. High C : N ratios are generally caused by organic materials poor in nitrogen such as cereal residues of wheat, paddy, *jowar, bajra,* maize, sugarcane trash stalks of cotton, jute and sawdust. Farm and urban wastes have been found to vary widely in C : N ratio ranging between 30 and 80. As the stock of available nitrogen gets exhausted, the activity of nitrogen fixing microorganisms predominates and thus there is a gain of nitrogen fixed from the atmosphere. With a C : N ratio of raw materials less than 30 : 1, the proportion of nitrogen is in excess of the requirements of microorganisms. Although the process of decomposition goes on uninterrupted the unassimilable nitrogen is lost in the form of ammonia gas.

Similarly, 20 to 40 per cent nitrogen is lost in the form of ammonia during preparation of organic manure from cattle-dung and night-soil. This implies that a thorough mixing of there fractions with the carbonaceous wastes cannot be attained in practice and there are always pockets of concentration of nitrogenous materials which cause an imbalance in the decomposition process and loss of nitrogen. It is obvious that simple nitrogenous compounds are more susceptible to loss which are easily decomposed by microorganisms and these compounds are of high utility to plants. Control of the loss and retention of this important plant nutrient in the final manure would be of practical value in obtaining a better quality manure.

The time of composting can be reduced by adding a nitrogen source of organic materials such as activated sewage sludge or by blending with organic residues richer in nitrogen such as legume residues and aquatic weeds (water hyacinth), slaughter-house wastes, green leaves, wastes of sugar, and wastes of antibiotic, yeast and paper industries.

Shredding and Blending

Composting process can be accelerated if raw materials are shredded into smaller pieces or ground since the material becomes more susceptible to bacterial attack due to exposure of greater surface area. The most desirable particle size for composting is less than 5 cm although larger sizes can be composted satisfactorily. However, the advantages of shredding may be uneconomical while composting on farms and in villages or by individual farmers.

The C/N ratio and moisture percentage are two important parameters which should be considered while mixing or blending different types of organic wastes for composting. There may not be any need of blending if the C/N ratio of organic materials is between 25 to 50, although 30 to 40 may be a good range. Substances poor in nitrogen such as sawdust, paper, straw, etc. can be mixed with comparatively nitrogen rich materials, *e.g.*, legume residues, water hyacinth, slaughter-house wastes, fish scrap, nightsoil, sewage sludge and biogas spent slurry to obtain a near optimum C/N ratio between 30 to 40. Similarly materials too dry and too moist can be blended together to obtain a desirable moisture level to avoid anaerobic conditions leading to bad odour.

Soil at 5 to 10 per cent is also sometimes added to compost to reduce the moisture content and to absorb ammonia in low C/N ratio materials. The dry soil may be added if sufficient dry materials are not available. Soils may be added to high C/N ratio organic matter to buffer acid conditions and to act as a diluent for retarding anaerobic process. It can also be used for improving the appearance of finished compost to give it more granular structure.

Moisture

If the amount of moisture in the compost pile is below 40 per cent (W/W), decomposition will be aerobic but slow. Aerobic decomposition can go on at any moisture content between 30 and 100 per cent if adequate turnings are provided. During aerobic composting high moisture content should be avoided so that anaerobic conditions are not created. The optimum moisture level of 50 to 60 per cent may be quite satisfactory for aerobic composting. However, it may be in range of 40 to 80 per cent depending on the nature of organic material to be composted. At IARI, when composting fibrous materials containing straw were composted aerobically. 80 to 85 per cent moisture was maintained. If anaerobic, composting is to be carried out, the maximum moisture content is not important. The moisture content between 80 to 90 per cent may be maintained.

Temperature

With the multiplication of microorganisms in the composting mass, the heat of exothermic biological reaction is retained due to sufficiently large mass and when the temperature rises above 40°C, the mesophilic microorganisms are replaced by thermophilic. High temperatures are essential for destruction of pathogenic organisms and weed seeds. This generally occurs within 2 to 5 days of the start of the composting. The temperature in the middle of the pile goes up to 55° to 70°C and after this it gradually cools to ambient temperature. Decomposition is fastest in the thermophilic stage. The optimum temperature based on oxidation of organic matter into carbon dioxide and water has been found by Willey and Pierce to be 60°C. Schulz showed that a maximum temperature of 71°C was optimal, but it was found that temperature should not exceed 70°C for long because decomposition will be slowed down by a thermal kill of microorganisms and only a few of the thermophilic organisms actively carry on decomposition above 70°C. However, while composting high C/N ratio materials such as wheat and paddy straw, *jowar* stalk and *jamun* leaf fall, etc. of C/N ratio 48 to 50 in cubic metre pits, it was observed by Gaur that temperature of the compost pile did not go beyond 52°C. This indicated that extent of rise of temperature in compost will depend on the type material being composted and probably also on the size of the heap.

Reaction

The initial pH in compost heaps is generally slightly acidic round pH 6 as is found in the cell sap of most of the plants. The production of organic acids during the early stages of composting causes further acidification (pH 4.5–5.0) but as the temperature rises, the pH increases to slightly alkaline (pH 7.5–8.5) reaction.

Microbiology

The microbial population changes during aerobic composting, Chang have described a typical pattern. The fungi and acid producing bacteria appear during the initial mesophilic stage. As the temperature increases above 40°C, these are replaced by thermophilic bacteria, actinomycetes and fungi. Spore-forming bacteria develop at temperature above 70°C. Finally, mesophilic bacteria and fungi reappear as the temperature falls down.

Many aerobic mesophilic bacteria initially present in the composting material multiply and show increased activity. As the temperature is raised, their numbers decrease due to change in environment. A minimum is reached at 55° to 65°C during composting of grass cuttings and straw but their number increases again as temperature drops below 45° to 50°C.

The role of bacteria may be involved in raising the temperature of the compost for the development of thermophilic microorganisms which colonise. Mesophilic bacteria which flourish during limited time, consume the most readily degradable carbohydrates. During the process compounds rich in nitrogen such as proteins, amino acids, blood-meal and peptones are bio-degraded especially by spore forming *Bacillus* spp. Actinomycetes degrade starch actively and also bring about large losses of water soluble fractions. Thermophilic bacteria do not appear to be important in degradation of cellulose and lignin but attack protein, lipids and hemicellulose fractions. Lipids are degraded to a great extent whereas degradation of celluloses and hemicelluloses is comparatively slow and intermediate. Lignin are the most resistant to decomposition and persist for longer period and tend to accumulate.

Actinomycetes (*Thermomonospora curvata*) may be important in cellulose decomposition. *T. curvata* was the most frequently occurring actinomycete in municipal and mushroom composts, because thermophilic actinomycetes can grow at higher temperatures than thermophilic fungi and they become dominant at the warmest stage.

Mesophilic fungi are present as the compost warms up moderately. These fungi are saprophytic sugar fungi which are quickly replaced by thermophilic fungi. The mesophilic fungi reappear in large number as the heap cools down below 40°C. Evidently they persist in outer layers during the thermophilic stage and reappear when the temperature drops. They can utilise cellulose and hemicellulose but not as efficiently as thermophilic fungi.

Thermophilic fungi occur when the temperature ranges between 40°C and 60°C. Like mesophilic fungi, they survive in the periphery of the heap when the temperature exceeds this range. The bacteria, fungi and actinomycetes isolated and reported from compost are presented in Table 5.1.

Table 5.1: Bacteria, Actinomycetes and Fungi Isolated from Compost

Bacteria

Mesophilic

- *Cellumonas folia*
- *Chondrococcus exiguus*
- *Myxococcus virescens*
- *M. fulvus*
- *Thiobacillus thiooxidons*
- *T. denitrificans*
- *Aerdbacter* spp.
- *Proteus* spp.
- *Pseudomonas* spp.

Thermophilic

- *Bacillus stearothermophilis*

Actinomycetes

Thermotolerant and Thermophilic

- *Micromonospora vulgaris*
- *Nocardia brasiliensis*
- *Pseudonocardia thermophila*
- *Streptomyces rectus*
- *S. thermofuscus*
- *S. thermoviolaceus*
- *S. thermophilus*
- *S. thermovulgaris*
- *S. violoceoruber*
- *Thermomonospora curvata*
- *T. fusea*
- *T. glaucus*
- *Thermopolyspora polyspora*

Fungi

Mesophilic

- *Fusarium culmorum*
- *F. roseum*
- *Stysanus stemonitis*
- *Coprinus cinereus*
- *C. megacephalus*
- *C. lagopus*
- *Clitopilus pinsitus*
- *Aspergillus niger*
- *A. terreus*
- *Geotrichum condidum*
- *Rhizopus nigricans*
- *Trichoderma viride*
- *T. (lignorum) harzianum*
- *Gospora variabilis*
- *Mucor spinescens*
- *M. abundans*
- *M. variens*
- *Cephalosporium acremonium*
- *Chaetomium globosum*
- *Glomerularia* sp.
- *Pullularia (Aureobasidium)*
- *Fusidium* spp.
- *Actinomucor corymbosus*
- *Mucor jansseni*
- *Taloromyces (Penicillium) variabile*
- Helminthosporium sativum
- *Aspergillus fumigatus*

Thermotolerant and Thermophilic

- *Humicola insolens*
- *H. griseus* var. *thermoideus*
- *H. lanuginosa (Thermomyces lanuginosus)*
- *Mucor pusillus*
- *Chaetomium thermophile*
- *Absidia ramosa*
- *Talaromyces (Penicillium) duponti*
- *T. emersonii*
- *T. thermophilus*
- *Sporotrichum thermophile*
- *S. chlorinum*
- *C.t. 6 (Mycelia sterilia)*
- *Stilbella thermophila*
- *Malbranchea putchella* var. *sulfurea (Thermoidium sulfureum)*
- *Dactylomyces cruslaceous (Thermoascus aurantiacus)*
- *Byssochlamys* spp.
- *Torula thermophila*

Inoculants and Other Preparations to Compost

If C/N ratio is too wide, adding nitrogen can hasten composting by lowering the ratio. Microbial inoculations may also be beneficial in a compost heap poor in microorganisms for speeding up decomposition of materials rich in cellulose and lignin.

It has been observed in sawdust compost that inoculation with spores of cellulolytic fungus, *Coprinus ephemerus* and addition of nitrogen, phosphorus and potash shortened composting of sawdust from usual 1–2 years to 3 months and produced a compost that did not immobilise nitrogen as compared to fresh sawdust.

Japanese patents for additives to accelerate decomposition of wood products have been reported recently. The addition of nitrohumic acid (humus or complex natural ligno-protein treated with nitric acid) to bark wastes and chicken manure (1 : 7 : 100) reduced the decomposition time from 180 days to 60 days. Decomposition of wood pulp was improved by urea and crude acetic acid distilled from wood. The recent work done in India on microbial inoculants for accelerating the compost-making has been discussed separately in this chapter.

Calcium Phosphates

Several investigators have reported that calcium phosphate increases the rate of decomposition and nitrogen conservation but more than 2 per cent inhibited the decomposition rate. Chang found increased rate of decomposition occurring mainly in cellulose and hemicellulose fractions. The number of cellulolytic microorganisms increased with the addition of phosphate and conserved nitrogen by decreasing the number of denitrifying bacteria. Although N fixation may be increased by adding calcium phosphate, there is generally sufficient phosphate in the organic wastes for meeting the requirements of microorganisms which need about 5 to 20 per cent only as much P as N. However, our experiments have shown that application of rock phosphate is important for nitrogen conservation and enrichment of organic manures, while composting wide C/N ratio organic materials (paddy, *jowar* straw, *jamun* leaves). Gaur recorded that *Azotobacter* numbers in compost were greater in presence of added rockphosphate.

Gaur reported that addition of rock phosphate at 1.0 per cent (W/W) increased the decomposition of paddy straw by cellulolytic fungi (*Aspergillus* spp. supplemented with urea nitrogen) (Table 5.2). The carbon content was decreased by 17.2 per cent and nitrogen augmented by 31.4 per cent. As a result C/N ratio of 23.1 was obtained within 8 weeks as compared to 37.7 in the control. Increased amounts of rock phosphate (2 and 3 per cent) showed slight reduction in decomposition intensity. Mathur from Ranchi Agricultural College reported that quality of compost prepared from mixture of paddy straw, grass and water hyacinth was improved when rock phosphate was applied to it with and without pyrite. The chemical composition of the compost is given in Table 5.3. The results showed that low dose of rock phosphate (5 per cent) increased the nitrogen content by 30 per cent of the compost over the control. The enrichment of compost with rock phosphate besides increasing the phosphorus content, increased the concentration of calcium and micronutrients particularly of iron, manganese and zinc. Beneficial effect of addition of rock phosphate at an optimum rate during the initial stage of compost making has been reported at IARI and ICAR Co-ordinated Project on "Microbial Decomposition and Recycling of Organic Wastes."

Destruction of Pathogens and Parasites

The destruction of pathogens and parasites of animals and humans is an important aspect and a problem when compost is prepared from materials containing sewage sludge and night-soil. Aerobic composting at high temperatures is effective in killing pathogens. Gotaas recommended 60°C for "Thermal Kill" of common pathogens listed in the Table 5.4

Table 5.2: Effect of Rockphosphate on the Decomposition of Paddy Straw by *Aspergillus* spp.

Treatments	*% Loss in Weight (2 months)*	*Organic Matter %*	*Carbon %*	*Total N %*	*C/N Ratio*
Urea (0.5% w/w) control	45.2	75.35	43.71	1.190	37.73
Urea + Rock phosphate (1.0% w/w)	55.0	62.36	36.17 (– 17.2)	1.564 (31.4)	23.12
Urea + Rock phosphate (2.0% w/w)	52.9	62.64	36.33 (– 16.9)	1.326 (11.24)	27.40
Urea + Rock phosphate (3.0% w/w)	52.8	64.60	37.47 (– 14.3)	1.275 (7.1)	29.39

Figures in parentheses show per cent loss or gain.

Table 5.3: Chemical Characteristics of Compost Prepared with Rock Phosphate

Components	T_1	T_2	T_3
Total Ca (%)	1.85	2.27	1.92
Total Mg (%)	0.37	0.51	0.38
Total Na (%)	0.06	0.19	0.09
Total K (%)	0.92	0.81	0.83
Fe (ppm)	5,080	12,720	28,200
Mn (ppm)	180	520	540
Zn (ppm)	60	95	80
Cu (ppm)	10	24	44
pH	4.40	5.55	5.80
Organic Carbon (%)	14.64	18.06	12.83
Total Nitrogen (%)	0.61	0.79	0.58
C/N ratio	24.00	22.80	22.10

T_1: 650 kg raw material (P_0)

T_2: 650 kg raw material + 32.5 kg Mussoorie rock phosphate (MRP)

T_3: 650 kg raw material + 32.5 kg (MRP) + 32.5 kg Amjhore Pyrite

The extent and duration of high temperatures in compost heaps as well as antibiosis as a result of mixed population of microorganisms provide reasons to believe that pathogens and parasites are not able to survive the composting process. To achieve this at least 2 to 3 turnings of compost heap is suggested. Knoll found that 65°C for one day ensured elimination of *Salmonella* spp. Stranch reported the destruction of *S. enteriditis, Erysipelothrix, Rhusiopathiae* and the *Psittacosis* virus under similar conditions. *Bacillus anthracis* was destroyed only at temperature above 55°C for 3 weeks or longer at a moisture content of 40 per cent or more. He concluded that higher temperature may not be the only way to kill pathogens during composting but some are killed by competition with other microorganisms.

Schraff found that resistant forms of parasites such as *Ascaris* eggs, cysts of *Entamoeba histolytica* and hookworm eggs were destroyed during aerobic composting. Scott found that at least 2 turns were necessary for the destruction of *Entamoeba histolytica* in 12 days and three turns were required to destroy the eggs of *A. lumbricoides* in 36 days. Anaerobic composting at the mesophilic temperature range does not effect satisfactory destruction of parasites in a relatively short time. Anaerobic composting when using sewage sludge or nightsoil should be preceded by aerobic composting at least for a week. The natural death of pathogens and parasites will occur in an anaerobic environment and the microbial antagonism will eventually eliminate them but this will generally take longer period of about six months.

Table 5.4: Temperature and Time of Exposure Required for Destruction of Some Common Pathogens and Parasites

Organism	*Observations*
Salmonella typhosa	No growth beyond 46°C; death within 30 minutes at 55°–60°C and within 20 minutes at 60°C; destroyed in a short time in compost environment
Salmonella spp.	Death within 1 hour at 55°C and within 15–20 minutes at 60°C
Shigella spp.	Death within 1 hour at 55°C
Escherichia coli	Most die within 1 hour at 55°C and within 15–20 minutes at 60°C
Entamoeba histolytica cysts	Death within a few minutes at 45°C and within a few seconds at 55°C
Taenia saginata	Death within a few minutes at 55°C
Trichinella spiralis larvae	Quickly killed at 55°C; instantly killed at 60°C
Brucella abortus or *Br. Suis*	Death within 3 minutes at 62–63°C and within 1 hour at 55°C
Micrococcus pyogenes var. *aureus*	Death within 10 minutes at 50°C
Streptococcus pyogenes	Death within 10 minutes at 54°C
Mycobacterium tuberculosis var. *hominis*	Death within 15–20 minutes at 66°C or after momentary heating at 67°C
Corynebacterium diptheriae	Death within 45 minutes at 55°C
Necator americanus	Death within 50 minutes at 45°C
Ascaris lumbricoides eggs	Death in less than 1 hour at temperatures over 50°C

Heaps versus Pits

When preparation of manure is carried out in over-ground heaps, the decomposition becomes mainly aerobic, with air freely accessible for microorganisms to carry out their activities. The rate of decomposition is accelerated if periodical turnings are given which also promotes physical disintegration of the raw materials. Owing to quick oxidation of organic matter, high temperatures (60°–70°C) develop, thereby killing weed seeds and pathogenic organisms. Loss of organic matter in this process often amounts to above 50 to 60 per cent and decomposition may continue even after the C/N ratio is reduced to the optimum. Loss of nitrogen from 20 to 50 per cent invariably takes place in this method.

Moisture content necessary for microbial activity is effected in the heap method, with losses occurring because of exposure to the atmosphere and the high temperatures that develop in the manure heap. In dry climate, this would affect the rate of decomposition and necessitate addition of extra water periodically. During the rainy season, on the other hand, there may be too much of water in the heap to slow down decomposition and even cause loss of nutrients and organic matter by leaching, unless manure preparation is carried out under sheds. This is not always feasible where large quantities of manure are required to be prepared.

Preparation of manure in pits has several advantages over method of preparation in. heaps over ground although it involves some initial expenditure on digging and permanent occupation of land for this purpose. The decomposition is mainly anaerobic, except in the surface few inches and proceeds slowly. Losses of organic matter as well as nitrogen are consequently reduced, amounting to about 25 and 20 per cent respectively, enabling a larger quantity as well as good quality of the manure to be obtained. However, the process of composting can be accelerated by providing a few turnings at periodic intervals which will create aerobic conditions as well. The materials are protected against loss of moisture by evaporation as well as excess of water in the rainy season and physical loss of nutrients by leaching. High temperatures do not develop in this case owing to the anaerobic nature and the slow rate of decomposition. But weed seeds and pathogenic organisms are destroyed because of the toxic products of this type of decomposition. Little physical disintegration of material such as straw and refuse, however, takes place in this system but this is not necessary from scientific standards of assessment of quality of manure of which a final C/N ratio of less than 20: 1 indicates its maturity and fitness for application to land. The manure may remain stored in pits until required without much loss due to very restricted decomposition after it is ready.

Utilisation of Urine Fraction

The liquid excreta of animals contains nitrogen in the form of urea, phosphorus, potash and several other nutrients in water soluble form which are of high utility to plants. Stress has always been laid on proper collection of this fraction and its utilization by adding urine-soaked litters or urine-earth to the manure-pit. However, in the process of its microbial fermentation with carbonaceous materials, the original simple compounds of nitrogen are converted into complex microbial proteins or into ligno-proteins by chemical reaction. There is also considerable loss of nitrogen in the form of ammonia due to the ease of decomposition of urea by numerous microorganisms. An overall reduction in the manurial value of this fraction thus takes place when it is incorporated in manure heaps. The low C/N ratio of urine and the simple nature of its nitrogenous compounds suggests that its separate collection and direct utilisation as manure would be a more effective alternative if practical methods of collection and conservation of this liquid manure are evolved. Work on this aspect carried out at the Indian Agricultural Research Institute and National Dairy Research Institute showed that urine-soaked earth can be treated with super phosphate or gypsum to control the decomposition of nitrogen and then quickly dried to stop bacterial activity or urine-soaked ground paddy husk or saw-dust can be composted along with dung with a good quality manure. The plant nutrient constituents thus remain conserved in their original form and the manure obtained is highly effective. Alternative materials like dry leaves, dry dung, etc. be used in conjuction with soil in cattle-sheds to soak urine more effectively and processed as above. In *pucca* cattle-sheds where urine and cattle-shed washings are collected in cess pits, the liquid, manure may be similarly treated with superphosphate or a mineral acid like hydrochloric acid and then soaked and dried in absorbing materials of the type mentioned.

Results obtained at the Indian Agricultural Research Institute have shown that the method can be effectively extended and applied to the collection and utilisation of large quantities of human urine which has not been exploited so far. A simple design of urinals evolved by the Institute (IARI) enables urine to be collected in absorbing materials contained in a drum or a pit dug for the purpose. The decomposition of urine is controlled by acidulating the material with superphosphate or hydrochloric acid. The material can be repeatedly soaked and dried to obtain concentrated manures. Dehydrated human urine manures have been found to be highly effective in stimulating growth and yield of crops. The proposition of separate collection and utilisation of cattle and human urine in the above manner offers a new, yet unexplored field for efficient utilisation of an inexhaustible and rich source of manure available in the country.

Table 5.5: Effect of Human Urine Dehydrated in Different Materials on Wheat (Manures applied at 112 kg N/ha)

Treatment	*Yield in g from 30 Plants*		
	Grain	*Straw*	*Total*
No manure	11.5	23.8	35.3
Urine in soil	26.6	33.9	60.5
Urine in saw-dust	27.4	37.5	64.9
Urine in charcoal-dust	28.0	37.3	65.3
Ammonium sulphate	22.8	51.6	54.4
C.D.	5.7	4.8	7.6

Preparation of Farmyard Manure

Organic waste materials that generally go in production of farmyard manure including dung, urine, soaked earth or litter, remnants of cattle food, ash and other household and kitchen wastes. The existing method of preparation by dumping the small daily collection in a heap or shallow pit leaves much to be desired and yields a low-quality manure.

Sectional Filling

One of the effective methods of preparation of farmyard manure that has been found to yield superior quality manure is the process called 'Sectional Filling of Trenches', which includes all the advantages of carrying out of decomposition in pits and suits well to the conditions existing in this country of having to deal with a comparatively small daily collections. The method consists of digging suitable-sized trenches depending on the number of cattle as shown in Table 5.6.

Table 5.6

No. of Cattle	*Length (m)*	*Breadth (m)*	*Depth (m)*
2 to 6	6.5	1.0	1.0
6 to 10	8.0	1.2	1.0
11 to 20	10.0	1.4	1.0
above 20	10.0	1.6	1.0

Filling of mixed wastes from the cattle-shed is recommended from one end of the trench in a section, one metre in length. A movable partition made from cotton stalks or such other material between two bamboo poles is used for making the section to be filled. The daily collection is dumped in this and covered with a thin layer of soil. This is continued until the section is filled to a level of about 50 cm above ground level. The top is then covered with earth and the partition is moved one metre further along the length to form the next section for filling as before. Leaves and other farm wastes like paddy and groundnut-husks, straw, etc. could be added to increase the quantity of manure if sufficient litter from the cattle-shed is not available. The manure is ready in about three months and contains about 2 per cent nitrogen on dry weight basis. The quantity of manure obtained per pair of animals amounts to 10 to 12 tonnes per year.

Composts

The principles of production of farmyard manure from cattle-shed wastes have been extended for the preparation of 'composts' in which a variety of carbonaceous and nitrogenous waste materials may be utilised. Thus, farm-wastes like straw, chaff, leaves, paddy and groundnut-husks, sugarcane-trash, bagasse and other agricultural as well as industrial and habitation wastes of all sorts like street-refuse, night-soil, slaughter-house wastes, sewage and sullage water, sludge, etc. can be pressed into service to yield manures.

In India, pioneering work in this field was carried out by Howard at Indore and Fowler at Bangalore. who standardized conditions for degradation of leaves, straw and town-refuse, utilizing dung and night-soil as starters for promoting bacterial activity.

Indore Method of Composting

In this method the waste materials are spread out and moistened with dung or night-soil slurry. These are then built up into heaps, four to six metres in length, one metre in breadth and one metre in height. Under these conditions, the rate of decomposition is rapid and high temperatures develop quickly. The process is accelerated if turnings are given, whereby aeration and mixing of the materials and moistening at this stage, if necessary, cause more or less total disintegration, yielding a brown homogenous manure in about three months. Under this aerobic process of decomposition losses of organic matter and nitrogen are heavy, amounting to 40 to 50 per cent of the initial. The manure resembles the traditional farmyard manure in appearance and properties and it has been shown that decomposable organic materials of any type can thus be processed to supplement the supplies or manure. This method of manure preparation, however, involves considerable labour in building up the heap to proper shape and periodical turnings that are given and becomes expensive and impracticable where 1arge quantities or materials have to be processed. However, the turnings of the mass are not necessary and decomposition can go on to the desired extent if adequate moisture level is maintained. The average composition of manure prepared by the Indore method has been round to be 0.8 per cent nitrogen, 0.3 per cent phosphoric acid and 1.5 per cent potash.

Fowler worked out the process of 'activated compost' in which fresh raw materials are incorporated m an already fermenting heap so that the established and large microbial population beings about quicker decomposition. The process is useful particularly where offensive materials like night-soil are to be quickly and effectively disposed of.

Compost from Habitation Wastes–Bangalore Process

Pioneering work in preparation of manure in pits has been carried out in India by Acharya, particularly on the utilisation of town-refuse and night-soil. What is called the hot fermentation method of manure production or the Bangalore process has by now been adopted in many towns in India, solving effectively one of the most difficult problems of sanitary disposal of these offensive wastes and yielding a high quality organic compost manure in the bargain. The operative part of the process can be adapted to the prevailing system of collecting refuse and night-soil. The method carried out in practice is as follows:

The compost production depot is located just on the outskirts for convenient transport of street-refuse and night-soil to the pits. In large towns, there may be two or three such depots to serve different localities. The compost depot should accommodate about 200 trenches of dimension as under, which can take the dumpings for one year:

Table 5.7

Population (in 000)	*Length (m)*	*Breadth (m)*	*Depth (m)*
Below 10	4.5	1.5	1.0
10 to 20	6.0	2.0	1.0
20 to 50	9.0	2.0	1.0
Above 50	10.0	2.5	1.0

The trenches are dug in rows with a space of 1 to 5 m between trenches. Roads of suitable width are provided between rows for the carts to approach and unload the materials inside the trenches. At first the refuse is dumped into the trench and spread out with rakes to make a layer of 15 cm. Night-soil is then discharged over this and, if necessary, spread out with long wooden spades in a layer of about 5 cm. This is then covered with a 15 cm layer of refuse. Night-soil and refuse thus follow in alternate layers until the pit is filled to 15 cm above ground level, with a final layer of refuse on the top. This may be dome-shaped and covered with a thin layer of soil. In some towns, night-soil comes mixed up with refuse in which case the material is simply dumped to fill the trenches. Sullage water, if collected in carts as in some towns, may also be emptied over the layers of refuse. In fact, the system provides a method of disposal of every kind of wast that the town may have, including slaughter-house, carcasses of animals, sewage, etc.

The decomposition of dumped materials in pits takes place largely in the absence of sufficient air except in the surface foot or so. This anaerobic decomposition is comparatively slow but markedly less wasteful. The difference in recovery of organic matter and nitrogen by this method as compared with the aerobic method in over-ground heaps is given below for manure prepared from town-refuse and night-soil.

Table 5.8: Percentage Recovery

	Heap	*Trench*
Organic matter	53	78
Nitrogen	54	74

In this method, no high temperatures develop in the lower layers and since the material does not receive turnings, disintegration into a homogeneous mass of manure does not take place. Even so, the C/N ratio is reduced to less than 20 : 1 in about six months and the manure is then ready for use. Deeper in the trenches, the material still possesses an unpleasant smell and it is desirable to dig it out and pile it up outside for some days before disposal.

In the absence of facilities for digging trenches, town-refuse and night-soil compost can be prepared in over ground heaps of one metre height, one metre width and any convenient length, with refuse and night-soil, placed in alternate layers, as in trenches and a final refuse layer on the top. Night-soil carts empty the material along the length of heap. No turnings need be given to the heaps and the material decomposes more quickly than in pits and can be used after three to four months.

The basic aspect of these processing treatments in manure production is that after the initial active decomposition and loss of excessive carbon has taken place, the proportion of nitrogenous night-soil and organic carbonaceous material is just proper and it can be used as manure. Often, the night-soil and refuse layers can be identified when the manure is dug out. The material gets mixed up only while it is being taken out, but because the mixture is in proper proportion, it is safe to apply as manure. Further decomposition is carried out in the soil itself and because of the presence of adequate nitrogen this causes no adverse effect and instead soon releases nitrogen for crops to assimilate.

In villages individual latrines are limited to only a few rich families, hence night-soil collection and its compost making is not popular. It can, therefore, be a great step in improving the existing insanitary conditions to provide suitable community or individual laterines, preferably Wardha-type trench latrines to villagers to utilise human-wastes. Before putting Wardha-type laterines in use, a 7.5 cm layer of refuse (grass, leaves, straw, weeds, trash, etc.) is spread at the bottom of the trench. A stock of refuse or dry earth is left to cover the faeces. After the material rises to a height below 15 cm of the top, the trench is sealed for decomposition and another trench put in use. Manure of good quality is ready in four to six months.

Preparation of Synthetic Compost

The basic principles underlying the microbiological process of cunversion of organic wastes into manure have been extended for preparation of synthetic composts, using inorganic fertilizers as sources of nitrogen for decomposition of carbonaceous materials. It has been found that organic nitrogen in the form of dung, required by microorganisms, can be completely substituted by inorganic nitrogen compounds like ammonium sulphate and urea, which are utilised equally effectively for decomposition and well fermented manure, resembling farmyard manure, can thus be obtained.

This opens up a vast field for utilisation of large quantities of various organic waste materials where supplies of dung are either short of requirements or not available at all, as on mechanized farms. The basic principles of C/N ratio in manure preparation can be applied to add nitrogenous fertilizers in sufficient quantity to reduce the ratio to about 30–40 : 1 and then allow the material to be decomposed and disintegrated until the C/N ratio is reduced to 20 : 1. The adco process of preparation of synthetic compost worked out by Hutchinson and Richards is based on this principle.

This process of production or compost from organic wastes of wide C/N ratio is worked with fertilizers like ammonium sulphate or urea. They provide complete or partial nitrogen for microbial decomposition in place of natural materials like dung or night-soil which may not be available at all or may be in short supply. Nitrogen is added to make up a total of 1.2 per cent on dry weight basis of the material. Thus straw containing 0.4 per cent nitrogen requires an additional 0.8 per cent nitrogen, *i.e.* about 4 kg of ammonium sulphate per 100 kg of straw. For neutralizing acidity, which may develop in this system, the process includes addition of lime at the rate of 5 kg to the above material. The material to be composted is spread out in layer in a heap or pit and sufficiently moistened. This is sprinkled with the fertilizer solution and then with lime. Superphosphate may be added to fortify the phosphatic contents of the manure. The treatment is continued layer-wise until the heap or pit is filled to size and allowed to ferment. The manure becomes ready for application in about four to six months and resembles farmyard manure in its action on soil and plant growth.

Mechanical Compost Plants

Although the major quantities of compost (310–350 million tonnes) at present used in agriculture are prepared in villages either by traditional or improved methods (Indore or Bangalore methods), these methods cannot be suitably used for processing of large quantities of organic refuse of bigger cities.

The mechanised/semi-mechanised plants will serve two pertinent objectives–sanitary disposal of city refuse and production of organic manure. It is estimated that by 1979 nearly 50 million tonnes of municipal wastes will be generated annually in India. Moreover, Indian city wastes contain about 50 to 80 per cent of compostable organic materials which can be processed conveniently in compost plants. During Sixth Five Year Plan, Ministry of Agriculture and Irrigation timely launched a programme to set up 35 composting plants in selected cities with a population of 3 lakhs and above. Five compost plants have already been set up one each at Ahmedabad, Baroda, Bangalore, Calcutta and Jaipur and the remaining 30 plants are expected to be set up during Sixth Five Year Plan (1978–83). The intake capacity of these plants already set up is about 125 to 200 tonnes garbage per day and can produce 60 to 70 tonnes of finished compost per day if the plant runs in shifts.

The costs of production is rather high due to mechanisation of the processes and at some places the plants were running at under-utilised capacity due to non-availability of required quantities of garbage. The selling price of finished compost has been kept between Rs 50 to 80 per tonne depending upon the composting system adopted and plant capacity. The cost may be reduced by installation of plants with less mechanisation.

The compost prepared from city wastes is bulky and low in plant nutrients. It contains 0.5 to 0.6 per cent N. 0.6 per cent P_2O_5 and 0.5 percent K_2O with a C/N ratio of 20 : 1. There is a need to enrich this manure in respect of nitrogen and phosphorus and reduce its bulkiness involving low-cost technology. The work carried by Gaur showed that by application of 10 kg low-grade rock phosphate containing 20 per cent P_2O_5 per tonne of organic wastes (crop wastes and leaf fall, etc.) and inoculation with microbial fertilizers improved the nitrogen, phosphorus and humus content appreciably. This can be adopted in improving the compost prepared by mechanical compost plants.

Use of Composted Versus Uncomposted Materials

Direct application of carbonaceous materials of wide C/N ratio to soils adversely affects plant growth. This is owing to the utilisation of all soil nitrogen by microorganisms, engaged in the process of decomposition. Mixed organic wastes containing adequate nitrogenous materials like dung and night-soil can, however, be directly applied to land and allowed to decompose *in situ* in soil without causing deficiency of nitrogen and the consequent depressing effect on plant growth. If the C/N ratio of the materials is restored to 30 : 1 or less, the decomposition in soil is faster owing to dilution and large variety of microbial population of the soil. Positive advantages that accrue from the decomposition carried out in the soil are: (*a*) no loss of nitrogen in the form of ammonia takes place in the presence of the large quantity of soil; (*b*) carbon dioxide evolved in large quantities during decomposition assists in enhancing the available plant nutrients of the soil; and (*c*) production of compounds possessing soil aggregating properties during decomposition by microorganisms, imparts good structure to the soil. The succeeding crop thus derives greater benefit from this manner of application than from the same raw materials after composting. It is, however, obvious that there are some limitations to the use of raw waste materials in practice. Materials like long-straw and street-refuse may be difficult to incorporate without prior disintegration. Offensive materials like night-soil cannot be handled in raw condition. Composting destroys pathogenic organisms and weed seeds that may be present in such materials. Under intensive cropping, the gap between one crop and the other may be too short for direct application of organic materials to soil and they may have to be composted before application. It has, however, been established that whenever possible, direct incorporation of materials like straw, leaves, etc. sufficiently in advance of sowing, with supplementary application of fertilizer nitrogen in adequate quantity, results in marked improvement in soil fertility with significant effect on crop yield.

Night-soil Poudrettes

The disposal of night-soil is often a serious problem in towns where facilities for its composting are not available or refuse in adequate quantity is not collected. Trenching the material as such takes a long time for it to dry and causes-nuisance of foul smell and proluse fly-breeding. Dehydration of night-soil by mixing it with any absorbing material like soil, ash, dry broken leaves, paddy-husk, etc. is a highly effective method of disposal of this offensive waste which yields poudrette manures of high manurial value. One of the most effective materials for this purpose is sawdust, which has high dehydrating capacity as well as the property of absorbing foul smell. The material can be used whenever available to treat night-soil and obtain an organic manure of high quality. Owing to its low C/N ratio, utilisation of night-soil as such without fermentation yields higher efficiency of response than when it is composted: The question of survival of pathogenic organisms in preparation and use of poudrette manures of this type is, however, a matter warranting consideration and further study.

Super-digested Compost

Recent trends in the preparation of farmyard manure and composts include the use of superphosphate which is sprinkled over the raw materials as they are being filled in pits, at the rate of about five per cent. It has been found that the treatment makes organic manures of this type more balanced and markedly reduces losses of nitrogen that take place during decomposition. The availability of phosphorus, thus intimately mixed with organic matter, is higher than when this fertilizer is used as such and this is of particular advantage to acidic, calcareous and heavy soils in which fixation of phosphords greatly reduces the efficiency of this plant nutrient. Results of work by Walunfkarhar and Acharya illustrated the effect of super-digested compost on crop growth are given in the Table 5.9. Yield of *marua* was significantly better with application of super-digested compost than with separate application of the two.

Table 5.9: Average Yield of *Marua* in g per Pot

Sl.No.	*Treatment*	*Grain*	*Straw*	*Total*
1.	No Manure	7.4	23.8	31.2
2.	Compost (2½ tonnes dry matter/hectare)	10.3	33.0	43.3
3.	Super-phosphare (P_2O_5 equal to No. 2)	8.9	30.8	39.7
4.	Super-digested compost	12.3	41.6	54.1
5.	Compost plus super-phosphate (as in No. 4)	10.3	32.6	42.9
	S.E.	**0.35**	**1.73**	**3.18**

Subbiah, by the use of tracer technique, showed that the movement of applied phosphorus ill soil generally increased when organic matter in the form of compost was also applied with the fertilizer arid this was further enhanced when applied as super-digested compost. The increase in movement due to super-digested compost over mere combination of super-phosphate and compost was higher in laterite, red and calcareous soils.

Control of Losses of Nitrogen During Manure Preparation

It has been mentioned that the process of decomposition in the preparation of organic manures results in significant losses of nitrogen in practice, which reduces the quality of final manure that is obtained. The losses ranging up to 20 per cent take place although they are heavy in heaps and are considerably reduced in pits. It is, therefore, always desirable to prepare manure in pits to obtain a better quality and quantity of manure.

Loss of nitrogen during fermentation takes place mainly in the form of ammonia gas and methods have been worked out to retain this in the manure by the use of acidic materials to fix nitrogen chemically or by absorbants which physically hold nitrogen. Thus addition of super-phosphate at the rate of about five per cent or a heap mineral acid like hydrochloric acid at one per cent have been found to conserve nitrogen in the manure. Materials like charcoal-dust, sawdust, gypsum and even soil added at the rate of five to ten per cent reduce nitrogen losses considerably. The inclusion of this simple treatment in the process of manure preparation would result in obtaining a superior quality of manure in which a substantial amount of nitrogen is held in forms that are highly available to plants.

Enriched Organic-Based Manures

Fermented organic manures are low grade materials and, under Indian conditions, have to be applied in large bulk to obtain significant effects on crop growth and yield. The total nitrogen content in them does not exceed 1.5 per cent at best and with the small quantity of dung used by farmers, it is of the order of 0.5 per cent. The percentage availability of nitrogen for plants in such manures ranges from 5 to 20 per cent with the result that low responses are obtained when applications are made which are inadequate to make good the primary deficiency of nitrogen in our soils.

At the Indian Agricultural Research Institute, a new approach to this problem has been worked out which enables such low-quality manures and other organic materials to be effectively utilized. It has been found that the rate and extent of liberation of nitrogen from applied organic manures is adequate for normal plant growth when their C/N ratio is less than 10 : 1 or the nitrogen content is more than 2.5 per cent. In practice, the C/N ratio of organic materials can be reduced by adding the requisite amount of nitrogen in the form of any nitrogenous fertilizer to it and thereby increasing the value of the denominator instead of decreasing carbon in the numerator by fermentation. Thus, a low-quality manure containing 0.5 per cent nitrogen and a C/N ratio of 20 : 1 can be treated with ammonium sulphate or urea solution to add 2 per cent more nitrogen and raise the total nitrogen content to 2.5 per cent. This would immediately reduce the C/N ratio to 8 : 1 and the enriched manure can then be utilized effectively. This method can utilise large quantities of other organic waste materials of wide C/N ratio like dry dung, paddy and groundnut-husks, sawdust, tobacco-waste, wheat-chaff and any material in condition to be easily incorporated in soil. This method enables the process of composting to be eliminated thus saving time and labour.

Enriched organic manures of the above type are processed and get ready in a short time and possess many advantages. Losses resulting from their composting are avoided and the total quantities of organic matter and nitrogen taken initially are conserved and utilised. The decomposition taking place in the soil in the presence of adequate nitrogen incorporated will cause no adverse effects and sufficient nitrogen, in excess of microbial requirements, is liberated for optimum growth of crops.

Variations in the preparation of these enriched manures include incorporation of more nitrogen to obtain concentrated manures containing 5 to 7 per cent nitrogen and supplementing this with super-phosphate and potassium sulphate to get mixed organic-mineral fertilizers or any desired grade to suit specific requirements of soil or crops.

Preparation of Enriched Manures

Enriched organic-based manures are prepared by taking requisite quantities of nitrogenous and other fertilizers and dissolving them in sufficient water. The solution is then treated with 100 ml of commercial hydrochloric acid per quintal of the final manure for controlling decomposition of the added nitrogen and its retention in the manure. This is then absorbed in the calculated quantity of the organic material, mixed thoroughly and dried in the sun. The dry enriched manure can then be bagged and stored for use.

Microbial Method for Hastening Compost Preparation

Compost is bulky in nature and contains on an average 0.5 to 1.0 per cent N, 0.6 per cent P_2O_5 and 0.5 per cent K_2O. There is a need to reduce its bulkiness and period of composting to make it more acceptable to the farmer. Moreover, the quality of manure should be improved to supply more nitrogen and phosphorus and humus content per unit weight. Researches conducted by Gaur in the Division of Microbiology, IARI have shown that it is feasible to hasten the period of composting by proper inoculation with cellulolytic microorganisms as well as improve nitrogen and phosphorus contents of organic fertilizers with low-cost technology. Gaur reported that mesophilic cellulolytic fungi when used as inoculant in composting of plant residues (straw, leaf fall) can hasten the composting process and improve the quality of compost. The carbon content of compost mass was lowered by 25 per cent of the initial raw material (48 per cent) organic carbon during the first month of composting in pits. The loss was more due to fungal inoculation, *Penicillium* spp. showing the maximum decomposition of 46 per cent followed by *Aspergillus* spp. with 37.5 per cent. As a result, C/N ratio within one month was decreased ranging from 19.6 to 23.6 due to inoculation with cellulolytic fungi whereas in uninoculated control, it was 31.4 (Table 5.10). Subsequently also the decreases in C/N ratio were greater duo to inoculation.

Table 5.10: Effect of Cellulolytic Culture in Hastening the Process of Composting

Treatment	*C : N Ratio or Compost Sampling Interval in Months*		
	I	*II*	*III*
Control	31.4	22.4	20.0
Aspergillus sp. (IF_1)	23.6	–	16.8
Trichoderma viride (IF_2)	23.1	19.5	18.9
Aspergillus sp. (RF_1)	20.6	16.9	14.3
Penicillium sp. (RF_2)	19.6	16.8	14.8

Initial C/N ratio 45.7; Weight of raw material: 32 kg/pit.

The data (Table 5.11) on recovery percentage of compost on oven dry basis indicated that bulkiness of organic manures decreased by 2 to 8 per cent due to inoculation. The efficient culture reduced bulkiness by 8 per cent. The total nitrogen content showed increase on the other hand, the increase being more in inoculated series. Maximum of 17 per cent increase was recorded with *Aspergillus* spp. (RF_1) followed by *Aspergillus* spp. (IF_1) and *Penicillium* spp. (RF_2) with 15.6 and 14.2 respectively as compared to uninoculated control.

The studies indicated that due to inoculation of compost heaps by cellulolytic cultures the period of composting of dry and wide C/N ratio materials can be reduced by 4 to 6 weeks. These findings have been confirmed by the studies carried out at other centres.

Table 5.11: Final Yield of Compost and Nitrogen Status in Compost Material

Treatment	Final Yield of Compost (kg)	% Recovery	Total N%	Total N in		% Increase Over Control
				Finished Compost Material (g)	Per Tonne of Finished Compost (kg)	
Raw material	–	–	1.05	(336)	(10.5)	–
Uninoculated	19.6	61.25	1.41	276	14.1	34.29
Aspergillus sp. (IF_1)	18.9	59.06	1.63	308	16.3	55.24 (15.6)
Trichoderma viride (IF_2)	18.1	56.56	1.33	241	13.3	26.67 (– 5.6)
Aspergillus sp. (RF_1)	17.0	53.12	1.65	281	16.5	57.14 (17.0)
Penicillium sp. (RF_2)	16.9	52.81	1.61	272	16.1	53.33 (14.2)

Weight or organic material composted 32 kg/pit.

Figures in parentheses show increase or decrease over uninoculated control.

Low-Cost Technology for Enrichment of Compost

Recent experiments conducted at IARI have shown that inoculation with *Azotobacter chroococcum* and phosphate solubilising microorganisms (*Aspergillus awamori* or *Bacillus polymyxa*) improved the manurial value of compost. Inoculation of compost with nitrogen fixer and P solubiliser was done when the temperature had stabilised around 30–35°C after one month of composting. Low-grade rock phosphate (10 kg/tonne of raw material) was applied at the initial steps of filling up of compost pits.

The results indicated that 31.5 per cent carbon was decomposed in control during the first month whereas, the loss in different treatments ranged between 33.3 to 41.5 per cent. This was further increased during second month showing maximum loss of 52.4 per cent of organic carbon in rock phosphate, *Azotobacter* and P solubiliser inoculated series (Table 5.12). In this treatment C/N ratio was reduced to 14.1 and 12.0 during 1st month and 2nd month after inoculation respectively.

Table 5.12: Changes in Carbon (%), Total Nitrogen (%), C/N Ratio during Enrichment of Compost by *Azotobacter* and P Solubilizing Cultures (Sampling interval in months after inoculation)

Treatment	1st Month				2nd Month			
	Carbon	Nitrogen	C/N Ratio	Humus %	Carbon	Nitrogen	C/N Ratio	Humus %
Control	31.4	1.36	23.0	7.8	28.5	1.38	20.6	4.85
Control + Rock Phosphate @ 10 kg P_2O_5/tonne	30.7	1.61	19.0	9.15	27.5	1.77	15.5	6.90
Azotobacter + P solubilizer	30.5	1.51	20.1	9.30	26.6	1.73	15.4	6.00
Rock phosphate + *Azotobacter* + P solubilizer	26.9	1.89	14.1	9.0	21.9	1.82	12.0	6.95

Initial sample: Organic carbon 46.0%, Total nitrogen 0.86%, C/N ratio 53.3.

There was appreciable gain in nitrogen content per tonne of finished compost in control as well as inoculated series ranging from 60.5 to 111.6 per cent over the initial raw material (Table 5.13). Addition of powdered rock phosphate alone resulted in an increase of 28.3 per cent over control. Inoculation with *A. chroococcum* and *A. awamori* amended with rock phosphate showed an increase of 32.0 per cent in nitrogen content over the control resulting in gain of 22 kg N/5 tonnes of finished compost. It was also noted that humus content increased in all the treatments including control (Table 5.12). The best effect in humus content 9.3 per cent was observed in case of *A. chroococcum* + *A. awamori* after 1 month of inoculation.

Table 5.13: Final Yield of Compost and Nitrogen Status in Compost Material

Treatment	Weight of Raw Material (kg/pit)	Final Yield of Compost (kg)	% Recovery	Total N %	Total N in		% Increase Over Control
					Finished Compost Material (g)	Per tonne of Finished Compost (kg)	
Raw material	32	–	–	0.86	275	8.6	–
Control	32	11.85	37.03	1.38	164	13.8	60.46
Rock phosphate @ 10 kg P_2O_5/tonne	32	12.30	38.44	1.73	213	17.3	101.16
Azotobacter + P solubilizer	32	11.45	35.78	1.77	203	17.7	105.81
Rock phosphate + *Azotobacter* + P solubilizer	32	13.00	40.63	1.82	236	18.2	11.63

In another study the number of *Azotobacter chroococcum* in compost was increased due to its inoculation and use of rock phosphate. The mean numbers of nitrogen fixers of nine weekly intervals per grain of dry compost in different treatments were as follows: Control 1,380, rock phosphate 3,590; *Azotobacter* 6,070; rock phosphate + *Azotobacter* 21,070. This also resulted in an increase in the nitrogen content of compost by 25 to 44 per cent over control. These findings clearly show that quality of compost can be improved by inoculation with *Azotobacter* and P solubilisers suitably amended with rock phosphate.

Fly-breeding in Manure Preparation and its Control

One of the disturbing aspects of manure preparation from organic-wastes particularly town-wastes, is the profuse fly brooding that takes place in this medium. In over-ground heaps, owing to the heat that develops inside, fly maggots migrate and concentrate on the surface few cm all over. If the material is given turnings, as in the Indore method, the maggots get substantially destroyed, otherwise flies develop profusely. In pits, the top layer of refuse or soil becomes the seat of migration of all maggots where they finally develop into flies.

The emergence of flies from such manure pits has been observed from the 6th day after filling and goes on till the 12th day or so, after which it ceases. Based on these observations, a simple and highly effective method of destroying flies, while they are resting in the maggot or pupae stage in the surface layer, has been worked out. The entire surface area of the pit is covered on the 6th day after filling with rags, refuse, straw, cotton or other stalks, sawdust, paddy or groundnut-husks or any such combustible waste and set on fire, which may continue to smoulder for about an hour. The maggots and pupae lodged on the few centimetres of upper surface are completely destroyed by the heat. The operation may be repeated on the 12th day to obtain absolute control of fly breeding. The treatment leaves a heat layer of ash on the top which is no more attractive to the flies.

Oil-cakes

Oil-cake is a valuable by-product of the oilseed extraction industry and contains significant quantities of proteins, of value as animal feed if edible and as nitrogenous manures otherwise. There is always a residue of oil present in them depending on the process of extraction as shown in Table 5.14.

Table 5.14

	Oil per cent
Country *ghani* Extracted	10–12
Hydraulic Press	8–10
Steam Expeller	5–8
Solvent Extract	1–2

The presence of oil slows down decomposition and the process of mineralization of their nitrogen in soil. Since this also gets decomposed in due course, the overall formation of mineral nitrogen is not materially affected in quantity. Their C/N ratio is narrow varying from 3 to 10 and about 60 to 75 per cent of the nitrogen is easily mineralizable and available for plants.

Effective Utilisation of Mahua-cake

Mahua-cake is the poorest of oil-cakes with a nitrogen content of 2.5 per cent and has been found to be much less effective than other even when applied in larger quantities. It has been found that nitrogen of this manure mineralizes to a small extent in the soil owing to the presence of toxic saponins. The efficiency of the manure can be increased by moistening it and allowing it to ferment for sometime to decompose the toxic substances. An effective treatment worked out at the IARI involves treatment of the material with ammonia solution which not only increases the nitrogen content of the manure but enables the original nitrogen to be effectively mineralized and utilized by plants. The treatment could be extended to enrich other oil-cakes and utilize them more effectively.

Utilisation of Slaughter-house Wastes

Blood

Blood is the most valuable of slaughter-house wastes which yields a rich proteinaceous meal on dehydration. It is useful as a feed or as a nitrogenous manure. The existing crude methods of its collection and processing in India result in wastage of a large part.

In slaughter-houses, blood is obtained partly in the form of spongy clots and partly as blood serum in liquid condition. These are dried by heating in large open pans over a slow fire, which is a time-consuming and highly offensive process. The Indian Agricultural Research Institute has worked out a simple method for quick treatment of blood to obtain blood-meal which can be used as feed or as a manure.

Blood-meal from serum is prepared by adding one per cent of lime and then heating it till the temperature rises to 80°C. At this stage, blood begins to coagulate by its reaction with lime, and further slow heating is continued with constant stirring until the flocculated blood separates out and the colour changes from red to ash. The remaining water is decanted off and the coagulated blood is passed through a quarter-inch mesh expanded metal sieve and dried in the sun. There is no loss of nitrogen in the process and the manure contains 10 to 12 per cent nitrogen.

Blood clots are processed separately by cooking them for a few minutes boiling in 0.5 per cent hydrochloric acid solution. The clots harden and rise up to the surface. They are then taken out, rubbed over an expanded metal seive and dried. The manure contains 14 to 15 per cent nitrogen.

Blood serum can also be absorbed as such in materials like dry dung powder, sawdust, dry leaf powder and such other materials and then dried in the sun. Such products may contain 4 to 5 per cent nitrogen. The use of these absorbing materials on the floor of slaughter-houses would enable blood to be directly absorbed and offers one of the simplest methods of effective collection and utilisation of this waste and keeping slaughter-houses clean.

Waste-meat

This waste in slaughter-house consisting of intestines, offal and other flesh cuttings is hardly utilized in our country. Their disposal and use in compost pits has been suggested where they can serve as a source of nitrogen required for the process of decomposition of other wastes. The proposition of utilization of meat wastes as such and obtaining concentrated meat-meal is more useful to consider. The method of cooking these in steam digesters under pressure can be adopted for the purpose but this requires expensive equipment. A simple method worked out at the IARI for treatment of such materials consists of cutting them into pieces and cooking in 5 per cent hydrochloric acid solution until they become soft. The material is then washed and rubbed over an expanded metal sieve and dried in the sun. The product may contain 8 to 10 per cent nitrogen and can be used as poultry feed or manure. The cost of processing works out to Rs. 2.50 per tonne.

The method can also be adopted to utilize the entire flesh of carcasses of all animals that die in towns and villages and can be organized on a countrywide scale for the exploitation of a valuable source of manure.

Bones

Bones from slaughter-houses, carcasses of all animals and from meat processing industry constitute a valuable waste which could be exploited more fully in the country. The existing methods of disintegration of bones involve crushing them in hammer mills by which raw bone-meal is obtained as a by-product or by digesting them in steam digesters, under pressure which yields steamed bone-meal in fine powdery condition. The methods require expensive equipment and are not always suite1 for adoption and treating small collection in villages and remote places.

Chemically Processed Bone-meal

The IARI has evolved a purely chemical process disintegrating bones and obtaining a sterilized bone-meal m fine condition which can be used as a feed or as manure. The process can be worked both as a cottage industry to treat small collections, suiting conditions in this country or on a large scale, if required.

Fresh or dry bones are taken in a suitably-sized container and soaked in sufficient water to cover them. Caustic soda is added to this to make a 5 per cent solution in the quantity of water taken. The bones are stirred once or twice daily until they disintegrate completely to a paste in 14 days. The supernatant is then decanted and utilized again and the disintegrated bones are washed twice with water. The material is finally taken up with water and treated with 10 per cent hydrochloric acid solution until it is neutral, to litmus paper. The supernatent is then decanted and the bone paste filtered through muslin cloth. This is then dried in the sun and it yields a sterilized bone-meal in fine condition. The product contains about one per cent nitrogen and 27 per cent phosphoric acid. The cost of production, excluding the value of bones works out to be Rs. 60 per tonne. The process is patented.

Fish-meal

Fish manure processed by drying non-edible fish or wastes from fish industry is a well-recognized, balanced organic manure containing both nitrogen and phosphorus in significant quantities. Fish-meal is also produced in India by cooking in steam digesters and then drying. The manure, however, has an extremely offensive smell which stands in the way of its wider use. At the IARI, it has been found that the objectionable smell can be completely smothered by the use of two litres or pine-oil emulsion per tonne. The long coast-line of the country offers a great scope for fuller exploitation of this highly valuable manure for increasing agricultural production.

Leather-wastes

Tanned and untanned leather wastes are obtained from tannery and leather processing industries where their disposal becomes a serious problem. Leather-wastes can be processed by heating in steam digesters and then grinding to powder. The product obtained-contains 7 to 8 per cent nitrogen and has been used as a nitrogenous manure. The IARI has a patented process for preparing leather-meal in which the material is soaked in water and then cooked in 5 per cent hydrochloric acid. The leather disintegrates into a plastic mass which is washed in water and dried and ground to powder. The acid treatment also helps in breaking down complex nitrogenous compounds and yields a manure containing 7 to 8 per cent nitrogen of which 70 to 75 per cent is mineralizable. The method can be adopted on a cottage industry scale, or large-scale according to the needs.

Hair and Wool-wastes

Large quantities of animal-hair wastes become available in tanneries, which have found no use so far. Human hair can be collected in a large quantity, from hair-cutting saloons. These have also not been exploited so far. Large quantities of wool-waste also become available from wool refining, cloth and carpet industries which are not properly utilized. These materials contain about 15 per cent nitrogen in the form of complex keratin and do not decompose in soil when used as such. No estimate is available of the quantities of these wastes that can be collected but it is considered that their utilization in agriculture as manure after suitable processing would provide a substantial additional source of a rich organic manure. A patented process worked out at the IARI involves treatment of these materials with a solution containing 8 per cent caustic soda on the weight of the material. This brings about their disintegration to a paste which is then treated with 10 per cent hydrochloric acid to neutralize the alkali and re-precipitate the proteins. The liquid is decanted and the spongy solid is rubbed over an expanded metal sieve and dried. The product contains 12 to 15 per cent nitrogen and has been found to be a good nitrogenous manure.

Chapter 6
Biogas Potential from Livestock Wastes and Human Excreta

According to the dung collection survey conducted by Institute of Agricultural Research Statistics, Delhi, it was estimated that in rainy season, 85 per cent of the collected dung was diverted into manure pits and only 13 per cent was converted into dung cakes, whereas during winter season 58 per cent was diverted into manure pits and 40 per cent converted into dung cake. The dung conversion into dried dung cakes amounted to 32 per cent during summer. However, the disposal of dung for other casual purposes like construction of *kucha* houses, etc. was only 2 per cent in all the seasons. The all-India average of the collected dung being used for preparing dung cakes is approximately 29 per cent which amounts to about 290 million tonnes of dung being burnt away by farmers as fuel for cooking. The solution to the problem of alternate fuel supply to rural farmers has to be tackled with more fire-wood production or other available sources, so that the dried dung cakes may not be burnt away into ash losing the manurial components in them. Acharya and Idnani have reviewed the work on gobar-gas production in India and recommended that the anaerobic digestion of cattle dung in biogas plants would yield fuel as well as manure. The rate of gas production from cattle dung during two months' digestion in gobar-gas plants at different temperature conditions is graphically represented in Figure 6.1. Assuming the above gas production rate for different States of India depending upon the temperature variations over a period of year, the statewise mean temperature and rate of gas production and annual production potential of biogas from bovine dung are presented in Table 6.1. The mean temperature in southern, eastern and western States and Islands in India vary with minimum temperature range of 18.1 to 28.9 °C during winter and maximum temperature of 28.8 to 33.7 °C during summer. The monthly mean temperature in northern States vary with minimum temperature range of 3.2 to 17.0°C during winter and maximum temperature of 21.0 to 34.2°C during summer. Hence, the rate of biogas production per tonne wet dung varied from 64.0 to 82.7 cu m in western, eastern and southern states and 23.2 to 66.3 cu m in northern States. The annual biogas potential works out to be 67,654.617 million cu m. The gas production from solid excreta of goat and sheep was calculated at the rate of double the gas yield from cattle dung due to roughly double the dry matter content in fresh dung. The biogas yield from excreta of poultry, pig, other livestock and human beings was also calculated at double the rate of biogas yield due to the higher digestible organic matter content in cattle excreta.

Table 6.1: Potential Biogas Production from Bovine Dung in Different States of India

Sl.No.	State/ Union Territory	Monthly Mean Temperature (°C)		Rate of Gas Production (cu m/tonne wet dung)	Animal Dung Availability (Million tonnes)	Annual Gas Production (Million cu m)
		Minimum	Maximum			
1.	Andhra Pradesh	22.8	33.7	75.0	77.389	5804.175
2.	Assam	16.8	29.4	64.0	25.376	1624.064
3.	Bihar	17.0	37.7	68.0	68.557	4661.876
4.	Gujarat	20.7	32.6	71.0	38.262	2716.602
5.	Haryana	13.5	34.2	66.3	32.213	2135.722
6.	Himachal Pradesh	10.7	28.7	52.3	19.181	1003.166
7.	Jammu and Kashmir	3.2	21.0	23.2	9.725	225.620
8.	Karnataka	21.9	29.4	69.3	54.800	3797.640
9.	Kerala	26.6	28.8	74.0	4.538	335.812
10.	Madhya Pradesh	17.7	33.3	67.2	80.732	5425.190
11.	Maharashtra	21.2	30.8	68.7	107.986	7418.638
12.	Manipur	16.8	29.4	64.0	1.285	82.240
13.	Meghalaya	16.8	29.4	64.0	1.779	113.856
14.	Nagaland	16.8	29.4	64.0	0.348	22.272
15.	Orissa	20.6	31.3	72.8	51.498	3749.054
16.	Punjab	13.5	34.2	66.3	51.190	3393.897
17.	Rajasthan	15.8	33.7	69.3	77.896	5421.562
18.	Tamil Nadu	24.7	31.6	77.3	70.027	5413.087
19.	Tripura	16.8	29.4	64.0	1.895	121.280
20.	Uttar Pradesh	15.1	32.9	61.1	179.032	10938.855
21.	West Bengal	18.1	29.7	66.3	45.687	3029.048
22.	Andaman Nicobar Islands	28.9	31.3	82.7	0.043	3.556
23.	Chandigarh	13.5	34.2	66.3	0.127	8.420
24.	Dadra and Nagar Haveli	28.9	31.3	82.7	0.070	5.789
25.	Delhi	13.5	34.2	66.3	1.372	90.964
26.	Goa, Daman and Diu	28.9	31.3	82.7	1.007	69.181
27.	Lakhadweep	28.9	31.3	82.7	0.006	0.496
28.	Mizoram	16.8	29.4	64.0	0.090	5.760
29.	Pondicherry	24.7	31.6	77.3	0.476	36.795
	All India Average	–	–	**67.1**	–	–
	All India Total	–	–	–	**1002.587**	**67654.617**

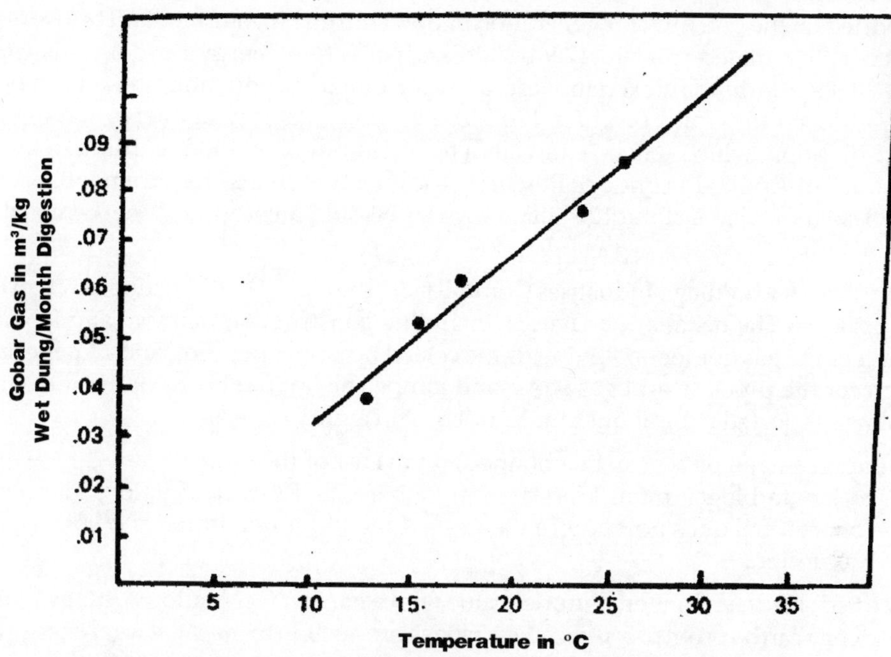

Figure 6.1: Rate of Gobar-gas Production at Different Temperatures

The biogas potential from livestock excreta other than cattle and buffalo and human excreta would be 7,447.583 million cu m (Table 6.2). The potential power generation from the biogas that can be obtained from all the livestock and human excreta was calculated on the basis of power generation of 1.605 kwh electricity produced from 1 cu m gas and presented in Table 6.3.

Table 6.2: Potential Biogas Production from Other Livestock Excreta and Human Excreta in India

Source	*Annual Excreta (million tonnes)*	*Annual Biogas Potential (million cu m)*
Sheep and goats	12.228	1640.998
Pigs	2.298	308.392
Other livestock	3.395	455.609
	6.024	808.420
Human beings	31.554	4234.541
Total	**55.499**	**7447.583**

Table 6.3: Potential Electrical Energy from Biogas per annum in India

Source of Excreta	*Biogas (million cu m)*	*Electrical Energy (billion kwh)*	*Effective Energy (billion kwh)*
Cattle	67654.617	108.586	86.869
Other livestock	3213.419	5.158	4.126
Human beings	4234.547	6.196	5.436
Total	**75102.583**	**120.540**	**96.431**

Rao reported on the potential for generation of power from biogas in India. The use of dung cakes in India has been estimated to yield 12.9 billion kwh effective energy for domestic fuel. If all the available livestock and human excreta were used for biogas production, the total biogas yield of 75,102.583 million cu m would produce 120.540 billion kwh electrical energy having 80 per cent effective consumption, which will be equivalent to 96.431 billion kwh. According to the estimate of National Council of Applied Economic Research, the effective energy requirement in rural sector of India would be of the order of about 125 billion kwh and would meet about 77 per cent of the rural fuel requirement of India.

According to Khadi Village Industries Commission, the cost of constructing a 6 cu m biogas plant is about Rs 4,200.00. The installation charges including construction material and labour costed 46 per cent whereas the gas holder and guide frame costed nearly 43 per cent, and 11 per cent of the cost was required for the pipeline and gas stove and lamp. The annual profit derived from the 6 cu m biogas plant was reported to be about four hundred and eighty rupees.

A higher percentage profit can be obtained, if the cost of the biogas plant can be reduced. The Chinese-based Janata biogas plant fabricated at Gobar Gas Research Station, Planning Research Action Division, Ajitmal does not require the gas holder and guide frame and costs only 50–60 per cent of the KVIC design.

The digested slurry has higher manurial value with a narrow C : N ratio. So, the available livestock and human excreta can be profitably utilized for biogas production to meet the rural energy requirement and to obtain better quality manure as well as minimize the environmental pollution.

Manurial Value of Digested Slurry

The digested dung slurry has a low content of N, P and K as compared to commercial fertilizer, but it is a valuable source of human substances and micronutrients. During anaerobic digestion of cattle dung 15 to 30 per cent protein-nitrogen is mineralised to ammonical form.

The average composition of digested cattle dung slurry is summarised in Table 6.4. The slurry or sludge can be composted with waste organic materials of high dry matter content for easy distribution and utilisation as manure.

Table 6.4: Composition of Digested Cattle Dung Slurry

Item	*D.M. %*	*O.M. %*	*Carbon %*	*Total N%*	*P%*	*C : N Ratio*
Digested cattle	8.0	72.57	42.1	2.07	0.35	20.47
dung	5.5	60.68	35.2	1.81	0.40	19.4
	8.2	71.80	37.5	2.12	0.41	17.6

The digested cattle dung slurry from a 1,000 cu ft biogas plant was composted in a battery of three manure pits with alternate layers of weeds, garden cuttings and other farm wastes. The samples of cattle dung and the two months old digested slurry and 4 months old compost were periodically analysed for chemical components in a year. It was found that on an average 13 per cent total carbon was digested during two months of anaerobic digestion of cattle dung.

The chemical analysis of the cattle dung, digested slurry and compost is presented in Table 6.5.

Table 6.5: Chemical Composition of Cattle Dung, Digested Slurry and Compost

Type of Manure	*Dry Matter (%)*	*Ash (%)*	*C (%)*	*N (%)*	*P (%)*	*C : N Ratio*
Dung	20.03	17.67	47.76	1.55	0.76	36.76
Digested slurry	5.73	22.60	41.56	1.99	1.02	23.19
Compost	30.82	38.32	36.36	2.24	1.20	16.73

Laura composted the nitrogen deficient wheat straw and sorghum fodder with spent (digested) slurry under anaerobic conditions. Addition of spent slurry accelerated decomposition of wheat straw during 45 days digestion from 10 to 30 per cent and that of sorghum fodder from 15 to 40 per cent of dry matter.

Acharya reported that the manure produced by anaerobic digestion showed good manurial value after sun-drying and was found to be better than farmyard manure (Table 6.6). In wet conditions, the manurial value of digested cattle dung slurry was less than that after sun-drying. Table 6.6 presents the yield of crops with wet and sun-dried digested slurry, farmyard manure and nitrogen application @ 100 lb N/ha.

Table 6.6: Crop Yield on Application of Wet and Sun-dried Digested Cattle Dung Slurry in Pot Culture Trial (g/pot)

Sl.No.	*Treatment*	*Wheat*		*Ragi*		*Sunnhemp (Dry wt)*
		Grain	*Straw*	*Grain*	*Straw*	
1.	No manure	8.84	13.46	10.1	31.4	93.4
2.	Wet slurry	10.32	15.26	12.0	33.8	106.3
3.	Sun-dried slurry	11.31	17.39	13.6	36.8	117.2
4.	FYM	10.02	16.28	12.4	34.6	104.4
5.	Ammonium sulphate	13.70	19.55	15.4	41.2	121.2

Ghosh reported that the digested slurry in wet and dried form gave significantly higher yield of rice and berseem as also with farmyard manure and groundnut cake. The residual effect of slurry was observed for two years, though the magnitude of response gradually narrowed down. Neelakantan reported on the effect of fresh and sun-dried digested cattle dung slurry @ 1 and 2 per cent level on oat in sand and soil under greenhouse conditions. The positive effect on fodder yield in soil culture was found to be of a lesser degree than sand culture. The results also showed that fresh slurry was better than dried slurry and the effect was more marked in sand culture (Table 6.7).

Maintenance of Organic Matter in Indian Soils

Controlled incubation studies conducted in different Indian soils for about 6 months have indicated that the application of FYM at the rate of 44 tonnes/ha (0.5 per cent carbon basis) effectively builds up the organic matter status of different soils whereas the control soils lost 15 to 40 per cent of the organic carbon. In case of cereal residues the quantity to be applied at the above rate for maintenance of soil organic matter comes to 24.5 tonnes/ha and higher doses than 25 tonnes/ha are needed for effective increase. The addition of farmyard manure and crop residues increased the soil carbon in different soil groups by 10 to 40 per cent. Those findings were supported by the data obtained from soil samples of various permanent manurial trials conducted at Sabour (Bihar), Chinsurah (West Bengal), Gurdaspur (Punjab) and Kanpur (U.P.) and analysed during 1972. The best results were obtained in the experiment conducted at Sabour where the application of 18.9 tonnes and 75.5 tonnes of FYM/ha/crop increased soil organic carbon by three times (1.230 per cent) and five times (1.930 per cent) respectively as compared to control (0.378 per cent). Humic and fulvic acid carbon percentages were almost doubled in case of 18.9 tonnes FYM/ha/crop as compared to control whereas this fraction was four times in case of 75.5 tonnes FYM/ha/crop. Organic nitrogen increased three times by addition of 75.5 tonnes FYM/ha/crop over control. However, the difference between 75.5 tonnes FYM/ha applied in one lot and 18.9 tonnes FYM/ha/crop as regards nitrogen content was not much different.

Table 6.7: Green Fodder Yield of Oat (g/pot)

Treatment	*Sand Culture*	*Soil Culture*
Control	19.2	14.7
Fresh dung (1%)	54.6	34.6
Fresh dung (2%)	66.0	40.0
Fresh slurry (1%)	73.3	43.3
Fresh slurry (2%)	89.3	56.7
Dried slurry (1%)	48.0	37.3
Dried slurry (2%)	53.3	45.1

Effect of Organic Matter on Soil Microorganisms

The results with different types of organic materials have clearly shown that living phase of soil was greatly stimulated which will be of consequence not only in biodegradation but also in nitrogen fixation, phosphorus solubilisation and in increasing the availability of plant nutrients to crops. Phosphorus is non-renewable asset and nitrogen is more labile and subject to losses. The augmented microbial activity can tap the inert nitrogen gas from the atmosphere, reduce the leaching losses and regulate the supply of phosphorus. The zymogenous response of microflora varied with the type and characteristics of soil and organics used. Bacteria due to application of organics was influenced to a greater extent than fungi and actinomycetes (Tables 6.8–6.10). Asymbiotic nitrogen fixers (*Azotobacter* and anaerobes) were stimulated by addition of straw in the field soil at 2, 5 and 10 tonnes/ha, Figure 6.2. The regulated application of organics as a practice will energise the living microorganisms of the soil, involved in biochemical activity of importance to soil fertility and plant nutrition. Addition of FYM and crop residues resulted in increase of total nitrogen varying from 10 to 70 per cent depending on the plant material and soil type used.

Immobilisation of nitrogen was usually associated with high C/N ratio material. Immobilisation was not observed when its C/N ratio was brought down to 40 and mineralisation of nitrogen was satisfactory and steady. The addition of chemical nitrogen or organic comparatively richer in nitrogen like non-edible cakes (*neem, karanj*), legumes and grasses can be advantageously utilised for lowering the C/N ratio of cereal straw/residues. Addition of rock phosphate or super-phosphate was found beneficial in hastening the *in situ* decomposition of organic material in soil and increased the available phosphorus. Farmyard manure application invariably increased the availability of phosphates.

Table 6.8: Average Bacterial Population in Differentially Treated Soils During Organic Matter Decomposition (Population expressed in 10^6/g soil)

Soils	Control (Soil)	Treatments					
		F.Y.M.		Wheat Straw/ Paddy Straw		Bajra Stalk/ Maize Stalk	
		0.5%C	2.0%C	0.5%C	2.0%C	0.5%C	2.0%C
Hilly soil (H.P.)	28.63	26.66	52.93	57.46	88.82	–	–
Alkali soil (Delhi)	3.39	8.79	29.60	17.37	29.17	–	–
Heavy clay soil (Patna)	11.66	16.54	17.56	24.69	28.71	20.09	28.39
Acid red loam (Kanke)	5.06	20.51	23.51	12.77	17.41	24.97	28.11
Sierozem soil (Hissar)	31.10	51.30	73.53	–	–	54.48	78.10
Alkali soil (Rajthal)	0.35	0.51	1.13	–	–	0.62	1.83
Alluvial soil (CRRI)	32.08	74.83	141.18	92.12 (88.41)	197.76 (193.75)	–	–
Hebbal red soil	16.38	20.52	39.20	–	–	228.89	367.41
Medium black soil	3.85	2.45	2.51	1.78	3.36	–	–
Acidic lateritic	5.12	4.56	3.07	1.63	3.61	–	–

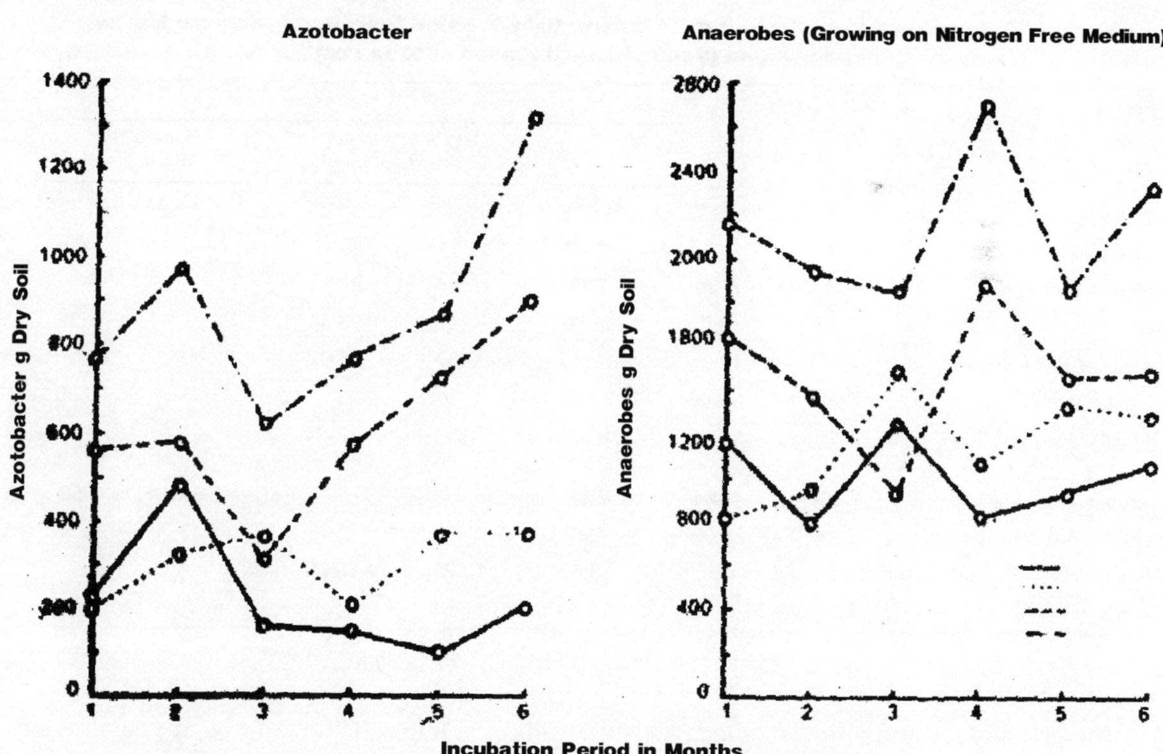

Figure 6.2: Effect of Incorporation of Wheat Straw on the Nitrogen Fixing Microorganisms

Table 6.9: Average Actinomycete Population in Differentially Treated Soils During Organic Matter Decomposition (Population expressed in 10^6/g soil)

Soils	*Control (Soil)*	*Treatments*					
		F.Y.M.		*Wheat Straw/ Paddy Straw*		*Bajra Stalk/ Maize Stalk*	
		0.5%C	*2.0%C*	*0.5%C*	*2.0%C*	*0.5%C*	*0.2%C*
Hilly soil (H.P.)	8.93	11.23	20.79	20.51	30.23	–	–
Alkali soil (Delhi)	0.67	2.60	16.27	3.74	6.56	–	–
Heavy clay soil (Patna)	2.26	3.10	3.46	4.41	5.06	4.21	4.79
Acid red loam (Kanke)	1.52	1.84	2.21	1.78	2.29	2.38	2.74
Sierozem soil	13.47	37.44	55.83	–	–	52.26	102.75
Alkali soil (10^3)	15.98	15.51	30.41	–	–	15.29	14.19
Alluvial soil (CRRI)	3.22	9.74	16.08	8.77 (9.35)	13.24 (12.38)	–	–
Medium black soil (Poona)	1.62	1.10	1.56	1.09	1.94	–	–
Acidic lateritic	3.29	4.26	3.60	2.98	1.72	–	–
Hebbal red soil (10^5)	8.44	5.94	9.39	–	–	12.47	11.69

Table 6.10: Average Fungal Population in Differentially Treated Soils During Organic Matter Decomposition (Population expressed in 10^4/g soil)

Soils	*Control (Soil)*	*Treatments*					
		F.Y.M.		*Wheat Straw/ Paddy Straw*		*Bajra Stalk/ Maize Stalk*	
		0.5%C	*2.0%C*	*0.5%C*	*2.0%C*	*0.5%C*	*2.0%C*
Hilly soil (H.P.)	11.57	14.54	17.57	17.11	28.96	–	–
Alkali soil (Delhi)	4.64	6.26	5.23	12.25	27.14	–	–
Heavy clay soil (Patna)	12.39	13.54	13.66	18.26	28.34	55.91	73.32
Acid red loam (Kanke)	10.82	12.35	12.84	29.57	57.02	72.69	99.39
Sierozem soil	28.69	25.46	24.98	–	–	43.36	40.30
Alluvial soil (CRRI)	27.15	39.00	48.23	75.54 (90.08)	134.15 (114.92)	–	–
Alkali soil (Hissar)	1.68	1.84	1.82	–	–	12.98	2.56
Hebbal red soil	6.74	4.06	4.89	–	–	42.43	95.55
Medium black soil (Poona)	13.29	11.71	13.38	13.21	12.68	–	–
Acidic lateritic	18.07	14.41	14.54	17.54	26.29	–	

Organic Mulch

This is an unique and simple method particularly for recycling of dry and nitrogen-poor organic materials since its incorporation in soil is not required at this stage and the costs of ploughing, water and inputs of chemical fertilizer for their favourable utilisation is not involved. Although decomposition of mulch is a slow process, its biomass and C/N ratio during the course of one crop season is appreciably reduced and this facilitates its incorporation in next season at a low cost. The experiments have shown that straw mulch besides controlling the weeds and conserving moisture and thereby saving irrigation, also increases crop yield and population of beneficial soil microflora.

Table 6.11: Effect of Organic Mulching with Paddy/Wheat Straw Yield of Field Grown Wheat and Pea Crops

Crop	*Per cent Increase in Yield in*						
	Bangalore Red Sandy Loam	*Barrackpore Alluvial Soil*	*Hissar Sierozem Soil*	*Poona Medium Black Soil*	*Acidic Soil*	*Loam (Kanke)*	*Delhi Alluvial Soil*
Wheat	31.6	67.0	66.0	–	14.3[a]	22.9[b]	16.1[e]
Pea	29.4	9.60	–	13.4	–	77.6[b]	–
Maize	–	–	–	–	5.00[c]	21.2[a]	95.7
Moong	–	–	–	–	–	–	31.7

[a]: Mulched with paddy straw.

[b]: Mulched with karanj leaves.

[c]: Mulched material incorporated in soil, previous crop-wheat.

[d]: Mulched material incorporated in soil, previous crop-peas.

[e]: Mulched material incorporated in soil, previous crop-maize.

The yields of wheat crop by mulching with straw was increased in acidic-red-loam, sierozem and laterite soils by 14.3, 24 and 29 per cent over unmulched treatment under different conditions. Mulching showed better effect on pea crop in acidic red-loam, medium black and laterite soils and increased the grain yield by 77.6, 43 and 40 per cent, respectively. Trials were repeated in different agro-climatic conditions (Bangalore red sandy loam, Ranchi acidic red, Poona medium black, Hissar sierozem, Delhi and Barrackpore alluvial soils) on wheat, pea, *moong* and maize crop. The previous year's findings were confirmed showing the beneficial effect of wheat or paddy straw or *karanj* leaves mulch at 10 tonnes/ha on the crops (Table 6.11). The grain yield of wheat was increased by 66 to 67 per cent in sierozem and Barrackpore alluvial soil, by 31.6 per cent in red sandy loam 14.3 per cent in-acidic red loam and 4.2 per cent in medium black soil. Pea grain yield was increased by 77.6, 29.4, 13.4 and 9.6 per cent in acidic red loam, red sandy loam, medium black and Barrackpore alluvial soil, respectively. The grain yield of *moong* and maize (Table 6.12) was increased significantly in Delhi alluvial soil due to wheat straw mulch by 31.7 and 110.5 per cent, respectively. The incorporation of mulched material used on previous crop–maize, in soil increased the grain yield of wheat crop by 16.1 per cent. Nitrogen uptake by both these crops was increased and mulching had favourable effect on the population of soil bacteria, fungi and actinomycetes (Table 6.13). The population of nitrogen fixers (Table 6.13) and phosphate solubilisers were increased several-folds. Organic mulch acts in several ways on soil and crop:

1. Enhances number of soil microflora
2. Increases plant nutrient availability
3. Conserves soil moisture
4. Controls weeds

5. Regulates soil temperature, and
6. Augments crop yields.

Table 6.12: Effect of Mulching on the Yield and Nitrogen Uptake by Grain in Crops

Treatment	*Grain Yield (g/plot)*	*% Increase Over Unmulched Treatment*	*Dry Matter Yield (kg/plot)*	*% Increase Over Unmulched Treatment*	*Grain Nitrogen %*	*Uptake of Nitrogen by Grain (g/plot)*
			Moong			
Unmulched	84.5	–	4.0	–	3.3	2.8
Mulched Significant at 5%	111.3	31.6	5.0	26.2	3.4	3.8
			Maize			
Unmulched	325.0	–	3.6	–	1.6	5.3
Mulched Significant at 5%	684.2	110.5	4.5	26.5	1.7	11.7

Table 6.13: Effect of Straw Mulch on Soil Microbial Population

Treatment	*Moong Crop (Average values of 3 intervals)*				*Maize Crop (Average values of 3 intervals)*			
	Bacteria	*Azoto-bacter*	*Actinomy-cetes*	*Fungi*	*Bacteria*	*Azoto-bacter*	*Fungi*	*Actinomy-cetes*
	1×10^6	1×10^2	1×10^5	1×10^4	1×10^6	1×10^2	1×10^4	1×10^5
Fallow unmulched	11.73	5.5	7.53	2.63	12.86	4.83	4.46	5.96
Fallow mulched	28.6	9.33	26.33	7.66	40.06	18.3	12.7	9.16
Cropped unmulched	17.66	5.03	13.26	3.63	20.2	3.86	5.3	9.16
Cropped mulched	32.96	8.0	24.66	9.36	38.93	13.63	11.0	25.76

Effect of Crop Residues on Yield of Legume Crops

The direct utilisation of organics low in nitrogen, *i.e.*, high C/N ratio material may not always be feasible in intensive agriculture and particularly in culture of cereal crops. The utilisation of organic materials in soil in case of cereals requires proper understanding of the technology by the farmer. While growing legume crops which have got host-*Rhizobium* symbiosis system *i.e.*, fixing of nitrogen in the nodules by rhizobia for the supply to plant, is not affected with nitrogen immobilisation processes which generally occurs when organic matter low in nitrogen is ploughed in the soil. On the other hand, this benefits the process of nitrogen fixation. Moreover, the beneficial effects can be obtained by improvement of soil conditions, nitrogen immobilisation and greater supply of carbon dioxide to plants for production of sufficient photosynthates. This method of recycling can be carried out by ploughing straw one week before sowing with small dosage of nitrogen (10–20 kg N/ha) and phosphate (90 kg P_2O_5/ha) as recommended for leguminous crops.

Nodulation in groundnut crop (Table 6.14) was improved due to application of wheat straw at 2,5 and 10 tonnes per hactare in Delhi alluvial soil. Pod yield (Table 6.15) was significantly increased by 66 and 95 per cent due to application of 2 and 5 tonnes straw per hectare by this method. Growth of plants was favourably influenced by 5 tonnes straw application. The residual effect of straw incorporation after groundnut increased the wheat grain yield by 3.9 to 36.8 per cent. Maximum residual effect was obtained with 10 tonnes straw applied per hectare. The practice can be followed in legume-cereal rotation. The straw gets decomposed during the growth period of legume crops and residual effect can be observed on the following crop. This has been confirmed with other legume crops (lentil, peas and *moong*).

Table 6.14: Effect of Straw on the Nodulation of Groundnut Crop

Treatment	*Sampling Period in Weeks*					
	4th Week			*7th Week*		
	No. of Nodules	*Dry wt of Nodules (mg)*	*Dry wt of Plants (g)*	*No. of Nodules*	*Dry wt of Nodules (mg)*	*Dr wt of Plants (g)*
Control	68	31	3.60	115	51	7.7
Wheat straw tonnes/ha	82	54 (74.2)	4.50 (25.0)	165	89 (74.5)	10.8 (40.0)
Wheat straw 5 tonnes/ha	125	103 (232.3)	5.87 (63.1)	223	148 (190.2)	16.9 (120.0)
Wheat straw 10 tonnes/ha	58	30 (–)	3.43 (– 4.7)	101	66 (29.4)	9.6 (25.2)
CD at 5% level	20	12	0.65	29	17	1.9

Values in parantheses indicate per cent increase over control.

Table 6.15: Effect of Straw on the Yield, Nitrogen Uptake and Oil Content of Groundnut Crop and Yield of Wheat Crop

Treatment	*Groundnut*						*Oil %*	*Wheat*	
	Pod Yield (g/plot)	*Dry Matter Yield (g/plot)*	*Nitrogen % in Pod*	*Nitrogen Uptake by Pod*	*Nitrogen % of Dry Matter (g/plot)*	*Nitrogen Uptake by Dry Matter (g/plot)*		*Grain Yield (g/plot)*	*Straw Yield (kg/plot)*
Control	336.7	840.0	3.95	13.3	0.36	3.0	47.6	760	1.53
Wheat straw 2 tonnes/ha	560.0 (66.3)	1170.0 (39.3)	4.03	22.6	0.39	4.6	48.2	790 (3.9)	1.86 (20.0)
Wheat straw 5 tonnes/ha	658.3 (55.5)	1473.3 (75.4)	4.10	27.0	0.40	5.9	48.4	890 (17.1)	2.13 (37.4)
Wheat straw 10 tonnes/ha	471.7 (40.1)	996.7 (18.7)	4.03	19.0	0.39	3.9	49.4	1040 (136.8)	2.41 (55.5)
CD at 5% level	53.7	136.6						196	0.265

Values in parenthesis indicate % increase over control.

Effect of Straw, *Neem* Cake and Farmyard Manure on Yield of Maize Crop

A field trial was undertaken to investigate the effect of organic materials including *neem* cake on the yield and nitrogen uptake by maize crop 'Ganga 5') over the normal package of practices. The treatments tested are given in Table 6.16. Phosphate and potash were applied at 80 kg P_2O_5/ha as super-phosphate and 50 kg K_2O/ha as murate of potash, respectively. All organic materials except farmyard manure were incorporated in soil at 18 tonnes/ha. 15 days prior to sowing and allowed to decompose. In case of straw and *neem* cake treatment, straw was applied @ 13.5 tonnes/ha and remaining organic matter 4.5 tonnes/ha was added through *neem* cake. Farmyard manure at 18 tonnes ha was mixed at the time of sowing. First instalment of nitrogen (50 kg N/ha) at both the depths was added right at the time of incorporation of straw in soil and the remaining 150 kg N was added in split doses. In remaining three treatment, nitrogen was applied in three split doses at 66.6 kg N/ha. Each treatment was replicated 5 times. Irrigation and inter-culture operations were given whenever necessary.

Table 6.16: Effect of Organic Materials on Yield and Nitrogen Uptake by Maize Crop

Treatment	Grain Yield (q/ha)	Straw Yield (q/ha)	Grains N (%)	Grain N Uptake (kg/ha)
Control	48.0	160.0	1.58	73.4
N (200 kg N/ha)	62.2	200.0	1.73	107.7
Straw (0–15 cm depth) + N	62.6	220.0	1.68	105.1
Straw (0–7.5 cm depth) + N	55.6	230.0	1.73	96.2
Straw + *neem* cake (0–15 cm depth) + N	76.0	270.0	1.75	133.5
FYM (0–15 cm depth) + N	70.4	240.0	1.63	114.8
CD at 5%	11.0	52.0		

The application of nitrogen at 200 kg N/ha increased the yield of maize crop significantly by 14 q/ha. Application of straw in 0 to 15 cm depth alongwith nitrogen maintained the grain yield obtained with 200 kg N/ha. When straw was mixed in 0 to 7.5 cm depth, a decrease in yield was observed showing nutrient immobilisation due to increased quantum of addition of organic matter in shallow layer. Significantly higher increases in yield were obtained when one-third of the organic matter has *neem* cake was incorporated alongwith straw. The next best treatment was farmyard manure incombination with nitrogen which also increased the yield significantly. Similar effect was obtained in the case of straw yields. It is interesting to record that maximum nitrogen uptake was obtained in straw + neem cake + nitrogen treatment and the second best was farmyard manure treatment. Straw application in 0 to 15 cm surface soil under tropical conditions can be done without any detrimental effects on crop yield. *Neem* cake application alongwith straw showed promising results.

Effect of Incorporation of Organic Matter on Paddy Crop

Organic matter (FYM and straw, 0.5 per cent w/w) was incorporated in sandy loam alluvial soil under field conditions one week before transplanting paddy seedlings. After one week of decomposition, it was found that application of farmyard manure and wheat straw increased the soil organic carbon by 72 and 79 per cent over control respectively (Table 6.17). The rate of decomposition of wheat straw was maximum till tillering stage and slower afterwards. The decomposition rate of FYM was faster as compared to wheat straw between tillering and harvesting stage. Application of organic materials increased both humic and humus carbon of soil.

Table 6.17: Effect of Incorporation of Organic Matter on Different Fractions of Soil Organic Matter Under Paddy Crop

Treatment	Before Transplanting				At Tillering				After Harvesting			
	Organic Carbon (%)	Humin Carbon (%)	Humus Carbon (%)	Humus Content (%)	Organic Carbon (%)	Humin Carbon (%)	Humus Carbon (%)	Humus Content (%)	Organic Carbon (%)	Humin Carbon (%)	Humus Carbon (%)	Humus Content (%)
Control	0.435	0.360	0.075	0.462	0.510	0.390	0.120	0.657	0.495	0.405	0.090	0.587
1 + 0.5'% FYM	0.750	0.630	0.120	0.750	0.720	0.510	0.210	1.085	0.645	0.525	0.120	0.760
1 + 0.5% wheat straw	0.780	0.675	1.105	0.502	0.660	0.405	0.195	0.747	0.570	0.435	0.135	0.670

Total nitrogen content in control soil decreased from 0.059 per cent before transplanting to 0.038 at harvest of crop showing a decrease of 35.6 per cent. Due to FYM and straw, appreciable increase in nitrogen content of soil was observed (Table 6.18). Availability of phosphate was favourably affected by FYM and a very marginal immobilisation of phosphate was observed in straw treated soil.

Table 6.18: Effect of Incorporation of Organic Matter on Soil Nitrogen and Available Phosphorus Under Paddy Crop

Treatment	Before Transplanting				At Tillering				After Harvesting			
	Total Nitrogen (%)	NH_4-N (ppm)	NO_3-N (ppm)	Available P (ppm)	Total Nitrogen (%)	NH_4-N (ppm)	NO_3-N (ppm)	Available P (ppm)	Total Nitrogen (%)	NH_4-N (ppm)	NO_3-N (ppm)	Available P (ppm)
Control	0.059	17.87	143.87	15.6	0.045	7.49	24.33	16.7	0.038	2.74	13.71	14.4
1 + 0.5% FYM	0.084	17.94	146.25	16.7	0.064	6.60	34.92	17.8	0.054	6.19	17.86	22.2
1 + 0.5% wheat straw	0.076	16.04	91.08	11.1	0.056	6.63	35.73	14.4	0.042	4.81	14.81	17.8

Farmyard manure significantly increased the grain yield of paddy by 41 per cent and nitrogen uptake by 10 per cent (Table 6.19). Straw incorporation increased the yield by 25 per cent and uptake of nitrogen by the crop by 8.2 per cent.

Table 6.19: Effect of Organic Materials on Yield and Nitrogen Uptake by Paddy Crop

Treatment	Grain Yield (g/plot)	% Increase Over Control	Straw Yield (g/plot)	% Increase Over Control	N % in Grain	Uptake of N by Grain (g/plot)
Control	666.7	–	831.7	–	0.95	6.3
FYM (0.5%)	941.7	41.2	1048.3	26.0	1.06	10.0
Wheat straw (0.5%)	836.7	25.2	923.3	11.0	0.98	8.2
C.D. at 5% level	76.3	–	78.4	–	–	–

Influence of Humic Substances on Crop Yields

The effect of humic substance extracted from farmyard manure and other sources was investigated on the growth of plants and microorganisms. The effect on crop yields is summarized in Table 6.20.

Application of sodium humate prepared out of farmyard manure @ 0.03 per cent to soil (w/w) significantly increased the yield of berseem, *dhaincha* and wheat crops by 47.0, 22.8 and 27.7 per cent over their control, respectively. Humus (Humic + fulvic fractions) when applied to sandy loam alluvial soil @ 0.025 and 0.05 per cent (w/w) increased the yield of paddy crop by 55.7 and 85.4 per cent respectively. Humus @ 0.05 per cent increased the yield of gram crop by 32.1 per cent.

Table 6.20: Influence of Humic Substances on Yield of Crops

	Per cent Increase Over Control							
	Berseem	*Dhaincha*	*Gram*	*Soyabean*	*Moong*	*Paddy*	*Tomato*	*Wheat*
Sodium humate (0.3%)	47.0	22.8	–	–	–	–	–	27.7
Rhizobium	23.0	18.7	–	–	–	–	–	–
Azotobacter	–	–	–	–	–	–	–	4.6
Rhizobium/Azotobacter + humate (0.0%)	52.8	61.5	–	–	–	–	–	35.4
Humus (0.025%)	–	–	–	–	–	55.7	–	–
Humus (0.05%)	–	–	32.1	–	–	85.4	–	–
Rhizobium	–	–	1.0	–	–	–	–	–
Rhizobium-Humus (0.05%)	–	–	41.0	–	–	–	–	–
Hydroquinone (10 ppm) sprayed	–	–	–	8.8	33.0	–	97.8	–
Humate (10 ppm) sprayed	–	–	–	23.2	77.0	–	109.0	–

Spraying of humates even in small dosages (10 ppm) 2 or 3 times during the growth of plants increased the yield of soybean, *moong* and tomato crops by 23.2, 77.0 and 109.0 per cent respectively. Similarly hydroquinone sprayed at the same rate increased the yield over control but the response was less as compared to humates.

Beneficial effect or humic substances on the growth of nitrogen-fixing microorganisms has been reported. In these studies (Table 6.20), it can be observed that the efficiency of *Rhizobium* and *Azotobacter* inoculants was increased due to application of humic materials. The response due to *Azotobacter* inoculation on wheat crop was 5 per cent increase in yield and due to humate it was 27.7 per cent and the combined effect (35.4 per cent) was greater than total of the two. Similar observation was recorded in case of gram and *dhaincha* crops. Nodulation was also improved substantially as recorded in case of gram and *dhaincha*.

Chapter 7

Response of Crops to Organic Manures

Response of crops to organic manures depends on several factors such as degree of decomposition and its carbon, nitrogen ratio, soil characteristics, moisture regime of the soil during the period of crop growth, time of its application, etc.

The major organic manures used in this country are the bulky manures, farmyard manure, compost and green manure. Concentrated organic manures such as cakes are not recommended now-a-days and in the past also whatever quantity of cake was used for manuring, was generally on certain crops like sugarcane, fruit crops and some other special crops such as betel-vine.

Numerous experiments have been carried out in the past at various research stations to study the response of crops to several organic manures. The main results obtained with farmyard manure or compost, oil-cakes and bone-meal as already reported by Garg together with results obtained later on are briefly given here. A good deal of research work on the response of farmyard manure to food crops under varied soil and agro-climatic conditions has been done in "All-India Coordinated Agronomic Research Project".

Farmyard Manure and Compost

Rice

Data of 341 experiments on rice crops are available for studying the response of rice to the application of farmyard manure and compost. No distinction could be made between farmyard manure and compost as neither was indicated in the various sources referred to, for the summarization of data. The actual doses tried-varied widely from centre to centre. Therefore, in order to summarize the results and give them in a concise form, the responses were standardized at a level of 12.6 tonnes per hectare (5 tonnes per acre). This was done by first fitting suitable response curves relating to the dose applied in those experiments where several doses were tried. Where the amount of manure was specified in terms of nitrogen content, it was assumed, in the absence or defimite knowledge on the actual composition of the manure, that the nitrogen percentage was 0.5 for conversion to the quantity of farmyard manure. The results obtained are given in the Table 7.1 (arranged State-wise). The response to application of 12.6 tonnes per hectare varied from about 100 kg of rice per hectare in Maharashtra and Bihar to 216 kg per hectare in Orissa. The average response based on 341 experiments distributed over 63 research stations in the country was 168 kg per hectare.

Table 7.1: Response of Rice to Farmyard Manure and Compost

State	*Response to 12.6 tonnes of Manure per hectare*		
	No. of Stations	*Number of Experiments*	*Response in kg per hectare*
Andhra Pradesh	10	64	194
Bihar	6	14	100
Gujarat	3	11	114
Kerala	3	20	128
Madhya Pradesh	6	31	212
Tamil Nadu	6	51	190
Maharashtra	8	33	99
Karnataka	6	13	168
Orissa	7	44	216
Uttar Pradesh	2	16	139
West Bengal	6	44	195
Average	**63**	**341**	**168**

N.B.: Average is taken by weighing the State figures by the area under the crop in the State.

Wheat

Results of 210 experiments on irrigated wheat and 71 experiments on unirrigated wheat were studied for finding out the response of wheat to farmyard manure or compost. The responses are given in the Tables 7.2 and 7.3 for irrigated and unirrigated wheats respectively. As in rice, the responses were worked out for a standard level since the actual dose varied widely from centre to centre. For irrigated wheat, the dose was taken as 12.6 tonnes per hectare whereas for unirrigated wheat, half of this dose was taken for estimating the response. The actual doses tested varied from 5 to 25 tonnes per hectare on irrigated crop.

Response of irrigated wheat to farmyard manure application at 12.6 tonnes per hectare ranged as low as 80 to 90 kg per hectare in Gujarat and Delhi to over 250 kg per hectare in Maharashtra, Madhya Pradesh and Bihar. The response of irrigated wheat, on the average, based on 210 experiments distributed over 31 research stations located in different States, was a little over 200 kg per hectare which is somewhat higher than the response obtained with rice. On unirrigated wheat, the response obtained was much lower being only 85 kg per hectare to an application of 6.3 tonnes per hectare. Unirrigated wheat has shown particularly good response in Rajasthan but this is based only on three experiments. In both Madhya Pradesh and Maharashtra, where the number of experiments was large, the responses were about 75 kg per hectare.

Sugarcane

Data of 258 experiments distributed over 8 States and 19 experimental stations were available for examining the response of sugarcane to farmyard manure or compost. The doses tried ranged between

5 to 50 tonnes per hectare. Therefore, the responses were standardized at 25 tonnes per hectare using techniques similar to those described in rice and wheat (Table 7.4). There is no large variation in the response of sugarcane in different States, the average response being of the order of 8 tonnes of sugarcane for an application of 25 tonnes of farmyard manure. The highest response has been obtained in Andhra Pradesh followed by Tamil Nadu. Only in Bihar and Gujarat, the response was somewhat low. It may be observed that the results for these two States are based on only two and one experiments respectively and, as such, cannot be relied upon much.

Table 7.2: Response of Irrigated Wheat to Farmyard Manure and Compost

State	*Response to 12.6 tonnes of Manure per hectare*		
	No. of Stations	*Number of Experiments*	*Response in kg per hectare*
Bihar	3	9	295
Delhi	1	21	92
Gujarat	1	1	82
Madhya Pradesh	3	46	259
Maharashtra	2	2	285
Punjab	7	67	213
Rajasthan	6	32	216
Uttar Pradesh	8	32	176
Average	**31**	**210**	**202**

N.B.: Average is taken by weighing the State figures by the area under the crop in the State.

Table 7.3: Response of Unirrigated Wheat to Farmyard Manure and Compost

State	*Response to 6.3 tonnes of Manure per hectare*		
	No. of Stations	*Number of Experiments*	*Response in kg per hectare*
Gujarat	1	2	87
Madhya Pradesh	7	56	74
Maharashtra	4	10	77
Rajasthan	2	3	140
Average	**14**	**71**	**85**

N.B.: Average is taken by weighing the State figures by the area under the crop in the State.

Cotton

Data for 71 experiments on irrigated cotton and 294 experiments on unirrigated cotton were available to study the response of cotton to farmyard manure. The doses tried on irrigated cotton ranged between 5 to 25 tonnes per hectare, whereas those on unirrigated cotton between 5 to 13 tonnes. The responses were standardized at 12.6 and 6.3 tonnes per hectare or farmyard manure on irrigated and unirrigated cotton respectively (Tables 7.5 and 7.6). The response of irrigated cotton shows large variation, it being nearly double in the southern States as compared with Punjab and Uttar Pradesh, in the north. Under unirrigated conditions the response was somewhat poor ranging between 12 to 22 kg of lint per hectare.

Table 7.4: Response of Sugarcane to Farmyard Manure and Compost

State	*Response to 25.1 tonnes of Manure per hectare*		
	No. of Stations	*Number of Experiments*	*Response in tonnes per hectare*
Andhra Pradesh	1	10	11.7
Bihar	1	2	3.7
Gujarat	1	1	3.7
Maharashtra	5	132	7.9
Karnataka	2	12	5.9
Punjab	5	30	8.0
Tamil Nadu	2	5	9.8
Uttar Pradesh	2	66	8.4
Average	**19**	**258**	

N.B.: Average is taken by weighing the State figures by the area under the crop in the State.

Table 7.5: Response of Irrigated Cotton to Farmyard Manure and Compost

State	*Response to 12.6 tonnes of Manure per hectare*		
	No. of Stations	*Number of Experiments*	*Response (lint) in kg per hectare*
Karnataka	2	16	80.9
Punjab	5	26	49.8
Tamil Nadu	1	5	96.1
Uttar Pradesh	2	24	48.9
Average	**10**	**71**	**56.4**

N.B.: Average is taken by weighing the State figures by the area under the crop in the State.

Table 7.6: Response of Unirrigated Cotton to Farmyard Manure and Compost

State	*Response to 6.3 tonnes of Manure per hectare*		
	No. of Stations	*Number of Experiments*	*Response (lint) in kg per hectare*
Andhra Pradesh	2	28	12.8
Gujarat	4	57	11.8
Madhya Pradesh	5	42	22.3
Maharashtra	10	113	15.9
Karnataka	4	54	20.2
Average	**25**	**294**	**16.2**

N.B.: Average is taken by weighing the State figures by the area under the crop in the State.

Potato

The results of more than 30 experiments conducted on potato are given in Table 7.7. As in other crops, there has been considerable variation in the quantity or farmyard manure used. Generally, there has been an incease in yield with the application of farmyard manure varying from 3 to 4 per cent to as much as 30 per cent of the control yield.

Table 7.7: Response of Potato to Farmyard Manure and Compost

State	*Station*	*Year(s)*	*No. of Experi-ments*	*Treatments*	*Control/ Yield (kg/ha)*	*Response (kg/ha)*	*S.E. of Res-ponse (kg/ha)*	*Response as Per-centages of Control*	*Remarks (Irrigated (Unirri-gated)*
Himachal Pradesh	Shillaroo	1951–53	7	Control F.Y.M. 36.88 tonnes/ha	15437	3894	1332	25.2	Unirrigated
Himachal Pradesh	Shillaroo	1953	1	Control F.Y.M. 9.21 tonnes/ha	8409	8589	–	102.1	Unirrigated
				F.Y.M. 18.42 tonnes/ha	–	9239	1213	109.9	
				F.Y.M. 36.85 tonnes/ha	–	13178	–	156.7	
Madhya Pradesh	Chhindwara	1948–51	2	Control T.C. at 37.66 tonnes/ha	8159	354	860	4.3	Irrigated
Orissa	Sambalpur	1955–56	2	Control F.Y.M. 8.96 tonnes/ha	2392	381	1384	15.9	Irrigated
Punjab	Jullundhar	1949–52	3	Control F.Y.M. 22.42 tonnes/ha	6437	235	236	3.7	Irrigated
Uttar Pradesh	Farrukhabad	1955–57	3	Control F.Y.M. 22.42 tonnes/ha	15892	536	710	3.4	Irrigated
Uttar Pradesh	Lucknow	1951	1	Control F.Y.M. 26.89 tonnes/ha	7806	2396	1781	30.7	Irrigated
Uttar Pradesh	Kanpur	1952–66	7	Control F.Y.M. 22.42 tonnes/ha	16820	2439	2200	14.5	Irrigated
Uttar Pradesh	Kanpur	1955	1	Control F.Y.M. 3892 tonnes/ha	11308	1502	987	13.3	Information regarding irrigation status not available
Uttar Pradesh	Kanpur	1953	1	Control F.Y.M. 28.02 tonnes/ha	24821	764	1568	3.1	Irrigated

Other Crops

Experimental results on the effect of farmyard manure on some crops of vegetables and chillies are also available (Tables 7.8–7.17). Application of farmyard manure has resulted in a significant increase in the yield of chillies. The response varied from 16 to 89 per cent. Results of similar experiments on *methi,* onion, sweet-potato, tomato and tapioca are also given. It will be seen that the response showed wide valuation. But in general, there has been a positive response, particularly to sweet-potato, tapioca and onion.

Table 7.8: Response of Sweet-potato to Farmyard Manure and Compost

State	*Station*	*Year(s)*	*No. of Experiments*	*Treatments*	*Control/ Yield (kg/ha)*	*Response (kg/ha)*	*S.E. of Response (kg/ha)*	*Response as Percentages of Control*	*Remarks*
Karnataka	Mangalore	1951–56	6	Control F.Y.M. 11.21 tonnes/ha	9007	2757	298	30.6	Unirrigated
Tami Nadu	Coimbatore	1951–53	3	Control F.Y.M. 11.21 tonnes/ha	7426	1084	261	14.6	Irrigated

Table 7.9: Response of Tapioca to Farmyard Manure and Compost

State	*Station*	*Year(s)*	*No. of Experiments*	*Treatments*	*Control/ Yield (kg/ha)*	*Response (kg/ha)*	*S.E. of Response (kg/ha)*	*Response as Percentages of Control*	*Remarks*
Tamil Nadu	Coimbatore	1951–53	3	Control F.Y.M. 11.21 tonnes/ha	3872	455	254	11.8	Irrigated

Table 7.10: Response of Onion to Farmyard Manure and Compost

State	*Station*	*Year(s)*	*No. of Experiments*	*Treatments*	*Control/ Yield (kg/ha)*	*Response (kg/ha)*	*S.E. of Response (kg/ha)*	*Response as Percentages of Control*	*Remarks*
Maharashtra	Niphad	1961–63	2	Control F.Y.M. 11.21 tonnes/ha	32062	2980	1190	9.3	Information regarding irrigation status not available
				F.Y.M. 22.42 tonnes/ha		5407		16.9	

Table 7.11: Response of Garlic to Farmyard Manure and Compost

State	*Station*	*Year(s)*	*No. of Experiments*	*Treatments*	*Control/ Yield (kg/ha)*	*Response (kg/ha)*	*S.E. of Response (kg/ha)*	*Response as Percentages of Control*	*Remarks*
Uttar Pradesh	Lucknow	1950	1	Control F.Y.M. 22.42 tonnes/ha	3384	508	191	15.0	Irrigated

Table 7.12: Response of Chillies (Dry) to Farmyard Manure and Compost

State	*Station*	*Year(s)*	*No. of Experi-ments*	*Treatments*	*Control/ Yield (kg/ha)*	*Response (kg/ha)*	*S.E. of Res-ponse (kg/ha)*	*Response as Per-centages of Control*	*Remarks*
Kerala	Taliparanha	1948–50	4	Control F.Y.M. 12.55 tonnes/ha	693	616	326	889	Unirrigated
Andhra Pradesh	Lam	1945–48	8	Control F.Y.M. 11.20	462	74	29	16.0	Information regarding irrigation status is not available
Andhra Pradesh	Lam	1950	1	Control	633				"
				F.Y.M. 26.89 tonnes/ha		235	97	37.1	
Andhra Pradesh	Lam	1955–57	3	Control	888				
				F.Y.M. 8.96 tonnes/ha		305	101	34.3	Unirrigated
				F.Y.M. 13.46 tonnes/ha		354		39.9	
				F.Y.M. 17.93 tonnes/ha		451		50.8	

Table 7.13: Response of *Bhindi* to Farmyard Manure and Compost

State	*Station*	*Year(s)*	*No. of Experi-ments*	*Treatments*	*Control/ Yield (kg/ha)*	*Response (kg/ha)*	*S.E. of Res-ponse (kg/ha)*	*Response as Per-centages of Control*	*Remarks*
Maharashtra	Poona	1961–63	2	Control	6558				
				F.Y.M. 11.21 tonnes/ha		235	269	3.6	Information on irrigation status not available
				F.Y.M. 22.42 tonnes/ha		357		5.4	Heavy rainfall is reported

Table 7.14: Response of Tomato to Farmyard Manure and Compost

State	*Station*	*Year(s)*	*No. of Experi-ments*	*Treatments*	*Control/ Yield (kg/ha)*	*Response (kg/ha)*	*S.E. of Res-ponse (kg/ha)*	*Response as Per-centages of Control*	*Remarks*
Maharashtra	Poona	1960–63	3	Control	13128				
				F.Y.M. 11.21 tonnes/ha		175	304	1.3	Information on irrigation status is not available
				F.Y.M. 22.42 tonnes/ha		866		6.6	Heavy rainfall is reported

Table 7.15: Response of Brinjal to Farmyard Manure and Compost

State	*Station*	*Year(s)*	*No. of Experiments*	*Treatments*	*Control/ Yield (kg/ha)*	*Response (kg/ha)*	*S.E. of Response (kg/ha)*	*Response as Percentages of Control*	*Remarks*
Maharashtra	Poona	1956–59	4	Control	31204				Unirrigated
				F.Y.M. 12.55 tonnes/ha		660	3906	2.1	
				F.Y.M. 25.11 tonnes/ha		370		1.1	–
Uttar Pradesh	Lucknow	1949	1	Control	15538	247		1.6	Irrigated
				F.Y.M. 7.83 tonnes/ha					
				F.Y.M. 15.69 tonnes/ha		710	2234	4.6	–
				F.Y.M. 31.38 tonnes/ha		148		0.9	–

Table 7.16: Response of Spinach to Farmyard Manure and Compost

State	*Station*	*Year(s)*	*No. of Experiments*	*Treatments*	*Control/ Yield (kg/ha)*	*Response (kg/ha)*	*S.E. of Response (kg/ha)*	*Response as Percentages of Control*	*Remarks*
Uttar Pradesh	Lucknow	1953	1	Control	559				
				F.Y.M. 6.72 tonnes/ha		163		29.2	
				F.Y.M. 8.96 tonnes/ha		308	63	55.0	Irrigated
				F.Y.M. 11.20 tonnes/ha		524		93.7	

Table 7.17: Response of Methi to Farmyard Manure and Compost

State	*Station*	*Year(s)*	*No. of Experiments*	*Treatments*	*Control/ Yield (kg/ha)*	*Response (kg/ha)*	*S.E. of Response (kg/ha)*	*Response as Percentages of Control*	*Remarks*
Maharashtra	Poona	1961–62	1	Control	3542				It is not known whether the crop was irrigated or not. But during crop season heavy rainfall was reported.
				F.Y.M. 5.60 tonnes/ha		– 118	220	– 3.3	
				F.Y.M. 11.20 tonnes/ha		348	2	9.8	
	Poona	1960–61	1	Control	5550				
				F.Y.M. 12.55 tonnes/ha		428	430	7.7	
				F.Y.M.25.11 tonnes/ha		588		10.6	

Oil-cakes

The principal oil-cakes used for manuring in India are groundnut-cakes and mustard-cakes. As already mentioned, oil-cake as manure for food-grains is not being encouraged for various reasons.

Sugarcane

Results of more than 70 field experiments conducted on sugarcane in different States are available. These experiments were generally conducted to compare the performance of oil-cakes with ammonium sulphate, the traditional fertilizer (Tables 7.18).

Table 7.18: Response to Sugarcane to Oil-cakes

State	*Station*	*Year (s)*	*No. of Experi-ments*	*Treatments*	*Control/ Yield (cane) in tonnes/ha*	*Response (cane) in tonnes/ha*		*S.E. of Response in tonnes/ha*	*Remarks*
						Cake Alone	*A/S : Cake*		
Andhra Pradesh	Ankapalle	1951–59	9	Control	49.48				A/S : cake in
				112.1 kg N/ha		42.79	47.06	1.96	the ratio of 1 : 0
	Ankapalle	1950–51	2	Control	96.36	N.A.		4.87	A/S : cake in
				112.1 kg N/ha			18.56		the ratio of 2 : 1
				224.2 kg N/ha			12.38		
				336.3 kg N/ha			17.20		
	Rudur	1957–59	3	Control	15.27				A/S : cake in the
				112.1 kg N/ha		58.12	66.03	5.05	ratio of 1 : 0
Bihar	Dehri-on	1954	2	Control	39.52				
	Sone			44.8 kg N/ha		6.12			
				89.6 kg N/ha		8.61	N.A.	3.49	
	Patna	1956–57	2	Control	28.52			2.74	A/S : cake in the
				89.6 kg N/ha		3.67	0.65		ratio of 1 : 0
				Control	42.63			2.56	A/S : cake in the
		1954–56	4			6.61	7.71		ratio of 1 : 1
	Warisali-	1954–56	3	134.5 kg N/ha					
	gang			Control	89.71		–		
				44.8 kg N/ha		4.94		4.67	
				89.6 kg N/ha		7.15	N.A.		
Maharashtra	Padegaon	1954–59	6	Control	61.91			2.84	A/S : cake in the
				336.3 kg N/ha		26.72	20.04		ratio of 1 : 0
Uttar Pradesh	Anandnagar	1958	1	Control	58.32			5.32	A/S : cake in the
				67.2 kg N/ha		0.33	2.21		ratio of 1 : 0
	Bhojapur	1958	1	Control	35.50			N.A.	A/S : cake in the
				67.2 kg N/ha		8.69	2.31		ratio of 1 : 0
Uttar Pradesh	Etawah	1957–59	3	Control	44.67			2.33	A/S : cake in the
				67.2 kg N/ha		2.05	10.36		ratio of 1 : 0
	Golgokura-	1959	1	Control	45.29				A/S : cake in the
	nath			67.2 kg N/ha		2.76	6.36	2.54	ratio of 1 : 0
	Khadda	1958	1	Control	22.67				A/S : cake in the
				67.2 kg N/ha		2.06	0.78	6.95	ratio of 1 : 0
	Kunraghat	1956–38	3	Control	34.85				A/S : cake in the
				67.2 kg N/ha		5.90	6.33	4.89	ratio of 1 : 0

Contd...

Table 7.18–Contd...

State	Station	Year (s)	No. of Experiments	Treatments	Control/ Yield (cane) in tonnes/ha	Response (cane) in tonnes/ha		S.E. of Response in tonnes/ha	Remarks
						Cake Alone	A/S : Cake		
	Masodha	1958	1	Control	52.17				A/S : cake in the
				67.2 kg N/ha		27.62	28.95	5.32	ratio of 1 : 0
	Modinagar	1955	1	Control		39.99			
				44.8 kg N/ha		0.70			
				89.6 kg N/ha		– 7.48	N.A.	7.13	
				134.5 kg N/ha		– 1.99			
	Muzaffar-nagar	1954–59	8	Control	43.99				A/S : cake in the
				134.5 kg N/ha		20.89	22.12	1.23	ratio of 1 : 0
		1956–58	3	Control	50.87			2.13	
				67.2 kg N/ha		1.96	3.91		A/S : cake in the ratio of 1 : 0
	Neeli	1959	1	Control	41.50			6.40	A/S : cake in the
				67.2 kg N/ha		11.25	6.35		ratio of 1 : 0
	Rampur	1959	1	Control	30.81				A/S : cake in the
				67.2 kg N/ha		5.62	7.38	5.65	ratio of 1 : 0
Uttar Pradesh	Raya	1959	1	Control	52.98	13.56	7.38	0.40	A/S : cake in the
				67.2 kg N/ha					ratio of 1 : 0
	Saini	1959	1	Control	58.55			15.77	A/S : cake in the
				67.2 kg N/ha		6.68	– 4.57		ratio of 1 : 0
	Seohara	1959	1	Control	58.20				
				67.2 kg N/ha		2.11	4.62	4.57	A/S : cake in the ratio of 1 : 0
	Shahjahan-pur	1954–59	10	Control	61.09				A/S : cake in the
				67.2 kg N/ha		8.53	– 8.34	1.28	ratio of 1 : 0
			1	Control	53.93				
				112.1 kg N/ha		16.39	N.A.	2.41	
	Varanasi	1957–59	3	Control	44.67			3.34	A/S : cake in the
				67.2 kg N/ha		0.19	2.06		ratio of 1 : 0
	Haldwani	1959	1	Control	42.98				
				67.2 kg N/ha		2.21	4.97	1.76	A/S : cake in the ratio of 1 : 0
Maharashtra	Padegaon	1954–59		8	336	1 : 0	97.26	2.39	
						0 : 1	76.40		
						1 :1	95.73		
						2 : 1	93.10		
						1 : 2	89.91		

In Andhra Pradesh, at the main Sugarcane Station, Anakapalle, trials conducted with groundnut-cake gave high response which was of the same order as for ammonium sulphate applied on equal nitrogen basis. At Rudrur in Andhra Pradesh also, the response to groundnut-cake was very high, but not higher than that of ammonium sulphate. At Padegaon in Maharashtra, the response to oil-cakes was higher than that of ammonium sulphate. In Bihar, the trials were conducted with castor-cakes. The response to cakes applied at the rate of 90 kg per hectare was 6.57 tonnes per hectare. In Uttar Pradesh, groundnut-cake was generally used. The responses were not particularly high, the average response to 57 kg per hectare over 14 centres being equal to 6.01 tonnes per hectare. The results of some trials with varying proportions of cakes and ammonium sulphate at different levels of nitrogen were also available in Maharashtra. The data generally showed that with the increase in the proportion of groundnut-cake, the response also increases.

A number of experiments comparing nitrogen application with varying proportions of groundnut-cake and ammonium sulphate were conducted in Maharashtra.

Groundnut-cake alone has given a higher response. The yield figures corresponding to other ratios of groundnut-cake and ammonium sulphate also show that the response is higher with higher proportion of cakes in the mixture, indicating that groundnut-cake is superior to ammonium sulphate.

Cotton

Data from 13 experiments on cotton were available to study the effect of application of groundnut-cake. The responses on irrigated cotton showed large variation (Table 7.19), the best response having been obtained at Warrangal in Andhra Pradesh. The responses in all the cases were not significantly different from those to ammonium sulphate applied at the same level.

Table 7.19: Response of Unirrigated Cotton to Oil-cakes

State	*Station*	*Year(s)*	*No. of Experi-ments*	*Treatments*	*Control/ Yield (lint) in kg/ha*	*Response (lint) in kg/ha*		*S.E. of Response in kg/ha*	*Remarks*
						Cake Alone	*A/S Alone*		
				(A)					
Andhra Pradesh	Mudhol	1948–49	2	Control	62	30		4.2	
				22.4 kg N/ha					
				44.8 kg N/ha					
	Warangal	1948	2	Control	142	44	N.A.		
				22.4 kg N/ha		50	109		
				44.8 kg N/ha		102	144	15.5	
				67.2 kg N/ha		105	140		
Madhya Pradesh	Amdaha	1955	1	Control	42				
				44.8 kg G.N. C/ha		21	8	8.1	
	Indore	1954–55	2	Control	58				
				22.4 kg N/ha		– 4	N.A.	5.5	
				44.8 kg N/ha		29			
Karnataka	Dharwar	1954	1	Control	243			14.3	
				67.2 kg N/ha		34	32		
	Hogari	1954–58	5	Control	125			4.5	
				33.6 kg N/ha		8	13		
				(B)					
Punjab	Rauni	1954	1	Control	381		A/S alone	44.9	
				28.0 kg N/ha		63	64		

Long-term Effect of Organic Manures

The knowledge on the effect of continuous application of organic and inorganic fertilizers on crop production and soil fertility is of considerable importance in formulating a sound and long-term manurial policy. We have earlier given the results of the direct effect of organic manures. Similar knowledge on the long-term effect has to be obtained from the results of carefully conducted manurial trials. Such long-term experiments have been very few in this country. Results of some of the long-term experiments are given below:

Rice

In West Bengal, long-term experiments with and without application of farmyard manure and with different fertiliur doses were conducted at Chinsura, Berhampoce and Suri farms. At Chinsura and Suri farms, experiments were started in 1948–49 and at Berhampore in 1949–50. The soil of Chinsura farm is typical Ganga low-land and is nearly neutral in reaction (pH 6.8). The soil of Suri farm is of lateritic origin and is acidic in reaction (pH 5.5). The soil at Berhampore is sandy-loam and alkaline in reaction (pH 7.7).

The response of rice to farmyard manure and trend of yield over the entire period of experimentation as compared with the application of fertilizers are given in Tables 7.20–7.22. Application of farmyard manure at Chinsura did not increase the yield, but at Suri and Berhampore there was a significant increase with the application of farmyard manure alone at the rate of 9 tonnes per hectare. Further, a study of trend in yield as indicated by the regression of yield over the years showed that plots treated with farmyard manure at Suri and Berhampore farms gained annually by 90 and 156 kg per hectare, respectively over the plots to which farmyard manure was not applied.

Table 7.20: Response of Rice to Farmyard Manure and Regression of Yield Over Years (kg/ha)

Station: Chinsurah			**State: West Bengal**			**Period: 1991–92 to 2001–02**	
Soil Characteristics	*Levels of N in kg/ha*	*Mean Cumulative Yield*			*Linear Regression of Yield Over Years*		
		F.Y.M. Levels			*F.Y.M. Levels*		
	2	*0*	*9220*	*Average*	*0*	*9220*	*Average*
Alluvium neutral in reaction	0	1430	1470	1450	43.0	45.8	44.4
	33.6	1540	1550	1545	31.4	31.9	31.7
	67.2	1440	1370	1405	31.2	17.1	24.1
	100.8	1290	1220	1255	21.1	13.2	17.2
	133.4	1190	1150	1170	17.2	13.0	15.1
	Average	**1378**	**1352**	**1365**	**28.7**	**24.2**	**26.5**

S.E. of difference
(*i*) between N means = 20.0
(*ii*) between F.Y.M. means = 9.0

S.E. of difference
(*i*) between N means = 4.09
(*ii*) between F.Y.M. means = 1.91

At the Central Rice Research Institute, Cuttack, experiments were conducted from 1949–50 to study the long-term effect of compost application with and without nitrogen. One experiment was conducted with combinations 0, 22.4. 44.8, 67.2 and 89.6 kg of nitrogen with and without compost at 9.2 tonnes per hectare. In another experiment, three levels each of nitrogen, phosphorus and potash were tried in combination with green manure, groundnut cake and farmyard manure applied at the rate of supplying 22.4 kg per hectare nitrogen. A third experiment was conducted at the same centre, with three levels of nitrogen and three levels of lime. After the first four years, the experiment was modified to include compost as one of the factors (Tables 7.23–7.26). In the experiment No. 1, compost applied alongwith nitrogen up to 22.4 kg per hectare has shown significant response. Plots in which compost was applied showed gain in yield for years as compared with the plots in which no farmyard manure was applied. In the experiment with different organic manures, fertilizers and application of compost has not shown a significant increase in yield either in the absence of other nitrogenous fertilizers or when nitrogen was applied. There were also no significant differences in the trend in yield but showed that compost-treated plots gained at the rate of 17 kg per hectare annually over plots receiving no compost.

Table 7.21: Response of Rice to Farmyard Manure and Regression of Yield Over Years (kg/ha)

Station: Suri | **State: West Bengal** | **Period: 1991–92 to 2001–02**

Soil Characteristics	*Levels of N in kg/ha*	*Mean Cumulative Yield*			*Linear Regression of Yield Over Years*		
		F.Y.M. Levels			*F.Y.M. Levels*		
		0	*9220*	*Average*	*0*	*9220*	*Average*
Lateritic acidic	0	1620	2020	1820	2.9	18.6	10.7
	33.6	1970	2150	2060	– 16.7	– 3.9	– 10.3
	67.2	2020	2020	2020	– 29.1	– 28.8	– 29.0
	Average	**1870**	**2063**	**1967**	**– 14.3**	**– 4.7**	**– 9.5**

S.E. of difference
(*i*) between N means = 35.0
(*ii*) between F.Y.M. means = 29.0

S.E. of difference
(*i*) between N means = 5.39
(*ii*) between F.Y.M. means = 4.40

Table 7.22: Response of Rice to Farmyard Manure and Regression of Yield Over Years (kg/ha)

Station: Berhampore | **State: West Bengal** | **Period: 1991–92 to 2001–02**

Soil Characteristics	*Levels of N in kg/ha*	*Mean Cumulative Yield*			*Linear Regression of Yield Over Years*		
		F.Y.M. Levels			*F.Y.M. Levels*		
		0	*9220*	*Average*	*0*	*9220*	*Average*
Laterite, sandy-loam	0	20	530	475	– 24.7	– 7.6	– 16.1
alkaline in reaction	33.6	640	750	695	– 24.7	– 6.5	– 15.6
	67.2	790	890	840	– 20.9	– 14.2	– 17.6
	100.8	830	930	880	– 39.7	– 17.4	– 28.6
	133.4	740	920	830	– 28.5	– 14.8	– 21.7
	Average	**684**	**804**	**744**	**– 27.7**	**– 12.1**	**– 10.9**

S.E. of difference
(*i*) between N means = 33.0
(*ii*) between F.Y.M. means = 11.0

S.E. of difference
(*i*) between N means = 4.4
(*ii*) between F.Y.M. means = 2.6

Table 7.23: Response of Rice to Compost and Regression of Yield Over Years (kg/ha)

Station: Cuttack **State: Orissa** **Period: 1991–92 to 2001–02**

Soil Characteristics	*Levels of N in kg/ha*	*Mean Cumulative Yield of Compost F.Y.M. Levels*			*Linear Regression of Yield Over Years of Compost F.Y.M. Levels*		
		0	*9220*	*Average*	*0*	*9220*	*Average*
N.A.	0	14.00	1720	1560	– 2.3	8.7	3.2
	22.4	16.30	1920	1775	2.7	3.3	3.0
	44.8	18.60	18.50	18.55	0.0	– 9.1	– 4.5
	67.2	1890	1710	1800	7.0	44	5.7
	100.8	1780	14.20	1600	14.7	13.0	13.9
	Average	**1712**	**1724**	**1718**	**4.4**	**4.1**	**4.2**

S.E. of difference
(*i*) between N means = 28.0
(*ii*) between compost means = 27.0

S.E. of difference
(*i*) between N means = 7.6
(*ii*) between compost means = 2.6

Table 7.24: Response Yield of Rice to Organic Manures (kg/ha)

Station: Cuttack **State: Orissa** **Period: 1996 to 2000**

Level of N (kg/ha)	*Organic Manures*		*Groundnut-cake*	*Compost*	*Average*
	No Manure	*Green Manure*			
0	1448	1615	1607	1533	1551
22.4	1553	1634	1578	1646	1603
44.8	1635	1644	1469	1692	1641

S.E. of difference
(*i*) between N means = 10.0
(*ii*) between manure means = 91.0

Table 7.25: Linear Regression of Yield of Rice Over Years (kg/ha)

Station: Cuttack **State: Orissa** **Period: 1996 to 2000**

Level of N (kg/ha)	*Organic Manures*		*Groundnut-cake*	*Compost*	*Average*
	No Manure	*Green Manure*			
0	13.41	8.34	– 65.28	33.20	– 2.58
22.4	– 16.53	21.23	– 63.58	15.37	– 10.88
44.8	– 10.56	9.28	– 62.88	23.07	– 10.27
Average	**– 4.56**	**12.95**	**– 63.92**	**23.88**	**– 7.91**

S.E. of difference
(*i*) between N means = 21.32
(*ii*) between manure means = 33.72

Table 7.26: Response of Rice to Compost and Regression of Yield Over Years (kg/ha)

Station: Cuttack			**State: Orissa**			**Period: 1991–92 to 2001–02**	
Soil Characteristics	*Levels of N kg/ha*		*Mean Cumulative Yield*			*Linear Regression of Yield Over Years*	
			Compost	*Average*		*Compost*	*Average*
N.A.	0	0	9220		0	9220	
	0	13.60	1780	1570	– 15.6	14.5	– 15.1
	22.4	1700	1990	1845	– 43.8	– 22.5	– 33.1
	44.8	1910	2050	1970	– 47.6	– 17.3	– 32.5
	Average	**1657**	**1933**	**1795**	**– 35.7**	**– 18.1**	**– 26.9**

S.E. of difference

(*i*) between N means = 25.0

(*ii*) between compost means = 31.0

S.E. of difference

(*i*) between N means = 3.8

(*ii*) between compost means = 1.9

Sahu reported the results of a long-term experiments on "Soil Fertility Investigations" conducted for ten years from 1956 to 1965. Investigations were carried out on a sandy loam lateritic soil of Bhubaneswar, Orissa to study the effect of continuous application of ammonium sulphate and organic manures, *viz.* FYM, green manure and groundnut cake (GNC) singly as well as in combination on the yield of lowland rice crop. Grain yield data (mean of 10 years) have been given in Table 7.27 which shows that the highest grain yield of 2.667 kg/ha was obtained with basal dressing of FYM to supply 45 kg N/ha together with 45 kg N/ha as ammonium sulphate the increase over control yielding 2,141 kg/ha being 520 kg/ha.

Table 7.27: Grain Yield of Rice Crop Due to Organic and Inorganic Manures (kg/ha)

Station: Bhubaneswar		**State: Orissa**		**Period: 1991–92 to 2001**	
Levels of N (kg/ha)	*Yield without Basal Dressing*	*Yield with basal dressing (45 kg N/ha)*			*Mean*
		F.Y.M.	*G.M.*	*G.N.C.*	
0	2147	2443	2511	2402	2376
22.5	2345	2468	2400	2364	2394
45.0	2361	2667	2496	2390	2479
67.5	2411	2396	2267	2179	2313
90.0	2231	2280	2021	2200	2183
Mean	**2299**	**2451**	**2339**	**2307**	

Padalia reported the results of a long-term experiment on a high-yielding variety of rice carried out at Central Rice Research Institute, Cuttack for seven years from 1967 to 1973. Two levels of compost to supply 0 and 80 kg N per ha were tried with and without inorganic nitrogen. The doses of inorganic nitrogen from ammonium sulphate were 0, 40, 80, 120 and 160 kg N per ha. Seven years mean grain yield data of rice crop, have been given in Table 7.28. Response due to compost alone to supply 80 kg

N/ha was 408 kg/ha of grains, the increase over control being 9.9 per cent. Application of compost alone was found to be inferior to ammonium sulphate at similar rate of application. However, application of compost at the rate of 80 kg N/ha in combination with ammonium sulphate to supply total quantity of 120 or 160 kg N/ha gave higher yields than the corresponding levels of nitrogen in the form of ammonium sulphate.

Table 7.28: Grain Yield of Rice Crop Due to Organic and Inorganic Manures (kg/ha)

Station: Cuttack | **State: Orissa** | **Period: 1996 to 2001**

Levels of N (kg/ha)	*Grain Yield kg/ha*		*Increase Due to Compost in kg/ha*
	Without Compost	*With Compost Providing 80 kg N/ha*	
0	4132	4540	408
40	4568	4972	404
80	4882	5037	155
120	4844	4831	Decrease
160	4769	4782	13
Mean	**4639**	**4832**	

Sugarcane

Apart from the long-term experiments in rice, results of a few experiments on sugarcane are also available. Long-term experiments on sugarcane were conducted at Padegaon, Kopergaon, Deolali and Akluj. At Padegaon, the experiment was started in the year 1939–40 on a three-course rotation of sugarcane, *jowar* and groundnut. The treatment consisted of combinations of two levels of compost with the six treatments made up of a control and 336 kg of nitrogen supplied by ammonium sulphate, groundnut-cake and mixtures of the two in three fixed rotations. Compost was applied at the rate of 25 tonnes per hectare, which contained approximately 135 kg N, 135 kg P_2O_5 and 720 kg K_2O. The experiment was modified in the year 1951–52. The 'no compost' plots were divided into two parts. To one of these, an additional treatment of nitrogen, phosphate and potash mixture in the form of commercial fertilizers and equivalent to a dose of 25 tonnes of compost per hectare was applied. The experiment was again modified in 1954–55 when the compost plots were also subdivided in two and the nitrogen, phosphate and potash mixtures were applied as above (Tables 7.29 and 7.30). Significant responses to the application of nitrogen and compost were obtained. The response was significantly higher with groundnut-cake as compared with ammonium sulphate. The results of plots treated with combinations of two fertilizers in different ratios also indicated the superiority of groundnut-cake. The response to nitrogen in the form of ammonium sulphate was enhanced when this fertilizer was applied in combination with compost. Considering the rates of change in yield with years, it is seen that there was a deterioration in the yield of all plots, but the application of compost resulted in a pronounced reduction in the rate of deterioration in nitrogen-treated plots. Considering the results for the first four crops and that for the entire period, it will be seen that the application of compost was effective in reducing deterioration in the soil fertility resulting from the continued application of nitrogen for short period but this could not, however, be maintained for a longer period.

Table 7.29: Response of Sugarcane to Organic Manures and Regression of Yield Over Areas (tonnes/ha)

Station: Pedegaon			**State: Maharashtra**			**Period: 2001–2002**
Treatments	*Mean Yield*			*Linear Regression of Yield Over Years*		
	No Compost	*Compost*	*Average*	*No Compost*	*Compost*	*Average*
Control	30.8	39.2	35.0	– 4.44	– 1.75	– 2.09
Groundnut-cake	102.7	111.3	107.0	– 5.86	– 8.16	– 7.01
Ammonium Sulphate	61.4	86.2	73.8	– 12.27	– 10.49	– 11.38
A/S : G.N.C. (1 : 1)	95.6	107.0	101.3	8.86	– 7.81	0.52
A/S : G.N.C. (1 : 2)	96.9	112.1	104.5	– 9.94	– 8.80	– 9.37
A/S : G.N.C. (2 : 1)	81.2	102.3	91.8	– 9.53	– 6.12	– 7.82
Average	**78.1**	**93.0**	**85.6**	**– 5.53**	**– 7.19**	**– 6.36**

S.E. of difference
(*i*) between compost means = 1.66
(*ii*) between treatment means = 2.88

S.E. of difference
(*i*) between compost means = 0.70
(*ii*) between treatment means = 1.22

Table 7.30: Response of Sugarcane to Organic Manures and Regression of Yield Over Years (tonnes/ha)

Station: Pedegaon			**State: Maharashtra**			**Period: 1980–2000**
Treatments	*Mean Yield*			*Linear Regression of Yield Over Years*		
	No Compost	*Compost*	*Average*	*No Compost*	*Compost*	*Average*
Control	37.2	43.9	40.5	1.98	5.97	3.97
Groundnut-cake	105.8	124.7	115.2	– 8.46	– 2.78	– 5.62
Ammonium sulphate	72.0	100.4	86.2	– 7.70	– 4.01	– 5.85
A/S : G.N.C. (1 : 1)	96.1	155.8	105.9	– 6.78	– 6.02	– 6.40
A/S : G.N.C. (1 : 2)	104.2	123.0	113.6	– 7.46	– 0.60	– 4.03
A/S : G.N.C. (2 : 1)	86.2	115.0	99.6	– 8.97	– 1.15	– 5.06
Average	**83.6**	**103.5**	**93.5**	**– 6.23**	**– 1.23**	**– 3.83**

S.E. of difference
(*i*) between compost means = 2.45
(*ii*) between treatment means = 4.24

S.E. of difference
(*i*) between compost means = 0.73
(*ii*) between treatment means = 1.26

The long-term experiments at Kopergaon, Deolali and Akluj commenced on identical plans from the year 1941–42. The three-course rotation of sugarcane, *jowar* and groundnut was adopted. The experiment consisted of three series corresponding to each of the three crop phases of the rotation.

The yield of sugarcane became available from 1941–42. The experiment was discontinued after 1949–50. The results pertaining to the sugarcane crop are briefly discussed below:

The experimental treatment consisted of all combinations of two levels of compost with four treatments made up of 336 kg nitrogen per hectare, supplied by groundnut-cake and mixture of groundnut-cake and ammonium sulphate in three fixed ratios. Compost was applied at 20 cart-loads per hectare.

At Kopergaon (Table 7.31), plots treated with combinations of ammonium sulphate and groundnut-cake indicated superiority of groundnut-cake over ammonium sulphate. Significant response to the application of compost was also obtained. The rate of change of yield with years showed a negative trend in the yields of plots under all the treatment. Application of compost did not bring in a significant reduction in the rate of deterioration.

Table 7.31: Response of Sugarcane to Organic Manures and Regression of Yield Over Years (tonnes/ha)

Station: Kopergaon **State: Maharashtra** **Period: 1980 to 2000**

Treatments	Mean Yield			Linear Regression of Yield Over Years		
	No Compost	Compost	Average	No Compost	Compost	Average
Groundnut-cake	106.5	115.9	111.2	– 6.28	– 7.56	– 6.98
A/S : G.N.C. (1 : 1)	100.3	113.2	106.7	– 6.09	– 8.21	– 7.15
A/S : G.N.C. (1 : 2)	99.0	115.3	107.1	– 14.18	– 11.48	– 12.83
A/S : G.N.C. (2 : 1)	91.1	108.4	99.7	– 11.12	– 6.35	– 8.73

S.E. of difference
(*i*) between two compost means = 1.64
(*ii*) between treatment means = 2.32

S.E. of difference
(*i*) between two compost means = 1.59
(*ii*) between two treatment means = 2.25

At Deolali (Table 7.32), groundnut-cake alone showed significantly higher response as compared with plots treated with combinations of oil-cakes and ammonium sulphate. It is interesting to note that in the plots treated with mixture of ammonium sulphate and groundnut-cakes, the yields were higher with higher content of groundnul-cakes. Application of compost also resulted in a significant increase in yields. As in Kopergaon, there was a significant deterioration in the yield of all plots. Addition of compost in all plots except in those treated with groundnut-cake alone resulted in significant reduction in the rate of deterioration. Even in the latter plots, there was a reduction in the rate of deterioration but it was not statistically significant.

As at the two centres, Kopergaon and Deolali, at Akluj (Table 7.33) also, plots treated with groundnut-cake alone or with higher percentage of groundnut-cake in the mixture gave higher response. Application of compost significantly increased the sugarcane yield. Although there was a negative trend in the rate of response under different treatments with years, application of compost had a beneficial effect in reducing the rate of deterioration.

The results of the treatments at the three centres, therefore, are generally quite similar. In all the cases, responses were higher when the percentage of nitrogen in the form of groundnut-cake was higher in the cake-sulphate mixture. Estimation of the difference in the effect of groundnut-cake and ammonium sulphate made from these experiments gave the following results:

Table 7.32: Response of Sugarcane to Organic Manures and Regression on Yield Over Years (tonnes/ha)

Station: Deolali **State: Maharashtra** **Period: 1991–92 to 2001–02**

Treatments	*Mean Yield*			*Linear Regression of Yield Over Years*		
	No Compost	*Compost*	*Average*	*No Compost*	*Compost*	*Average*
Groundnut-cake	89.6	98.5	94.0	– 21.05	– 16.71	– 18.88
A/S : G.N.C. (1 : 1)	76.7	86.0	81.3	– 23.60	– 12.09	– 17.84
A/S : G.N.C. (1 : 2)	83.7	88.4	86.0	– 21.54	– 13.40	– 17.47
A/S : G.N.C. (2 : 1)	72.5	86.6	79.5	– 17.54	– 8.38	– 12.96
Average	**80.6**	**89.9**	**85.2**	**– 20.93**	**– 12.64**	**– 16.79**

S.E. of difference
(*i*) between compost means = 2.94
(*ii*) between treatment means = 4.16

S.E. of difference
(*i*) between compost means = 2.80
(*ii*) between treatment means = 3.94

Table 7.33: Response of Sugarcane to Organic Manures and Regression on Yield Over Years (tonnes/ha)

Station: Akluj **State: Maharashtra** **Period: 1991–92 to 2001–02**

Treatments	*Mean Yield*			*Linear Regression of Yield Over Years*		
	No Compost	*Compost*	*Average*	*No Compost*	*Compost*	*Average*
Groundnut-cake	115.1	122.5	118.8	– 12.40	– 4.96	– 8.68
A/S : G.N.C. (1 : 1)	100.3	109.8	105.0	– 22.25	– 10.39	– 16.32
A/S: G.N.C. (1 : 2)	105.8	119.3	112.5	– 22.33	– 8.57	– 15.45
A/S : G.N.C. (2 : 1)	83.8	104.6	94.2	– 17.54	– 10.34	– 13.94
Average	**101.2**	**114.0**	**107.6**	**– 18.63**	**– 8.56**	**– 13.59**

S.E. of difference
(*i*) between compost means = 2.94
(*ii*) between treatment means = 4.16

S.E. of difference
(*i*) between compost means = 2.80
(*ii*) between treatment means = 3.94

As can be seen, the difference between the true effect of nutrients applied in the form of groundnut-cake and ammonium sulphate is statistically significant at all the three centres.

Table 7.34: Mean Difference Between True Effects of Groundnut-cake and Ammonium Sulphate in Tonnes per Hectare

Centre	*Mean Difference*	*S.E. of Mean Difference*
Kopergaon	15.4	3.32
Deolali	22.7	3.23
Akluj	35.5	5.97

A long-term experiment to determine the relative value of organic and inorganic manures applied to sugarcane crop was conducted for 13 years from 1989 to 2002 at Sugarcane Research Sub-station, Muzaffarnagar, U.P. and the results were published by Ambika Singh. The trial was conducted in two adjacent fields in alternate years with a sugarcane-fallow-sugarcane rotation so that a crop of sugarcane was available every year. So in one field, sugarcane crop was taken for 7 years from 1980 to 2000 and in the other field for 6 years from 1980 to 2000. The experiment consisted of seven treatments, *viz.*, (1) control (no nitrogen); (2) farmyard manure (132 kg N/ha); (3) groundnut cake (132 kg N/ha); (4) ammonium sulphate (132 kg N/ha); (5) farmyard manure + ammonium sulphate (each at 66 kg N/ha); (6) ammonium sulphate + groundnut cake (each at 66 kg N/ha); (7) farmyard manure + groundnut cake + ammonium sulphate (each at 44 kg N/ha).

Mean yields are presented in Table 7.35. There was significant response to the application of nitrogen from all sources. Farmyard manure applied alone gave significantly lower response than the other five nitrogen treatments. The average percentage responses to the dressings of farmyard manure, groundnut cake and ammonium sulphate were 24, 41 and 47 per cent respectively.

Table 7.35: Response of Sugarcane to Organic Manures

Station: Muzaffarnagar		**State: Uttar Pradesh**	**Period: 1990 to 2000**
Sl.No.	*Treatments*	*Mean Yield Tonnes/ha*	*Mean Yield as % of Control*
1.	Control	45.4	100
2.	F.Y.M. (132 kg N/ha)	56.7	124
3.	G. Cake (132 kg N/ha)	64.0	141
4.	A/S (132 kg N/ha)	66.8	147
5.	A/S F.Y.M. (132 kg N/ha)	61.9	136
6.	A/S + G. Cake (132 kg N/ha)	64.6	142
7.	A/S + G. Cake + F.Y.M. (132 kg N/ha)	64.1	141

On the basis of soil analysis, it was concluded by the author that continuous application of ammonium sulphate did not result in a fall in the pH or in the content of organic carbon or of nitrogen in the soil, to an extent that would make the soil infertile. On the contrary, the yields on these plots were higher than those for any of the other treatments and these differences have been maintained.

Other Crops

In Tamil Nadu, long-term experiments were started in the year 1909 at Agricultural College and Research Institute, Coimbatore in red sandy loam soil with organic manure (FYM) and inorganic fertilizers. From the commencement of the experiment, various crops like Great millet (*cholam*), Finger millet (*ragi*), Common millet (*panivaragu*), Italian millet (*tanai*), cotton, etc. had been raised under irrigated conditions till September, 2001 after which the crops were raised under rainfed conditions. From the results (Table 7.36) reported by Krishnamoorthy yield data of 5 crops due to some selected manurial treatments have been taken and converted into metric system.

Application of FYM @ 12.55 tonnes/ha every year gave the highest average yield of Great millet (*cholam*), the increase over control being 1,323.9 kg/ha. This yield was 3 times higher than the control but was at par with NPK and NP treatments. Regarding grain yield of Finger milet (*ragi*) average responses of 1,194.1, 1,128.7 and 932.9 kg/ha were obtained with NPK, NP and FYM treatments respectively, the mean yield in control plot being 559.6 kg/ha. In Italian millet (*tanai*), the highest grain yield was obtained with FYM, the response over control yielding 219.5 kg/ha being 343.7, 275.5 and 184.1 kg/ha with FYM, NP and NPK treatments respectively. The order of response in common millet (*panivaragu*) grains were 541.3, 526.9 and 502.9 kg/ha with NP, FYM and NPK treatments respectively, the control yield being 700.6 kg/ha. In cotton crop the response was the highest with FYM (346/7 kg/ha) followed by NP and NPK giving response of 235.4 and 210.1 kg/ha, the yield of control being 286.8 kg/ha. In all these crops, the effects of FYM, NP and NPK fertilizers, were significant over control. However, nitrogen alone gave non-significant increase over control in all crops.

Table 7.36: Response to Various Crops to Organic and Inorganic Manures at Coimbatore (Tamil Nadu) During 1980 to 2000

Sl.No.	*Crop*	*Treatment*	*Control Yield kg/ha*	*Response kg/ha*	*Percentage of Control*	*S.E.*	*C.D.*
1.	Great millet	Control	601.9	–	–	134.5	372.6
	(*Cholam*)	N		106.9	117.8		
		NP		1038.5	272.6		
		NPK		1041.6	273.1		
		FYM		1323.9	320.3		
2.	Finger millet	Control	559.6			78.6	221.5
	(*Ragi*)	N		91.8	116.4		
		NP		1128.7	301.7		
		NPK		1194.1	313.4		
		FYM		932.9	266.7		
3.	Italian millet	Control	219.5			47.6	135.8
	(*Tanai*)	N		34.7	115.8		
		NP		275.5	225.5		
		NPK		184.1	183.9		
		FYM		343.7	256.6		
4.	Common millet	Control	700.6			92.0	260.1
	(*Panivaragu*)	N		111.9	116.0		
		NP		541.3	177.3		
		NPK		509.2	172.7		
		FYM		526.6	175.2		
5.	Cotton	Control	286.8	–		39.4	111.7
	(*Kapas*)	N		26.7	109.3		
		NP		235.4	182.1		
		NPK		210.1	165.8		
		FYM		346.7	220.9		

N @ 25.2 kg/ha; P @ 67.8 kg/ha; K @ 60.5 kg/ha; FYM @ 12.55 tonnes/ha.

Effect of Organic Manures in Rotation

Pusa (Bihar)

A long-term experiment was started in *kharif*, 1982 at the Indian Agricultural Research Institute (Botanical Sub-station), Pusa, Bihar, with a view to determining the effect on soil fertility of organic and inorganic manures, under sub-tropical climatic conditions. The treatments were applied to the following four-year eight-course rotations:

Table 7.37

Year	*Name of the Crop*	
	Kharif	*Rabi*
First year	Maize	Oats
Second year	Maize	Peas
Third year	Maize	Wheat
Fourth year	Maize	Gram

The experimental treatments consisted of all the combinations of two levels of ammonium sulphate, 0 and 44.8 kg N per hectare, two levels of super-phosphate, 0 and 89.6 kg P_2O_5 per hectare and two levels of potassium sulphate, 0 and 56.0 kg K_2O per hectare plus two additional treatments, *viz.*, rape-cake at 44.8 kg N per hectare and farmyard manure at the rate of 9 tonnes per hectare (to supply 44.8 kg N per hectare).

The inorganic fertilizers alone and in various combinations were applied, half before *kharif* sowing and the other half before *rabi* sowing. The total amount of 9 tonnes of farmyard manure per hectare was applied in the last week of April or first week of May, whereas rape-cake was applied, half before maize sowing and the other half at the last interculture in maize.

The examination of the data showed that farmyard manure treatment and nitrogen in contribution with phosphate and potash gave higher response than the remaining treatments. A similar finding was observed with oats and wheat also. On peas and gram, farmyard manure gave the highest response. In general, under all the-treatments, there was a deterioration in the rate of response with years. On the whole, plots treated with farmyard manure and rape-cake gave high response for all the crops whereas NP and NPK gave high responses on maize, oats and wheat only.

The results of a few long-term experiments with organic manure and fertilizers discussed above show that although there are some variations in the results obtained, by and large, compost has shown a moderate response and its continued application has been beneficial in reducing deterioration in soil fertility and crop response to fertilizer application.

Shinde conducted field experiments on manurial requirements of a fixed crop rotation of rice followed by gram from 1955–56 to 1966–67 at Bagwai in Madhya Pradesh in order to study the effect of continuous use of FYM and fertilizers on the yield and fertility status of the medium black soil. Application of FYM at 5.6 tonnes/ha significantly increased the grain yield of rice crop (Table 7.38) and also resulted in significant increase of organic matter and available phosphorus in the surface layer.

Table 7.38: Grain Yield of Rice Crop (q/ha) Due to FYM (5.6 tonnes/ha) and Phosphorus Levels

Station: Bagwai			**State: Madhya Pradesh**				**Period: 1990 to 2000**
Levels of P_2O_5 kg/ha	*FYM × Phosphate*		*Mean*	*Levels of N kg/ha*	*FYM × Nitrogen*		*Mean*
	F_0	*F_1*			*F_0*	*F_1*	
0	8.11	15.77	11.94	0	9.50	13.98	11.74
33.6	15.63	16.88	15.75	33.6	13.49	17.25	15.37
67.2	15.75	19.29	17.52	67.2	15.50	20.71	18.10
Mean	12.83	17.31		Mean	12.83	17.31	

CD at 5%; FYM = 0.64 q/ha; N and P = 0.79 q/ha; FP and FN = 1.12 q/ha

In an experiment with wheat-pearl millet rotation conducted at Hissar during 1971 to 1973 in soil having pH 8.2 (1 : 2). Poonia found that application of FYM at 50 tonnes/ha once in a year before wheat crop gave significant response in wheat and pearl millet crops. Although direct response was low in wheat crop in both years yet response due to residual effect of FYM was comparatively higher on pearl millet being 9.0 and 4.4 q/ha of grains in 1972 and 1973 respectively (Table 7.39).

Table 7.39: Yield of Wheat and Pearl Millet Crops Due to FYM at 50 tonnes/ha (q/ha)

Station: Hissar				**State: Haryana**				**Period: 1997 to 2000**
FYM	*Wheat*				*Pearl Millet*			
	1971–72		*1972–73*		*1972*		*1973*	
	Grain	*Straw*	*Grain*	*Straw*	*Grain*	*Straw*	*Grain*	*Straw*
F_0	47.8	79.0	34.9	72.7	33.0	105.3	30.1	100.0
F_1	49.6	78.0	36.5	83.7	42.0	126.8	34.5	114.1
Response	1.8	–	1.6	11.0	9.0	21.5	4.4	14.1
C.D. at 5%	1.0	NS	1.5	5.0	2.5	5.0	0.6	4.6

Manurial Requirements of a Fixed Crop Rotation

To study the direct, residual and cumulative effect of phosphorus, potassium and farmyard manure on a fixed, single year, two crop rotations with high-yielding varieties of rice, wheat, *jowar*, *bajra* and maize, experiments were conducted all over India in Coordinated Agronomic Research Project. Treatments included all combinations of three levels of phosphorus (0, 30, 60 kg P_2O_5/ha) two levels of potassium (0, 30 kg K_2O/ha) and two levels of farmyard manure (0, 15 tonnes/ha) in three phases over a basal dressing of 100 kg N/ha to each crop.

The results obtained with farmyard manure have been briefly given and discussed rotation wise and State wise.

Rice-wheat Rotation

Uttar Pradesh

The data of trials conducted at four stations of Uttar Pradesh from 1980 to 2000 have been given in Tables 7.40 and 7.41.

Table 7.40: Direct, Residual and Cumulative Responses (kg/ha) to FYM at 15 tonnes/ha Rotation : Rice–Wheat

Sl.No.	State and Station	Year(s)	Crop	Phase I Application in every Season			Phase II Application in Kharif only			Phase III Application in Rabi only		
				Control Yield	Res-ponse	C.D. 5%	Control Yield	Res-	C.D. 5%	Control Yield	Res-ponse	C.D. 5%
Uttar Pradesh												
1.	Varanasi	1970–71 to	Rice	4494	153	190	4696	52	211	4530	63	215
		1972–73	Wheat	4181	352	130	3791	331	142	4026	303	158
		Total Over 2	Crops	8681	505	232	8487	383	260	8556	366	233
		1975–76	Rice	5502	806	315	4768	681	311	5487	902	244
		1975–76	Wheat	3991	245	220	3766	158	258	4204	241	251
		1976–77	Rice	4669	392	270	4829	480	224	5004	294	249
		1976–77	Wheat	2706	419	230	2444	590	186	2740	704	205
2.	Bichpuri	1969–70 to	Rice	2701	265	119	2581	236	155	2696	152	119
		1972–73	Wheat	3816	633	89	3824	406	83	3900	460	87
		Total over 2	Crops	657	898	147	6405	642	154	6596	612	172

Table 7.41: Direct, Residual and Cumulative Responses (kg/ha) to FYM at 15 tonnes/ha Rotation : Rice–Wheat

Sl.No.	State and Station	Year(s)	Crop	Phase I Application in every Season			Phase II Application in Kharif only			Phase III Application in Rabi only		
				Control Yield	Res-ponse	C.D. 5%	Control Yield	Res-	C.D. 5%	Control Yield	Res-ponse	C.D. 5%
Uttar Pradesh												
1.	Masodha	1972–73 to	Rice	4532	445	97	4456	359	109	4268	352	99
		1974–75	Wheat	2811	918	82	2290	470	104	2708	897	77
		Total over 2	Crops	7343	1363	124	6746	829	161	6976	1249	125
		1975–76	Rice	2697	216	171	2635	179	226	2779	185	245
		1975–76	Wheat	2608	820	165	2175	381	268	2234	594	285
		1976–77	Rice	2288	593	179	2700,	660	140	2833	810	154
		1976–77	Wheat	3329	358	133	2990	260	266	3494	417	142
2.	Pura Farm	1975–76	Rice	3701	240	128	3477	261	149	3728	522	193
		1975–76	Wheat	4654	232	281	4297	192	155	4621	452	122
		1976–77	Rice	3592	317	123	3475	195	121	3877	363	138
		1976–77	Wheat	4821	202	166	4473	220	181	4811	289	140

Varanasi

Trials conducted from 1970–71 to 1972–73 showed that application of FYM at 15 tonnes/ha was significantly effective in *rabi* crop of wheat only. When averaged over 3 years, response of 5 q/ha of total grain yield was obtained when FYM was applied in every season. In 1975, direct, residual and cumulative response of 8.1, 6.8 and 9.0 q/ha were observed in rice crop while response was meagre in wheat in 1975–76. In 1976–77 response to FYM was moderate in rice crop ranging from 2.9 to 4.9 q/ha in different series while in wheat crop responses were fairly high being 4.2, 5.9 and 7.0 q/ha in direct, residual and cumulative series respectively.

Bichpuri

Trials conducted from 1969–70 to 1972–73 brought out that application of FYM was significantly effective both in rice and wheat crops. When averaged over 4 years, the total grain yield of the two crops increased by 9 q/ha when FYM was applied in every season. This was reduced by one-third when the application of FYM was limited to only one season, *kharif* or *rabi* (Table 7.40).

Masodha

Data of 1972–73 to 1974–75 (Table 7.41) showed that total response of annual grain yield to FYM when applied during *rabi* only was of the order or 12.5 q/ha. The additional response was 1.1 q/ha when FYM was applied in every season. In 1975–76, direct response of 8.2 q/ha was found for wheat crop. In *kharif* rice of 1976, response to FYM was 6.6 and 8.1 q/ha in the residual and cumulative series while in wheat 1976–77, response to FYM ranged from 2.6 to 4.2 q/ha in different series.

Pura Farm

In rice crop of 1975 and wheat crop of 1976–77, cumulative response of 5.2 and 4.5 q/ha respectively were obtained when FYM was applied in *rabi* only. The trend of results was similar in 1976–77 having cumulative response of 3.6 q/ha in rice and 2.9 q/ha in wheat when FYM was applied in *rabi* only (Table 7.41).

Madhya Pradesh

Trials were conducted from 1969–70 to 1976–77 at Kathulia Farm, Raipur and Jabalpur and the data has been presented in Table 7.42.

Kathulia Farm

FYM was found effective in stepping the yield of rice as well as wheat. The total annual grain yield increased by 8.2, 4.7 and 4.4 q/ha respectively (average of 4 years–1969–70 to 1972–73), by the application of 15 tonnes of FYM in every season, in *kharif* only and in *rabi* only.

Raipur

Data of 1971–72 to 1973–74 showed that application of 15 tonnes/ha of FYM during kharif only was adequate and this increased the annual grain yield by 9.2 q/ha. In the year 1976–77 the response was of the order of 10.7 q/ha in rice and 7.9 q/ha in wheat when FYM was applied in rabi only (Table 7.42).

Jabalpur

The data of 1972–73 to 1974–75 brought out that total response of the annual grain yield to FYM when applied in *kharif* only was 7.1 q/ha. An additional, response of 2.6 q/ha was obtained when F.Y.M was applied in every season (Table 7.42).

Table 7.42: Direct, Residual and Cumulative Responses (kg/ha) to FYM at 15 tonnes/ha Rotation : Rice–Wheat

Sl.No.	State and Station	Year(s)	Crop	Phase I Application in every Season			Phase II Application in Kharif only			Phase III Application in Rabi only		
				Control Yield	Res-ponse	C.D. 5%	Control Yield	Res-	C.D. 5%	Control Yield	Res-ponse	C.D. 5%
Madhya Pradesh												
1.	Kathulia	1969–70 to	Rice	6190	247	154	6173	309	168	5798	185	135
	Farm	1972–73	Wheat	2819	569	123	2591	157	117	2671	251	145
		Total over 2	Crops	9009	816	200	8764	466	217	8469	436	208
2.	Raipur	1971–72 to	Rice	2351	528	212	2340	618	239	1769	396	207
		1973–74	Wheat	1476	452	161	951	402	159	1505	314	202
		Total over 2	Crops	3827	980	266	3291	920	309	3274	710	273
		1976–77	Rice	1824	747	340	1618	664	222	1313	1075	450
		1976–77	Wheat	2099	632	353	1501	645	281	2026	786	737
3.	Jabalpur	1972–73 to	Rice	3016	946	107	2984	827	113	2795	687	112
		1974–75	Wheat	1942	1021	86	1719	882	72	1945	768	92
		Total over 2	Crops	4958	1967	139	4703	1709	150	4740	1455	163

West Bengal

The data of trials conducted at Kharagpur from 1972–73 to 1974–75 has been given in Table 7.43. The total responses of the annual grain yield to FYM was 7.2 q/ha when FYM was applied in *kharif* only and 8.2 q/ha when FYM was applied in every season.

Table 7.43: Direct, Residual and Cumulative Responses (kg/ha) to FYM at 15 tonnes/ha Rotation : Rice–Wheat

Sl.No.	State and Station	Year(s)	Crop	Phase I Application in every Season			Phase II Application in Kharif only			Phase III Application in Rabi only		
				Control Yield	Res-ponse	C.D. 5%	Control Yield	Res-	C.D. 5%	Control Yield	Res-ponse	C.D. 5%
West Bengal												
1.	Kharagpur	1972–73 to	Rice	3749	640	61	3711	520	71	3069	274	56
		1974–75	Wheat	2261	183	91	1752	198	60	2095	234	72
		Total over 2	Crops	6010	823	117	5463	718	98	5164	508	98
Jammu and Kashmir												
1.	Bagtisheroo	1976–77	Rice	4523	542	348	4275	171	296	4540	456	255
		1976–77	Wheat	3363	560	410	2910	515	419	3356	660	199

Jammu and Kashmir

At Bagatishero in 1976–77, the response was 4.6 q/ha in rice and 6.6 9/ha in wheat when FYM was applied in *rabi* only (Table 7.43).

Rice-rice Rotation

Andhra Pradesh

Trials conducted at Tirupati during 1972–73 to 1974–75 showed that application of FYM in every season was beneficial giving annual response of 4.8 q/ha (Table 7.44).

Table 7.44: Direct, Residual and Cumulative Responses (kg/ha) to FYM at 15 tonnes/ha Rotation : Rice–Wheat

Sl.No.	State and Station	Year(s)	Crop	Phase I Application in every Season			Phase II Application in Kharif only			Phase III Application in Rabi only		
				Control Yield	Response	C.D. 5%	Control Yield	Res-	C.D. 5%	Control Yield	Response	C.D. 5%
Andhra Pradesh												
1.	Tirupati	1972–73 to	Rice	5013	268	102	4998	151	99	5020	87	77
		1974–75	Rice	5253	216	76	5208	127	83	5329	183	101
		Total over 2	Crops	10266	484	124	10206	278	118	10349	270	140
Karnataka												
1.	Mangalore	1972–73 to	Rice	3993	1802	171	3960	1740	143	3949	518	156
		1974–75	Rice	3655	453	172	3569	104	140	3616	431	177
		Total over 2	Crops	7648	2255	265	7529	1844	233	7565	949	261
		1975–76	Rice	3529	688	285	3357	280	224	3526	1044	320
		1975–76	Rice	3381	1147	233	2122	62	92	2068	320	126
		1976–77	Rice	5194	1608	364	5187	618	259	5249	1204	200
		1976–77	Rice	5735	1448	194	5688	720	220	5886	1521	194

Karnataka

At Mangalore during 1972–73 to 1974–75 response to FYM was spectacular (Table 7.44). FYM was applied in *kharif* only increased the annual grain yields by 18.4 q/ha. When applied in every season, the annual grain yield was further increase by about 4.1 q/ha. During 1975–76 response to *kharif* rice was 10.4 q/ha when FYM was applied in *rabi* only and response to *rabi* rice was 11.5 q/ha when FYM was applied in every season, in *kharif* only and in *rabi* only respectively.

Kerala

Experiments were conducted at Karamana. Application of FYM in every season gave annual response of 8–8.5 q/ha during 1972–73 to 1974–75. In 1975–76 response was 6.7 q/ha in *kharif* rice and 5.1 q/ha in *rabi* rice when FYM was applied in *rabi* only (Table 7.45).

Orissa

At Bhubaneswar during *kharif* rice of 1976, response to FYM was significant in direct (5.4 q/ha), residual (2.8 q/ha) and cumulative series (4.2 q/ha) while in *rabi* rice response was 3.9, 4.8 and 5.7 q/ha in direct, residual and cumulative series respectively (Table 7.45).

Table 7.45: Direct, Residual and Cumulative Responses (kg/ha) to FYM at 15 tonnes/ha Rotation : Rice–Wheat

Sl.No.	State and Station	Year(s)	Crop	Phase I Application in every Season			Phase II Application in Kharif only			Phase III Application in Rabi only		
				Control Yield	Response	C.D. 5%	Control Yield	Res-	C.D. 5%	Control Yield	Response	C.D. 5%
Kerala												
1.	Karamana	1972–73 to	Rice	5490	473	242	5616	329	256	5621	10	231
		1974–75	Rice	2870	410	182	2960	93	149	2997	117	182
		Total over 2	Crops	8360	883	358	8576	422	365	8616	187	338
		1975–76	Rice	5339	484	263	5840	425	361	5370	672	280
		1975–76	Rice	2704	210	270	2748	227	275	2478	507	302
Orissa												
1.	Bhuban-	1976–77	Rice	3216	537	203	3248	277	204	3330	417	342
	eshwar	1976–77	Rice	3282	389	239	3052	481	254	3208	572	298

Tamil Nadu

At Karairuppee, when averaged over 3 years (1971–72 to 1973–74) the annual response to FYM was 4.0, 2.7 and 2.8 q/ha when applied in every season, in *kharif* only and in *rabi* only respectively (Table 7.46). In 1975–76, response was better when FYM was applied in *rabi* only being 2.9 q/ha in *kharif* rice and 2.5 q/ha in *rabi* rice. Response to FYM was poor in the coastal alluvium soil of Thanjavur.

Table 7.46: Direct, Residual and Cumulative Responses (kg/ha) to FYM at 15 tonnes/ha Rotation : Rice–Rice

Sl.No.	State and Station	Year(s)	Crop	Phase I Application in every Season			Phase II Application in Kharif only			Phase III Application in Rabi only		
				Control Yield	Response	C.D. 5%	Control Yield	Res-	C.D. 5%	Control Yield	Response	C.D. 5%
Tamil Nadu												
1.	Karaiyi-	1971–72 to	Rice	3246	106	69	3320	115	79	3189	8	64
	ruppu	1973–74	Rice	4143	292	118	4138	159	100	4358	277	116
		Total over 2	Crops	7429	398	139	7458	274	117	7540	285	143
		1975–76	Rice	5457	257	98	5344	198	112	5520	295	95
		1975–76	Rice	3782	243	60	3716	197	64	3883	248	70
2.	Thanjavur	1972–73 to	Rice	4667	119	78	4760	70	89	4453	70	79
		1974–75	Rice	3563	94	71	3435	41	57	3689	57	37
		Total over 2	Crops	8330	213	105	8195	111	121	8142	127	88

Maize-wheat Rotation

Punjab

At Ludhiana, FYM immensely benefited maize and wheat crops in all phases and years (Table 7.47). Taking both crops together, application of FYM in *kharif* season only could be considered as adequate giving a response of 25.5 q/ha (average of 4 years 1969–70 to 1972–73).

Table 7.47: Direct, Residual and Cumulative Responses (kg/ha) to FYM at 15 tonnes/ha Rotation : Maize–Wheat

Sl.No.	State and Station	Year(s)	Crop	Phase I Application in every Season			Phase II Application in Kharif only			Phase III Application in Rabi only		
				Control Yield	Res-ponse	C.D. 5%	Control Yield	Res-	C.D. 5%	Control Yield	Res-ponse	C.D. 5%
Punjab												
1.	Ludhiana	1969–70 to	Maize	3016	1545	154	3085	1358	145	3001	740	149
		1972–73	Wheat	3299	1117	93	2610	1195	113	2883	1195	131
		Total over 2	Crops	6315	2662	220	5695	2553	208	5884	1935	239
		1975–76	Maize	1394	1786	430	1726	1243	231	1344	2414	1495
		1975–76	Wheat	1599	2604	614	1543	2552	633	1665	2571	597
		1976–77	Maize	1896	784	252	2183	1093	339	2337	955	325
		1976–77	Wheat	2370	2140	781	2295	1949	498	3628	2188	671
Jammu and Kashmir												
1.	Talab Tilloo	1971–72 to	Maize	2954	275	293	3060	38	186	2910	188	248
		1973–74	Wheat	3148	621	193	3137	437	235	3101	508	202
		Total over 2	Crops	6102	896	386	6196	475	320	6011	696	362
		1975–76	Maize	3418	751	301	2917	461	292	3156	410	313
		1975–76	Wheat	2451	208	169	2267	774	231	2686	300	354
		1976–77	Maize	1862	1165	232	1896	889	231	1940	1249	346
		1976–77	Wheat	1032	361	145	764	731	202	907	787	248

In *kharif* maize, response to FYM was 17.9, 12.4 and 24.1 q/ha in 1975 and 7.8, 10.9 and 9.5 q/ha in 1976 in direct, residual and cumulative series respectively. In wheat, response was 26.0, 25.5 and 25.7 q/ha in 1975–76 and 21.4, 19.5 and 21.9 q/ha in 1976–77 in direct, residual and cumulative series respectively.

Jammu and Kashmir

At Talab Tilloo as per average data for 1971–72 to 1973–74 (Table 7.47) response to FYM was significant only in the *rabi* crop or wheat, the direct, residual and cumulative effects being 6.2, 4.and 5.1 q/ha respectively. However, in maize crop of 1975 and 1976, responses were significant being 7.5, and 4.6 and 4.1 q/ha in 1975 and 11.6, 8.9 and 12.5 q/ha in 1976 in direct, residual and cumulative series respectively. In wheat direct, residual and cumulative responses were 2.1, 7.7 and 3.0 q/ha in 1975–76 and 3.6, 7.3 and 7.9 q/ha in 1976–77 respectively.

Jowar-wheat Rotation

Madhya Pradesh

At Indore, application of FYM during *kharif* only was found to be adequate, the increase in the annual grain yield being of the order of 7.2 q/ha (Table 7.48).

Table 7.48: Direct, Residual and Cumulative Responses (kg/ha) to FYM at 15 tonnes/ha Rotation : *Jowar*–Wheat

Sl.No.	State and Station	Year(s)	Crop	Phase I Application in every Season			Phase II Application in Kharif only			Phase III Application in Rabi only		
				Control Yield	Res-ponse	C.D. 5%	Control Yield	Res-	C.D. 5%	Control Yield	Res-ponse	C.D. 5%
Madhya Pradssh												
1.	Indore	1969–70 to	*Jowar*	2436	305	135	2403	186	144	2083	267	166
		1973–74	Wheat	4314	369	160	3986	430	186	4242	290	136
		Total over 2	Crops	6750	674	155	6389	716	126	6325	557	153
Karnataka												
1.	Siruguppa	1972–73	*Jowar*	2151	1957	261	1685	1821	258	1829	1560	256
		1974–75	Wheat	2336	601	154	1783	779	115	2280	570	129
		Total over 2	Crops	4487	2558	321	3468	2600	332	4119	2130	335

Karnataka

At Siruguppa, the annual response of FYM applied in *kharif* only was as high as 26.0 q/ha. Further application of FYM in *rabi* did not increase the annual yield (Table 7.48).

Bajra-wheat Rotation

Haryana

At Hissar, taking both crops together, response to FYM was nearly the same ranging between 5.4, 5.9 and 6.0 q/ha whether applied in *rabi* or in *kharif* or in every season respectively (Table 7.49).

Rajasthan

At Jodhpur, FYM applied during *kharif* season only increased the annual grain production by 6.8 q/ha (Table 7.49).

Gujarat

At Jagudon, there was no response of FYM to *kharif* crop of *bajra*. However, in *rabi* crop of wheat, direct response to FYM was 3.9 q/ha (Table 7.49).

Rotation-Jowar in Kharif-*bajra* in Rabi

Tamil Nadu

At Bhavanisagar, during 1971–72 to 1973–74 effect of FYM was phenomenal in all phases (Table 7.50). Its application in every season increased the annual grain yield by 19.8 q/ha and in one season only either in *khaif* or *rabi* by 16.0 and 17.6 q/ha respectively. In the year 1975–76, direct, residual and cumulative responses were 8.8, 3.6 and 4.9 q/ha in *jowar* and 20.3, 10.5 and 18.9 q/ha in *bajra* respectively. In 1976–77 application of FYM in *rabi* only was round more beneficial, the response being 13.7 q/ha in *jowar* and 7.3 q/ha in *bajra* (Table 7.50).

Table 7.49: Direct, Residual and Cumulative Responses (kg/ha) to FYM at 15 tonnes/ha Rotation : *Bajra*–Wheat

Sl.No.	*State and Station*	*Year(s)*	*Crop*	*Phase I Application in every Season*			*Phase II Application in Kharif only*			*Phase III Application in Rabi only*		
				Control Yield	*Res-ponse*	*C.D. 5%*	*Control Yield*	*Res-ponse*	*C.D. 5%*	*Control Yield*	*Res-ponse*	*C.D. 5%*
Haryana												
1.	Hissar	1970–71 to	*Bajra*	2856	225	81	2792	288	85	2816	197	81
		1972–73	Wheat	4188	372	154	4093	298	158	4070	340	166
		Total over 2	Crops	7044	597	193	6885	586	176	6886	537	185
Rajasthan												
1.	Jodhpur	1971–72 to	*Bajra*	1719	319	194	1581	265	191	1743	125	184
		1973–74	Wheat	2442	204	189	2222	415	178	2291	285	141
		Total over 2	Crops	4161	523	290	2803	680	281	4034	410	250
Gujarat												
1.	Jagudan	1971–72 to	*Bajra*	2052	112	210	2366	114	143	1903	43	202
		1973–74	Wheat	3199	387	322	3408	247	249	3027	240	339
		Total over 2	Crops	6252	499	428	5774	361	291	4930	283	460

Table 7.50: Direct, Residual, and Cumulative Responses (kg/ha) to FYM at 15 tonnes/ha Rotation : *Jowar* in *Kharif*–*Bajra* in *Rabi*

Sl.No.	*State and Station*	*Year(s)*	*Crop*	*Phase I Application in every Season*			*Phase II Application in Kharif only*			*Phase III Application in Rabi only*		
				Control Yield	*Res-ponse*	*C.D. 5%*	*Control Yield*	*Res-*	*C.D. 5%*	*Control Yield*	*Res-ponse*	*C.D. 5%*
Tamil Nadu												
1.	Bhavani-	1973–74	*Jowar*	3105	496	365	3114	1069	855	2587	1099	459
	sagar	1973–74	*Bajra*	1904	785	254	1524	526	253	1760	657	187
		Total over 2	Crops	5009	1981	569	4665	1595	546	4347	1756	580
		1975–76	*Jowar*	1549	796	304	927	358	295	1734	488	301
		1975–76	*Bajra*	459	2029	372	957	1055	305	434	1890	369
		1976–77	*Jowar*	2390	512	448	1258	297	370	1564	1366	655
		1976–77	*Bajra*	1354	329	291	1262	497	319	1375	733	318

Response of Crops to Bone-meal

A large number of experiments were conducted at agricultural research stations and in cultivators' fields with bone-meal on rice and wheat. A few experiments were also conducted on other crops. In the majority of these experiments, bone-meal was compared against phosphorus applied in the form of super-phosphate. In the agricultural research stations, by and large, there was no response to phosphate applied either in the form of bone-meal or super-phosphate. Therefore, the results or these trials are not included. In the cultivators' fields, however, generally there was a good response to phosphate applied in the form of bone-meal. The results of or these trials conducted on rice are given in Table 7.51 and on irrigated wheat in Table 7.52. The fertilizer was tried at the levels of 22.4 and 44.5 kg P_2O_5 per hectare in these trials. Bone-meal gave moderate to good response on rice crop. On an average, the response was only a little less than that to the application of super-phosphate. However, on the acidic soils of Kerala and Tripura, bone-meal gave even a better response than super-phosphate.

Table 7.51: Comparison of Responses of Rice to Bone-meal and Super-phosphate Experiments on Cultivators' Fields–T.C.M. Trials

State	*Centre*	*No. of Experiments*	*Bone-meal Dose in kg/ha (P_2O_5)*	*Response in kg/ha*	*Triple Dose in kg/ha (P_2O_5)*	*Super-phosphate Response in kg/ha*	*S.E. of Response in kg/ha*
Tripura	Agartala	17	22.4	184	22.4	215	123
			44.8	369	44.8	123	
Assam	Darrang	10	22.4	166	22.4	197	154
			44.8	105	44.8	381	
Karnataka	Mangalore	34	22.4	155	22.4	141	55
			44.8	172	44.8	203	
Andhra Pradesh	Samalakota	15	22.4	240	22.4	289	49
			44.8	264	44.8	320	
Orissa	Kalahandi	83	22.4	117	22.4	178	31
			44.8	98	44.8	326	
Kerala	Chalakudy	43	22.4	191	22.4	105	49
			44.8	338	44.8	184	
Madhya Pradeh	Raipur	48	22.4	86	22.4	105	43
			44.8	105	44.8	178	
Andhra Pradesh	Bodhan	61	22.4	123	22.4	86	49
			44.8	234	44.8	215	
Average		311	22.4	148	22.4	166	31
			44.8	209	44.8	240	

The response of wheat to application of bone-meal was somewhat lower than in rice. However, the results were available only from Rajasthan.

It appears from the results given above that the bone-meal can be very effective in the acidic areas.

Table 7.52: Comparison of Responses of Irrigated Wheat to Bone-meal and Super-phosphate Experiments on Cultivators' Fields–T.C.M. Trials

State	Centre	No. of Experiments	Bone-meal		Triple Super-phosphate		S.E. of Response in (kg/ha)
			Dose in kg/ha (P_2O_5)	Response in kg/ha	Dose in kg/ha (P_2O_5)	Response in kg/ha	
Rajasthan	Raisingnagar	9	22.4	65	22.4	203	191
			44.8	83	44.8	175	
	Sumerpur	18	22.4	92	22.4	111	200
			44.8	138	44.8	129	
	Average	27	22.4	83	22.4	137	86
			44.8	111	44.8	157	

Chapter 8

Response of Crops to Organic Materials in Salt Affected Soils

The applications of various organic manures are very useful practices for reclamation of salt-affected soils. Besides source of plant nutrients, organic manures produce favourable effect on soil physical properties. It counteracts the unfavourable effect of exchangeable sodium. The decomposition of cattle manure and plant residues liberate carbon dioxide and organic acids which help to dissolve any insoluble calcium salts in soil solution and neutralize the alkali present. Decomposition of organic matter improves soil permeability and bacteria present in it increase water stable aggregate.

Organic materials which have been tried for reclamation of salt-affected soils are farmyard manure, compost, green manures, molasses, press-mud, paddy straw, crop residues and different weeds, particularly *Argeinone mexicana.* However, in trials where green manures have been tried with other organic materials, their data has also been incorporated for the sake of comparison. Organic materials have been tried singly or in combination with other organic and inorganic amendments.

The data of past experiments have been converted into metric system for the sake of uniformity.

Response of Crops in Salt-affected Soils of Punjab and Haryana

In Punjab at Nissang experimental farm, application of 22.2 tonnes/ha of FYM to highly sodic soil doubled the yield of rice as would be evident from Table 8.1.

Table 8.1: Response of Rice to Organic Materials in Sodic Soil at Nissang, Punjab

Sl.No.	*Treatments*	*Control Yield (kg/ha)*	*Response (kg/ha)*
1.	Control (leaching)	1310	–
2.	Press mud @ 222 q/ha	–	520
3.	FYM @ 222 q/ha	–	1370
4.	*Dhaincha* @ 222 q/ha	–	1600

Kanwar carried out trials in alkali soils at Kamma (Punjab) from 1998to 2000 and found high response to FYM applied annually at the rate of 38.2 tonnes/ha (Table 8.2).

Table 8.2: Response of Crops to FYM @ 38.2 tonnes/ha and Effect on Soil pH at Kamma, Punjab

Crop	Year(s)	Control Yield (kg/ha)	Response (kg/ha)	pH (1 : 2) 0–15 cm	
				Control	FYM
Rice	1996	177	143	9.7	9.4
Rice	1997	795	1180		
Rice	1998	961	1313		
Rice	1999	2467	441		
Senji (grain)	1999	283	360		
Senji (fodder)	1999	12,357	14,385		
Wheat (grain)	2000	427	272		
Wheat (straw)	2000	1132	1287	9.4	8.4

Singh reported that on an alkali soil in Punjab, FYM at 10.2 tonnes/ha plus *dhaincha* green manure at 13.9 tonnes/ha proved superior to green manure with 27.7 tonnes/ha (Table 8.3).

Table 8.3: Response of Rice Crop to FYM and Green Manure in Alkali Land in Punjab

Sl.No.	Treatments	Control Yield (kg/ha)	Response (kg/ha)
1.	Leaching (7.5 cm per week)	251	–
2.	Green manure (27.7 tonnes/ha) plus leaching	–	530
3.	FYM 10.2 tonnes/ha		
	+ Green manure 13.9 tonnes/ha	–	999
	+ leaching		

Dargan initiated some experiments on the use of FYM and gypsum on a sodic soil and found that in the first crop of berseem, sugarcane and rice grown in separate fields having pH 10.5 and above, there was high order interaction of combined application of FYM and gypsum (Table 8.4).

In the experiment started with berseem as the first crop, rice crop variety 'IRS-68' was taken in *kharif* 1991 and berseem in *rabi* 1991–92 without adding any more FYM or gypsum. Results reported by Dargan showed that the effect of F.Y.M. added to the first crop of berseem was significant on two succeeding crops of rice and berseem (Table 8.5). In the same fixed layout, there was no further response of FYM in rice of 1992 and berseem of 1992–93. In *kharif* 1993, FYM at 0, 25 and 50 tonnes/ha was added again as per original treatments and maize variety 'Vijay' was taken. Dargan *et al.* (1976) reported that the direct effect of FYM @ 25 tonnes/ha was highly significant on maize grain yield (Table 8.6).

In field experiments at C.S.S.R.I. Karnal initiated in 1970 on sodic soil with pH above 10.0, Yadav found that the growth of *Eucalyptus hybrid* after treating the soil with gypsum @ 50 per cent G.R. and FYM in planting pit was almost as good as in the original alkali soil replaced with good soil. Yadav made similar observations on height increment of different species during the period of September 1974 to September 1975 in the same experiment (Table 8.7).

Table 8.4: Yield of Crops in q/ha Due to FYM and Gypsum at CSSRI, Karnal (Haryana)

Sl.No.	*Treatments*	*Berseem Fodder*	*Sugarcane Stripped Cane*	*Paddy Grain*
1.	Control	1.5	–	24.5
2.	FYM at 25 tonnes/ha	8.3	4.9	35.9
3.	FYM at 50 tonnes/ha	17.4	4.7	37.4
4.	Gypsum at 25% G.R.	–	143.2	42.3
5.	FYM @ 25 tonnes + 25% G.R.	–	213.7	52.9
6.	FYM @ 50 tonnes + 25% G.R.	–	357.8	54.6
7.	Gypsum at 50% G.R.	94.9	286.5	57.0
8.	FYM @ 25 tonnes + 50% G.R.	294.8	374.6	57.6
9.	FYM @ 50 tonnes + 50% G.R.	318.9	391.1	68.9
	C.D. at 5%	47.7	92.9	8.9
	Soil pH at start	10.5	10.6	10.6
	Gypsum tonnes/ha @ 25% G.R.	–	9.0	9.0
	Gypsum tonnes/ha @ 50% G.R.	11.0	18.0	18.0

Table 8.5: Residual and Cumulative Effect of FYM on Rice Grain and Berseem Fodder at CSSRI, Karnal (Haryana)

Sl.No.	*Treatments*	*Rice 1971*		*Berseem 1971–72*	
		Control Yield (q/ha)	*Response (q/ha)*	*Control Yield (q/ha)*	*Response (q/ha)*
1.	FYM @ 0 tonnes/ha	70.5	–	878	–
2.	FYM @ 25 tonnes/ha	–	8.0	–	39
3.	FYM @ 50 tonnes/ha		13.0	–	136
	C.D. at 5 per cent		5.7		83.0

Table 8.6: Direct Effect of FYM on Maize Grain at CSSRI, Karnal

Sl.No.	*Treatments*	*Control Yield (kg/ha)*	*Response (kg/ha)*
1.	FYM @ 0 tonnes/ha	23.4	–
2.	FYM @ 25 tonnes/ha	–	5.8
3.	FYM @ 50 tonnes/ha	–	5.1
	C.D. at 5 per cent	3.5	

Yadav in a separate field experiment at CSSRI, Karnal observed that the height growth of *Ecualyptus hybrid* during 1971 to 1974 in combined application of gypsum and FYM treatment was as good as in the normal soil and was much superior to that obtained in treatments of gypsum or FYM applied alone (Table 8.8).

Table 8.7: Height Increment (cm) of Different Forest Species on Saline Sodic Soil at Karnal Under Different Treatments

Soil Treatment	*Acacia arabica*	*Albizia lebbeck*	*Eucalyptus hybrid*	*Prosopis juliflora*	*Terminalia arjuna*
Control (sodic soil)	40	15	33	46	25
F.Y.M.	76	46	62	81	68
Gypsum	88	62	91	92	85
Gypsum + F.Y.M.	102	71	120	102	95
Normal soil	107	87	119	100	98

Table 8.8: Change in Height of *Eucalyptus* Hybrid Under Different Treatment of Sodic Soils

Soil Treatments	*Height in Metres*			
	1971	*1972*	*1973*	*1974*
Control (sodic soil)	0.90	–	–	–
F.Y.M.	0.90	2.20	3.50	4.90
Gypsum	0.90	2.50	4.00	5.10
Gypsum + F.Y.M.	0.90	2.70	4.40	6.20
Normal soil	0.90	3.10	4.40	6.10

Mendiratta reported that in *jowar*-wheat and fallow-wheat rotation, gypsum in combination with either FYM and/or *dhaincha*, appeared to be the best treatment in increasing the yield of wheat grain in Ballop area (Kota) of Chambal Command affected with salinity and alkalinity (Table 8.9).

Table 8.9: Response of Wheat to Organic Manures

Ballop Area (Kota)–Rajasthan–1998–99

Sl.No.	*Treatments*	*Control Yield q/ha*		*Response q/ha*	
		After Jowar	*After Fallow*	*After Jowar*	*After Fallow*
1.	Control	19.85	16.00	–	–
2.	F.Y.M.			1.78	2.91
3.	Gypsum			5.44	4.46
4.	*Dhaincha*			1.72	2.32
5.	F.Y.M. + *Dhaincha*			6.53	5.84
6.	F.Y.M. + Gypsum			6.28	5.62
7.	Gypsum + *Dhaincha*			4.08	3.70
8.	F.Y.M. + Gypsum + *Dhaincha*			6.54	6.98
	C.D. at 5 per cent			1.75	2.20

Chapter 9
Nitrogen Fixation

Introduction

A logarithmic expansion in our understanding of the biochemistry, physiology and genetics of nitrogen fixation has occurred in the last 20 years. Although few agricultural or industrial uses have resulted to date, considerable effort is being expended on model systems as a prerequisite to commercial application. This article will first review nitrogen fixation and then explore how some of this recently acquired knowledge may be applied. The rapid advances that have been witnessed in the field of nitrogen fixation lead us to believe that direct application of this knowledge will be forthcoming.

Biochemistry

Historical Review

Ancient peoples acknowledged the benefits of annually rotating leguminous and non-leguminous crop plants. The first explanation for this effect was made in the 1830s by the French scientist Boussingault, who suggested that leguminous plants could utilize atmospheric nitrogen for growth. Boussingault's work met the harsh criticism of the distinguished German organic chemist, Liebig, and consequently Boussingault's claim fell into disregard with the scientific elite of the period. However, during the next 50 years others sought to verify Boussingault's results.

The three decades following the experiments of Hellriegel found researchers seeking to identify not only the organisms responsible for symbiotic nitrogen fixation but also free-living N_2-fixing microbes. Winogradsky demonstrated fixation by the strictly anaerobic *Clostridium* spp., Beijerinck showed the obligately aerobic *Azotobacter* spp. were capable of nitrogen fixation and Drewes reported fixation by the algae *Nostoc* and *Anabaena.* Fred, Baldwin thoroughly reviewed the work on the leguminous symbioses in their treatise. That monograph describes the morphological, cultural and physiological characteristics of the rhizobia, cross-inoculation groups, the nodule-formation process, inoculation methods, the quantification of fixation rates under various conditions, and so on.

In the late 1920s, investigations on the biochemistry of the nitrogen fixation process began in earnest. Progress was made in several major areas:

1. The effects of different partial pressures of N_2 on the fixation process were investigated.
2. H_2 was reported to be a specific competitive inhibitor of nitrogen fixation.
3. Ureides were found to be the major nitrogen assimilate in certain leguminous plants.
4. The inter-relationships of carbon and nitrogen metabolism during the N_2 fixation process were actively investigated.

It was during this period, before active, cell-free extracts or pure enzyme became available, that the first stable product of nitrogen fixation was identified. Virtanen supported hydroxylamine, whereas Burris considered ammonia to be the first stable product. In the early 1940s, a new technique was introduced which aided in establishing ammonia as the first stable product of nitrogen fixation. Burris developed the use of the stable isotope, $^{15}N_2$, for nitrogen fixation research. Not only did this technique resolve the hydroxylamine *versus* ammonia controversy, but it continues to provide the most definitive and accurate method of quantifying nitrogen fixation. By the early 1950s, it was firmly established from many additional lines of evidence that ammonia was the first stable product of nitrogen fixation.

In 1960, both the group at Wisconsin and at Dupont reported reproducible, active, cell-free extracts from blue-green algae and *Clostridium pasteurianum*, respectively. These reports were quickly followed by others describing active extracts from *Azotobacter vinelandii, Klebsiella pneumoniae, Bacillus polymyxa, Azotobacter chroococcum, Rhodospirillum rubrum, Chromatium* spp. and *Rhizobium* bacteroids. These extracts were used to define the requirements for the nitrogen-fixing reaction and also pointed out why most previous attempts had failed. Nitrogenase, the enzyme catalyzing the reduction of N_2 to NH_4^+ was found to require large amounts of ATP, to be strongly inhibited by ADP, to require a constant supply of low potential electrons and to be irreversibly inactivated by O_2.

In these early studies on the cell-free systems, the ATP was supplied *via* the phosphoroclastic metabolism of pyruvate and was later replaced by an exogenous ATP-regenerating system. The utilization of $Na_2S_2O_4$ as an electron donor, as suggested by Bulen, coupled with the ATP-regenerating system, provided a readily available and reliable *in vitro* method for measuring nitrogenase activity. These surrogate systems permitted investigation of the ATP and electron requirements, the H_2 evolution activity of nitrogenase, the discovery of alternative substrates and the purification of nitrogenase.

In the 1970s, highly purified nitrogenase component proteins suitable for protein chemistry and, enzymological characterization were obtained from *C. pasteurianum, K. pneumoniae, A. vinelandii, A. chroococcum, B. polymyxa* and *R. rubrum*. The investigations on the nature of nitrogenase structure and function advanced rapidly, utilizing such classical tools as enzyme kinetics and protein chemistry, and such state-of-the art physical methodologies as electron paramagnetic resonance (EPR). Mossbauer spectroscopy and extended X-ray absorption fine structure.

Nitrogenase requires the functioning of two distinct proteins for the reduction of atmospheric nitrogen. Neither of these component proteins separately displays any activity characteristic of nitrogenase itself. The smaller of the two proteins, dinitrogenase reductase (also called Fe protein, component 2), is reduced by an appropriate reductant and binds two molecules of ATP. Dinitrogenase reductase then transfers one electron to dinitrogenase (also called MoFe protein, component 1) with the concommitant hydrolysis of both ATP molecules. The electrons are transferred within dinitrogenase to the various iron-sulfur centers and the iron-molybdenum cofactor (FeMoCo). The latter is believed to be the substrate-reduction site.

Hageman recently proposed a nomenclature for nitrogenase based on a functional role of the component proteins (dinitrogenase and dinitrogenase reductase), rather than on a physical property or an arbitrary designation. This nomenclature. though not accepted by all, will be utilized here since it does describe the major functional role for these proteins (as far as our present understanding permits) and it provides the reader unfamiliar with the field with a way of more easily comprehending the biochemistry of the process.

Molecular Properties of Nitrogenase

Dinitrogenase

Dinitrogenase is the larger and more complex component of nitrogenase. Dinitrogenase is an $\alpha_2\beta_2$ tetramer ranging between 210,000 daltons for *C. pasteurianum* and 245,000 daltons for *A. vinelandii*. The molar masses for the two subunits are approximately 50,000 and 60,000 daltons. Metal analyses indicate two molybdenum atoms and 24 to 32 iron atoms per molecule. The acid-labile sulfur content is approximately equivalent to the iron content. The metals appear to be arranged as four Fe_4S_4 clusters, two iron-molybdenum (FeMo) cofactors and possibly an Fe_2S_2 center. Dinitrogenase as normally isolated yields a characteristic electron paramagnetic resonance (EPR) spectrum with g values of approximately 4.3, 3.7 and 2.01 which originate from the FeMo cofactor. Reduction of the paramagnetic form of dinitrogenase by dinitrogenase reductase results in the loss of the EPR signal. The EPR signal returns upon catalytic reoxidation (substrate reduction). Orme-Johnson monitored the change in the EPR signal during oxidative titration of reduced *A. vinelandii* dinitrogenase. Removal of four electrons, one from each of the Fe_4S_4 centers as monitored by Mossbauer spectroscopy, did not alter the EPR signal. Removal of the fifth and sixth electrons (from the FeMo cofactor) resulted in the disappearance of the signal. These two distinct phases are not observed with dinitrogenase from either *C. pasteurianum* or *K. pneumoniae,* implying subtle differences among these nitrogenases in the spatial arrangements of these metal centers.

FeMo Cofactor

Shah and Brill were able to extract the labile FeMo cofactor of dinitrogenase utilizing *N*-methylformamide and have subsequently purified the molecule. The isolated FeMo cofactor

1. Contains 7–8 iron atoms per molybdenum atom.
2. Contains no amino acids.
3. Possesses an EPR signal with g values of 4.6, 3.4 and 2.0.
4. Is able to reconstitute FeMo cofactor-less mutants of *A. vinelandii* and *K. pneumoniae.*
5. Is capable of binding and reducing acetylene in the presence of an appropriate reductant.

X-ray absorption spectroscopy (XAS) and extended X-ray absorption fine structure (EXAFS) have been used to probe the environment around the molybdenum atoms in the FeMo cofactor and in dinitrogenase. Data from both materials indicate that three to four sulfur atoms and two to three iron atoms are at an average distance of 0.235 and 0.266 nm, respectively from the molybdenum atoms. The only discrepancy is the appearance of oxygen or perhaps nitrogen in the FeMo cofactor, but not in the dinitrogenase spectra. This difference may be an artifact of the isolation procedure. The isolation and analysis of the FeMo cofactor has initiated considerable activity in synthesizing inorganic iron-molybdenum model compounds.

Dinitrogenase Reductase

Dinitrogenase reductase is composed of two identical subunits. containing a single Fe_4S_4 cluster, with a molar mass between 55,000 and 67,000 daltons depending upon the source of the protein. Reported molar masses of the subunits range between 27,500 and 34,600 daltons. Dinitrogenase reductase from *C. pasteurianum,* which has been completely sequenced, has 273 ammo acid residues per subunit, yielding a molar mass of 57,674 daltons. The amino acid sequence had no sequence homology with other known iron-sulfur proteins. *C. pasteurianum* dinitrogenase reductase has a higher content of glycine-glycine sequences than any other protein previously examined. The twelve cysteine residues were randomly distributed, unlike the clustered sequences of ferredoxins. Thus, it was not possible to deduce the liganding structure of the Fe_4S_4 cluster between the two subunits. The Fe_4S_4 cluster can be extruded readily with the aid of thiol ligands and characterized.

Dinitrogenase reductase possesses two catalytically active binding sites for MgATP. MgDAP, which strongly inhibits catalysis, binds strongly to only one of the MgATP sites. The binding of MgATP causes a number of changes in dinitrogenase reductase indicative of a conformational change of the protein. MgATP binding:

1. Decreases the midpoint potential from about –250 mV to –400 mV.
2. Increases the accessibility of the Fe_4S_4 center to iron chelators.
3. Increases the sensitivity to O_2.
4. Alters the –SH titer of the protein in the presence of dithionitrobenzoate or iodoacetamide.

Dinitrogenase reductase has an EPR signal with g values of approximately 2.04, 1.94 and 1.88. The binding of MgATP to dinitrogenase reductase induces a transition of the EPR spectrum from a rhombic to an axial-type signal. Orme-Johnson have questioned the significance of the MgATP-induced EPR spectral change for *C. pasteurianum* dinitrogenase reductase because the spectral change can be observed only at pH values above 7.5 and not at values close to the pH optimum for sub-strate reduction (pH 6.5).

Substrates

Nitrogenase is capable reducing a large number of doubly- and triply-bonded molecules. The substrates are reduced in two-electron steps or multiples thereof:

Nitrogen: $N_2 + 6H^+ + 6\,e^- \rightarrow 2\,NH_3$

Acetylene: $C_2H_2 + 2\,H^+ + 2\,e^- \rightarrow C_2H_4$

Protons: $2\,H^+ + 2\,e^- \rightarrow H_2$

Cyanide: $HCN + 6\,H^+ + 6\,e^- \rightarrow CH_4 + NH_3$

and $HCN + 4\,H^+ + 4\,e^- \rightarrow CH_3NH_2$

Nitrous oxide: $N_2O + 2\,H^+ + 2\,e^- \rightarrow N_2 + H_2O$

Azide: $N_3 + 3\,H^+ + 2\,e^- \rightarrow N_2 + NH_3$

and $N_3^- 7\,H^+ + 6\,e^- \rightarrow N_2H_4 + NH_3$

and $N_3^- + 9\,H^+ + 8\,e^- \rightarrow 3\,NH_3$

Hydrazine: $N_2H_4 + 2\,H^+ + 2\,e^- \rightarrow 2\,NH_3$

Cyclopropene: $C_3H_4 + 2\,H^+ + 2\,e^- \rightarrow C_3H_6$ (cyclopropane)

and $C_3H_4 + 2\,H^+ + 2\,e^- \rightarrow C_3H_6$ (propane)

N_2 is the natural substrate for the enzyme, but the acetylene reduction technique is commonly employed as an index of nitrogenase activity. Although rapid and convenient, and acceptably adequate for routine applications, the acetylene reduction method may yield misleading results if not used judiciously. Michaelis and inhibitor constants for substrates and inhibitors (competition for electrons in the presence of two or more substrates) have been calculated from steady-state kinetics with nitrogenases from *C. pasteurianum, A. vinelandii, A. chroococcum* and *K. pneumoniae*. These studies are the basis of a hypothesis predicting five distinct substrate-reducing sites. The physical relationship between these kinetically determined sites and FeMo cofactor or the other metal sites is not known.

Hydrazine, long suspected as a bound intermediate of N_2 reduction, was shown by Bulen to be a substrate for nitrogenase. The reaction was dependent upon pH, indicating that the protonated form of hydrazine ($N_2H_5^+$) was the preferred substrate. The reactivity of hydrazine provided the first evidence for its existence as a bound intermediate. Further evidence was reported by the ARC unit at Sussex, UK. By quenching an actively N_2-reducing preparation of nitrogenase with ethanolic HCl, Thorneley captured a hydrazine-like intermediate with *p*-dimethylaminobenzaldehyde. The time course for the production of this compound is consistent with its existence as an intermediate of N_2 reduction. Recently, Dilworth while investigating the reduction of azide to ammonia discovered that hydrazine was a product of azide reduction. Addition of $^{15}N_2H_4$ during azide reduction did not result in $^{15}NH_3$, indicating lack of equilibration between enzyme-bound N_2H_4-like intermediates and added N_2H.

Recently, two new substrates have been reported that show promise as active site probes. McKenna have demonstrated that cyclopropene is reduced by nitrogenase to a mixture of propene (1/3 of total) and cyclopropane (2/3 of total). Orme-Johnson reported that diazirine is a substrate for nitrogenase.

During the reduction of N_2, HD is produced from reaction mixtures containing either $H_2 + D_2O$ or $D_2 + H_2O$. Burgess *et al.* have proposed a mechanism whereby HD formed by the N_2-dependent process originates from a bound, reduced N_2 intermediate. Thus, H_2 inhibition of N_2 reduction and N_2-dependent HD formation are believed to arise from the same molecular process.

Steric effects are a major factor determining which unsaturated molecules will serve as substrates and also their effectiveness in competing for available electrons. For example, the Michaelis constants for $CH_2CHCHCN$, CH_2CN and CH_3CHCCH_3 are three orders of magnitude higher than N_2, C_2H_2 and HCN. Rates of substrate reduction have been reported to be equivalent for all substrates, although these have primarily been comparisons between N_2 and C_2H_2. Recently, Hageman and Burris have demonstrated that the electron flux through dinitrogenase controls the relative effectiveness and thus the rates at which various substrates are reduced by nitrogenase. Furthermore, the relative electron allocation is temperature dependent.

Energy Requirements

The, nitrogen-nitrogen triple bond is exceedingly strong requiring 941 kJ mol^{-1} to completely Issoclate the molecule. The large energy requirement for the reduction to the double-bonded Intermediate (523 kJ mol^{-1}) may explain the difficulty in reducing dinitrogen even though the free energy for the overall reaction is negative.

$$N_2(gas) + 3\,H_2(gas)\ \hat{}\ 2\,NH_3(aq),\ 25°C\ \Delta G = -53.34\ kJ\ mol^{-1}$$

The energy to drive biological nitrogen fixation is supplied by ATP. Since two ATPs are hydrolyzed for each electron transferred from dinitrogenase reductase to dinitrogenase, a minimum of 12 ATPs are needed to reduce N_2. However, for each mole of N_2 reduced, a minimum of one mole of H_2 is evolved from nitrogenase. The evolution of H_2 requires the same energy input (two ATPs per electron transferred) as other substrates, thus raising the minimum number of ATPs to 16. Also, the eight low potential electrons (six for N_2; two for H^+) can be assumed to possess energy as they could alternatively be used for oxidative phosphorylation or some other energy yielding reactions. Evans assumed each pair of electrons is, equivalent to three ATPs and, therefore, the apparent minimum requirement for N_2 reduction is approximately 28 ATPs.

The ATP for biological nitrogen fixation must originate from the metabolism of carbon compounds (although photosynthetic microorganisms may derive energy directly from light). Theoretically, 0.11 mol of glucose is needed to produce 1 mol of ammonia. Commonly, in the leguminous symbioses, 12 g of glucose must be metabolized to produce the energy needed to reduce 1 g of N_2. This results in an overall efficiency of 12 per cent. The efficiencies of the free-living heterotrophs are considerably lower.

Electron Donors

The efficacy of $Na_2S_2O_4$ as an *in vitro* reductant has deferred the need to identify the endogenous electron carriers to nitrogenase. Thus, many donors have not been identified nor have the metabolic pathways from which the electrons originate.

The *in vivo* reductants for dinitrogenase reductase that have been identified are ferredoxins or flavodoxins. Ferredoxins reduced by the phosphoroclastic pathway have been shown to be the source of electrons for dinitrogenase redutase from *C. pasteunanum* and *B. polymyxa.* Carter reported that a ferredoxin purified from *R. japonicum* bacteroids could supply electrons to nitrogenase utilizing an assay system containing 5-deazariboflavin or heat-treated chloroplasts. Hageman have demonstrated that *Azotobacter vinelandii* flavodoxin is the likely *in vivo* reductant because of the significant stimulation of activity compared to the *in vitro* reductant, $Na_2S_2O_4$. A flavodoxin in reduced *via* pyruvate metabolism has been shown to be the electron donor in *K. pneumoniae via* biochemical and genetic analysis of the *nifJ* and *nifF* gene products.

Catalytic Mechanism

Currently the most detailed mechanisms of nitrogenase catalysis are those offered by Thorneley. The flow of electrons through nitrogenase is well established; reduced flavodoxin/ferredoxin transfers an electron to dinitrogenase reductase, then the single electron is transferred to dinitrogenase with the hydrolysis of two molecules of MgATP to MgADP. The mechanism by which the electrons are transferred or stored within dinitrogenase and the manner by which the electrons are allocated to substrates is poorly understood.

The oxidation/reduction of dinitrogenase reductase can be monitored by EPR, Mössbauer, potentiometry and visible absorption spectroscopy. The Fe_4S_4 center operates between the $[Fe_4S_4(Cys)_4]^{2-}$ state and the $[Fe_4S_4(Cys)_4]^{3-}$ state. The binding of MgATP can be followed by EPR, thiol group reactivity, O_2 sensitivity, chelation susceptibility and, of course, ligand binding methods.

It is not known if there is an obligatory *in vivo* order in which dinitrogenase reductase is charged with MgATP and then reduced or if these processes occur independently. The rate of reduction by $Na_2S_2O_4$ of oxidized dinitrogenase reductase in the absence of MgATP is approximately 1000 times faster than the rate observed during catalytic turnover. Thus, dissociation of the MgADP-oxidized dinitrogenase reductase complex may greatly affect the rate of re-reduction. However, $Na_2S_2O_4$, which is normally utilized in these investigations, is not a very effective reductant for dinitrogenase reductase and the use of the natural electron donors may yield different results. Theoretically, reduced dinitrogenase reductase, free in solution, may transfer electrons to a molecule of oxidized dinitrogenase reductase that is complexed to dinitrogenase.

The reduced, MgATP-complexed dinitrogenase reductase rapidly associates with dinitrogenase. The rate constant for this association has been estimated at greater than 10^7 m^3 mol^{-1} s^{-1} with a dissociation constant of the order of 0.5 μmol m^{-3}. Ratios of dinitrogenase reductase to dinitrogenase of both 2 to 1 and 1 to 1 have been reported as optimal for nitrogenase activity. Hageman reported that a 1 : 1 ratio of dinitrogenase reductase : dinitrogenase produced full activity of the ATP hydrolysis and electron transfer reactions for *A. vmelandii* nitrogenase. A report on an heterologous nitrogenase complex (*A. vinelandii* dinitrogenase plus *C. pasteurianum* dinitrogenase reductase) indicates that *A. vinelandii* dinitrogenase does have two binding sites for dinitrogenase reductase. *C. pasteurianum* nitrogenase requires a 2 to 1 ratio of dinitrogenase reductase to dinitrogenase for full catalytic activity. These differences in binding ratios may reflect subtle distinctions in the electron transfer reactions between various nitrogenases.

Although nitrogenase is frequently referred to as a complex, there is little evidence to suggest a given dinitrogenase reductase molecule remains bound to a particular dinitrogenase molecule for a complete catalytic cycle (in terms of substrate reduction). Recently, Hageman and Burris have presented persuasive evidence that during proton reduction dinitrogenase reductase and dinitrogenase associate and dissociate with each electron transfer event. Mortenson and Thorneley suggest that perhaps for acetylene or N_2 reduction a longer-lived complex may be necessary.

Inhibitors

Classical Inhibitors

The so-called 'classical' inhibitors are those compounds that also serve as substrates for nitrogenase. These inhibitors can be classified into five different groups depending upon their inhibition patterns *versus* the other substrates:

1. H_2, N_2O and cyclopropene: these compounds are competitive *versus* N_2.
2. CN^-, N_3 and CH_3NC: these three are mutually competitive but non-competitive *versus* N_2.
3. C_2H_2: acetylenc is non-competitive *versus* N_2 and N_3^-. N_2 is competitive *versus* acetylene. H_2 evolution is completely suppressed by the presence of substrate levels of acetylene.
4. CO: carbon monoxide is competitive with all substrates but is unable to block H^+ reduction.
5. H_2 evolution: H^+ reduction is not inhibited by the presence of H_2 nor blocked by CO.

Originally these results were interpreted as five different sites or perhaps modification of a single site. There is no active site data to support this interpretation and these results may simply reflect differences in the allocation of electrons to substrates.

Recently, several new substrates have been investigated to pursue the conformation of the substrate-reducing active site(s) of dinitrogenase. McKenna *et al.*, have synthesized and utilized cyclopropene and 3,3-difluorocyclopropene as substrates and active site probes. 3.3-Di-fluorocyclopropene inhibits H_2 evolution, acetylene reduction and N_2 reduction. Diazurine, a strained-ring diazene analog, is a substrate and also an inhibitor of acetylene reduction. Dilworth reported that N_2, CO and N_2O inhibited the reduction of azide to hydrazine. H_2 was not an inhibitor of hydrazine formation from azide.

Regulatory Inhibitors

The regulation of nitrogenase is governed by energy, nitrogenous compounds and oxygen. Normally, nitrogen fixation activity is expressed when it is required for the growth of organisms. Once the nitrogen fixing system has been genetically turned on, it can either be genetically or biochemically turned off. In this section we will deal only with the biochemical considerations.

Energy

Nitrogen fixation imposes a considerable energy drain on the energy metabolism of an organism. Nitrogenase requires ATP for catalysis and is inhibited by ADP. The ratio of ADP/ATP affects electron transfer and thus allocation of electrons to substrates as well as total activity. *In vivo,* N_2-fixing cells have ADP/ATP ratios of 0.3–0.5, whereas organisms under non-N_2-fixing conditions have ratios of 0.8–0.9. The *in vivo* data, if considered exclusively, would imply nitrogenase is 80 per cent inhibited under normal conditions. Mortenson utilized $p/2e^-$ ratios as a measure of the efficiency of nitrogenase *in vitro* and found that the best apparent efficiency is at an ADP/ATP ratio of 0.3–0.5. Thus many other factors in addition to *ADP/A* TP ratios contribute to optimize rates of N_2 fixation *in vivo.*

Ammonia

The effect of added nitrogenous compounds, particularly ammonia, to cultures of actively fixing cultures is perhaps the most often studied aspect of nitrogenase regulation. Addition of ammonia to the culture media causes a rapid 'switch-off' of nitrogenase, a simple repression of nitrogenase synthesis or a combination of these effects. The observed effects depend upon the organism, the carbon and fixed nitrogen sources and the degree of membrane energization. The mechanism by which ammonia represses nitrogenase synthesis is not known. However, during depression of *K. pneumoniae*, the *nif* mRNA possess a half-life of about 20 min. After the addition of ammonia to a derepressed culture, the half-life of these mRNAs was reduced to approximately 9 min. Apparently, there is no need for post-transcriptional control of nitrogenase by ammonia in *K. pneumoniae.*

Covalent Modification

The nitrogenases of the photosynthetic bacteria *Rhodospirillum rubrum, Rhodopseudomonas capsulata* and *Rhodopseudomonas palustris* exist in one of two different forms depending upon the nitrogen source. The two different forms are actually interchangeable by the addition or removal of a group covalently bound to dinitrogenase reductase containing phosphate, a sugar (presumably ribose) and an adenine-like molecule. Cultures grown on glutamate or N_2 are poised so that modification occurs during harvesting and express little or no nitrogenase activity upon isolation. Cultures grown under NH_4^+-limited conditions lack this potential and possess active cell-free N_2-fixing preparations. The inactive dinilrogenase reductase can be converted into the active form by removal of the covalently linked adenine-like molecule. The deactivation is not well understood, but during this process ribose, phosphate and the adenine-like molecule are attached to dinitrogenase reductase. The inactive form of dinitrogenase reductase is incapable of transferring electrons to dinitrogenase. Glutmine synthetase may play a role in the interconversions of dinitrogenase reductase.

Oxygen

In vitro, oxygen irreversibly inactivates both nitrogenase components. The half-lives of dinitrogenase and dinitrogenase reductase in air are approximately 10 min and less than one min, respectively. *In vivo,* it is not known if O_2 denatured nitrogenase proteins can be reactivated or must be completely resynthesized. Based on investigations of putaredoxin and other iron-sulfur proteins, O_2 denaturation of the nitrogenase proteins is most probably due to the oxidation of the labile sulfide of the Fe_4S_4 clusters and/or FeMo cofactor to the zero oxidation state.

The physiology of each nitrogen-fixing organism must have the capacity to maintain the activity of the O_2-sensitive nitrogenase proteins as well as provide adequate sources for energy and reducing equivalents. The physiology of each N_2-fixing organism must be compatible with its respective ecological niches. For example, the *Azotobacter* possess a unique mechanism to protect nitrogenase components during O_2 stress. A complex is formed between an Fe_2S_2 protein and dinitrogenase reductase. This Fe_2S_2 protein has a molar mass of 14,000 daltons in *A. chroococcum* and 23,000 in *A. vinelandii*. O_2 stress may initially oxidize the Fe_2S_2 protein, thereby increasing its affinity for dinitrogenase reductase. Reduction of the Fe_2S_2 protein (removal of O_2 stress) causes dissociation of the complex and restores nitrogenase activity.

Conversely, sub-optimal partial pressures of O_2 can severely affect nitrogenase activity in those organisms which generate ATP *via* oxidative phosphorylation. Low O_2 tensions reduce ATP levels and thus lower the nitrogenase activity.

Ammonia Assimilation

Ammonia, the first stable product of nitrogen fixation, must be assimilated for transport to symbionts and/or for conversion into proteins. Classical enzyme studies, ^{15}N and more recently ^{13}N have been employed to trace the assimilatory pathways of N_2-fixing organisms. Several reviews appeared recently describing the ammonia assimilation process in prokaryotes, plants and legume nodules.

There are two possible primary routes for the assimilation of ammonia into amino acids: (1) glutamate dehydrogenase and (2) glutamine synthetase-glutamate synthase. Glutamate dehydrogenase provides the major assimilatory pathway when ammonia is abundant. However, during nitrogen fixation when nitrogen-limiting conditions prevail, it operates predominately in a catabolic mode rather than a synthetic one. Glutamate dehydrogenase has a high Michaelis constant for ammonia and furthermore the specific activities in crude extracts of nitrogen limited cells are usually quite low. Conversely, Lees have reported the kinetic parameters for soybean and lupine glutamate dehydrogenase, respectively, and suggest that this enzyme may have a more important role in ammonia assimilation than implied by the Michaelis constant for ammonia. Other amino acid dehydrogenase activities, such as alanine or aspartate, are too low to be of physiological importance.

The glutamine synthetase-glutamate synthase pathway is the major assimilatory pathway under nitrogen-fixing conditions in prokaryotes. The Michaelis constants for all the substrates are at physiological levels and the specific activities can account for the observed rates of ammonia assimilation. This pathway requires ATP in addition to reductant and thus may represent a mechanism for regulatory control in free-living organisms.

Once glutamine and glutamate are formed, all the other amino acids and other nitrogenous compounds can be formed by transamination or similar processes. However, in some symbiotic associations more complex pathways are required. In nodulated leguminous plants, the ammonia produced *via* nitrogen fixation within the *Rhizobium* bacteroids is excreted into the plant cytosol. Although two glutamine synthetases have been reported in free-living cultures of rhizobia only one has been reported within bacteroids. Present evidence suggests glutamine synthetase of *Rhizobium* bacteroids performs an insignificant role in symbiotic ammonia assimilation.

Glutamine synthetase of the plant cytosol is the major pathway of ammonia assimilation in nodulated leguminous plants. In certain leguminous plant species such as lupine, the principal form of nitrogen assimilates that are translocated to other plant tissues are amides. In other leguminous species, for example soybeans, allantoin and allantoic acid, commonly referred to as the ureides, are the major transport forms of nitrogen. The ureides are synthesized entirely in the plant host cell,

Schubert has shown that the synthesis of the purines takes place entirely within the plant cytosol, whereas the catabolism of the purines involves sequential steps within the cytosol, peroxisome and endoplasmic reticulum.

Genetics

Introduction

Until the 1970s, the molecular genetics of N_2 fixation was essentially unexplored. Earlier progress was confined to the isolation and biochemical characterization of mutants unable to use N_2 as a nitrogen source. The primary impediment to further analysis was the absence of genetic exchange systems for those diazotrophs which had received the most biochemical attention, *i.e. Clostridium* and *Azotobacter.* Because many of the genetic tools available for *Escherichia coli* could be applied to *K. pneumoniae,* it became clear that this nitrogen-fixing cousin of *E. coli* offered the greatest advantages for genetic analysis. As a result of this recognition, two major advances were made. First, Streicher *et al.* reported on the intrastrain transfer of *K. pneumoniae* genes for nitrogen fixation (*nif*) by bacteriophage PI and showed their linkage to genes for histidine biosynthesis. Almost simultaneously, Dixon demonstrated that genes for nitrogen fixation could be transferred from *K. pneumoniae* to *E. coli* by conjugation and, amazingly, expressed there to form a functional nitrogenase system. These advances established the approximate chromosomal location of *nif,* identified an easily-selected, linked marker and demonstrated that some *nif* genes were clustered on the chromosome. These successes, a political climate favorable to the support of this research and major advances in molecular technology, provided the basis for the exponential acquisition of genetic and molecular information which has taken place in the ensuing decade.

The genetic tools and current results of analyses will be outlined for several types of diazotrophs after a general discussion of the available approaches and methods.

Approaches and Techniques Available

An excellent review of the techniques used to explore the *nif* genes of *K. pneumoniae* has recently been published. Because it is written for use in the laboratory and is part of a more comprehensive methods volume, it is certainly a critical reference for those working in the area of nitrogen fixation. Here we will attempt to point out approaches which can be used with diazatrophs not as genetically accessible as *K. pneumaniae* and initially confine our remarks to free-living organisms.

First, a bank of independent mutations blocking the ability to fix dinitrogen can be generated. Although NTG (*N*-methyl-*N*-nitro-*N*-nitrosoguanidine), a powerful mutagen, is often used, it readily results in multiple lesions which may complicate the interpretation of pleiotropic pheno-types. Other mutagens such as alkylating agents, hydroxylamine, nitrous acid or UV may offer fewer problems. Because all strains are not equally sensitive to these agents, a dose-response curve should be established for each before use.

After mutagenesis of an appropriately genetically marked strain, Nif^- mutants are isolated as those unable to grow with N_2 but able still to grow well with ammonium salts. Next some phenotypic characterization of the *nif* mutant strains can be made. If the wild-type phenotype(s) is restored spontaneously at a frequency of 10^{-9} or higher, the original lesion is likely to be a point mutation, although the possibility of suppression should be considered. Generally loss of a plasmid, deletions or multiple mutations do not revert at these frequencies. In addition, any pleiotropic effects of the mutations should be examined, *e.g.*, altered growth rates on nitrogen sources other than N_2.

A biochemical and physical analysis of the *nif* components of the mutants can proceed without regard to the genetic capabilities of the organism. However, the establishment of the number of genes represented among the mutations greatly decreases the amount of biochemistry which must be done. Residual *in vivo* and *in vitro* nitrogenase activity should be measured to determine suitability for further biochemical or genetic studies. Furthermore, if activity is present *in vitro* but not *in vivo*, a block in endogenous electron flow to the nitrogenase complex is likely. Additional functional characterization can be made if antibodies to purified protein components of the nitrogenase complex are available. Purified components can sometimes be used to restore activity to inactive extracts of mutants, thus identifying the altered protein.

Because the three polypeptides of the dinitrogenase and the dinitrogenase reductase can be easily visualized by a comparison of the protein patterns of repressed and derepressed cells on denaturing polyacrylamide gels (SDS-PAGE), a rapid screen of mutants representative of each linkage group or gene will assess the gross integrity of the nitrogenase complex. Additional *nif*-specific polypeptides can be identified by pulse-labelling proteins during derepression of nitrogenase, separation by SDS-PAGE, followed by autoradiography. Finally, as a further refinement, pulse-labelled extracts can be subjected to two-dimensional gel electrophoresis and subsequent autoradiography. These experiments can give gene-protein relationships as well as operon structure when used to analyze the appropriate mutations.

If a genetic exchange mechanism such as transduction, transformation or conjugation is available, mapping can proceed. For Gram-negative organisms lacking indigenous transferrable plasmids, conjugation has often been obtained with the promiscuous drug resistance plasmids of the PI incompatibility group. Two factor crosses of *nif-l* × *nif-2* can be made to determine the relative linkage of the *nif* lesions. This procedure, called the ratio test, can be performed using an unlinked marker to standardize the results of each cross. Thus the closer two mutations are in the chromosome, the fewer prototrophic recombinants will be obtained relative to the number of recombinants for the unlinked marker. If a linked marker is available, such as *hisD* in *K. pneumaniae*, three point crosses can be carried out in a similar manner. It should be remembered that the size of the DNA transferred, *e.g.*, the capacity of the phage head in transduction, determines the physical limits for establishing linkage between markers. In addition, fine structure mapping requires a relatively small genetic vector.

If a linked marker is not available for mapping purposes, it is possible to generate such a mutation by the use of NTG. A Nif^- strain is mutagenized with NTG and plated for Nif^+ revertants. Most of these revertants will have been the result of the action of NTG and, therefore, will have a high probability of having an additional mutation(s) nearby. Temperature sensitive mutations can be isolated and used, even without a knowledge of the genes altered, or specific phenotypic alterations can be sought.

To Investigate the number of genes represented by the mutational clusters, complementation analysis is required. The complementation test simply determines whether two mutations affect the same gene. For this analysis it must be possible to establish a relatively stable merodiploid, *i.e.*, a strain with two copies of the chromosomal segment covering the mutations. This is generally brought about by the introduction (conjugation or transformation) of a plasmid containing the appropriate region of DNA into a recombination-deficient recipient strain (Rec^-). Although complementation has been successfully performed in a Rec^+ background, an absolute interpretation of results is questionable.

The construction of a plasmid containing *nif* may be accomplished either *in vivo* or *in vitro*. To form an F-prime or an R-prime *in vivo* carrying *nif*, conjugation into a Nif Rec^- recipient can be carried out and a Nif^+ recipient selected. Since recombination is prevented, the incoming chromosomal DNA can be stably maintained only if it is a part of the plasmid replicon. Other ways of restricting recombination are to use a recipient with a large deletion of the area coding for *nif* proteins, or to use another, closely related species as recipient so that DNA homology is low but one in which *nif* expression can be monitored.

In vitro construction of plasmids carrying *nif* genes can be carried out using recombinant DNA techniques, which have been adequately reviewed elsewhere. The highly conserved sequence of *nifH* and *nifD* genes among diazotrophs allows the identification of nitrogenase structural genes from almost any species with an appropriately labelled probe, *e.g.*, pSA30 from *K. pneumaniae.*

Once a plasmid containing the *nif* genes has been obtained, mutations can be introduced either by mutagenesis of the plasmid, by homogenotization with chromosomal mutations or by cotransduction of *nif* lesions with a selectable marker into the plasmid. Following the derivation of a bank of plasmid and chromosomal mutations, all pair-wise crosses can be constructed and the *Nif* phenotypes observed.

Information concerning operon structure can be obtained from insertion mutations induced by transposable genetic elements such as drug resistance transposons or bacteriophage Mu. These insertions have been shown to produce strongly polar effects on genes operator-distal to the point of the insertion. Therefore, noncomplementarity between an insertion mutation and point mutations in several clustered genes suggests an operon or transcriptional unit.

Another advantage of insertion mutations is their usefulness in the construction of deletions. Since these elements excise imprecisely, deletions are often observed among strains selected for loss of the insert.

A third type of genetic insertion resulting in the fusion of the gene for β-galactosidase to the operator-promoter region of the gene of interest has proved to be especially valuable in studying *nif* regulation. Because of the extreme O_2 sensitivity of some *nif* products and the lack of discriminatory assay systems for most of these products, observations of the control of expression are quite difficult. By fusing the gene for β-galactosidase (which is easily assayed) to the control region of the various *nif* operons, fluctuations in expression are readily followed. The limitation to this procedure, as for the other transposable elements, is to obtain a suitable vector for introducing the element into the organism of choice. Unless the bacterium of interest is sensitive to λ or μ bacteriophages, a conjugable plasmid or transformation must be used which may have limited efficiency.

nif Genes in *Klebsiella pneumoniae*

The results of numerous investigations aimed at the elucidation of the number, arrangement and control of the genes essential for nitrogen fixation by *K. pneumoniae* have been reviewed well and frequently in the recent literature. Here we will attempt to summarize briefly the information with updates from the primary journals.

Figure 9.1 shows the order, operon organization, direction of transcription and approximate molar mass of the products of the genes presently identified in the *nif* regulon of *K. pneumoniae.*

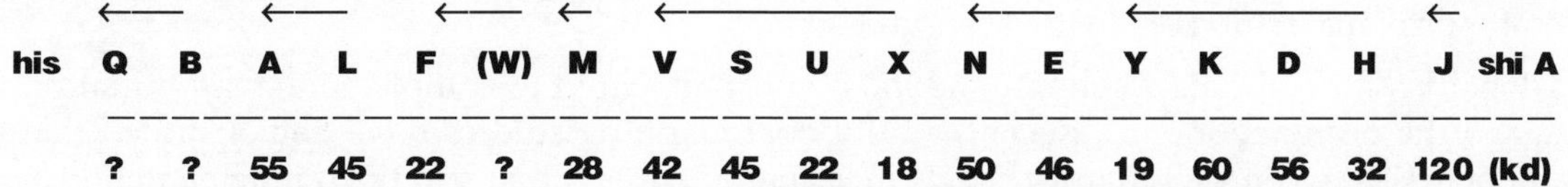

Figure 9.1: Order, Operon Organization, Direction of Transcription and Approximate Molar Mass of the Products of the *nif* Regulon Genes of *Klebsiella pneumoniae*

The order was established relative to *hisD* by three point crosses, deletion mapping and physical analysis with cloned fragments. The additional use of transposon mutagenesis of the cloned fragments has allowed a more detailed physical map to be established. It is now apparent that the *nif* regulon is contained in a 23 kb segment and that all the essential genes in this segment have been identified.

The number of operons and the direction of transcription have been identified by complementation between insertion mutations and point mutations. Present data suggest seven or eight transcriptional units read in the leftward direction toward the genes for histidine biosynthesis (Figure 9.1).

Although the protein products of only five of the *nif* genes have been purified to homogeneity, the functions of most are known but not the actual enzymatic reactions catalyzed. The dinitrogenase reductase and the dinitrogenase, which have been purified, are coded by *nifH* and *nifKD*, respectively.

The protein products of both genes involved in electron transport to the nitrogenase complex, *nifF* and *nifJ*, have been purified, *nifF* codes for a flavoprotein which is essential for physiological electron transport. In contrast, the J protein is a dimer of *ca.* 245,000 daltons containing 30 mol iron and 24 mol labile sulfur/mol protein and is probably an oxidoreductase.

Three of the genes whose products have not been purified are involved in the FeMo cofactor synthesis, *nifE*, *nifN* and *nifE*.

Because extracts of strains with mutations in *nifM* which have inactive nitrogenase can be restored to activity by the addition of dinitrogenase reductase, it appears that M protein is involved in a post-transcriptional modification of the product of *nifH*. A recent analysis of mutations in *nifV* indicated that the product may also be involved in post-transcriptional modification but of the dinitrogenase protein. Although dinitrogenase in a *nif* V^- background reduces some substrates, it is not capable of reducing N_2 with physiological electron fluxes. Two other gene products, those of *nifS* and *nifU*, also appear to be involved in the maturation of dinitrogenase since strains containing Mu insertions in *S* or *U* were shown to lack 'normal' levels of this protein.

The last two genes for which functions are known are those comprising the *nifLA* operon. The *nifA* product is required for the expression of all *nif* genes except for its own operon. In contrast, the *nifL* product is not essential for *nif* expression and acts as a repressor for all transcripts (except its own operon) in response to O_2, NH_4^+ and temperature.

The remaining genes, *nifQ*, *nifX*, *nifY* and *nifW* have not been assigned a specific function. The gene designated *nifQ* was inferred from the observation that a deletion from *hisD* which recombined with all known *nif* mutations resulted in a leaky Nif-phenotype. In addition, the strain containing this deletion was not dramatically altered in acetylene reduction, suggesting that protein may influence substrate selection. By analysis of the protein products of cloned fragments of the *nif* regulon, genes *X* and *Y* were identified. These two genes have not been shown to be necessary for nitrogen fixation.

Only limited analysis has been carried out on *nifW* (Roberts and Brill, 1980). It was assigned on the basis of reversion of a Mu induced mutation which was polar onto *nifF* and may represent an operator proximal region of *nifF* that is non-essential.

Finally a mutation between *nifJ* and *nifH* which appeared to complement a *nifJ* mutation was assigned the designation *nifC*. Recently additional mapping studies have demonstrated that this mutation lies between well characterized *nifJ* lesions and that intracistronic complementation occurs which confused the interpretation of earlier results.

In summary, 18 to 19 gene assignments have been made in the *nif* regulon. All except *nifW* have been shown to produce a protein product, 15 by genetic experiments and two to three by examination of cloned fragments. It is clear that the commitment to nitrogen fixation is a major investment for a bacterium.

Regulation of *nif*

A large number of environmental factors are involved in the regulation of expression of nitrogen fixing activity, among which are ammonia concentration, dissolved oxygen tension, presence of amino acids, availability of molybdenum and temperature. It appears that control may operate at transcription, the stability of transcripts and the stability of the enzyme complex. In most cases, evidence for regulatory effectors still rests with the physiological descriptions rather than detailed molecular mechanisms.

Ammonia, the end product of nitrogen fixation, is assimilated by the successive action of glutamine synthetase (GS) and glutamate synthase. As a result, the investigations of the mechanism of ammonia repression of nitrogenase have focussed on these enzymes. Only recently has it been directly proven that this control occurs at the level of transcription.

Because the levels of GS and the state of its covalent modification were seen to fluctuate in response to the supply of fixed nitrogen available to the cell; attention began to focus on the importance of this enzyme in a generalized nitrogen control system. Concomitant with the changes in GS activity and content, changes in the levels of other enzymes capable of supplying NH_4^+ or glutamate to the cell were also observed and a cause and effect relationship was proposed. The description of two types of mutants mapping in the region of the structural gene for GS, (a) those leading to glutamine auxotrophy, Gin^- which also resulted in a Nif^- phenotype and (b) those designated Gln^c which were constitutive for nil expression in the presence of NH_4^+, confirmed the involvement of the assimilatory system. The model which grew out of these studies suggested that the activation of the synthesis of the enzymes of nitrogen fixation and nitrogen metabolism required an increase in the cellular level of GS and its conversion to the unadenylylated form, which could then function as a positive regulatory element for transcription.

Since this model was proposed, there has been an accumulation of evidence showing that there is no apparent correlation between the adenylylation state of GS and the level of other enzymes of nitrogen metabolism. The discovery of additional genes involved in nitrogen regulation tightly linked to *glnA* has made a re-evaluation of nitrogen regulation necessary. Genetic and physical data presently support the model proposed by McFarland for *Salmonella* and elaborated on by Merrick, illustrated in Figure 9.2 in a modified form showing an interaction with *nif* genes.

Loss of *ntrA* or *ntrC* results in loss of ability to express nitrogen controlled genes at high levels. Thus, when nitrogen is limiting, *ntrA* protein or an enzymatic product of that protein interacts with *ntrC* protein to form an activator. The *ntrC* product also can interact with the *ntrB* protein to form a repressor when ammonia levels are high. As a consequence of the loss of *ntrC*, both repression and activation of *glnA* are eliminated resulting in a low *glnA* expression insensitive to nitrogen availability. In contrast, mutations in *ntrB* allow a high constitutive level of expression of *glnA*. This model now appears to hold for several enteric bacteria.

In *E. coli* and *K. pneumoniae* the P_{II} protein, a component of the adenylylation system for glutamine synthetase, has been reported to be a corepressor capable of interacting with the *ntrBC* complex. Mutants incapable of converting P_{II} to its undylylated form in response to low levels of ammonia were unable to activate nitrogen assimilation genes including nitrogenase.

Mutations occurring within the *nif* regulon which render the expression of that regulon insensitive to fixed nitrogen are at the operator-promoter site of *nifLA* or within the *nifLA* operon. These results

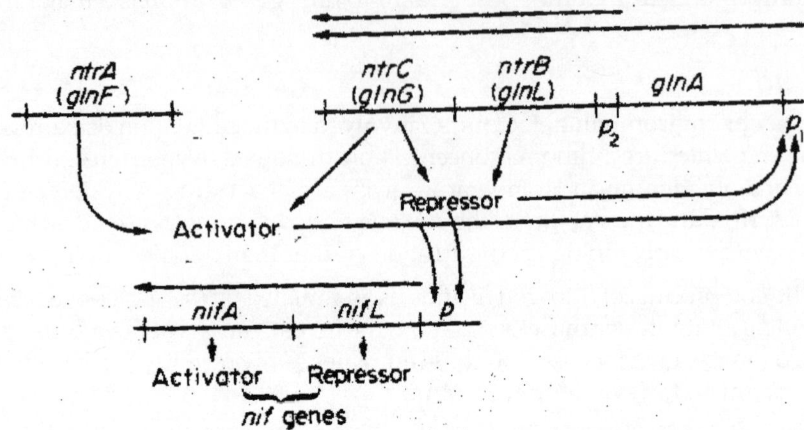

Figure 9.2: Model for Regulation of Nitrogen Metabolism. Gene Designations in parentheses are those Used in *Eschtrichia coli.* The *ntr* designation derives from nitrogen regulation ← indicate transcriptional units and directionality ← indicate regulatory functions.

suggest that the transcription of *nifLA* is controlled by the generar nitrogen-regulatory system and the products of that operon in turn regulate the remaining *nif* operons.

Recent experiments have pointed out the remarkable similarities between the *nifA* and *ntrC* protein products. Both are approximately 55,000 daltons, have similar isoelectric points, and require a functional *ntrA* gene for regulatory activity. Both have been shown to be capable of activity with the *nifLA* operon, *glnAntrBntrC* operon, and the *Rhizobium meliloti nifH* promoter in addition to those genes governing the catabolism of certain amino acids. These similarities have led to the suggestion that the two genes are evolutionarily related. Significantly the *K. pneumoniae nifH* promoter is responsive only to *nifA* protein.

As yet the physiological meaning of the ability of *nifA* protein to substitute for *ntrC* protein is unclear. Perhaps it is simply a sparing effect for *ntrC* or the vestiges of the evolutionary antecedent.

The *nifL* product which is the *nif* regulon repressor affords a more immediate response to fixed nitrogen than is possible through nitrogen regulation. Since *nifL* and *nifA* coding for the activator protein are transcribed as a single operon, the *nifL* protein must be maintained in an inactive state during active nitrogen fixation and be converted into a repressor state by molecular signals sensing fixed nitrogen or O_2.

Synthesis of most *nif* polypeptides in *K. pneumoniae* has been shown to be temperature sensitive. It is as yet unresolved whether A protein and/or L protein are responsible for thermosensitivity.

The mechanisms involved in the control mediated by amino acids or by nitrate remain to be elucidated. However, both repressive effectors are still functional in mutants of *K. pneumoniae* derepressed for nitrogenase biosynthesis in the presence of ammonium.

The involvement of molybdenum (Mo) in the regulation of biosynthesis of nitrogenase in *K. pneumoniae* has been brought into question with results from more recent studies. Under Mo deprivation some dinitrogenase apoprotein is made and can be activated by the addition of molybdate to the cells. Thus Mo must not be essential for *nit* poly-peptide synthesis in this bacterium. Only in *A. vinelandii* does Mo appear to be required for conventional dinitrogenase synthesis.

A form of control at the level of mRNA destabilization operates in *Klebsiella* for *nif* expression. Under nitrogen fixing conditions, *nif* mRNA appears to be remarkably stable with a half-life of 18 min or longer. When these diazotrophs were shifted to repressive conditions, the half-life became significantly shorter. However after addition of rifampicin plus NH_4^+, mRNA hybridization experiments suggested that *nif* mRNA was detectable well after *nif* protein synthesis was terminated. This result would suggest that a translational control was superimposed on the transcriptional regulation and mRNA stability in *Klebsiella* or that a specific nuclease was required to shorten the half-life of *nif* mRNA upon repression.

Azotobacter Species

The members of the Gram-negative *Azotobacter* genus are nitrogen-fixing obligate aerobes. They are widely distributed in soil and some species have been found associated with the rhizosphere of tropical grasses. There are reports of increased plant growth following inoculation with *Azotobacter*, possibly due to the production of plant growth hormones. Because of the potential for coupling ammonia excretion from N_2 reduction with plant-hormone production in an organism that has an associative growth mode, this genus deserves considerable attention.

Although *A. vinelandii* has been reported to have 10 times the amount of DNA that *E. coli* contains, mutants lacking the ability to fix nitrogen were obtained rather readily. While purine and pyrimidine auxotrophs of *A. vinelandii* have been isolated, amino acid auxotrophs have not been reported even though considerable effort has been made to obtain them. Therefore, a very limited array of genetic markers are presently available for extensive mapping efforts in this genus.

Several *nif* mutations have been characterized with respect to activities for dinitrogenase and dinitrogenase reductase, antigenic cross-reacting material (CRM) and electron paramagnetic resonance signals. Among the mutants described were all the predictable classes for a system of two proteins which could be assayed separately. Two additional classes of special interest were found: one which was Nif^- but hyperproduced the reductase and one which produced nitrogenase in the presence of ammonia.

Although a transformation system was described 10 years earlier, a reliable procedure was not established until 1976. Refinements of this procedure have been published. The plate-transformation system was used for the determination of a rough linkage map for several Nif^- mutant strains. Because the size of DNA transferred was not determined and congression (simultaneous transfer of markers on separate fragments of DNA) was relatively high, no physical interpretation of distance could be made from the ratio test crosses. However, the results did demonstrate that the *nif* genes in *A. vinelandii* do not fall into one cluster. Subsequent DNA-DNA hybridization studies with cloned *K. pneumoniae nif* genes have shown that the highly-conserved structural genes for nitrogenase *nifH* and *D* do occur on a single, small fragment.

Regulation of the nitrogen-fixing complex of *Azotobacter* is similar to other free-living diazotrophs in being ammonia repressible with the involvement of the ammonia assimilatory system. Two classes of regulatory lesions have been described: (1) presumed point mutations resulting in the loss of activity and CRM for both nitrogenase proteins, as well as (2) a mutation overproducing the dinitrogenase reductase and simultaneously not producing dinitrogenase. The existence of these phenotypes has been interpreted to mean that a common regulatory gene is required for expression of the structural genes for the nitrogenase complex.

Surprisingly, an extended analysis of *A. vinelandii* Nif^- mutants which included two-dimensional PAGE of the proteins of several Nif^+ revertants revealed that the 'conventional' dinitrogenase polypeptides were missing and four additional, ammonia-repressible proteins were present. Thus the Nif+ revertants were pseudorevertants. Further investigation has led to the conclusion that this diazotroph may possess an alternative nitrogen fixing system which is Mo repressed. While the description of this alternative system awaits further confirmation, the survival potential for a bacterium forced, to fix nitrogen in an environment depleted of Mo is clear. The biochemistry of N_2 reduction without the involvement of Mo should prove extremely interesting..

Cyanobacteria

The cyanobacteria can be roughly divided into three main groups, the unicellular, filamentous nonheterocystous and filamentous heterocystous forms. Although nitrogen fixing species occur in all classes, all the filamentous heterocystous species have been shown to possess this capacity. The heterocyst is the differentiated cell that functions as an anaerobic site specifically designed for N_2 reduction and release of fixed nitrogen to vegetative cells.

The only reliable and efficient genetic exchange system for the cyanobacteria to date has been developed for *Anacystis nidulans,* a non-N_2 fixing species. The consequences of the absence of a genetic exchange system in the diazotrophic cyanobacteria was brought into clear focus by their omission from a recent review of the genetics of nitrogen fixation by Roberts and Brill. However, progress is being made towards genetic analysis through the generation of defined mutations and cloning technology.

Although a few mutant strains had been isolated in several species of filamentous cyanobacteria, until the work of Currier there had not been a concerted effort to obtain well characterized mutants in one species. Now a bank of mutant strains has been isolated in *A. variabilis,* including auxotrophs and those altered in nitrogen fixation and heterocyst development. To accomplish mutant isolation from a filamentous organism, it was necessary to separate the rare mutant cell from the filament. To rupture cells randomly and fragment the filaments, cavitation was used and the short filaments or single cells were then subjected to penicillin enrichment and selection for auxotrophs.

The strongly conserved nature of the structural genes for nitrogenase, *nif H* and *D,* was demonstrated by heterologous hybridization of *K. pneumoniae nif* genes carried on the plasmid pSA30 with fragments of *Anabaena* DNA. In addition, homology between *nif* genes other than the structural genes for the nitrogenase was reported for DNA from *Anabaena* and *K. pneumoniae* which may be a unique feature among the diazotrophs. These studies have indicated that the *nif* structural genes in *Anabaena* are rearranged from the order found in *K. pneumoniae* and are probably separated on the chromosome.

During the course of the hybridization experiments, what appears to be a second copy of the *nifH* gene was found. Because the restriction pattern was different from that of the *nifH* gene located near *nifD,* it is not known whether this is a second copy of *nifH,* a nonfunctional pseudogene, or another gene which by chance has homology. A role for such a gene in a hypothetical alternative nitrogen-fixing system such as that found in *A. vinelandii* was also suggested.

The regulation of the capacity for nitrogen fixation by cyanobacteria differs from other free-living diazotrophs in a few areas and is discussed in several reviews. Ammonia repression (and probably O_2 repression) operates in the filamentous cyanobacteria. Supporting the involvement of a generalized nitrogen regulatory system in nitrogenase control were results which showed that the GS inhibitor MSX (methionine-*SR*-sulfoximine) caused a relief of inhibition of heterocyst formation by exogenous ammonia and the excretion of newly fixed ammonia from cells of N_2-fixing *Anabaena cylindrica*. The majority of the nitrogenase in filamentous cyanobacteria occurs within the heterocysts of a photosynthetically active culture. The elegant experiments of Wolk and coworkers have demonstrated that ammonia generated by nitrogenase in *A. cylindrica* is assimilated *via* the glutamine synthetase-glutamate synthase enzyme couple. Therefore, as might have been expected, the GS activity, which is the result of a single enzyme in *Anabaena,* is slightly higher in heterocysts, although the glutamate synthase activity is lower as compared to vegetative cells. These findings support the contention that glutamine is produced from newly fixed N_2 within the heterocyst and exported to the vegetative cell where glutamate is made.

The GS in the cyanobacteria does not enjoy large fluctuations in activity in response to nitrogen source nor is it covalently modified. In addition, the excretion of ammonia from *Anabaena azollae* when in association with the water fern *Azolla caroliniana* has been shown to be the result of the absence of GS activity and antigen from the cyanobacterium. Thus some question arises concerning the role of GS or *gln* associated genes in the regulation of nitrogenase. As a result, molecular biological techniques have been applied to the problem and the *glnA* gene from *Anabaena* 7120 has been cloned into *E. coli* where it complements a *glnA* deletion. Interestingly the *Anabaena* gene is not repressible in *E. coli.* Additional studies are under way to elucidate the extent of the *glnA* region cloned and the regulatory properties encoded.

Photosynthetic Bacteria

The purple, non-sulfur phototrophs have been most intensively studied among the free-living photosynthetic bacteria, excluding the cyanobacteria. Within the Rhodospirillaceae, the nitrogenase system of *R. rubrum* has received most biochemical attention, while genetic transfer systems have been established in *R. capsulata* and *Rhodopseudomonas sphaeroides.* Although some metabolic and morphological differences exist, it is assumed here that the nitrogen fixing function is conserved in these organisms, an assumption which has received support from a number of physiological studies.

Interest in the nitrogen fixing capacity of these bacteria has been rekindled by the demonstration of regulation of nitrogenase activity by covalent modification of the dinitrogenase reductase. This mechanism of inactivation and subsequent activation suggest that the number of genes essential for nitrogen fixation may be increased in these organisms.

Several genetic tools are available in the Rhodospirillaceae and a recent review has an excellent description of each. Here we will briefly summarize the transfer mechanisms and the strains in which they have been successfully applied.

Conjugation mediated by PI incompatibility group plasmids appears to be ubiquitous among these Gram-negative phototrophs. Plasmid transfer was first used to promote chromosomal transfer and demonstrate linkage in *R. sphaeroides.* Subsequently, transfer of several P and W group plasmids into this species has been shown as well as low chromosomal mobilization. Similar experiments of conjugational transfer have been reported for *R. capsulata.* The introduction of RP4::Mu *cts* has been accomplished in both species. Although Mu was not thermoinducible in *R. sphaeroides,* phages were produced. In contrast, *R. capsulata* strains differ in Mu expression; 37b4 containing RP4::Mu *cts* was reported to be thermoinducible while no Mu expression was observable in B100. These experiments suggest that *in vivo* engineering with Mu may be practical in *R. sphaeroides* and *R. capsulala* 37b4. In addition, the introduction of R751::Mudlac (Ap^R, *lac*) into *R. sphaeroides* has been accomplished and should allow operon fusions in this organism.

By the introduction of the mercury resistance transposon Tn501 into RP1, Pemberton and Bowen have been able to demonstrate that this plasmid promoted a high frequency of chromosomal transfer in *R. sphaeroides*. As a result, they have published the first map of auxotrophic markers in the photosynthetic bacteria, which opens the way for additional mapping and strain construction. A single *nif* mutation has subsequently been mapped to a position very near the genes for the photosynthetic apparatus.

A derivative of RP1, pBLM2, which has enhanced chromosome mobilization ability in *R. capsulata* was isolated by Marrs by screening rare exconjugant clones for donor ability. A frequency of chromosome transfer of 6×10^{-4}/donor was obtained for some markers and R-primes bearing the genes for photosynthesis were found. Similar constructs should be possible for the *nif* genes of the organism.

These preceding genetic tools have been adaptations of systems first described in the enteric bacteria, *R. capsulata* enjoys, in addition, an elegant endogenous system of generalized gene transfer discovered by Marrs in 1974. The agent (GTA) that serves as the vector for gene transfer appears to be a small phage-like particle. DNA extracted from the GTA particles is linear, double-stranded of *ca.* 3×10^6 daltons and is randomly packaged from the chromosome. Mapping results with the GTA system have yielded map distances that are remarkably close to the physical distances obtained by restriction endonuclease mapping.

This transfer agent has been used to show *nif* transfer between *R. capsulata* mutant strains and to begin to construct a linkage map of *nif* genes. By applying the ratio test markers of the same phenotype less than 2500 base pairs (bp) apart can be reliably demonstrated to be linked with this agent. Although 13 Nif^- mutations have been shown to fall into five linkage groups, the question of overall clustering must be settled by different exchange techniques employing larger pieces of the chromosome, Preliminary indications from cloning studies using the heterologous pSA30 plasmid to identify *R. capsulata nif* structural genes indicated that these genes are clustered.

Regulation of the repression of synthesis of nitrogenase in the Rhodospirillaceae is assumed to be similar to that in other diazotrophs. No activity is measurable in cultures grown in ammonia, complex medium or air. In addition, the polypeptides corresponding to the nitrogenase proteins are absent under these conditions.

Enzymatic studies with *R. capsulata* and *Rhodopseudomonas palustris* demonstrated that GS and glutamate synthase are the key enzymes of ammonia assimilation. The involvement of a generalized nitrogen regulatory system in the control of nitrogenase was indicated by the derepression of the enzyme complex in the presence of the GS inhibitor, MSX. Support for the involvement of common regulatory elements for *gln* and *nif* has been derived from, the isolation of glutamine auxotrophs lacking GS activity which are derepressed for nitrogenase in the presence of ammonia.

Recently, an interesting class of spontaneous Nif^- mutants has been isolated after senal sub-culture of *R. capsulata* cells in derepressing medium. Because the only substrate for nitrogenase was protons, these cells produced large quantities of H_2. After 10 to 12 subcultures, a substantial portion of the population was Nif^-. This result points out one of the difficulties which may be encountered with applications of diazotrophs for commercial H_2 or fertilizer production.

Rhizobium Species

The agricultural economic importance of the symbiotic associations involving the rhizobial species has focussed attention on these organisms for many years. The mechanisms of recognition, infection and nodulation have been investigated extensively; however, the innate difficulties of a developmental symbiosis has slowed the acquisition of knowledge which might be used to improve the process.

Renewed interest has arisen because of the recent advances in understanding nitrogen fixation in the free-living bacterial systems. Much of the genetic research with *Rhizobium* has been thoroughly reviewed in the recent literature. Earlier genetic studies were critically summarized by Schwinghamer and Roberts. Here we will summarize the mechanisms available for analysis and some of the recent findings.

Fast Growing Species

Because of the faster doubling times, *R. meliloti, R. leguminosarum. R. trifolii* and *R. phaseoli* have been easier to manipulate by standard genetic techniques than the slower growing species. Mutants have been isolated after chemical mutagenesis and penicillin enrichment techniques *via* classical procedures. Although transformation procedures have been available for most species for many years, little linkage data has resulted from its use. In addition, generalized transduction in *R. meliloti* was reported as early as 1967; however, only one instance of cotransduction of markers has been demonstrated with this system. It is expected that these genetic exchange processes will begin to be re-investigated since the establishment of circular linkage maps of the chromosomes has been accomplished *via* conjugation. Transductional analysis will be essential for fine structure mapping and the ability to transform opens the way for additional cloning manipulations. Indeed, the recent literature reflects this interest since transduction has now been reported for *R. leguminosarum* and *R. trifolii* and improved methods for plasmid transformation of *R. meliloti* have been developed.

Systems of conjugation have played the most prominent role in the genetic investigations of the rhizobial species. Early endogenous conjugational systems have either not been rigorously pursued or the *Rhizobium* species used has been questioned. More recently, the existence of conjugative plasm ids in three strains of *R. leguminosarum* which code for bacteriocin production and have chromosome mobilization ability (Cma) has been demonstrated.

In contrast with the endogenous plasmids which are interesting because of their own genetic content, the plasmids of the Pl incompatibility group originally from *Pseudomonas aeruginosa* have been the most productive for genetic analysis. As a result of their use, chromosome linkage maps now exist for *R. meliloti* 2011, for *R. meliloti* 41 and for *R. leguminosarum*. Recombination and linkage between markers in crosses of *R. leguminosarum* with *R. trifolii* or *R. phaseoli* were essentially the same as results obtained from crosses within *R. leguminosarum*. Therefore, the map derived for *R. leguminosarum* is believed to represent all three species. A recent comparison of these maps shows the similarities and complementation obtained among the *R. meliloti* strains and *R. leguminosarum*.

Use of the promiscuous plasmids has also made the introduction of transposons into *Rhizobium* a fairly straightforward procedure. Although Pl plasmids are generally stable in *Rhizobium* when bacteriophage Mu is present in the plasmid, the plasmid is no longer stably maintained after its introduction by conjugation. When a transposon conferring drug resistance is also included on such a plasmid, selection for that drug resistance after transfer of the plasmid to *Rhizobium* selects for those cells in which transposition has occurred. Two such 'suicide' plasmids, one containing Tn7 and one containing Tn5, have been constructed and used.

An additional procedure for transposon mutagenesis has been described by Ruvkun. Any cloned gene can be introduced into *E. coli* and there mutagenized with transposons. If it is not in a transmissable vector, it may be inserted into a broadhost-range cloning vehicle and conjugated back into the original host. The cloning vehicle can then be 'chased' from the bacterial cytoplasm by the introduction of an incompatible plasmid while continuous drug resistance selection is made. The majority of drug resistant exconjugants will have undergone a homologous recombinational event, or homogenotization, such that the transposon now resides in the chromosomal gene.

In *R. trifolii*, *R. phaseoli* and *R. leguminosarum*, the genes for nodulation lost specificity and nitrogen fixation have been reported to be present on large plasmids. In contrast, experiments with *R. meliloti* 102F34 designed to examine the location of *nif* genes, suggested both a chromosomal position and that at least three discrete units necessary for nitrogen fixation were contained in an 11.2 kb fragment. Most intriguing was the recent observation of the occurrence of reiterated *nifD* and *nifH* genes of *R. phaseoli* with at least one copy on a larger plasmid.

Slow Growing Species

Included in the slower growing group of rhizobia strains are *R. lupini*, *R. japonicum* and *Rhizobium* sp. or cowpea rhizobia. The genetic analysis of these species lags behind that of the faster growers primarily because of the greater difficulty encountered in the microbiological manipulations. In contrast to the fast growers, these strains express high nitrogenase activities in defined culture; therefore, these bacteria may ultimately be more amenable to the analysis of *nif* functions. Although most of the mutants isolated early were drug resistant, more recently Nif^- mutants of *R. japonicum* have been obtained after mutagenesis and screening *in planta* or *ex planta*. Glutamine auxotrophs of cowpea strain 32H1 showing impaired nitrogenase activity have also been isolated. Few other auxotrophs have been reported in these strains.

Although an endogenous conjugation system was reported for a non-nodulating *R. lupini* strain as early as 1968, methods for gene transfer in other strains of slow growers have only now begun to be developed. Using an uncharacterized bacteriophage. Shah *et al.* have now generated a linkage map of *R. japonicum* by transduction. This map can now be used to locate and manipulate the genes essential for nodulation and nitrogen fixation.

In preparation for a more extensive genetic analysis, Kennedy have reassessed the ability of several strains of cowpea rhizobia to transfer, maintain and express P group plasmids. Strains which have a reasonable frequency of transfer and good plasmid stability during nodulation were identified but no Cma was reported. Obviously Cma is the next step in the analysis of the slow growers.

Regulation

From studies with cloned DNA fragments containing *nif* genes, it has been shown that the control for expression of these genes in the fast-growing species operates at the level of transcription during nodulation. Similar experiments with slow growers have not yet been done.

The involvement of the ammonia assimilatory system in the regulation of nitrogenase of *Rhizobium* remains an open question. In contrast to *K. pneumoniae*, results with bacteroids have suggested that the assimilation system of the bacteria is essentially non-functional during greatest N_2-fixing activity. Studies with *R. japonicum* bacteroids and free-living bacteria supported this finding in that the ammonia produced from N_2 was found in the medium. In addition, mutants lacking glutamate synthase and strains naturally lacking glutamate dehydrogenase were normal in their abilities to nodulate and fix-N_2.

On the other hand, studies with glutamine auxotrophs showed that nitrogenase activity was lacking in the mutant bacteroids. These results must be interpreted with care, since it is now known that all *Rhizobium* species so far examined have two distinct glutamine synthetase enzymes. The two enzymes have different physical properties; GSI undergoes adenylylation in response to a nitrogen signal while GSII does not. The description of an auxotrophic revertant of *Rhizobium* sp. 32H1 in which (1) GSI was constitutively adenylylated, (2) GSII was still missing and (3) nitrogenase was constitutively synthesized implies common regulatory elements for GSI and nitrogenese. Because the auxotrophy of GSII mutants could be satisfied by either glutamine or purines, a role for this enzyme in assimilation has been questioned.

Applications

Physiology or Organisms

Application of the biological nitrogen fixation process requires a knowledge of the physiology of these organisms. The broad range of the ecological adaptations of nitrogen-fixing organisms permits an even broader range of applications. The capacity to fix atmospheric nitrogen occurs in a larger number of diverse bacteria but does not naturally occur in eukaryotes except within symbiotic associations.

Aerobes

Aerobic nitrogen-fixing bacteria are found mainly in the family Azotobacteriaceae, which includes the taxonomically similar genera *Azotobacter, Azomonas, Azotomonas, Azotococcus, Beijerinckia* and *Derxia.* Members of the genera *Azotobacter, Beijerinckia* and *Derxia* have been demonstrated to fix N_2, but Parejko could find no evidence for fixation in strains of *Azotomonas* that were examined. Most of these aerobic bacteria actually fix nitrogen only under microaerophilic conditions. It has been proposed that these organisms adapt physiologically by producing copious amounts of polysaccharide which hinders diffusion of oxygen thereby permitting the functioning of nitrogenase. Of these organisms, the *Azotobacter* is best adapted to aerobic nitrogen fixation. *Azotobacter* possess very high rates of respiration; indeed, the highest respiratory rates ever measured are those of *Azotobacter.* These high respiratory rates have been called respiratory protection since they maintain intracellular partial pressures of oxygen that are compatible with the functioning of nitrogenase. However, even the *Azotobacter* strains are susceptible to oxygen at partial pressures beyond 20 per cent O_2, or when forced into phosphate limitation. Under these conditions, the nitrogenase is 'switched off' or protected by an iron-sulfur protein that forms a complex with dinitrogenase reductase. Nitrogenase is switched on when favorable conditions return.

Although *Azotobacter* is generally thought of as a free-living nitrogen fixing organism, several species have been reported to form associations with the roots of plants. For example. *Azotobacter paspali* forms a loose association with the roots of *Paspalum notatum* and other grasses.

Azospirillum fixes atmospheric nitrogen microaerophilically forming characteristic pellicles in culture. The discovery of associations between *Azospirillum* and many plant roots. including some agriculturally important plants, has spawned much promise and controversy.

Aerobic nitrogen-fixing organisms also include iron- and methane-oxidizing bacteria. Three strains of *Thiobacillus ferrooxidans* have been shown to incorporate ^{15}N when growing in culture media at pH values of 2. Also nitrogen fixation among the methane-oxidizing bacteria is common, Dalton has shown that the methane mono-oxygenase enzyme of *Methylococcus capsulata* provides respiratory protection for the organism such that cultures can be conditioned to withstand elevated partial pressures of oxygen without increased respiration.

Facultative Anaerobes

The majority of the physiological and biochemical investigations of nitrogen fixation have been performed with facultatively anaerobic bacteria. This class includes species of *Bacillus, Klebsiella,*

Rhodospeudomonas, Rhodomicrobium and *Rhodospirillum.* These bacteria, although able to grow aerobically on fixed nitrogen, cannot fix nitrogen in air. Only under strict anaerobic, or extremely low oxygen tensions are these organisms capable of nitrogen fixation. The non-sulfur phototrophic bacteria are able to utilize light energy directly for the fixation of atmospheric nitrogen anaerobically and recently Madigan have demonstrated that *Rhodopseudomonas capsulata* fixes nitrogen microaerophilically in the dark.

Certain cyanobacteria, most notably the filamentous *Anabaena* species fix atmospheric nitrogen aerobically. This would appear to be a dilemma since the cyanobacteria conduct plant-type photosynthesis and thereby evolve O_2. *Anabaena* has overcome this incongruity by morphologically altering approximately every tenth cell along its length into a thick walled 'heterocyst' devoid of photosystem II but possessing the nitrogen-fixing apparatus. The heterocyst supplies fixed nitrogen to the vegetative cells, which in turn provide reduced carbon compounds for the fixation process.

The coccoid cyanobacteria *Gleocapsa* also fixes nitrogen aerobically. *Gleocapsa* forms large dense colonies in which the outermost cells restrict O_2 from diffusing into cells at the center of the colony mass. Thus, a diffusion gradient is formed and cells at the center of the colony are able to fix N_2 under microaerophilic conditions.

Anaerobes

The most notable anaerobic diazotrophs are in the genus *Clostridium. Clostridium* is one of the most primitive of organisms and thus its physiology has attracted considerable attention. These organisms are strict anaerobes, but will tolerate oxygen to varying degrees in culture without irreversible denaturation of their nitrogenase. By comparison the nitrogenase of the sulfate-reducing bacteria is impaired by any level of oxygen present in the culture.

Symbionts

The complexity of many of the symbioses has hindered progress toward elucidating the physiology and biochemistry of the prokaryotic symbiont. Although the free-living state of these organisms is well defined, in many cases the symbiotic state is poorly understood.

The *Rhizobium* species are defined as heterotrophs able to grow both aerobically and anaerobically in culture. Recently, Hanus have demonstrated certain *R. japonicum* and cowpea rhizobia are capable of chemolithotrophic growth if provided with a fixed nitrogen source. *Rhizobium,* the endophytes of leguminous plants, have been demonstrated to fix measurable levels of N_2 when grown in culture on well-controlled, low partial pressures of oxygen. Whereas nitrogen fixation can be obtained on defined media in the slow-growing rhizobia (*R. japonicum, R. lupin;* cowpea miscellany), the fast growers (*R. melitoti, R. leguminosarum, R. trifolii, R. phaseoli*) can fix nitrogen *ex planta* only in media containing liquid extract from plant callus growth media. The rates of fixation are extremely low compared to those obtained when these organisms are participating in their respective symbioses.

The genus *Frankia* has been designated to encompass those endophytes of symbioses between non-leguminous plants and the actinorhizii. Callaham were the first to successfully isolate the endophyte of these associations, obtaining the organism from *Comptonia peregrina* nodules in pure culture. Subsequently, the endophytes from *Alnus* and *Eleagnus* have also been isolated. These organisms are actinomycetes with a characteristic sporangia. The doubling time of these strains on defined media is of the order of several days.

The cyanobacteria participate as the nitrogen-fixing endophyte in lichens and with the water fern *Azolla.* The lichens are associations between cyanobacteria and a fungus. In the case of *Peltigera aphthosa,* a complex symbiosis occurs consisting of an ascomycetous fungus, a species of the cyanobacteria *Nostoc* and a green alga *Coccomyxa* sp. The *Nostoc* in the symbiosis contains a much higher number of heterocysts than is normally found in the free-living culture (20 per cent *versus* 5–10 per cent). *Anabaena azollae,* the endophyte found in the leaf cavities of the water fern *Azolla,* also has an increased number of heterocysts. The factors controlling the increased heterocyst frequency in these cyanobacterial symbionts have not been determined. The *Anabaena-Azolla* symbiosis is a major contributor to rice production in China and Southeast Asia. Talley report nitrogen accretion rates between 30 and 100 kg N ha^{-1} $year^{-1}$ in replicated experiments conducted in California.

Agronomic Applications

The most grandiose application will be the integration of the Nif genes directly into plant cells such that the transformed eukaryotes now are capable of providing their own fixed nitrogen directly from the atmosphere. Whereas the transfer of genes may certainly be possible, the expression of the genes at a significant level, provision of sufficient energy for nitrogen fixation without severely affecting plant growth or development, protection of the O_2 labile nitrogenase proteins and so on, represent intractable barriers that will require extraordinary effort and time to surmount. The most logical intermediate step to achieve nitrogen fixing eukaryotes is *via Agrobacterium. Agrobacterium,* taxonomically related to the *Rhizobium,* is able to infect a wide variety of plants, rather than being restricted to the legumes as are the rhizobia, and form crown galls. This infection and host-bacteria relationship require the transfer of a plasmid from the bacteria into the plant genome. Thus, *Agrobacterium* may be a very useful model system to solve the problems that the nitrogen fixation process may impart upon the host plant. This section will address some of the more practical, short term applications.

Rhizobium

The *Rhizobium* spp. have attracted considerable attention as the primary approach to increasing nitrogen fixation potential in crop legumes. The features that make *Rhizobium* symbioses attractive are (1) their symbiotic relationship with major economic/agricultural crops and (2) that the *Rhizobium* and many of the important legumes can be manipulated genetically. Compared to the associative systems, the intimate symbiosis between *Rhizobium* and leguminous plants afford a more definitive approach toward improvement of its efficiency and effectiveness.

During the early stages of plant development a period of nitrogen limitation is endured before the required nodule mass has been synthesized. The high energy demand of nitrogen fixation then imposes a carbon stress upon the metabolic processes of the plant. This and a multitude of other inter-relationships govern the yield potential of the leguminous crops. To increase yields in the short term, the proper *Rhizobium* strain-plant cultivar combination as well as the necessary disease resistance traits should be utilized. The use of the proper *Rhizobium* strain–plant cultivar can markedly increase yields, but the farmer often does not have access to the desired information. Moreover, new cultivars are introduced more rapidly than *Rhizobia* strain-plant profiles can be completed.

Typically, fields in which leguminous crops are planted contain large indigenous populations of rhizobia. For example, the majority of these indigenous *R. japonicum* in midwestern soils consist of the 123 serotype, a group which is not conductive to optimal yields with most soybean cultivars. This serogroup is quite competitive *versus* applied strains and thus forms a high percentage of the bacteroids found within the nodules of plants grown in these soils. Applying strains of rhizobia possessing superior biochemical and genetic traits is futile if they eventually form only a minority of the nodule bacteroid population. Thus, improved strains of rhizobia need to contain the necessary characteristics for competitiveness as well as any beneficial biochemical or genetic characteristics. Thus, strains of rhizobia should be selected for competitiveness before genetic/biochemical alterations are made or, conversely, the genes for competitiveness should be transferred to improved strains of rhizobia. However, the parameters controlling competitiveness are not understood at this time, although lectins are believed to be a major factor.

The problem of competitiveness may be circumvented by sterilization of the soil or by producing rhizobia strains resistant to those agrichemicals deleterious to rhizobia. At present, soil sterilization methods are impractical on a large scale. A short term solution may be 'reagent-selective' inoculants. Legume seeds can be coated with various agents deleterious to the indigenous rhizobia. A desired *Rhizobium* strain which can be made resistant to this agent also can be incorporated Into the seed coating. The nodules resulting from this treatment contain a high proportion of the desired *Rhizobium*. Thus, full expression of the beneficial traits are obtained.

The list of desired traits is endless and with the development of *Rhizobium* genetics the near future promises an abundance of improved strains. Among these traits is the hydrogen recycling mechanism as exemplified by certain strains of *R. japonicum*. During the nitrogenase reaction, a minimum of 1 mol of hydrogen is evolved for each mole of atmospheric nitrogen reduced. Under less favorable conditions. more hydrogen is evolved than nitrogen reduced. In the extreme case, that is in the absence of nitrogen, only hydrogen is evolved. Evans and associates have identified strains that do not evolve hydrogen during the fixation of atmospheric nitrogen. These strains possess an uptake hydrogenase which consumes all of the hydrogen evolved, *via* nitrogenase, in an oxygen-dependent reaction and concomitantly produces ATP. This ATP can then be recycled for the fixation of nitrogen. Use of these strains in field tests has increased the nitrogen content of soybeans by 10–13 per cent. The transfer of this trait to other rhizobia possessing additional beneficial parameters will be possible in the very near future.

At the present time, identification of the biochemical processes that limit symbiotic nitrogen fixation is more difficult than the genetic transfers. The biochemistry and physiology of the symbioses have not been adequately characterized, so that the limitations of the symbiotic fixation process can be defined. During the peak of nitrogen fixation activity, the symbiotic process is thought to be carbon limited. There is considerable physiological information supporting this concept and it has been calculated that if the peak of nitrogen fixation activity could be lengthened by several days the nitrogen content of soybean seed theoretically could be doubled. The interdependence of carbon metabolism and nitrogen metabolism in *Rhizobium*-leguminous plant symbioses suggest that any increase in the availability of carbon compounds to the root nodules, or alternatively more efficient utilization of available carbon compounds within root nodules, will increase the nitrogen fixation potential. As an index of photosynthate availability many workers have reported differences between cultivars in the rate of CO_2 uptake in soybeans. However, the rate of CO_2 uptake is not a reliable indicator of net photosynthate production nor photosynthate transport to root nodules. More systematic methodologies for screening cultivars and cultivar-*Rhizobium* strain combinations are required.

Mutations affecting the metabolism of the citric acid cycle or the uptake of these metabolites significantly alters the nitrogen fixation capacity of the symbiosis. Compared to other carbon compounds, the citric acid cycle intermediates support greater rates of nitrogen fixation in suspensions of anaerobically isolated bacteroids. However, It is not known how the photosynthetic compounds supplied to the root nodule, primarily glucose or sucrose, are metabolised or in which compartment of the symbiotic tissue (infected plant cell *vs.* non-infected plant cell *vs.* bacteroid) the metabolism occurs.

The genetics and biochemistry of the host plant must also be considered and a number of investigators are focusing on this area. For example, in *Pisum sativum*, where two genes have been found that control nodule number, a correlation exists between nodule number and seed yield in peas have described a soybean plant line that does not undergo senescence like normal plants. Hopefully these plants will continue to fix N_2 for prolonged periods and thus lead to increased yields.

Perhaps the greatest application of *Rhizobium* research will be in the tropics and subtropics where the majority of the 20000 species of leguminous plants thrive. These regions coincide with the majority of the world's underdeveloped population. Because of the poor soils in many tropical regions, applications of nitrogen fixation technology have the greatest potential for helping mankind. However, agricultural research on legumes has been conducted primarily in the temperate climates of the more developed countries.

Azospirillum

A large number of diazotrophs have been reported in association with plants, usually found near, on or within the roots. These associations are not nearly as specialized in their symbiotic morphology as the *Rhizobium*, but their biochemistry may be equally complex. These associations have generated considerable excitement, since their wider range of association indicates they may be adaptable to many non-leguminous, agriculturally important plants. In the forseeable future, the expected benefits in terms of plant productivity are considerably less than the *Rhizobium-leguminous* plant symbioses, but they may provide significant crop improvement in soil with poor or moderate nitrogen fertility.

Azospirillum has been the most widely studied associative organism because (1) it can form associations on a large variety of plants often in high numbers (10^7 per g of root) and (2) some reported rates of fixation can significantly affect crop yields. *Azospirillum* has been reported associated with maize, sorghum, sugar cane, rice, millet, oats, rye, barley, forage grasses and the water plant *Spartina alternifora*. Dobereiner and Day have reported rates of fixation of *Azospirillum* on *Digitaria decumbens* as high as 1 kg N ha^{-1} d^{-1}. Although some of the early nitrogen fixation measurements were overestimates due to experimental conditions, more careful work by Okon and others has shown that in Israel, *Azospirillum* association can contribute the majority of the nitrogen required by the plant. In the more temperate regions, particularly the midwestern and northern United States, it appears that *Azospirillum* may contribute little to the nitrogen economy of typical crop plants. However, it may be practical to select cell lines (mutants or genetically improved strains) of *Azospirillum* as inoculum for specific cultivars to form productive associations under a particular set of environmental or ecological conditions.

Cyanobacteria

In the United States, the cyanobacteria are usually associated with the eutrophication of lakes, rivers and streams and thus are treated as a scourge. However, they are treated as a boon in the rice growing regions of the world since the cyanobacteria are beneficial to the crop. The cyanobacteria are nurtured and subsequently used as a green manure to fertilize the rice field after it is drained. The extent of the contribution of cyanobacteria to rice is not yet known, but a typical rice crop will remove about 50 kg N ha^{-1}. Wetland rice can be grown continuously with reasonable yield levels without N-fertilizer additions if adequate cyanobacteria populations are maintained.

Cyanobacteria frequently dominate the phytoplankton population of eutrophic lakes and account for the major source of fixed nitrogen. In non-eutrophic lakes and rivers and in the open ocean, N_2-

fixation rates may be imperceptible. Burris has reported that cyanobacteria attached to rocks in the intertidal zone of the Great Barrier Reef could fix 6.8 to 30.6 kg N ha^{-1} of rock surface $year^{-1}$. Stewart reported rates of fixation of 25 kg N ha^{-1} $year^{-1}$ on the supralittoral fringe of temperate shores. Much higher rates of fixation have been reported in association with coral reefs have isolated a rapid growing blue-green alga from the warm shallow coastal area of Port Aransas, Texas capable of fixing 40 kg N ha^{-1} $year^{-1}$.

The contribution of cyanobacteria to soils has not been estimated. Although contributions to the nitrogen economy may be quite substantial in certain environments, the contribution on a global scale is rather small.

Cyanobacterial Associations

Associations of cyanobacteria with eukaryotic cells are rare. The two primarily studied cyanobacterial associations are *Azolla* and the lichens. *Azolla* spp. are some of the few vascular plants capable of forming an intimate association with a blue-green alga. *Azolla* spp. are fast-growing water ferns that contain the cyanobacteria, *Anabaena azollae,* as an endophyte within their leaf cavities. In the symbiotic state, the *Anabaena* express an elevated number of nitrogen-fixing heterocysts, as high as 20 per cent of the total cell. In southeast Asia, *Azolla* are nurtured simultaneously with rice or as a green manure crop in fallow paddies. When grown simultaneously with rice, some species of *Azolla* serve only as a future green source of nitrogen, but other *Azolla* species may leak or excrete ammonia into the aquatic environment which may then be available to the rice crop immediately. Thus, the nitrogen input of *Azolla* blooms toward a particular rice crop is difficult to determine; when *Azolla* are considered solely as a green manure crop, values as high as 250 kg N ha^{-1} have been estimated, but values between 30 to 100 may be more common. *Azolla* have provided a readily available inexpensive source of fertilizer nitrogen to underdeveloped regions of the world, where the cost and technology of applying commercial fertilizer would be prohibitive.

Lichens have little direct agronomic importance; however, their ability to fix atmospheric nitrogen and occupy habitats not suitable for other plants make them important long term investments of fixed nitrogen. Lichens can colonize poor and/or acidic soils, survive prolonged periods of desiccation and fix N_2 under severe temperature extremes. Lichens are composed of diverse classes of organisms. The fungal partner is usually an ascomycete but may also be a basidiomycete or in the Fungi Imperfecti group. The cyanobacterial partner may have to provide the photosynthetically fixed carbon in addition to the fixed nitrogen although this depends upon the composition of the lichen. The reported ranges for fixation by lichens is 0.2 to 12.0 kg N ha^{-1} $year^{-1}$.

Photosynthetic Bacteria

The photosynthetic bacteria, although ubiquitous, probably do not provide meaningful agronomic levels of fixed nitrogen. The photosynthetic bacteria contribute in anaerobic and microaerophillic environments, especially those rich in H_2S but nutrient-poor and illuminated. They have been reported to contribute fixed nitrogen in rice paddies, salt marshes, estuarine muds and sulfur-spring ditches.

New Associations

The discovery of *Azospirillum* in association with the roots of grasses initiated a renewed interest in associative systems. Although most of this interest has centered around *Azospirillum,* a number of other associative systems have been reported. The best defined new associations are (1) sugar cane–*beijerinckia;* (2) wheat–*Bacillus* spp. and/or *Erwinia herbicola;* (3) rice–*Achromobacter;* (4) *Paspalum notatum–Azotobacter paspali;* and (5) *Azospirilium.* Because of the importance of wheat as a major agronomic crop, the reports of associations with *Bacillus* spp. and *E. herbicola* have received particular interest and scepticism. Klucas and Pedersen have used *K. pneumoniae* or *E. herbicola* as inoculum on winter wheat and grain sorghum. They reported significant differences between plant cultivars and inoculants, pointing out that plant genotype, species of microorganism and environmental conditions all play a major role in determining yield. Depending upon the various treatments, inoculation produced increases of up to 44.4 g dry wt. or decreases of as much as 24.7 g dry wt. Rennie and Larson have shown that inoculation of wheat with *Bacillus* in sterile leonard jars resulted in increased plant nitrogen. As discussed by Jensen a great number of difficulties remain before the associative systems become agronomically feasible. Although the yields of most of these associative systems appear small when judged against yields in typical agricultural soils, their benefits are significant when compared to the poor soils (and crop yields) in many underdeveloped countries.

Industrial Applications

Chemical Catalysts

The Haber-Bosch process, the principal industrial process for ammonia production since 1913, requires pressures of 350–1000 atm, temperatures of around 350°C plus elementary hydrogen for the reduction of N_2. The usual catalyst is composed of iron. Ruthenium and osmium catalysts are available that perform at lower temperature and which push the equilibrium further toward ammonia, but their greater costs discourage their use. The high energy costs of the Haber process means an entirely different procedure must be developed to provide cheaper ammonia. Even the biological process requires large inputs of energy, but unlike the industrial process it occurs at room temperature and atmospheric pressure. Hopefully, new catalysts and procedures will not only provide cheaper sources of ammonia but will require a minimum amount of equipment and capital Investment so that ammonia can be produced where it is needed and the costs of transportation can be reduced.

The three types of chemical reactions being considered as possible alternative procedures are (1) nitriding reactions, (2) reactions with transition metals in aqueous or alcoholic solutions, and (3) formation of coordination compounds with concomitant reduction to hydrazine or ammonia. None of these reactions has as yet yielded a suitable new method. Some of the metal-nitrogen complexes have yielded nearly quantitative yields of hydrazine or ammonia, but have not been catalytic. Those methods that have been capable of turnover or catalysis exhibit rates that are extremely low and not suitably reproducible. A large majority of the research on nitrogen complexes has been done in non-aqueous solutions due to the hydrophobic nature of the metal ligands. In these solutions, production of ammonia has been favored when tungsten or vanadium are used as the coordinating metal rather than molybdenum. The most promising systems utilizing aqueous or alcoholic media have been those based on vanadium. The most effective metal in the nitriding systems has been titanium.

Recently, the emphasis has switched to molybdenum-containing compounds modelled after the FeMo cofactor of dinitrogenase, Shah have demonstrated that isolated FeMoCo is capable of catalytically reducing acetylene to ethylene in the presence of sodium borohydride. Also, Thorneley have demonstrated a hydrazine-like intermediate which they believe is bound end-on to the molybdenum of dinitrogenase. A number of cubane iron-sulfur clusters of the sum formula $[Mo_2Fe_6S_9(SC_2H_5)_8]^{3-}$ or $[Mo_2Fe_6S_9(SC_2H_5)_9]^{3-}$ have been synthesized in which one Mo atom replaces one Fe atom in a corner of a typical Fe_4S_4 cluster. Analysis of these clusters by EXAFS show that they closely resemble native dinitrogenase. A large number of other structures which bear resemblance to the FeMo cofactor of dinitrogenase have also been reported. At present there have been no reports that these compounds function catalytically.

Nitrogenase, in addition to reducing nitrogen to ammonia, can reduce acetylene to ethylene, evolve H_2 from protons and reduce a number of other triple-bonded compounds. Developments toward creating better procedures for producing ammonia could also be applied for other oxidation-reduction reactions. Although the same catalyst may not be appropriate for all the various reactions needed, the variety of new Mo-Fe compounds synthesized indicates there should be no shortage of specialized catalysts. Possible applications include producing methylamine from cyanide, removing nitrogen oxides from emission gases, forming doubly- or triply-bonded carbon and nitrogen reagents, preparing benzene or unsaturated heterocyclic ring systems and catalyzing specific substitutions in unsaturated ring systems. A large number of chemical feedstocks will be prepared with these newly developed catalysts.

Ammonia Production

Ammonia production from continuously cultured organisms or organisms bound to solid supports has been widely discussed, but application has been difficult. The requirements for microbial production of ammonia are:

1. Derepression of the genes for nitrogen fixation.
2. Repression of ammonia assimilation enzymes.
3. Availability of large quantities of energy for nitrogen fixation and cell viability.
4. Export of the ammonia produced by nitrogenase into the surrounding media.
5. Extraction of ammonia from undesirable components in the effluent.

There has been considerable variability and instability of genetic alterations described in points (*i*) and (*ii*) above during continuous culture experiments. The best strains in terms of longevity of ammonia export are those capable of maintaining low rates of protein synthesis.

The energy sources utilized most frequently for culturing nitrogen fixing organisms are expensive reduced carbon sources such as sucrose, glucose, mannitol and/or organic acids. Thus, provision of an inexpensive energy source has been a major limitation. The transfer of cellulase genes into a nitrogen-fixing organism has not yet been reported. but utilization of cellulose or any other cheap energy source (industrial or commercial wastes) is a necessary prerequisite for microbial production of ammonia to be cost-effective.

If anhydrous ammonia is needed, extraction of ammonia from the media would require available technologies to achieve the desired product purity. This step may require considerable expense. Should the desired, stable genetic alteration of a nitrogen-fixing organism be achieved, the extraction procedure may become the major cost limitation. If an effective extraction procedure can be devised with minimal capital investment and furthermore can be automated along with the culture production, then on-site ammonia generation may be available for the average farmer or small industrial consumer.

The overall cost-effectiveness depends upon the rate of ammonia production per culture, the density at which the organisms can be cultured or attached to a matrix and the extent or type of extraction procedure needed. The actual ammonia production per culture or per microbe depends upon the genetic alteration achieved, assuming the rate of ammonia production equated with the doubling time of the faster growing nitrogen-fixing organisms. The rapid advances of gene cloning and gene regulation coupled with the rising costs and dwindling supplies of fossil fuels indicate biological ammonia production may be cost-effective within the next few decades.

Hydrogen Production

Like biological ammonia production, hydrogen production *via* nitrogen-fixing organisms requires derepressed strains capable of utilizing cheap, available energy sources. Unlike ammonia production, cellular export and extraction of H_2 from the media is not a problem. Considerable volumes of H_2 are generated under certain culture conditions. For example, the build-up of hydrogen from *Rhodopseudomonas capsulata* can rupture sealed gas culture tubes. H_2 evolution from *Rhodospirillum rubrum* has been reported at rates of 20 ml h^{-1} g^{-1} dry wt. of culture.

The culturing conditions required for H_2 production (continuous or matrix) and enzyme source are similar to those described above for ammonia production. The same genetically altered diazotroph could probably be utilized for both ammonia production and hydrogen production, since in both cases nitrogenase regulation and cloning are required plus utilization of a cheap energy source. Culturing problems for both products would be very similar. The major technological difference would be product extraction. Since the hydrogen gas will be mixed with whatever gas (or gas mixture) has been used to support cellular metabolism, the extraneous gases will have to be removed. However, the methodologies for purification by liquefaction are available. The extraction requirement would limit small site applications and add the cost of transportation for the consumer. Again the cost-effectiveness depends upon the genetically altered rate of production that can be achieved as well as capital purchases, maintenance and other production and transportation costs.

Biomass Conversion

The nitrogen fixation process has often been linked with biomass conversion. Since nitrogen is quite often limiting in diets of people in the less developed countries, biomass conversion with nitrogen supplementation (via nitrogen fixation) has been a popular theme. Utilization of neutralized pulp mill wastes, spoiled grains and wastes from food processing could serve as energy and carbon sources for nitrogen-fixing organisms. Coculturing of diazotrophs and other microbes has been suggested as a means of providing additional vitamins or minerals not necessarily found in diazotrophs. The resulting biomass could then be processed as food for humans and/or animals. Biomass production rates of blue-green algae in waste water treatment ponds have been reported as high as 100 tons ha^{-1} $year^{-1}$ with the average centering around 50 tons ha^{-1} $year^{-1}$. Major drawbacks have been production costs, nutritional quality, palatability and, when dealing with less developed countries, transportation. The technology required would limit its availability in those areas where these products would be most needed. Alternatively, the biomass produced could be converted into nitrogen fertilizers, either by utilizing the biomass directly or by extraction of the nitrogen-containing components.

Timber Production

Nitrogen is the most common limiting nutrient for timber production as well as for agronomic crops. The contribution of nitrogen fixation to forests can occur *via* lichens–cyanobacteria, leguminous plants–*Rhizobium* symbioses and woody plant–actinomycete symbioses. The most often used biological method of forest fertilization has been the use of actinomycete-nodulated plants, such as *Myrica, Alnus, Purshia* and others. Actinomycetous root nodules have been reported in more than 140 species of plants, most of which are woody shrubs or trees. Many of these trees provide excellent lumber for building materials, furniture, pulp wood, or fuel.

In the Pacific Northwest, red alder has the highest rate of nitrogen fixation (up to 480 kg ha^{-1}) lowed by snowbush (up to 100 kg ha^{-1}), Scotch broom (160 kg ha^{-1}) and lupines (loo kg ha^{-1}). Most reports show mixed stands of alder and Douglas fir increase total timber production, in some cases up to 100 per cent, compared to pure stands of Douglas fir. However, management of these mixed forests can require additional labor and capital. Alder and other native woody nitrogen-fixing plants are considered nuisances since they will out-grow the Douglas fir during the first several years after planting and consequently shade and reduce the growth of the fir. Prohibiting the growth of these nitrogen-fixing plants is difficult, since many are pioneer species and rapidly colonize selective or clean-cut areas. Proper management requires intercropping of alder several years after the fir have been planted to avoid shading. This demands cultivation of the Douglas fir for the first few years, perhaps additions of chemical nitrogen fertilizers and then planting of the alder. The uses of alder as a second lumber product from this intercropping system are pulp, furniture and fuel. With regard to the application of biological nitrogen fixation to forests. DeBell has stated 'In many instances the most limiting factors are not the need for more scientific research or breakthroughs regarding N_2–fixation *per se*, but rather some fundamental biological information as well as appropriate demonstration of benefits.'

Phytochemical Production

Many of the legumes and non-leguminous plants produce valuable biochemicals such as dyes, fibers, flavorings, odors, substitutes for chocolate, citrus, coffee, garlic, licorice, tea, tobacco and vanilla medicines pharmaceuticals, oils for fragrances, cooking and machinery and a host of other chemicals. Cultivation of these plants on a commercial scale would require large initial expenditures. However, since the symbionts for many of these plants are known or are obtainable, biological nitrogen fixation is the logical alternative to chemical at fertilizer to reduce the cost of applying nitrogen. Bio-Alionetics has recently announced that it will use an algal strain to produce commercially 3.8 to 7.6 million 1 of ethanol per year. The algae residue after processing will be sold as fertilizer.

Chapter 10

Mycorrhizae in Agriculture

The evolution of the eukaryotic cell, which made possible the great plethora of living things that now inhabit the earth, is thought to have resulted from the repeated development of mutualistic symbioses between prokaryotes. This ancient tradition of mutualism has been maintained among the eukaryotes, and when green plants first invaded the land more than four hundred million years ago, their ultimate success in taking this giant step may well have been due to their establishment of an intimate and mutually beneficial relationship with fungi.

While at first sight this may seem improbable, on reflection it is not such an unlikely partnership; the early land plants could photosynthesize effectively even before the evolution of leaves, but they undoubtedly had difficulty in accumulating the water and mineral nutrients they needed to sustain this activity. The filamentous fungi, which had themselves only recently emerged from the water, were perfectly adapted for ramifying through the soil and scavenging for those very things, but had an absolute requirement for energy-rich carbon compounds of the kind produced by the plants. We now have some proof of the antiquity of this relationship. Fossils of Devonian plants have been found to contain well-preserved fungal structures virtually indistinguishable from those that can be seen today in the roots of many healthy modern plants. Similar structures occur in the rhizomes of a Carboniferous fern and have in fact been recorded sporadically in the underground parts of Palaeozoic, Mesozoic and Cenozoic plants.

Man first became aware of this long tradition of plant-fungus interaction about a century ago, when several biologists noticed that some plant roots, while extensively invaded by fungi, did not become diseased. The name 'mycorrhiza', literally translated 'fungus root', was coined by Frank in 1885. Although the beneficial nature of the relationship was not established until much later, there is now an overwhelming body of evidence showing that in many situations (particularly in infertile soils) mycorrhizal plants grow better than non-mycorrhizal plants. It has been demonstrated that the hyphae of the fungal symbionts permeate the soil and obtain scarce and relatively immobile elements, especially phosphorus, but also nitrogen, potassium, copper and zinc, more effectively than the root hairs on a non-mycorrhizal plant.

But long before the true nature of the mycorrhizal association was suspected, the constant association of particular fungi with certain trees had been noted. In the third century BC, Theophrastus commented on fungi which grew from the roots of oaks and other trees, and more recently, unsung naturalists coined such suggestive names as *dubovik* or 'oak mushroom' for *Boletus luridus* in Russia, and *Larchenmilchling*, or 'larch milky cap' for *Lactarius porillsis* in Germany. From the early 1800s the French had been encouraging the growth of truffles, which were (and still are) in great demand for *haute cuisine*, by planting oak trees in particular kinds of soil.

Interest in this symbiosis has escalated dramatically in recent years, partly because of what we have learned about the benefits of mycorrhizae, and partly because of economic and geopolitical events.

On the geopolitical and economic fronts, several factors emerged which stimulated interest in mycorrhizae.

The World-wide Energy Crisis

The large-scale manufacture, transportation and application of fertilizers are energy-intensive processes, so drastic increases in the price of oil have made fertilizers more expensive.

The Population Explosion

This has necessitated increased agricultural production, which can be achieved either through the 'green revolution', using new rates of crop plants that require heavy and repeated applications of fertilizer, or by bringing marginal land into cultivation, which is again often assumed, according to conventional wlsdum, to call for extensive fertilization.

The Enormous Requirements of Industrial Societies for Raw Materials

These have resulted in widespread destruction of natural soils and vegetation.

The Vast Quantities of Waste Products Generated by Industrial and Domestic Consumption

These have led to massive pollution of various kinds.

Among other things, these undesirable aspects of unregulated private (and also, sometimes, of government) enterprise eventually engendered the environmental movement, which has successfully lobbied for efforts at restoration of many damaged ecosystems. Each of these phenomena has helped to increase interest in mycorrhizae.

On the academic front, we learned that some plants cannot become established or grow normally without an appropriate mycobiont, and that as compared to non-mycorrhizal plants, those of the same species but with a suitable fungal partner need less fertilizer; withstand heavy metal and acid rain pollution better; grow better on the infertile soils of marginal lands, or on mine spoils and other areas in need of revegetation, and at high elevations; are more resistant to soil-borne diseases; withstand high soil temperatures better; grow better in soils of high salinity; can tolerate greater extremes of soil pH; and survive transplant shock better. We will explore each of these areas in some detail in the appropriate section of the chapter. As these positive features emerged, the rapidly accumulating data permitted other generalizations. It became apparent that mycorrhizal fungi are ubiquitous, and it is now estimated that over 90 per cent of all higher plants are normally mycorrhizal.

The dramatic increase in research effort being devoted to mycorrhizae can be demonstrated by two simple statistics. At the first North American Conference on Mycorrhizae in 1969, 26 communications were presented. At the fifth NACOM in 1981, the number had risen to 263. In 1970, the abstracting service of the Commonwealth Agricultural Bureaux listed six papers on mycorrhizae. By 1979, the CAB listed 314 mycorrhizal references.

There are several different kinds of mycorrhiza, but this chapter will consider only the two most important and widespread forms, which are those constantly found in association with our agricultural and forest crops. These mycorrhizae differ widely both in their structure and in the. systematic position of the fungi involved. By far the commoner of the two is the vesicular-arbuscular endotrophic mycorrhiza (VAM). This kind of symbiosis involves what appears to be a relatively small number of fungi which will grow only in association with plant roots (*i.e.* they are obligate biotrophs) and seem not to reproduce sexually. These attributes made experimental work with V AM difficult, and left the fungi in something of a taxonomic limbo, although they are grouped in a family called the Endogonaceae, which is tentatively placed in the phylum Zygomycota.

Nevertheless, they do form mycorrhizae with hundreds of thousands of plant species, including almost all our field crops, with such exceptions as sugar beet and the brassicas, so investigators persisted. At first spores for experimental work had to be painstakingly sieved and picked out by hand from naturally infected soil. Later a variety of flotation and density-gradient techniques were devised in order to speed up the collection process. Now even more efficient ways of producing inoculum have been found, and are described in the appropriate section of this chapter. We anticipate that these will make the application of VA mycorrhizal fungi in agriculture and revegetation a practical process.

The second kind of mycorrhiza is the ectotrophic mycorrhiza. This kind of symbiosis involves a fairly large number of fungi which, although they are usually found associated with tree roots, can in most cases be grown in pure culture, and are almost all known to produce sexual fructifications *in viva,* if not universally *in vitra.* These fungi are mainly Basidiomycetes, though a few are Ascomycetes: both belong to the phylum Dikaryomycota. Since most of them can be grown in axenic culture, they are much easier to work with, and it is therefore not too surprising that their application in forestry is further advanced than that of the VA mycorrhizae in agriculture.

The mycorrhizal symbiosis, whether ectotrophic or endotrophic, must have three basic functioning components: (1) fungal mycelium exploring large volumes of soil and retrieving mineral nutrients; (2) a fungus-plant interface where the exchange of chemicals occurs: and (3) plant tissues which produce and store carbohydrates. The success of mycorrhizae is probably due to three factors: (1) their long evolutionary history: they predate the root hair and have had a long time to become finely tuned to their environment; (2) their economy: a delicate but extensive fungal mycelium requires less investment than a macroscopic system of fine roots and root hairs (3) their efficiency: the fungi may have surface phosphatases that enable them to obtain soil phosphates more rapidly than non-mycorrhizal roots can. What is abundantly clear is that we are just now on the three hold of important developments in the exploitation of both VA mycorrhizae and ectomycorrhlzae. We hope this chapter will make more scientists aware of that potential.

We note that the synergistic attributes of mycorrhizal fungi and their hosts cannot be pooled by gene cloning and transformation. The properties of mycorrhizal fungi that make them valuable to their hosts are inherent in the morphology of the fungi, especially in their diffuse, tubular, nutrient-absorbing thalli.

Comparison of Ectotrophic and Vesicular-Arbuscular Mycorrhizae

The two kinds of mycorrhizae have essentially the same functions. Acting as an interface between plant and fungus, they are the site of an ongoing exchange. Plant photosynthates are translocated to the fungus in return for phosphorus and other inorganic nutrients obtained by the fungal hyphae from a larger volume of soil than that to which the roots of non-mycorrhizal plants have access. But once we begin to consider other aspects of these phenomena, their numbers, their relationships, their morphology and development, their host ranges, we find little congruence between them.

The physical contrast is exemplified in Figure 10.1, which shows a vertically sectioned root with diagrammatic representations of characteristic endo- and ecto-mycorrhizal structures. Vesicular–arbuscular mycorrhizal (VAM) fungi cause no macroscopic changes in the roots they inhabit, and obviously represent a relatively minor investment on the part of their host plants. Even their fructifications are on a microscopic scale, with the exception of those species that form sporocarps. The ectomycorrhizal (EM) fungi, on the other hand, produce macroscopic changes in the morphology of their host roots, and their visible mantles and their numerous, large and often colourful fruit bodies just as obviously represent a considerable investment on the part of their host plants. This is partly explained by the fact that thc roots which become ectomycorrhizal are usually perennial. Since they

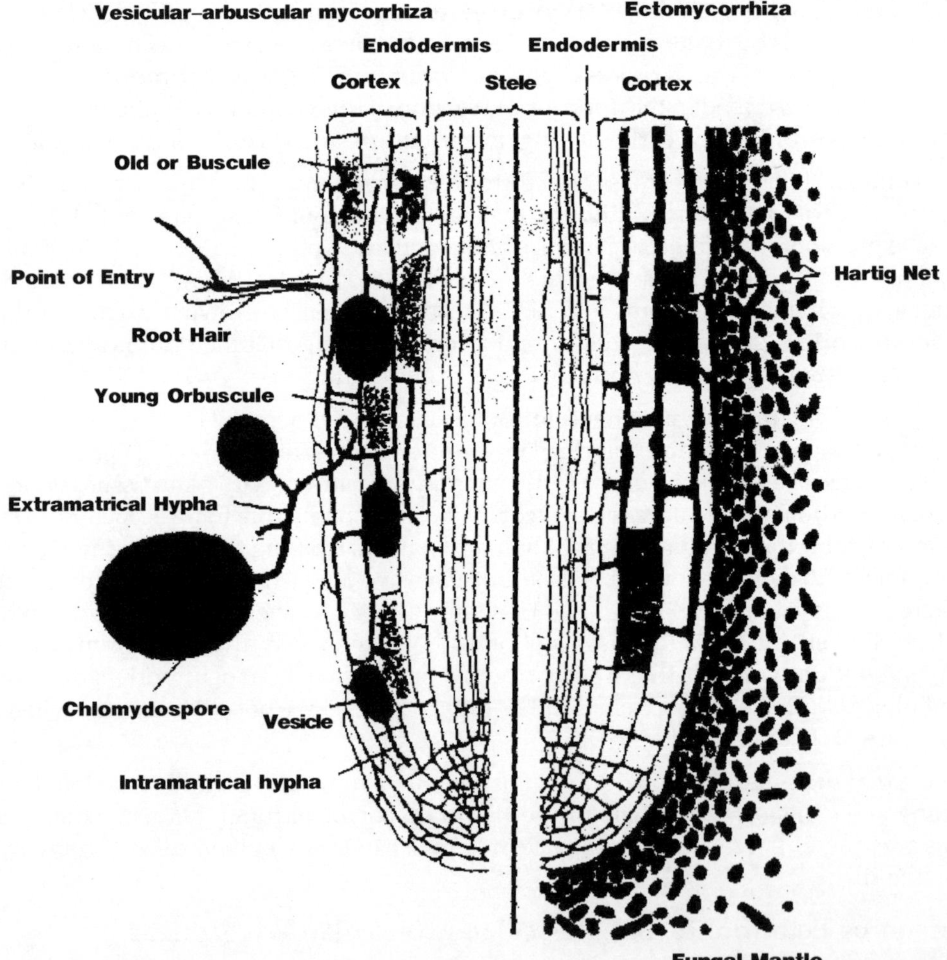

Figure 10.1: Vertical Section of a Root with a Schematic Representation of Vesicular-arbuscular Endotrophic Mycorrhiza (left) and Ectomycorrhiza (right)

are expected to be a long-term, re-usable investment they can involve much more biomass than a short-lived phenomenon like the VAM arbuscule, which is often associated with the roots of annual herbaceous plants. We have evidence that ectotrophic mantles act as nutrient sinks for both fungus and host, but no such function is ascribed to the VA mycorrhiza.

Most EM fungi produce sexual fructifications only during a short season, which is determined by the climate rather than the time of infection, though asexual sclerotia are produced by some species throughout the growing season. VAM fungi may produce spores or sporocarps whenever their host plants are growing, but there tends to be a population peak toward the end of the season. VAM spores may germinate at any time if ambient conditions are suitable and, in some species, if dormancy has been broken, so we must assume that they are also being continuously replaced.

While both kinds of mycorrhiza involve filamentous fungi, the two groups of fungi are taxonomically unrelated. All VAM fungi are still treated as members of a single family having affinity with the phylum Zygomycota, though this is by no means fully established. Most EM fungi are members of a completely different evolutionary line, the phylum Dikaryomycota, mostly hymenomycetous or gasteroid holobasidiomycetes with a few operculate, unitunicate ascomycetes, though a few Zygomycetes form ectomycorrhizae with conifers and eucalypts. Individual hyphae of the VAM fungi penetrate the root cells of their host plant, and the interfaces at which exchange of materials takes place, the finely branched arbuscules, develop within individual cells of the root cortex. This is why VA mycorrhizae are often called endomycorrhizae. By definition, hyphae of EM fungi mass around the feeder roots of their host to form a macroscopic sheath, and grow between and around, but never within, the cells of the root cortex to form the Hartig net, which is the functional interface between host and fungus. The extracellular nature of this interface gives the mycorrhiza the name ectomycorrhizae (though under some conditions. limited intracellular penetration may occur).

The different degrees of intimacy inherent in the two types of mycorrhiza may be reflected in the growth requirements of the two kinds of fungi involved. VAM fungi are obligately biotrophic, and can be grown only in dual culture, with a suitable host plant. Many, probably most, EM fungi can be grown in axenic culture, though a variety of vitamins and other growth substances must often be added to the culture medium.

Trees in boreal forests are usually ectomycorrhizal, while those in tropical rainforests usually have vesicular-arbuscular mycorrhizae. The reasons for this contrast are worth considering. The tree flora in tropical rainforests is extremely diversified. It is common to find 100 different species of trees per hectare and the total number of woody species may approach 400 ha^{-1}. As many as 208 species of trees have been recorded in 1000 m^2, and individual specimens of the same tree species may be 800 m apart. This diversity may not be reflected in the VA mycorrhizal fungi associated with the roots of the trees. However, the VAM fungi of tropical forests are largely unknown, and recent observations suggest that many taxa await discovery and description. New techniques may be necessary to study these, since turnover is high and production may be seasonal. The described world mycota comprises about 100 species at present, and even if, as Trappe predicts, this number doubles by the year 1990 (largely as a result of the renascent interest in these fungi), they will still not even remotely approach the diversity of fungi involved in ectotrophic mycorrhizae.

The tree flora of the boreal forest is extremely restricted. This great band of forest that stretches completely across North America, Europe and Asia is essentially made up of representatives of only six genera; spruce, hemlock, fir, pine, willow and birch. Sometimes, pure stands of a single coniferous species extend for many kilometres. Within this extremely uniform plant community, there are hundreds of different species of ectomycorrhizal fungi. We can emphasize this comparison by giving approximate world totals for the number of species involved on each side of the mycorrhizal equations: VA mycorrhizae, 200 fungi and 300 000 plants; ectomycorrhizae, 5000 fungi and 2000 plants. This seems to show which partner has undergone adaptive radiation to cope with habitat variability. It has been suggested that a tropical rainforest is like a boreal. Forest standing on its head. In the tropical forest the diversity is above ground, in the boreal forest it is below ground. If this is true, it represents an enormous reversal of roles, and is worthy of further study for what it will tell us about the evolution of mycorrhizae and of trees.

Vesicular-arbuscular mycorrhizae are, as we mentioned in the introduction, by far the oldest-established and commonest form of mycorrhiza. They have been detected in some of the oldest land plants, *Rhynia* and *Asteroxylon,* they occur scattered throughout the subsequent fossil record, and are present in the roots of the great majority of living plants. From the sample of plants so far examined, it is estimated that they will be found to occur in about 90 per cent of all extant vascular plants. In fact, the most logical way of discussing their host range is to list the groups of plants that do not normally have VA mycorrhizae.

The only woody family among the entirely non-mycorrhizal plants is the Proteaceae, which is characterized by finely branched roots and extensive roothairs. Many of the rest are weeds, vigorous pioneer herbaceous annuals with highly opportunistic lifestyles. They germinate quickly in disturbed or depauperate soils, grow rapidly, and in some cases flower and set seed in a phenomenally short period. This means that they have no time to sit around and wait for the local VAM fungi to find and colonize their roots–indeed in some of the habitats occupied by weeds there may be little or no VAM inoculum–so the plants have evolved systems of roothairs that enable them to get along without mycobionts. The endomycorrhizal relationship is extremely old, and it is hardly surprising that some plants may now, after hundreds of millions of years, be evolving alternate strategies. That this should happen mainly among the herbaceous annuals is even more understandable when we consider that they are the most recently evolved group of plants. In addition to their weedy habit, members of the Brassicaceae and related families may have evolved chemical defences to repel herbivorous animals, and thereby discouraged their now inessential mycorrhizal fungi. Glenn suggested that these families lack functional mycorrhizae because of the presence of glucosinolates and their hydrolysis products, isothiocyanates, in and around their roots.

This leads to some speculations on the reasons for the emergence of the ectotrophic mycorrhiza. Pirozynski suggests that, as land plants and insects evolved, the effects of leaf-eating insects became more and more severe. This selection pressure favoured those plants which contained unpalatable or toxic substances such as tannins, phenols and resins. These substances, however, if they became disseminated throughout the plant, as they did in the Pinaceae, Myrtaceae, etc., could be inimical to the continued presence of endomycorrhizal fungi. The less intimate association involved in ectomycorrhizae may well have evolved to fill the void left when the VAM fungi were effectively expelled from some of their hosts.

It is well known that EM fungi produce plant growth hormones, notably auxins, in axenic culture. In addition, they produce phenols that have been demonstrated to have a regulatory action on auxins. This suggests that the EM fungus may have a vital role in controlling root development in its host, a function presumably absent from the VAM symbiosis, This may correlate with the obviously modified structure of ectomycorrhizal roots, and the lack of such modification in VA mycorrhizal roots.

Another instructive comparison can be drawn between the distributions of the two kinds of mycorrhiza. Although our knowledge of the agarics is far from complete, we have considerable biogeographical data on them, information that has been supplemented by our attempts to plant conifers in parts of the world to which they were not indigenous. In many cases, these attempts failed initially, simply, we now know, because no appropriate ectomycorrhizal fungi were present. The dependence of the fungi on their tree hosts is so complete that despite their long-range airborne propagules, they were obviously unable to establish themselves on many islands, or in many areas of the southern hemisphere, in advance of their hosts. Thus the only way to establish either partner is to introduce both simultaneously. The VAM fungi suffer under no such disability. Although they are obligately biotrophic, their host range, at least *in vitro*, is so enormous, and their history so long (predating, for example, the breakup of Pangaea) that wherever there is soil capable of supporting plant growth, there are likely to be VAM fungi.

Ectomycorrhizal hosts tend to produce small, dry, wind-dispersed fruits, which are most suited to short-range dispersal. This is in marked contrast to the large, fleshy and appetizing fruits of many tropical endomycorrhizal trees, which stand a good chance of becoming established even if transported over fairly long distances by animal or bird vectors.

Ectoritycorrhizal tree species tend to occur at timberline in both Northern and Southern Hemispheres (*viz.* the *Nothofagus* forests in the Southern Alps of New Zealand) and to grow better than endomycorrhizal plants on soil that is poor or disturbed. The disturbance may be anthropogenic, due to mining, or natural, due to landslides or even glaciation. The implication is that ectomycorrhizal trees have a competitive edge in marginal conditions, and that major climatic shifts in the recent geologic past (especially the ice ages) have been instrumental in the spread of ectomycorrhizal trees, while the endomycorrhizal trees retreated to areas of warmer climate. However, this generalization cannot be applied to the herbaceous plants. In the recolonization of areas devastated by the 1980 eruption of Mount St. Helens, the first mycorrhizal plants were endomycorrhizal species.

Ectomycorrhizae

Systematics of Ectomycorrhizal Fungi and their Hosts

In the preceding section, we pointed out that the species of ectomycorrhizal fungi outnumber those of their hosts. An estimated 5000 fungi can establish ectotrophic mycorrhizae with about 2000 woody hosts. We obviously cannot give any detailed coverage of the very large number of fungi so far known to be ectomycorrhizal. We think it is appropriate, however, to sketch in the broad taxonomic outlines. All ectomycorrhizal fungi, with only one or two exceptions, belong to the mainstream fungal phylum, Dikaryomycota. The great majority are Basidiomycetes. Miller records mycorrhiza-forming activity in representatives of 73 basidiomycete genera distributed among 27 families in nine orders. These orders, families, and the number of mycorrhizal genera known for each family are recorded in Table 10.1. These fungi are holobasidiomycetes: agarics (mushrooms, toadstools); agarics which have evolved, or are evolving, into a gasteroid habit and have in many cases become hypogeous (puffballs, false truffles); and non-gilled hymenomycetes from among the club fungi, chanterelles, tooth fungi and what are called the resupinate bymenomycetes (paint fungi). There are also ectomycorrhizal representatives of 16 unitunicate ascomycete genera from eight families and two orders. All but one of the ascomycete genera involved are hypogeous, and all but one are related to the operculate Discomycetes (cup fungi).

Table 10.1: Taxonomic Distribution of Ectomycorrhizal Fungi

Taxon	Genera
Phylum: Dikaryomycota	
Subphylum: Basidiomycotina	
Order: Agaricales	
Family: Amanitaceae	(2)*
Hygrophoraceae	(1)
Tricholomataceae	(6)
Entolomataceae	(1)
Cortinariaceae	(5)
Paxillaceae	(2)
Gomphidiaceae	(5)
Boletaceae	(13)
Strobilomycetaceae	(3)
Russulales	
Russulaceae	(5)
Elasmomycetaceae	(4)
Gautieriales	
Gautieriaceae	(1)
Hymenogastrales	
Octavianinaceae	(4)
Hymenogastraceae	(1)
Rhizopogonaceae	(2)
Hydnangiaceae	(1)
Phallales	
Hysterangiaceae	(1)
Lycoperdales	
Mesophelliaceae	(1)
Melanogastrales	
Melanogastraceae	(2)
Leucogastraceae	(2)
Sclerodermatales	
Sclerodermataceae	(2)
Astraceae	(1)
Aphyllophorales	
Cantharellaceae	(3)
Clavariaceae	(?)
Corticiaceae	(3)
Hydnaceae	(?)
Thelephoraceae	(2)
Subphylum: Ascomycotina	
Order: Pezizales	
Family: Pezizaceae	(1)
Balsamiaceae	(3)
Geneaceae	(1)
Helvellaceae	(1)
Pyronemataceae	(3)
Terfeziaceae	(4)
Tuberaceae	(2)
Elaphomycetales	
Elaphomycetaceae	(1)

*: Number of confirmed ectomycorrhizal genera.

Some of the evidence that particular fungi are mycorrhizal is rather circumstantial, based on field observation, though enough is now known about the patterns of growth and fruiting of ectomycorrhizal fungi to make these putative connections fairly reliable. For example, although no representatives of the Clavariaceae or Hydnaceae have yet been confiemd as EM fungi by synthesizing mycorrhizae in culture, field observations suggest that this confirmation may not be long delayed.

Many EM fungi belong to cosmopolitan, highly speciated agaric genera such as *Russula, Lactarius, Cortinarius, Amanita, Tricholoma, lnocybe* and *Laccaria,* which, despite much attention from agaricologists, are still incompletely known. Moser estimates that *Cortinarius* has 2000 species. One of the features of the great expansion in our knowledge of EM fungi that has taken place in recent years is the increasing number of ectomycorrhizal hypogeous Basidiomycetes, which are frequently, and perhaps logically, gasteroid relatives of the conspicuous epigeous agarics. Some families are almost entirely ectomycorrhizal–Boletaceae, Gomphidiaceae, Russulaceae, Strobilomycetaceae, Cantharellaceae–as are all or most species of the genera *Amanita, Armillaria, Astraeus, Cortinarius, Hebeloma, lnocybe, Laccaria, Pisolithus, Ramaria, Rozites, Scleroderma, Thelephora* and *Tricholoma,* as well as almost all hypogeous Basidiomycetes.

At the other end of the scale, although a few scattered members of the Aphyllophorales are ectomycorrhizal, the vast majority of the members of this diverse order are vigorous saprobes, with the enzymatic capability to degrade lignin and/or cellulose. In the Ascomycetes, members of the hypogeous Tuberaceae are probably all mycorrhizal. One phylum-wide generalization may be in order: if a fungus produces macroscopic, hypogeous fruit bodies, then it is very likely to be mycorrhizal, whether it is a basidiomycete or an ascomycete.

So many different fungi are involved in ectomycorrhizal partnerships, and ectomycorrhizal hosts are so relatively few, that distributional patterns of these interdependent organisms are bound to be rich in biogeographic significance, and important to would-be mycorrhiza synthesizers. It is unlikely that the conifers, for example, could spontaneously make the leap across wide water barriers to oceanic islands, since they would have to be accompanied by their mycobiont(s), and the evidence is that they have not been able to achieve this. Since many of the EM fungi coevolved with their hosts, they will be found only within the distribution range of the hosts. They may not occur throughout that range, however, because they may have been replaced in some areas by other EM fungi.

The Gomphidiaceae and Rhizopogonaceae were originally restricted to the Northern Hemisphere, with their conifer hosts. Species of the bolete genera, *Suillus, Fuscoboletinus* and *Boletopsis* are associated only with Northern Hemisphere Pinaceae, but other bolete genera are more cosmopolitan. *Mesophellia,* the sole genus of the Mesophelliaceae, and the only known mycorrhizal member of the Lycoperdales, is restricted to Australia, where its phytosymbiont is *Eucalyptus.* The group of Agaricales which evolved mycorrhizae with the Pinaceae, Betulaceae, Fagaceae and Salicaceae in the Northern Hemisphere seem also to have done so with the ancestors of *Nothofagus* before these reached the Southern Hemisphere. Conceivably, the mycobionts associated with *Eucalyptus* and other Myrtaceae in the tropics and the Southern Hemisphere are derived from those which arrived there with *Nothofagus.*

The uniformity of phytobionts in the boreal forest is reflected also in forests of Caesalpinioid legumes in the tropics, and of *Eucalyptus* in Australia. In each case, the diversity lies in the mycobionts. The dominance of ectomycorrhizal trees over vast areas of the Northern Hemisphere was promoted by various episodes of cold or dry climate during the Tertiary or the Cenozoic, since the ectomycorrhizal symbionts are often well adapted to climatic stress and to poor soils.

Some EM fungi have a wide host range. Typical examples of this group are *Amanita muscaria, Boletus edulis, Cantharellus cibarius, Cenococcum geophilum, Laccaria laccata, Pisolithus tinctorius* and *Thelephora terrestris.* These would appear to be likely candidates for exploitation. Others appear to be more selective, and some are virtually host-specific, *e.g., Suillus grevillei* fruits only under *Larix, Suillus lakei* only with *Pseudotsuga menziesii, Lactarius obscuratus* with *Alnus, Gomphidius vinicolor* with *Pinus* and *Cortinarius hemitrichus* with *Betula.*

A single tree may have several or many different ectomycorrhizal partners on its roots at the same time; Trappe gives a range of from five to dozens of different fungi, and these mycobionts may give place to others as the tree ages. Thus, a single tree species may have a very large number of potential ectomycorrhizal partners. Trappe estimated that Douglas fir (*Pseudotsuga menziesii*) might be able to form ectomycorrhizae with as many as 2000 different species of fungi. The potential diversity in these partnerships is even greater than that number would indicate. Different isolates of the same fungus may also behave very differently asmycobionts with the same tree species. Marx reported that of 20 isolates of *Pisolithus tinctorius,* one was much better than all the rest as a mycorrhizal partner of southern pines, while, some isolates did not form mycorrhizae at all with these hosts.

This leads us to a consideration of the full host spectrum of the ectomycorrhizal fungi. Briefly, this can be described as follows: Gymnosperms–Pinaceae and some Cupressaceae; Angiosperms–a few monocots; *Kobresia bellardi, Kobresia myosuroides, Euterpe globosa* and *Festuca rubra*; all or some members of 21 dicot families; all members of the Fagaceae, Betulaceae, Salicaceae, and Dipterocarpaceae subfamily Dipterocarpoideae, and most Myrtaceae; also the tribes Amherstieae and Detarieae of the family Caesalpinioideae, some Mimosoideae and Papilionoideae. *Coccoloba* of the Polygonaceae, *Neea* and *Pisonia* of the Nyctaginaceae, scattered representatives of the Aceraceae, Bignoniaceae, Combretaceae, Euphorbiaceae, Juglandaceae, Rhamnaceae, Rosaceae, Sapindaceae, Sapotaceae, Tiliaceae, Ulmaceae, and even some ferns.

The 2000 plants comprising the above list have several features in common. They are almost all woody and perennial. Many of them grow in extensive pure stands. Many are indigenous to the Northern Hemisphere, and some are the main constituents of the boreal forest. The Pinaceae are the single most important ectomycorrhizal family, since they cover vast areas of the globe, and are harvested and replanted in astronomical numbers each year. It is the hope of enriching that harvest, and improving the success of the replanting, that has kindled our interest in the biotechnological exploitation of ectomycorrhizal fungi.

Morphology and Development of Ectomycorrhizae

In describing the development of vesicular–arbuscular mycorrhizae we are basically concerned only with the activities of the fungus, since there is little visible change in the roots. But the formation of ectomycorrhizae involves morphological responses by both host and fungus, so this process must be described in terms of both partners. The various morphological types of ectomycorrhizae have been described in detail by Chilvers and Zak. Zak gave a number of excellent photographs which should make it easy for anyone to recognize an ectomycorrhiza on sight.

Ectomycorrhizae normally begin to develop 1 to 3 months after the tree seed germinates. Although EM fungi may colonize the long roots, they occur most commonly on the 'short' or 'feeder' roots that develop from the sides of those long roots in the upper, humic layers of the soil. In many cases the fungus is spreading through the soil from a nearby mycorrhizal root and can call upon ample reserves of energy derived from its host. If, however, isolated inoculum of spores or mycelium is involved, it can apparently subsist in the rhizosphere (the area around the roots) on root exudates before actual infection occurs. Mycorrhization occurs when an EM fungus encounters a feeder root of a hospitable tree species. Colonization will occur only in a specific unsuberized zone of the root, behind the root tip, and before the region in which the primary cortex has begun to deteriorate. Although the root tip is not invaded, it eventually becomes completely covered by the fungal mantle.

Fortin using the root pouch technique, observed that mycelial inoculum of *Pisoilthus tinctorius* initiated the mycorrhization of short root primordia of *Pinus strobus* within 5 days. This involved: (1) the penetration of hyphae between the cells of the root cortex and the formation of the characteristic Hartig net, in which hyphae completely surround cortical cells (Figure 10.1); and (2) the establishment of a mantle of hyphae around the outside of the root. A *Cenococcum geophilum* isolate took 2 weeks to reach a comparable stage. The growth rate of the short roots was increased, and they eventually branched dichotomously. We should point out that changes in growth rate and pattern are typical of ectomycorrhizae, and are caused by the liberation of growth hormones from the fungus. These changes will occur if auxins are applied in the absence of an EM fungus.

Marks described two patterns of mycorrhization. In primary colonization, typical of an uncolonized root in colonized soil, the mantle (a mass of fungal hyphae surrounding the root) begins to form at about the time the first leaves appear. When the xylem vessels begin to differentiate in the vascular cylinder at the centre of the root, hyphae penetrate between the cells of the cortex and form a Hartig net (a single layer of closely packed fungal hyphae). In most, cases, this net spreads slowly inward until it reaches the endodermis, which effectively bars any further penetration, though in some angiosperms the Hartig net may not penetrate beyond the, first layer of cortical cells. In secondary colonization, typically derived from mycelia already established on other parts of the same root system, the mantle often simply spreads along a long root and rapidly, envelops any emerging short roots, sending hyphae into their cortex to set up the vital Hartig net interface.

The penetration of hyphae between the cells of the root cortex does not lead to plasmolysis or other deleterious cytological alterations in the host cells. As the hyphae insinuate themselves between the cortical cells, the latter simply separate at the middle lamella, and an almost complete single layer of fungal hyphae eventually separates and virtually encapsulates each cell (Figure 10.1), though plasmodesmatal connections may remain between cortical cells. Far from being deleterious, the presence of this fungal net actually prolongs the life of the cortical cells and of the root as a whole. The fungal mantle, which develops concomitantly around the outside of the root, varies from a relatively loose weft of hyphae to a thick, dense, pseudoparenchymatous layer which may account for nearly half the biomass of the mycorrhiza. Mycelial strands or hyphae often extend from the mantle into the surrounding soil, while the formation of roothairs by the plant is suppressed.

It has been shown that sugars are translocated from the root *via* the Hartig net to the fungal mantle, where they tend to accumulate. As sugars pass from plant to fungus, they are converted into trehalose (a disaccharide), mannitol (a polyhydric alcohol) and glycogen, all three being typical fungal carbohydrates. The glycogen is insoluble, and therefore unavailable for possible reabsorption by the plant. More surprisingly, although the mannitol and trehalose remain in solution in the fungus, the plant is incapable of reabsorbing them. Thus, the fungal mantle acts as a sink where reserves of carbohydrates derived from the phytobiont are stored. This fact is emphasized when, as autumn approaches, many of the fungi mobilize the stored carbohydrates and produce flushes of their large, fleshy basidiomata near the tree. If we add up the various parts of the fungus, the conspicuous fruit bodies, the extensive but usually inconspicuous mycelium ramifying through the soil, and the rootlet mantles, we can calculate that the tree often invests at least 10 per cent of its total production of photosynthates in its mycorrhizal fungus. This investment is clearly more than compensated for by the increased efficiency of mineral absorption provided by the EM fungus.

Visual inspection will show that, compared to non-mycorrhizal roots, mycorrhizal roots are: (1) a different colour, because of the enveloping fungal mantle; (2) thicker, because of the presence of the mantle and because their cells are larger and less liable to collapse; and (3) branched much more often, pinnately and racemosely in *Abies, Fagus* and *Eucalyptus,* and dichotomously in *Pinus.* Thus, the ectomycorrhiza is characterized by its form, colour and texture; by the presence of a sheathing fungal mantle (sometimes not very well developed); and by the presence in the outer layer(s) of the root cortex of a Hartig net, however restricted in development, which is the truly diagnostic structure. Although several different morphologies have been described among ectomycorrhizae, and the black monopodial mycorrhizae of *Cenococcum geophilum* are unique, it is widely thought impossible to determine the identity of the mycobiont unless this could be persuaded to produce fruit bodies (basidiomata or ascomata). Zak (1973) gives a photograph of a basidioma of *Hysterangium separabile* attached by a mycelial strand to a mycorrhiza of *Pseudotsuga menziesii.* This kind of evidence is unusual, but has been used to establish the mycorrhizal nature of a number of fungi. Today, there is renewed interest in the taxonomic value of mycorrhizal morphology.

Ectomycorrhizae remain active for periods ranging from several months to 3 years. Roots and mantles often extend at the same rate, but root extension sometimes outpaces mantle extension, and the root breaks through and extends beyond the mantle. The root may then be colonized by other EM fungi.

Sources of Ectomycorrhizal Inoculum

Ectomycorrhizae may be initiated by several different kinds of inoculum, which can be categorized as: (1) natural inoculum in the form of airborne spores (usually basidiospores, but in some cases also conidia); (2) soil already colonized by an EM fungus or fungi; (3) seedlings already colonized by an EM fungus or fungi, that is, bearing mycorrhizal roots; (4) fungal sporomata, spores or sclerotia specifically collected for the purpose; and (5) fungal mycelium produced in axenic culture. It is worth comparing the merits of these different kinds of inoculum.

Natural Airborne Spore Inoculum

This is, of course, one of the prime dispersal mechanisms for these fungi in nature, but there are a variety of reasons why it is basically unsuited to forestry applications.

1. It is available only during a relatively short period of the year, since most agarics fruit in late summer or early fall.
2. Many ectomycorrhizal fungi are hypogeous, and their spores may not be aerially transmitted in significant numbers.
3. Even when spore inoculum is being produced, its availability in a specific area where tree seedlings are being produced may be sporadic or quantitatively inadequate, especially if the nursery is a long way from the nearest stand of ectomycorrhizal trees.
4. We have no effective control over the nature of the fungal panners being introduced.
5. If the seedlings are being started at a low elevation for high elevation outplanting, they may acquire local mycobionts unsuited to conditions at the ultimate growth site. Such erratic, patchy and possibly unsuitable inoculation is basically unacceptable in modern forestry practice. Since spontaneous inoculation, even where it is a possibility, seems to be so unreliable, we will consider all available alternatives, dealing with them in increasing order of complexity and cost.

Soil Already Colonized by an EM Fungus or Fungi

This has been fairly widely employed, especially when attempts were being made to establish conifers in new areas. But it has a number of drawbacks.

1. Soil inoculum is very bulky and heavy, and although it can be used in small-scale operations, it is unsuited for any large-scale afforestation projects.
2. Soil is a complex system, usually containing a large number of living organisms, some of which may be pests or pathogens. It would be most unfonunate to introduce any of these to anew area.

3. As in the case of spontaneous inoculation, the nature of the mycobiont(s) in the soil will often be uncertain or unknown. The desirability of using precisely identified fungal panners is clearly established in next paragraphs.

The Introduction of Seedlings with Established Mycorrhizae

Although this implies that the mycobiont(s) being introduced are compatible with the desired tree species, this method shares several disadvantages with the previous one.

1. Pests or pathogens may be introduced with the mycorrhizal seedlings.
2. The nature of the fungal symbiont(s) will often be uncertain.
3. The spread of infection within a nursery or plantation may be slow and possibly uneven, unless inordinately large numbers of infected seedlings can be used.

The Deliberate Introduction of Spores, Sporocarps or Sclerotia

This would seem to be an obvious way of improving upon nature, and there is no doubt that it has potential, especially since the identity of the desired mycobiont could be established at the outset. But several difficulties stand in the way of its widespread adoption.

1. Since the only major source of this kind of inoculum is currently the naturally occurring fruit bodies of the fungi concerned, it must be apparent that the availability of most of these structures is seasonal.
2. This also means that in most cases the quantities of inoculum available will be limited, and that they will inevitably fluctuate from year to year as the fruiting of the fungi is affected by the climate.
3. Since the basidiomata of most agarics, although they may be locally abundant, are usually sporadic in occurrence and scattered over large areas, the collection of the quantities that would be required for large-scale forestry applications would be extremely labour-intensive (but see the case of *Pisolithus tinctorius*, below).

   Also, in most instances, a certain level of taxonomic expertise would be required of the collectors, since they would have to be able to discriminate between desirable and undesirable species.
4. Even if an appropriate quantity of basidiomata of a suitable fungus could be collected, the question of how these could be stored would still remain. Although it is conceivable that in some cases the inoculum would be applied immediately after collection, this is likely to be the exception rather than the rule, since most fumigation and seeding are carried out in spring. The basidiomata of agarics are a notoriously perishable commodity, and if they are not immediately lyophilized, dried, frozen, or treated in some other way to preserve the viability of the inoculum, the natural processes of decay will frustrate the entire operation. At best, the long-term storage of such inoculum is problematical.
5. Initiation of mycorrhizae by basidiospore inoculum takes 3 to 4 weeks longer than when mycelial inoculum is used (see below). This delay gives pathogens a greater chance to attack the roots, and the later-developing mycorrhizae also provide less growth stimulation during the crucial early stages of growth.

There is another fungal reproductive structure that is much less perishable than either spores or basidiomata, since it has obviously evolved as a long-term survival mechanism. We refer, of course, to the sclerotium. Unfortunately, most agarics do not produce these structures, but one widespread mycobiont is characterized by them. The sclerotia of *Cenococcum geophilum* occur naturally in the soil in huge numbers, making up a not inconsiderable biomass that might be harvested and used as inoculum. Fogel estimated that the A_0 and A_1 horizons of soil, *i.e.* the top 5 cm. in a 35 to 50 year old Douglas fir stand (*Pseudosuga menziesii*) contained 2785 kg ha^{-1} dry wt. of *Cenococcum geophilum* sclerotia.

Mycelial Inoculum Derived from Pure Cultures of known Mycobionts

Here the identity of the chosen fungus will be known, pests and pathogens will be absent, inoculum will be relatively compact, and it should be available on a year-round basis. However, it too has some inherent problems, since it must be easily the most expensive of the alternatives.

1. Some ectomycorrhizal fungi are difficult to isolate in pure culture, the process also calling for highly trained personnel.
2. Cultures are expensive to maintain, and they tend to grow relatively slowly, taking a long time to produce the quantity of biomass required for large-scale applications.
3. We still do not know how well such inoculum survives in the soil in face of predation and competition from indigenous organisms.
4. We have not yet defined the best possible fungus-host combinations for many soil-climate combinations. It is hardly worth going to the expense of mass-producing mycelium of single species until we are sure that the results will be economically worthwhile.

Pure cultures of ectomycorrhizal fungi can be derived from fruit body tissue, surface-sterilized mycorrhizal roots and sclerotia. Isolation from rhizomorphs or mycelial strands is also possible. Isolation of EM fungi from basidiospores is difficult and is rarely attempted.

Isolation from fruit bodies is usual, because this allows precise identification of the fungus at the outset. Members of some genera are often fairly easy to isolate. Among these are *Amanita, Astraeus. Boletus, Cortinarius. Fuscoboletinus, Hebeloma, Hymenogaster, Hysterangium, Laccaria, Lactarius, Leccinum, Melanogaster, Paxillus, Pisolithus, Rhizopogon, Scleroderma, Suillus* and *Tricholoma*. Fortunately, these include some of the better mycorrhizal partners with the broadest host ranges, *e.g., Pisolithus*. Other genera are more recalcitrant; only a few species of *Russula*, and none of *Gomphidius*, have been cultured. We believe, that most EM fungi can ultimately be grown in axenic culture when the rather stringent nutritional requirements they have developed as a result of their more or less obligately biotrophic lifestyle have been worked out.

Cultures should be made from young basidiomata or ascomata, though mature specimens should also be collected in order to facilitate identification and the preparation of voucher specimens. When collected, fruit bodies should be kept in waxed or brown paper bags until isolation can be attempted, since storage in plastic bags often causes rapid loss of viability. Extensive field notes, photographs and spore prints should be taken and the advice or assistance of an experienced fungal taxonomist may well be invaluable, since the accurate identification called for by mycorrhizal research is no easy matter for the uninitiated. If there is any uncertainty about the name to be applied to the organism, voucher specimens should be sent to a competent taxonomist, preferably one who has published recently on the group in question. If the fungus is to be used in research, voucher specimens should be deposited in an established, internationally recognized herbarium where they will be available for study by other researchers. Such herbaria are listed in the Index Herbariorum, and the accession number and the code letters of the Herbarium (*e.g.* DAOM 129643) should be cited in all publications concerning the fungus. If this is not done, the culture is in many ways an orphan, since essential information concerning its identity, and some of the value of subsequent research, will have been permanently lost.

Isolations should be carried out in still, and preferably sterile, air. A laminar-flow bench is ideal, but any draught-free place may be adequate, and isolations can even be made in the field, if necessary, in a portable isolation chamber. The work area should be sterilized with chlorine bleach; we have found that this is much more effective than alcohol. The isolation is made with flame-sterilized scalpels. Shallow slits are made in the surface tissues of the fruit body, then these are opened up to expose the inner tissues by gently bending the tissue on each side away from the slit. Now, small cubes of the freshly exposed tissue can be cut out with another flamed scalpel and transferred to modified Melin-Norkrans nutrient agar. Tubes containing agar ('slants') are usually much less susceptible to random atmospheric contamination than Petri plates, especially if the neck is flamed before and after the tissue is introduced. Ten to twenty isolations should be attempted from several different locations in the fruit body. Molina and Palmer give many useful hints for dealing with particular genera, but experience is the best teacher.

The isolates can be incubated at room temperature for 3 to 4 days, then examined under a dissecting microscope for signs of mycelial growth. Note that some fungi are slow starters and may require up to 6 weeks to establish themselves. Fast-growing fungi should be transferred after 2 to 4 weeks; slow-growing species not for up to 4 months. Really slow-growing isolates will probably not be of much use, and may well fail to survive subculturing. Contaminating bacteria and molds (usually hyphomycetes) will be encountered, but with a little experience these will be easily recognized and rejected. The ultimate test of any cultures obtained is their ability to establish ectomycorrhizae with aseptically grown tree seedlings.

Cultures may also be derived from freshly collected ectomycorrhizae, if basidiomata or ascomata are not available: The mycorrhizal roots are washed in tapwater, shaken vigorously in a dilute detergent solution (*e.g.* Tween 20), washed again, soaked in a surface-sterilizing agent (*e.g.* 100 mg l^{-1} mercuric chloride for 4 min, or 30 per cent hydrogen peroxide for 5–20 s), rinsed with sterile water, and finally transferred aseptically to nutrient agar.

The sclerotia of *Cenococcum geophilum,* a possible anamorph of *Elaphomyces*, are present in very large numbers in the soil under many conifers. They will be readily extracted from the soil and used to initiate cultures. Large, clean sclerotia are washed, surface sterilized with 30 per cent hydrogen peroxide, rinsed in sterile water, then aseptically transferred to nutrient agar. The wide, black hyphae of this fungus are unmistakeable.

Stock cultures of EM fungi should be maintained at 3 to 4 °C and transferred three or four times a year. Marx found that many cultures of EM fungi lost their ability to induce mycorrhizae after prolonged culturing, though some could be successfully stored as plugs cut from growing cultures and kept in cold sterile water in darkness. We must emphasize that not all culturable EM fungi will behave in the same way, and that some experimentation may be necessary to obtain the best response from many isolates. Carefully kept records will prevent loss of hard-won information.

Melin pioneered thc techniques which permitted the synthesis of ectomycorrhizae *in vitro,* and thus established the true identities of many EM fungi. As a result of his work, we can now use field observations to predict many functional host-fungus combinations. Mycorrhizae synthesized in this way have also permitted investigations of fungal and host physiology, and interactions, particularly as these concern the uptake and translocation of water and nutnents by the fungus, the movement of photosynthate from plant to fungus, the protection from pests and pathogens afforded the plant by the fungus, temperature effects, and the various degrees of host-fungus compatibility. Molina give details of the various steps necessary to establish aseptic tree seedlings and to inoculate them with EM fungi.

We must emphasize that before the considerable expense of producing large quantities of mycelial inoculum *in vitro* is undertaken, as many as possible of the factors just listed should be investigated, so that the final choice from among so many potential mycobionts may be made intelligently. In fact, since we know that the average ectomycorrhizal tree has several to many different symbiotic fungi on its roots at any one time, and may also change partners over the years, there is almost certainly no single ideal mycobiont.

Nevertheless, it is also certain that inoculation of young conifer seedlings with an appropriate fungal partner will greatly increase their chances of survival during the first year or so of life. The subsequent replacement of that partner by others may be regarded as part of the normal course of events.

Therefore, attempts are being made to mass-produce and market mycelial inoculum of several broad-spectrum EM fungi, most notably *Pisolithus tinctorius,* but also several others, including *Thelophora terrestris* and *Cenococcum geophilum.* These fungi not only meet most mycorrhiza-forming criteria, but are also easy to isolate, in most cases grow relatively quickly, and can withstand the various manipulations involved in preparation, storage, transportation and application of inoculum. *Pisolithus tinctorius* can grow at 42°C, and also at 7°C, can tolerate a pH range of 2.6–8.4, and can survive prolonged freezing. It also forms abundant mycelial strands. *Cenococcum geophilum* is extremely drought tolerant and will form ectomycorrhizae from pH 3.4 to 7.5.

It is easy to produce enough mycelial inoculum for small-scale research projects, but much more difficult to generate enough to inoculate the many millions of seedlings routinely produced each year, while preventing losses due to microbial contamination.

Moser grew *Suillus plorans* in liquid medium in small flasks, then added this starter inoculum to 10 l tanks of the same medium. These tanks were aerated for 2–3 h daily over 3–4 months. The mycelium thus generated was used to inoculate 5 l flasks containing sterile peatmoss soaked with liquid medium. Over the next few months the mycelium ramified throughout this substrate. This inoculum was then packaged in sterile plastic bags, taken to the nursery, and used within 3 days. As a result of the various manipulations involved, inoculum often became contaminated by common molds and bacteria. Moser went to all this trouble because he found that neither mycelium growing on solid agar media, nor a mycelial suspension in the liquid medium, made effective inocula. Moser used the same technique to produce inoculum of seven other EM fungi from the genera *Suillus* (3), *Paxillus* (1), *Phlegmacium* (1), *Amanita* (1) and *Lactarius* (1).

Gobi grew mycelia in sterilized cereal grains (wheat, millet), often with an addition of calcium sulfale, in 11 bottles. Shaken weekly, these substrates became thoroughly permeated by mycelium after about 4 weeks, and could be stored at 4°C for up to 9 months. The actual inoculum was produced by adding the colonized grain to peatmoss variously amended with inorganic nutrients plus ammonium tartrate, asparagine, soybean meal, blood meal, malt extract and glucose, depending on the requirements of the species at hand. One or more bottles of the grain were added to plastic bags containing 10-151 of peatmoss, the bags were plugged with cotton and periodically shaken over 3–6 weeks.

Takacs used a similar approach to produce inoculum of seven EM fungi for the establishment of conifers on formerly treeless land in Argentina, which had no indigenous EM fungi. Inocula of four different fungi were mixed with soil or litter at the nursery site, and incubated for a few weeks in small heaps. In this way, twenty 200 ml starter cultures were used to generate 100–200 kg of 'inoculum', though in the absence of ectomycorrhizal host plants and special nutrients, it is unlikely that the starter cultures can have done any more than survive, in a considerably diluted form.

Marx tried various methods of producing mycelial inoculum of *Pisolithus tinctorius, Thelephora terrestris* and *Cenococcum geophilum.* Functional inoculum could not be produced on sterilized grains, but vermiculite plus peatmoss moistened with modified Melin-Norkrans, (MMN) nutrient solution gave good results. The MMN solution contained; 0.05 g $CaCl_2$, 0.025 g NaCl, 0.5 g KH_2PO_4, 0.25 g $(NH_4)_2PO_4$, 0.15g $MgSO_4.7H_2O$, 1.2 ml 1 per cent $FeCl_3$, 100 µg thiamine HCl, 3 g malt extract, 10 g glucose and distilled water to make 1 l.

Peatmoss-vermiculite in a ratio of 28 : 1 is moistened with half its volume of MMN solution (*e.g.*, 1400 ml vermiculite, 50 ml peatmoss, 750 ml MMN solution). The initial inoculum can be in the form of plugs cut from an agar culture, or blended mycelium grown in liquid culture. Using blended mycelium mixed throughout the substrate, *Thelephora terrestris* and *Pisolithus tinctorius* will thoroughly colonize the substrate in 1–2 months at room temperature. *Cenococcum geophilum* may take 4–5 months to achieve the same result. If this inoculum is mixed directly with fumigated nursery soil, it will soon be overgrown by saprophytic microorganisms which exploit the unused nutrients. This problem can be reduced or eliminated by leaching the inoculum in tap water for 2–3 min to remove those nutrients. This leaves a bulky, sticky paste with 90 per cent water. Much of the water is normally removed by drying the inoculum on wooden frames at 20–26°C and 35–45 per cent relative humidity for about 4 days, mixing every few hours to reduce excessive surface drying. The final inoculum weighs about 400 g and its water content is 20–65 per cent. Inoculum of *Pisolithus tinctorius* can be stored at 5°C for 9 weeks without much loss of activity, but the sooner it is used, the better.

The USDA and Abbott Laboratories have developed a commercial formulation of *Pisolithus tinctorius* mycelial inoculum, called 'MycoRhiz'. This is grown in the vermiculite-peat-moss-MMN medium. The starter mycelium is grown in MMN solution in large fermenters, with continuous agitation and aeration, at 28–32°C for 7–14 days. The vermiculite-peatmoss-MMN substrate is steam-sterilized in deep-tank or drum fermenters, then inoculated with the starter culture at a rate of 5–20 per cent of fermenter volume. The medium is mixed, and then incubated with or without agitation for 1–4 weeks. The final inoculum is harvested by flotation on water, then dried in an Aeromatic fluid bed dryer to a moisture content of 20–25 per cent. 'MycoRhiz' weighs about 250–300 g l^{-1}, is packaged in 50 I units, and is stored at 5°C, when it has a shelf life of 5–6 weeks. A tractor-drawn combined seeder-inoculater has recently been developed.

However 'MycoRhiz' was withdrawn from the market in September 1983, due to quality control problems. Abbott Laboratories anticipated that it would return in 1984; but at the time of publication this had not occurred. The Butler Mushroom Farm in Pennsylvania is at present willing to produce inoculum for any viable EM under contract. Mycelial inoculum of *Suillus, Thelephora* and *Laccarla* is being field tested in Oregon. Sylvan Spawn Laboratories, PA are now producing mycelial inoculum of *Pisolithus tinctorius* in breathable plastic bags. One factor in the acceptance of inoculum such as 'MycoRhiz' is the high total cost of application at around \$15 m^{-2}.

Evaluation and Selection of Ectomycorrhizal Fungi

Thc thousands of different ectomycorrhizal fungi are probably necessary evolutionary responses to the diverse needs of many hosts in a multiplicity of habitats. Trappe addressed the question of how many host-fungus-soil-climate combinations we may expect to find. He reported that a 250 km transect running east from the coast of Oregon or Washington, USA, can pass through 17 major forest zones, hundreds of habitat types and at least 10 genera of ectomycorrhizal hosts that are grown in nurseries for forestation, erosion control or as ornamentals.

How, then, are we to select the best possible mycobionts for our trees? The first step is to define the range of fungal symbionts available for the chosen tree species. Even this may be difficult, since some trees are compatible with a wide spectrum of mycobionts, and the success of the symbiosis often varies with the provenance or strain of either partner. It will often be found, however, that one or two of the characteristics discussed below will override the rest in importance. For example, if a fungus cannot be grown in pure culture for the large-scale production of inoculum, it will be effectively excluded from consideration for biotechnological applications.

All potential host-fungus pairs should ideally be tested for all of the following characteristics: (1) rapidity and extent of mycorrhization; (2) host response; (3) inorganic nutrient uptake; (4) water relations (keeping in mind the conditions under which the pair must operate after outplantling); (5) tolerance of extremes of temperature (*cf.* field conditions); (6) tolerance of extremes of pH (*cf.* field conditions); (7) tolerance of natural or anthropogenic soil toxicity; (8) stability of the partnership (one expression of the competitive ability of the fungus); (9) disease resistance (this might be tested only for diseases known to exist in the outplanting site); (10) mycelial strand formation by the fungus; (11) ease of isolating the fungus in pure culture; (12) ease and rapidity with which large quantities of inoculum can be produced; and (13) edibility of the fruit bodies of the fungus.

The potential range of mycobionts for a given tree in a given area may initially be checked by field observations of basidiomata associated with that tree, combined with estimations of the degree of mycorrhizal infection on the roots, though this should not rule out the possibility of introducing new and efficient EM partners to the area. Field observations showed Moser that *Suillus plorans* was the predominant EM fungus associated with *Pinus cembra* at treeline in the Alps. Other obvious pairs are listed in previous section, but the choice is not usually so simple. The process of testing host-fungus pairs involves the isolation of the fungi in pure culture, and their inoculation onto seedlings grown individually and aseptically in tubes (Hacskaylo, 1953) or in root pouches, where the development of mycorrhizae can be visually checked. Molina inoculated containerized conifer seedlings with 15 potential mycobionts. Only two of the fifteen fungi formed abundant mycorrhizae with all the conifers tested.

It is important to design experiments with the inherent variability of the material in mind. In the host plant, much less variability may be expected in material derived from cuttings than from seed (though conifers are not usually propagated by cuttings), while the geographic origin of the host can be important. In the mycobiont, geographic origin, original host and age of the culture may all introduce variation, Marx suggested that before the mycorrhizal potential of any fungal species can be properly assessed, several isolates representing the range of variation expected should be tested. Appropriate conditions for experiments with EM fungi and their hosts are discussed in detail by Reid. We will discuss our 13 criteria in sequence.

Rapidity and Extent of Mycorrhization

Knowledge of the rate at which, and the degree to which, a tree's root system is colonized by EM fungi is an essential part of any comprehensive understanding of ectomycorrhizae. Ectomycorrhizae are distinguishable by the naked eye, and therefore require no special treatment before they can be quantified in various ways. Entire root systems of seedlings can be examined, but in older plants only a representative sample obtained by soil coring or local excavation can be studied. Soil cores reported in the literature vary in diameter from 1.2–10 cm, and in depth from 7,5 to 90 cm, with 30 cm as the norm. The number of samples required must be determined statistically for each study. Soil cores are soaked for up to 36 h before being agitated in water to clean the roots of soil particles. The percentage of mycorrhizal short roots can be determined visually. Grand discuss other assessment techniques, such as direct counts of entire root systems, of selected roots and of ectomycorrhizal tips, in tandem with determinations of the weights of these structures. Results may be expressed as number and weight of ectomycorrhizal structures per unit area, or per unit volume of soil.

If the amount of mycorrhizal material in core samples is very large, subsampling may be necessary. Marks used a 10 inch-square plexiglass tray divided into 1 in squares and 0.1 in subdivisions. Sieved material from cores was spread out evenly over the tray in 200 ml water, and the ectomycorrhizae and tips counted in seven randomly selected squares under a dissecting microscope.

Host Response

The response of the host seedling or tree to mycorrhizal colonization can be measured in various ways. The non-destructive, and therefore easiest, method is to determine seedling survival, expressed as percentages of the initial uninoculated and inoculated populations. Such data can be gathered at various ages, before and after outplanting. Other non-destructive measures are height of plant, thickness of stalk at ground level, number of leaves, leaf length and leaf area. More definitive measurements involve determining the dry weight of the whole plant, or of separate root and shoot systems. Roots, of course, will often have a significant fungal component, but this is usually acknowledged or assessed without measurement, though it can be determined by a glucosamine assay. Details of these parameters are given by Sinclair and Marx. Measurements of stem height and stem diameter at soil line are eventually replaced by diameter at breast height (1.4 m) in older trees.

Sinclair propose a 'mycorrhizal influence value' (MIV), which can be determined for any of the parameters mentioned above by expressing the mean value for non-inoculated plants as 100, and calculating the value for mycorrhizal plants as an integer relative to that 100. Thus the MIV would be a percentage of the control value in each case.

Inorganic Nutrient Uptake

Phosphorus uptake, and levels achieved in the mantle and in the plant (determined using radio-tracer techniques by Mejstfik are among the most important reflections of the effects of EM fungi on their hosts. Ectomycorrhizal plants also absorb many other minerals, *e.g.* calcium, potassium, copper, molybdenum, magnesium and zinc, from the soil much more efficiently than non-mycorrhizal plants. The fungal mantle can store inorganic nutrients, *e.g.*, chloride, ammonium and especially phosphate and release them to the plant during periods of deficiency or active growth. *Pisolithus tinctorius* thrives in soils of extremely low fertility, such as mine spoils, while *Paxillus involutus* does well only on sites with relatively abundant available nitrogen. But since it is in the uptake of phosphorus, often a limiting nutrient in poor soils, that EM fungi make their greatest contribution to the symbiosis, evaluation of rate and amount of phosphorus accumulation must be one of the most important criteria in selection.

Water Relations

Theodorou found that isolates of some mycorrhizil fungi were better able than others to survive drought while still contributing directly to the welfare of the plant partner. To assess the response of isolates of *Cenococcum geophilum, Suillus luteus* and *Thelephora terrestris* to induced water stress. Mexal grew them in artificial media with the water potential controlled at various levels by the osmoticum, poly(ethylene glycol) 4000. Of the three fungi, *Cenococcum geophilum* was found to be especially tolerant of low water potential, which correlates well with its propensity for forming ectomycorrhizae in dry areas. In fact, because *Cenococcum* grows best at a water potential of –1.5 MPa, it can be difficult to establish this fungus in irrigated nurseries, where it may be replaced by *Thelephora terrestris*. Other work demonstrating the superior performance of ectomycorrhizal plants under conditions of water stress has been carried out by Dixon.

Temperature Tolerance

Moser was careful to inoculate *Pinus cembra* destined for high-altitude outplanting with a cold-adapted strain of *Suillus plorans*. Other fungi, especially *Pisolithus tinctorius*, have been found to be adapted to high temperatures. *Cenococcum geophilum* appears to tolerate both extremes relatively well. Hacskaylo *et al.* (1965) examined the temperature responses of six EM fungi commonly associated with *Pinus virginiana*.

pH Tolerance

Marx reported that pine seedlings with *Pisolithus tinctorius* ectomycorrhizae survived and grew better on acid coal spoils than did non-mycorrhizal seedlings. Dale found that ectomycorrhizae improved the growth of pines in an alkaline soil. Marx investigated the effects of pH on formation of ectomycorrhizae on pine in aseptic culture.

Tolerance to Soil Toxicity

Zak demonstrated the destruction of heat-formed phytotoxins in the soil by EM fungi. Bowen pointed out that in view of the selective absorption of various ions by mycorrhizal fungi, and their capacity for storing ions in the mantle, they may be active in ameliorating marginal soil toxicities. There is still little published work in this area, but we are aware of studies in progress on the spoils derived from nickel mining at Sudbury. Ontario, which indicate the ability of some EM fungi to tolerate fairly high levels of some heavy metals in the substrate. Powell *et al.* treated soil around pecan trees with a variety of nematodes and fungicides, and observed a large increase in mycorrhiza formation by *Scleroderma bovista* L.

Stability of the Partnership

The stability of the partnership need only be established in the short term, since the choice of a mycorrhizal partner for a tree should probably be based on the immediate benefits it bestows. Although the initial mycobiont has in many cases been shown to be supplanted or supplemented by other EM fungi after the seedling has been outplanted, its presence in the early days may well make the difference between death and survival of very young seedlings. Once again, selection of a mycobiont adapted to the conditions of the outplanting site, and preferably already established there as determined by the occurrence of its basidiomata or ascomata, may produce the best results.

Disease Resistance

The presence of EM fungi on the roots of trees has repeatedly been shown to confer some protection against the effects of several important root-pathogenic fungi. Hyppel found that *Boletus bovinus* helped to protect *Picea abies* from *Fomes annosus*. Wingfield observed that *Pisolithus tinctorius* increased the survival rate of *Pinus taeda* seedlings exposed to *Rhizoctonia solani*. Corte reported that mycorrhizae formed by *Suillus granulatus* seemed to protect seedlings of *Pinus excelsa* from a root-rotting *Rhizoctonia* sp. Marx reviewed the antagonism of mycorrhizal fungi to root-pathogenic fungi. Ross found that seedlings of *Pinus clausa* were protected against *Phytophthora cinnamomi* by mycorrhizae of *Pisolithus tinctorius*, while Marx found that *Cenococcum geophilum* or *Pisolithus tinctorius* reduced the impact of the same pathogen on *Pinus echinata*. The effects of the pathogen *Mycelium radicis atrovirens* on *Picea mariana* and *Pinus resinosa* were markedly reduced by the presence of *Suillus granulatus*.

Strand Formation

Mycelial strands, associations of parallel hyphae, serve both as effective agents for the spread of the fungi through the soil and in the long-distance translocation of phosphate and other nutrients to the mycorrhizae. Different species, and different isolates of the same species of EM fungus, may have different capacities for mycelial strand formation.

Other things being equal, it would seem reasonable to choose a strand-forming fungus, such as *Pisolithus tinctorius*, over one that did not produce these structures.

Ease of Pure Culture Formation

Many of the fungi responsible for ectotrophic mycorrhizae can be isolated in axenic culture without much difficulty, but most will not fruit in culture, grow slowly, and are heterotrophic for vitamins like thiamine, for some amino acids, and for other normally root-derived substances, as well as for simple carbohydrates. Most of them are incapable of degrading cellulose or lignin, though these substances are the principal diet of many other basidiomycetes.

Ease and Rapidity of Production

The various attempts to produce inoculum of ectomycorrhizal fungi on a large scale are fully documented in previous paragraphs. Since *Pisolithus tinctorius* has been shown to establish mycorrhizae with almost 50 different tree species, thrives over a wide range of soil pH, tolerates high temperatures well, and can establish mycorrhizae in the poorest soils, it has been touted as a panacea for all our ectomycorrhizal problems. It is also the first EM fungus to be made available in the form of commercially produced mycelial inoculum. We applaud this initiative. but we would encourage researchers to continue working on other fungi, because it seems unlikely to us that *Pisolirhus* can be all things to all EM trees. It is probably at its best coping with heat and drought stress. It has, to the best of our knowledge, been collected only once or twice in Canada, and we suspect that it will not turn out to be the perfect partner for boreal conifers.

Edibility of the Fruit Bodies

If it should transpire that several potential partners are more or less equivalent, then the ultimate choice may be dictated by secondary, though not negligible, factors such as the edibility or otherwise of the sporomata of the fungi being considered. For example, if a hypothetical choice lay between a species of *Amanita* known to have highly toxic basidiomata, and another agaric whose basidiomata are edible and choice, the decision would be obvious. Less obvious, but also important, is the caution that species known to have toxic fruit bodies should not be introduced to new areas as mycorrhizal partners, even if they might seem otherwise desirable. One of the most toxic of all agarics, *Amanita phalloides*, was inadvertently introduced into South America as a mycobiont of oak seedlings imported from Europe during the early part of this century. The cyclopeptide toxins (amanitins) in this fungus have caused many fatalities. One example in which the conscious dissemination of poisonous agarics has been avoided is provided by the Australian government, which refused to allow the importation of cultures of *Amanita panrherina*, a good mycorrhizal partner, but a species producing basidiomata containing dangerous levels of ibotenic acid.

At the other end of the scale are the French experiments with 'trufficulture', the deliberate use of *Tuber melanosporum* as a mycorrhizal partner with an eye to the production of truffles, an extremely valuable crop. The first steps toward the culture of other choice edible fungi have been made, again by the French. Using pure cultures of the famous 'cepe' (*Boletus edulis*) and three other boletes, as well as *Lactarius deliciosus* and *Tricholoma flavovirens,* ectomycorrhizae have been established on *Pinus pinaster* and *Pinus radiata* in tubes and in greenhouse pots. It remains to be seen whether outplanted seedlings bearing mycorrhizae of these species will ultimately produce basidiomata, thereby providing an interesting and perhaps valuable by product of afforestation.

Natural Inoculum: Airborne Spores

Although spontaneous inoculation by airborne spores is still operative in some cases, this is largely by default, in the absence of any active human intervention. It will usually-be regarded only as a supplementary source when mycorrhizal fungi are already known or presumed to be present in the soil, and will play a role chiefly in conditions where natural regeneration is being allowed or encouraged, and significant investment of time or money is unavailable or is considered to be unnecessary.

Soil Colonized by EM Fungi

In Western Australia, pine seeds planted in 14 new nurseries germinated and produced seedlings that grew relatively normally for a few months, then began to decline and die. A few seedlings remained healthy and when these were examined it was found that they had developed ectomycorrhizae, structures conspicuous by their absence from the sickly seedlings. When soil from around the healthy seedlings was used to inoculate other seedbeds, the seedlings in those beds began to recover, and ultimately thrived. Soil from beneath established ectomycorrhizal trees is a fairly reliable source of inoculum. About 10 per cent by volume of infected soil is often added to a new nursery bed. Although the mycorrhiza-forming fungi are often unknown, and there is some risk of introducing pests and pathogens, this has rarely caused difficulty, because the location from which the soil is derived can be carefully scrutinized for such problems. Infected soil has been used to establish exotic pines in various parts of the Southern Hemisphere, and soil transfer is still a regular procedure in many third-world countries.

Seedlings Colonized by EM Fungi

The planting of 'nurse' seedlings that carry established mycorrhizae has also worked relatively well despite the drawbacks we mentioned earlier. This technique was first applied on a large scale in Indonesia, and is apparently still in use there. Mycorrhizal seedlings are planted in seedbeds at 1–2 m intervals. When the plants are lifted, some are left behind to infect the next crop. Attempts to extend the range of the indigenous *Pinus insularis* from the Philippine highlands to lowland sites in the Philippines. Hawaii and South Africa failed completely until seedlings potted in highlands soil were introduced to the lowland sites. This was a very clear-cut example, because there were no indigenous potential ectomycorrhizal partners in the lowland soil to confuse the issue.

Fungal Sporomata or Sclerotia

Spores or chopped basidiomata or ascomata have also been used with some success, though only experimentally. On occasion it can be feasible to collect sufficient quantities of basidiomata of particular species for large scale inoculation. Over 450 kg of basidiomata of *Pisolithus tinctorius* were collected on mine spoils in about 75 man-days. If it is assumed that 1 mg of spores is required for successful inoculation of a single plant, this collection represented enough inoculum for 225 million pine seedlings. Trappe suggests that in most years he could collect enough basidiomata of *Rhizopogon* spp. to inoculate many millions of conifer seedlings. The fruit bodies of gasteromycetous fungi are a particularly concentrated form of spore inoculum, and it would clearly be impossible to collect spores of any Agaricales or Aphyllophorales on this scale.

Mycelial Inoculum

Axenically grown mycelial inoculum of EM fungi. This is now considered to have the greatest potential for general application. Nevertheless, this kind of inoculum presents some problems. Since EM fungi are, to all intents and purposes, obligately biotrophic in the natural habitat, mycelial inoculum will not be able to grow far through the soil to find and colonize a hospitable rool. Therefore, an initial inoculum must be delivered in a very precise manner-it must be placed in contact with, or in the path of, the young roots. Several methods of application have been tried.

Broadcast Inoculation

A known quantity of inoculum is spread out over a given area of seedbed and mixed into the top 10–20 cm of soil before the bed is seeded. Inoculum of *Pisolithus tinctorius* broadcast at a rate of 1 l m^{-2} gave results equivalent to those obtained with higher levels of inoculum. Marx found that inoculum incorporated in container growth media at a rate of 6 per cent by volume produced effective mycorrhization in many tree species. With containerized seedlings, inoculation and container filling processes can be combined. Marx obtained good results by inoculating *Pinus taeda* nursery beds with cultures of *Pisolithus tinctorius.* Laboratory grown inoculum was leached under running tap-water, cool-dried to about 20 per cent moisture and kept cold, but not frozen, until used. Nursery beds previously fumigated with methyl bromide-chloropicrin were inoculated with about 100 cm^{-3} of the dried preparation, dug into the top 7–10 cm of soil. The bed was then machine sown.

Banding of Inoculum Below Seeds

This concentrates inoculum in a zone that will be penetrated by the growing roots. Seeds and inoculum can now be simultaneously dispensed by a modified nursery seed planter. One important advantage of this method over the broadcast technique is that it needs only about a third as much inoculum. Riffle reviewed recent developments in this field.

Slurry Inoculum

With slurry inoculum, bare-root or container-grown seedlings can be inoculated by dipping before transplanting.

Shaw showed that mycelial inoculum of EM fungi can be safely transported over long distances without loss of viability. Mycelial inoculum was grown in Oregon, leached, placed in plastic bags and cooled to 5°C, air freighted in an ice chest to Juneau, Alaska, and stored at 5°C for up to 3 weeks. Container-grown Sitka spruce seedlings were successfully inoculated with this material when it was mixed with the growth medium.

Endomycorrhizae (Vesicular–Arbuscular Mycorrhizae)

Systematics of Vesicular–Arbuscular Mycorrhizal Fungi and their Hosts

In the light of what we say elsewhere in this chapter about the enormous differences in efficiency among the various fungus-host-soil-climate combinations. it should be apparent that the identity of the mycobionts to be used in agricultural systems must be established with the greatest precision possible.

The taxonomy of the VA mycorrhizal fungi is still in a state of active ferment. Where thiny species of VAM fungi were recognized in 1974, seventy-eight species had been described by 1982, and this number is expected to more than double in the next decade. None of these apparently obligately biotrophic fungi have ever been seen to undergo sexual reproduction, and they are sufficiently different from all other known fungi to constitute something of a taxonomic enigma. They are all grouped in a single family, the Endogonaceae, which has usually been tentatively placed in the phylum Zygomycota of Kingdom Fungi. Of this family, only the non-endomycorrhizal genus *Endagone* has been seen to form zygosporangia (the kind of meiosporangia peculiar to the Zygomycota).

The disposition of the VA mycorrhizal fungi in the Zygomycota was challenged by Pirozynski, who suggested that these fungi might well be members of the phylum Oomycota In the Kingdom Protoctista. That challenge led to some investigations of wall chemistry by Weijman, who found that the non-chitinous glucans which characterize the walls of oomycotan fungi are absent from the walls of the Endogonaceae. This effectively ruled out the Oomycota as a home for the VA mycorrhizal fungi, and strengthened the likelihood of a connection with the Zygomycota. Recent observations of *Endagane pisijarmis,* the type species of *Endogane,* type genus of the Endogonaceae, by Berch have revealed arbuscule-like and vesicle-like structures in this non-VAM, saprophytic fungus, and have reinforced the cohesion of the Endogonaceae, making it even more likely that at least some VAM fungi are zygomycetous.

The Endogonaceae is composed of six genera. Four of these are unequivocally mycorrhizal– *Acaulospara, Gigaspara, Glomus* and *Scleracystis.* Another little-known genus. *Entrophaspara,* may also be endomycorrhizal, but is too infrequently collected to concern us here. *Glaziella.* formerly placed in the Endogonaceae, now has been demonstrated to be an ascomycete with uni-spore asci. *Acaulospora* and *Gigaspora* form what have been called 'azygospores', which are supposed to be parthenogenetic zygospores, though there is little or no karyological evidence to support this hypothesis. *Glomus* and *Sclerocystis* produce what are called chlamydospores: these are supposed to be asexual relating spores, a perfectly reasonable assumption in our opinion. For the sake of simplicity, we usually refer to all these structures as spores.

Mosse produced the first modern key to nine spore types associated with VA mycorrhizae. Although their coverage was limited, they attempted to focus on diagnostic features of size, shape and wall structure of spores, and on the mycelium giving rise to the spores. They gave helpful line drawings photomicrographs of thesatures.

The main features of VA mycorrhizal fungi are neatly summarized by Trappe and Schenck, with black-and-white and coloured figures of spores and sporocarps. We are optimistic enough to discern a trend toward better illustration of VA fungi, in terms of both scope and quality, in some recent papers. Although we have neither space, nor a mandate, to make our own contribution toward improving the situation, good illustrations should be a priority for taxonomists working in this group.

Relative to the number of genera of ectomycorrhizal fungi; there are very few genera of VAM fungi, which allows us to characterize them here.

Glomus, the most common genus of VAM fungi, has over 50 species which form globose, ellipsoid or rather irregularly shaped spores that range from 20–400 μm in their largest dimension. These spores are thick-walled (up to 30 μm), and hyaline, yellow, red-brown, brown or black. They are attached to a single subtending hypha (or to two, three or more in exceptional species). They are produced in the soil of the rhizosphere (near plant roots), at the soil surface, or occasionally in roots, and they may form singly, in groups of a few to many, or in large aggregates called sporocarps, which may range from 1 mm to more than 20 mm in their largest dimension. The sporocarps of certain species, such as *Glomus versiforme* (= *Glomus epigaeum*), are formed at the surface of the soil (epigeous), while those of other species, such as *Glomus macrocarpum* are formed in the soil or in leaf litter (hypogeous). Sporocarps are most commonly found in undisturbed forest communities with perennial plants and accumulated organic material. Sporocarpic species. when grown on plants in pot culture, usually produce spores singly or in small aggregates. Some *Glomus* species (*e.g., Glomus lacteum*) are not known to form spores in sporocarps, while others form spores outside the plant root on frequently (*e.g. Glomus intraradices*).

Sclerocystis species produce chlamydospores that are very similar to those of *Glomus,* except that they tend to be clavate rather than globose. All *Sclerocystis* spp. form spores only in sporocaros that are up to 700 µm in diameter. The spores, which may, be up to 125 µm long, develop in single layer, radially arranged around central plexus of sterile and sporogenous hyphae. The sporocarps may group together in masses up to several cm in diameter on the soil surface, or on leaves, twigs and mosses.

The so-called 'azygospores' of *Acaulospora* spp. form laterally on thin-walled terminal vesicles. When mature, the spore does not display a subtending hypha because the vesicular structure that gives rise to it loses its cytoplasm and collapses during development of the spore. The spores are globose or ellipsoid, from under 100 to over 400 µm in diameter, and hyaline, yellow or reddish-brown. The surface of the spore wall may be ornamented with pits, projections of various shapes, folds, spines or reticulations, and the wall is up to 12 µm thick. One species of *Acaulospora* is now known to produce sporocarps, but in the other species, spores are normally found in the, soil of the rhizosphere.

Individual spores of *Gigaspora* spp. can be over 600 µm in diameter, and the smallest ones are just under 200 µm. Spores develop terminally on a bulbous hyphal suspensor, which remains attached at maturity and may bear short lateral projections that are all that remain of collapsed fine hyphal branches. Spores are hyaline, white, yellow, pinkish, grey-green, brown or black, and the wall may be ornamented with minute pits, warts, spines or reticulations and is up to 20 µm thick. These spores form singly in rhizosphere soil. Additional structures that have some importance in the identification of species of *Gigaspora,* and are not reported to be formed by any other fungi, are called 'ornamented vesicles' or 'accessory cells'. These are usually 20–50 µm in diameter, borne singly or in clusters of 12 or more, and are typically formed on spiralling hyphae. Vesicle walls may be smooth, or have rounded lobes or knobs, digitate or coralloid projections, or spines. These structures have not been seen to germinate, so their function is obscure.

Another fungus, or group of fungi, that forms endomycorrhizae has been referred to in the literature as 'the fine endophyte', since its hyphae are less than 4 µm wide, compared to 10–20 µm for typical VAM fungi. This anomalous endophyte has been described as *Glomus tenue,* though it does not resemble the other species of *Glomus.* Its spores are 10–15 µm in diameter, and when young are hyaline and indistinguishable from the vesicles that form in the roots: when mature, however, they are pigmented. Because of their small size when compared to those of other VAM fungi, the spores of *'Glomus tenue'* are not often recorded in nature, though mycorrhizae of this type are common, particularly in dry regions.

Trappe pointed out that VA mycorrhizae have actually been observed in about 1000 genera of plants representing some 200 families. This is a relatively small sample of the 350 000 extant species of higher plants, but, as in opinion polls, a small sample is often perfectly adequate to establish general trends, sometimes with surprising accuracy. In this case, we now confidently forecast that about 900;0 of vascular plants normally establish symbiotic relationships with V AM fungi.

The taxonomy of the plants with which the V AM fungi form symbioses is a vast subject which lies far beyond the scope of this chapter. In fact, relatively few families are known to be largely non-VA mycorrhizal; the Brassicaceae, Commelinaceae, Cyperaceae, Juncaceae and Proteaceae, as well as some members of the Capparaceae, Polygonaceae, Resedaceae, Urticaceae and herbaceous members of the Caryophyllales (Amaranthaceae, Caryophyllaceae, Chenopodiaceae, Portulacaceae), plus, of course, most of the 2000 woody plant species that are ectomycorrhizal.

Since Cronquist lists just under 400 families of higher plants, we estimate that all or most members of over 380 of these families form VA mycorrhizae. In addition, there are about 50 families of non-flowering vascular plants (Gymnosperms, Pteridophytes, etc.), many of which are also VA mycorrhizal. The total number of plant species involved is astronomical. It is estimated that there are about 350,000 species of higher plants, so we suggest that over 300,000 of these will be receptive hosts for the incomparably smaller number of VA mycorrhizal fungi (78 taxa recognized as of February 1983). The implications of these figures for the host range of the VAM fungi are profound. If the hosts were divided up evenly among the fungi, with no overlap in host range, each fungus would have over 3000 potential partners. In fact, we know that host ranges overlap very extensively, and we can suggest that some individual VAM fungi may well have access to tens of thousands of hosts.

Present information suggests that the distribution of VAM fungi is essentially world-wide, since they could occur wherever more than 300,000 host plants grow. This forecast may well change, as host ranges and ecological specificities of individual fungal species are investigated in the field. A few distributional patterns are already emerging. A recent taxonomic redefinition of *Glomus macrocarpum* has reduced the apparent distribution of this species. The same study showed that *Glomus australe,* a segregate from *Glomus macrocarpum* sensu lato, may be restricted to Tasmania and mainland Australia.

Morphology and Development of Vesicular–Arbuscular Mycorrhizae

The vesicular-arbuscular mycorrhiza is a much more subtle and less obvious phenomenon than the ectotrophic mycorrhiza. The presence of a VA mycobiont in a root is usually undetectable by the naked eye. There is no overt morphological change in the roots, no mycelial sheath, no exuberant eruption of macroscopic fungal fructifications. Yet, as appropriate clearing and staining. will demonstrate, the absorbing roots may be extensively colonized. The technique most frequently used is that described by Phillips or some modification thereof. Since anyone who wishes to work with VA mycorrhizae will have to examine colonized roots for various purposes, we will give a complete staining technique.

Roots, fresh or fixed in FAA, are heated in 10 per cent (w/v) KOH at 90°C for 1–2 h to clear the cytoplasm from the cortical cells. If roots are darkly pigmented, it may be necessary to immerse them in an alkaline solution of 10 vol. H_2O_2 at 20°C for 10–60 min until bleached. The roots are then washed in several changes of water, the last change acidified by addition of a few drops of lactic acid or HCI. Roots are then stained in 0.5 per cent acid fuchsin or 0.05 per cent trypan blue in lactoglycerol heated to fuming. Excess stain is subsequently removed by heating the material in lactoglycerol. Segments (0.5–1.0 cm) of the stained roots are mounted in lactoglycerol or poly(vinyl alcohol) –lactophenol. Kormanik described a technique for handling large numbers of samples simultaneously.

We may outline the life cycle of the vesicular–arbuscular mycorrhizal fungi as follows. Spores in the soil germinate even in the absence of a suitable partner, and often in the absence of any apparent stimulus, but usually in conditions that are also appropriate for plant seed germination and root growth. If a receptive root is not encountered, the cytoplasm may be retracted from the germ tube and the spore may retain the capacity to germinate on several subsequent occasions. This multiple germination may reduce the vulnerability of the fungus to grazing by mycophagous soil microfauna, as well as increasing the likelihood that the fungus will successfully colonize a plant.

The interaction of fungus and plant may begin well before the hypha and the root make physical contact. Germ tubes of *Gigaspora gigantea* were attracted through the air toward roots of bean or corn, presumably by volatile substances that were active over at least 10 mm. Response of the VAM fungus to volatile attractants emitted by roots would increase the probability of eventual mycorrhiza formation. In addition, germinating spores of *Glomus mosseae* liberated substances having the same kind of biological activity as gibberellins and cytokinins. The role of these substances in recognition and colonization under natural conditions deserves investigation.

If the germ tube, or any part of the network of hyphae ramifying through the soil, encounters a receptive root or root hair, an appressorium forms at the point of contact, and penetration of the root cortex follows. Penetration occurs into or between epidermal cells in the zone of differentiation and elongation of the fine feeder roots. As is the case in other root symbioses, such as ectomycorrhizae and actinorrhizae, the symbiosis is initiated in juvenile tissues or cells. Hussey and Peterson observed that the preferred site of penetration of *Asparagus vulgaris* roots by *Glomus* sp. was the 'short cells', cells which had not yet developed the secondary wall thickening characteristic of mature epidermal cells.

Within the epidermal and outermost cortical cells, the VAM fungus may form loops of hyphae, or it may traverse these cells to penetrate the next deeper layer of cells. Subsequently, the hyphae grow between or within the cortical cells, although the meristematic cells and the endodermal cells are never entered. Specialized branches enter individual cortical cells and form finely arborescent structures called arbuscules, which remain enclosed by the host plasmalemma, and are apparently the sites of exchange between the fungus and the plant. After some time, the arbuscules gradually break down and their fine branches disappear. It was formerly assumed that the plant was digesting the fungal structure, and that phosphorus passed to the plant as a result of this digestion. However, Cox demonstrated that digestion of the arbuscule could not provide the plant with more than 0.065 per cent of the phosphorus that enters the mycorrhiza. This seems to indicate that there must be an active mechanism transferring phosphorus to the plant throughout the duration of the mycorrhizal relationship. Poly-phosphate granules. involved in P transport within the fungal cytoplasm, have been seen in vacuoles in the inter- and intra-cellular hyphae but are no longer observed in the finest branches of the arbuscules. Those branches have been found to contain acid and alkaline phosphatases.

Individual arbuscules seem to function for a period or 4–15 days, and the cytoplasm of the colonized host cell remains alive, and seems in fact to be greatly stimuated, throughout this period. The nucleus, in particular, of the arbuscule-containing cells is much larger than that of non-colonized cells, and the volume of the cytoplasm is also greatly increased. Ultimately, retraction septa form in the arbuscule as the cytoplasm is drawn back, beginning with the fine branches and progressing toward the main trunk. The fine branches then collapse and the cytoplasm of the root cell eventually returns to normal, though traces of the fungal wall may persist long after the active fungus has decamped.

Many though not all, endomycorrhizal fungi that are characterized by the formation of arbuscules also form terminal and/or intercalary vesicles within the mycorrhiza. These are thin-walled, expanded structures which are not delimited by a septum, and often contain large quantities of lipids. We have recorded up to 500 vesicles cm^{-1} of root in leek, the root cortex taking on the appearance of an almost solid mass of vesicles. This occurs when the root zone concerned is no longer as actively involved in absorption as in the earlier stages of infection, and it seems probable that the cortex is no longer functional as a pathway for nutrient uptake by the plant. Since the stele is not infected, it can of course still translocate substances to and from the active feeder roots.

Vesicles, which may be involved in temporary storage of modified photosynthates received from the plant, are not formed by the species of *Gigaspora*. Daft proposed that the term 'arbuscular mycorrhiza' or 'AM' be adopted to describe the kind of endomycorrhiza formed by members of the Endogonaceae, since the arbuscule is the only structure common to them all. Although we agree that 'arbuscular mycorrhiza' is indeed more accurate and therefore more appropriate than 'vesicular-arbuscular mycorrhiza' as a unifying descriptive term for this phenomenon, we have generally followed the established convention in this chapter, to permit easier access to the literature, which is usually keyed to the traditional term, vesicular-arbuscular mycorrhiza or 'VAM'.

In conjunction with the development of the fungus inside the root, which is generally described as the intramatrical phase, the fungus proliferates in the soil outside the root, this being described as the extramatrical phase. A network of extramatrical hyphae ramifies through the soil, extending as far as 8 cm from the root. This means that a mycorrhizal plant can exploit several times the volume of soil available to a non-mycorrhizal plant, since active translocation of minerals along the extramatrical hyphae to the host plant has been convincingly demonstrated.

In exchange, the fungus receives photosynthates, some of which are transported to the extramatricar hyphae, and to the spores that develop on those hyphae, as has been established through ^{14}C tracer studies.

The large, rounded, presumably a sexual spores of the fungus are formed in the soil singly, or in aggregations up to 1 cm in diameter called sporocarps. They are also occasionally found in roots, empty seed coats, insect carapaces, rhizomes and other available, relatively protected spaces. These spores range in size from 50 to 600 μm, and may be very thick-walled (up to 30 μm), darkly pigmented, and filled with storage lipids, all features that emphasize their role in long-term survival when host plants are absent or dormant. These spores may be dispersed in flood waters or in wind-driven soil, and are well equipped to survive desiccation and UV radiation. Soil fauna, including earthworms and small rodents, may contribute to the dispersal of the spores, which are known to remain viable after passage through the gut of such animals.

The spores will eventually germinate, producing hyphae which will once more grow through the soil and perhaps encounter another plant. The identity of the plant may not matter much, since VA mycorrhizal fungi can usually relate successfully to a very large number of host species, though there is no doubt that some pairings are more efficient than others, as has been demonstrated by Fairweather. This aspect of the relationship is discussed elsewhere in the chapter.

Sources of VAM Inoculum

The inoculation of experimental plants can be carried out with colonized soil, or colonized roots, or spores, or a mixture of all three. The first two kinds of inocula have some profound disadvantages. Soil may contain more than one mycorrhizal fungus, and it may also contain pathogenic organisms. Root inoculum can be used if it has been grown under more or less aseptic conditions, and if the original inoculum used to infect the roots was of a single, named species. The structures developing in roots, although diagnostic of VAM fungi as a group, are not usually diagnostic of a particular species. The features diagnostic of individual species are present only in the spores, which develop primarily on extramatrical hyphae, so these are perhaps the best inoculum for laboratory experiments.

Since the fungi cannot be grown in axenic culture, the only source of spores is the soil near the roots of infected plants. Agricultural soil has been found to contain varying numbers of VAM fungal spores. Sutton recorded up to 70±17 spores g^{-1} dry soil under maize, and up to 86±11 under strawberry, all in the uppermost 24 cm of soil. Hayman and Kruckelmann recorded much lower numbers, possibly because their extraction techniques were less effective. Kessler estimated that the upper 10 cm of soil in an undisturbed Acer-dominated hardwood forest in Michigan contained nearly 7,000,000 sporocarps ha^{-1} of VAM species.

Although inoculum for experiments could be derived from naturally colonized soil, this kind of inoculum will usually contain several different VAM fungi, and is also present at a fairly low level. It was discovered that individual VAM fungi could be propagated in a dual culture system, fungus plus host plant, in what is called pot culture. Furlan and Fortin noted that soil in which onions inoculated with *Gigaspora calospora* had been grown for 19 months contained 325 spores g^{-1} dry soil.

Gerdemann devised the first useful technique for extracting spores from soil. A soil, sample was suspended in four times its volume of water, heavier particles were allowed to settle for a few seconds, then the liquid was decanted through a sieve with 1 mm mesh. Whatever, passed through this sieve was then poured through another sieve with 0.25 mm (250 μm) mesh. Material retained by this sieve was washed and transferred to a Petri dish, and the spores picked out by hand under a dissecting microscope. This technique was slightly refined by Gerdemann, who used the following series of sieves: 1 mm, 710 μm, 420 μm, 250 μm, 149 μm, 105 μm, 74 μm and 44 μm. They found that most of the desired spores fell in the 420–149 μm range, and they used this fraction for their study. Since it is recommended that from 50 to 500 spores are required for inoculation of a single pot, and a well-designed experiment might need scores of pots, an investigator was condemned to spend an excessive amount of time simply to collect enough spores for each experiment. The initial sieving could, however, be speeded up by nesting the set of sieves in a vertical series.

Ohms reported that Gerdemann's technique yielded only about 100 spores per operator per day, and tried to make spore recovery less laborious by centrifuging the organic fraction of soil on a sucrose gradient. He sieved the soil initially to obtain the fraction retained by a 177 μm mesh, then suspended this material in water. Lighter material was decanted after the spores and healthier soil particles had settled, or had been centrifuged for 3 min at 3100 r.p.m. (presumably in a clinical centrifuge). A suspension of the spores and the remaining debris was added to a 50 ml centrifuge tube containing a sucrose gradient established by placing 10 ml of 50 per cent sucrose at the bottom, 15 ml of 25 per cent sucrose above it, and 10 ml of water at the top. The upper layers were added very carefully with a hypodermic syringe equipped with a curved extension that allowed the liquid to be placed directly on the side wall of the vertically held tube. After centrifuging for 5 0 min at 3100 r.p.m., the spores were largely to be found in the middle layer, while the soil particles had settled to the bottom layer. The spores in the middle layer were pipetted into a beaker, and the sucrose solution diluted with water to decrease the specific gravity of the solution and allow the spores to sink to the bottom. The solution was then decanted and the spores were washed several times with water. They could then be readily hand-sorted. This method yielded over 1500 spores per man-day. One disadvantage of this technique was the osmotic pressure exerted by the sugar solution, which could apparently damage young spores.

To avoid this problem, Mosse separated the spores by differential sedimentation on gelatin columns. Gelatin was dissolved in distilled water to make solutions of 5 per cent, 15 per cent and 20 per cent (w/v), which had specific gravities of 1.02, 1.04 and 1.05, respectively, at 32 °C. Liquid 20 per cent gelatin at 30°C was poured into glass tubes 30 cm long and with an internal diameter of 37 mm to a depth of about 5 cm and allowed to solidify at 4°C. A 5 cm layer of 15 per cent gelatin was then added and allowed to solidify, then a similar layer of the 5 per cent solution, and finally a 10 cm layer of water. Columns were stored at 4°C until used. For spore separation, the columns were placed upright in a water bath at 32°C and allowed to liquefy. Then the fraction of a 50 g soil sample that passed through a 250 µm sieve but was retained on a 100 µm sieve was washed into the water layer at the top of a column. After 30 min of sedimentation, the columns were cooled to 4°C again. Spores were extracted from the columns as follows. The water layer was discarded. The column was then held under a hot water tap until the gelatin became free from the wall of the glass tube, and could be slid out of the tube. The core could now be cut into segments. It was found that VAM spores tended to sediment near the 5–15 per cent interface. A 1 cm segment of the column (0.5 cm on either side of the interface) was cut out, gently melted, then strained through a piece of finely woven material supported on a 100 µm sieve. This gave good recovery of four different kinds of VAM spores. Mosse claimed that their gelatin columns had the following advantages over the sucrose solutions used by Ohms (1957): (1) high viscosity and relatively high specific gravity; (2) low osmotic pressure; (3) solidification below 30°C, which makes it easy to set up columns, to maintain sharp boundaries between layers, even in very wide columns, and to segregate portions of the resolidified columns for recovery of the material they contain.

Sutton extracted spores from soil by a flotation-adhesion technique that depended on the rapid flotation of spores on water, the hydrophobic nature of their surface layer, and their tendency to adhere to glass surfaces. A 10 g subsample of soil was placed in a 50 ml cylinder which was then filled with water to within about 2 cm of the brim, and shaken for 10–15 s. The suspension was allowed to settle for about 2 min, while the VAM fungal spores accumulated in the scum floating at the meniscus. Most of the aqueous layer was then decanted into a separatory funnel, carrying most of the spores with it. Further water was added to the cylinder, and the shaking, sedimentation and decanting processes repeated twice more. The accumulated liquid in the separatory funnel was then allowed to settle for 2–3 min. This allowed the spores to collect around the margin of the meniscus in the funnel. The water was then drained at 75–100 ml min^{-1} from the funnel into a second 250 ml separatory funnel, where it was allowed to, settle once more before being drained at a similar rate. The scum and spores adhering to the insides of the two funnels were then washed onto a filter paper. Preliminary experimentation showed that this technique recovered 94–98 per cent of the spores in any given soil sample, unfortunately including spores that are dead.

In an effort to simplify and speed up the process of collecting the thousands of spores needed for experimental purposes. Furlan introduced a flotation-bubbling collection technique. The soil was first sieved to concentrate the spores, then the material was added to a column containing a 50 per cent aqueous solution of glycerol, which was found to have a higher specific gravity (1.13) than the spores without exerting any deleterious osmotic effects on them. Compressed air was injected at the bottom of the column through a fritted disc, so that the small bubbles would both agitate the suspension, helping to separate the spores from soil particles, and propel the spores toward the surface, where they could be recovered. Furlan found that the 74–500 µm soil fraction was most appropriate for this technique. The apparatus consisted of a glass tube of 6 cm internal diameter and 55 cm long, held firmly in the vertical position, and with a fritted disc with pore size 4–5.5 µm fused in position 5 cm above the bottom. Compressed air (monitored by a pressure gauge) was supplied to the bottom of the column at 1 approximately 83 kPa. A litre of 50 per cent glycerol solution was added to the tube, the air pressure raised to 103.4 kPa, and 10 g of soil fraction added. After the bubbling had continued for 1–2 min the air was turned off, the foam allowed to subside, soil particles adhering to the wall of the tube above the liquid were carefully washed down, and the suspension was allowed to settle. After 30 min the supernatant was drawn off, the tip of the aspiration tube being held just below the surface of the liquid to pick up all the spores. All supernatant was aspirated to within 1 cm of the deposit at the bottom. The supernatant was now sieved, and the spores were washed thoroughly in tap water and stored in Ringer's solution (adjusted to a pH of 7.4 with NaOH, filtered and autoclaved). This technique was found to give 94.5 per cent recovery of spores. If virtually total recovery was required, a second treatment would increase the yield to 99.5 per cent.

Mertz used discontinuous sucrose gradients to recover large numbers of spores from massive soil samples. They decanted and wet-sieved 18 kg of soil with cold water, and found that most spores were present in the 425–250 µm fraction. Spores were separated from most of the remaining debris using discontinuous 30 per cent (w/v) aqueous sucrose gradients. The sieved material was layered on 600 ml water over 200 ml sucrose in all beaker. After settling, the spores and debris that collected at the interface were removed by vacuum aspiration, rinsed in cold water, and centrifuged for 1–5 min at 1600 × g in a clinical centrifuge on a second gradient (15 ml water over 20 ml sucrose in a 50 ml tube), the duration of centrifugation being determined by the kind and amount of debris present. Fresh pot culture material needed 5 min of centrifugation to ensure complete separation of the finest root fragments. The spores were then aspirated from the gradient interface, rinsed, and stored in cold water. In this manner, up to 10,000 spores d^{-1} could be recovered by a single operator. The authors claimed that their method was faster than the bubbling-glycerol technique of Furlan.

Furlan *et al.* used several radiopaque media (substances designed to be injected intravenously into humans for a variety of radiographic diagnostic procedures), instead of sucrose, to establish density gradients. These media are free from the deleterious osmotic effects associated with prolonged immersion in strong sucrose solutions. Also. being innocuous in human tissues, they should have no harmful effects on the fungal spores. This assumption turned out to be correct. The radiopaque media used were Renografin-60, Telebrix-30 and Telebrix-38 and Radioselectan. These substances are of generally similar composition: Renografin-60 contains 52.6 per cent of the meglumine (*N*-methylglucamine) salt and 7.5 per cent of the sodium salt of diatrizoic (3,5-diacetamido-2.4,6-triiodobenzoic) acid. It also contains 0.32 per cent sodium citrate as a buffer, and 0.04 per cent disodium edetate as a sequestering agent. It has a pH of 7.0–7.6.

The 53–250 µm or 53–500 µm soil fraction was initially subjected to the flotation-bubbling technique of Furlan, then the spores and soil particles derived from the supernatant were centrifuged in concentration gradients of the radiopaque media. Each of the gradients used contained a series of four dilutions of the medium, 60 per cent, 40 per cent, 20 per cent and 10 per cent (v/v). These gradients were assembled in 50 ml centrifuge tubes by placing 10 ml of 600;0 solution at the bottom of the tube, followed by 5 ml of each of the 40 per cent, 20 per cent and 10 per cent solutions, and lastly 5 ml of water or Ringers solution. Two ml of the spore concentrate was added to the top of the gradient and the tube centrifuged at 500 × g for 10 min. The spores usually concentrated just above the various gradient interfaces, with most above the 10 per cent and 20 per cent step gradients. After extraction, spores were rinsed in tap water and stored in Ringer's solution at 4°C. This technique is very effective for cleaning up the spore concentrate derived from the flotation-bubbling process. Furlan also found that another medium, Percoll (Pharmacia), could be used, and that a continuous gradient could be formed by centrifugation at high speed, 40 40,000 × g for 20 min.

An alternative approach to the separation of VAM spores from soil has been developed by Tommerup and Carter from a technique designed originally to measure the particle size of powders. Soils, collected dry, were fractionated in a dry particle size analysis elutriator. The basic principles are: (1) mobilization of particles into an airstream; (2) regulation of the airstream velocity to suspend particles of equivalent diameters; (3) collection of the fractions on filter papers; and (4) separation of spores in a sucrose gradient. Propagules of fungi remained viable after undergoing this treatment, as evidenced by their ready germination. This technique greatly reduces the exposure of the spores to wetting, and the sucrose gradient could probably be eliminated for rough bulk separation of spores. Since wetting may induce germination of many VAM spores, the older wet-sieving and decanting technique could result in some loss of inoculum potential. The dry particle size analysis elutriator is a simple and inexpensive piece of equipment that can be adapted to either small- or large-scale spore separation.

It must be noted that spores collected by the methods described above are not sterile. Mertz devised a gentle but effective surface sterilization process as a logical corollary of their spore-collecting technique. Within 24 h of being collected, spores were washed twice for 1 min by shaking in an 0.05 per cent (w/v) aqueous solution of Tween 20, a surfactant, and rinsed in sterile water after each washing. Then the spores were immersed twice in 2 per cent (w/v) Chloramine T for 10 min. 5 min of each treatment being under a vacuum of up to 80 kPa. Spores were rinsed in sterile water after each treatment. More rigorous sterilization was achieved by storing the spores for at least a week in the dark at 4°C in a tilter-sterilized solution of 200 mg l^{-1} streptomycin and 100 mg l^{-1} gentamicin. Just before use as inoculum, spores were treated for 20 min in Chloramine T, rinsed with sterile water, then rinsed in sterile streptomycin-gentamicin solution. This procedure yielded spores that were 100 per cent surface sterile, yet still 90 per cent germinable. Spores could be stored for up to 3 months (and possibly longer) at 4°C without loss of viability.

All the techniques discussed above recover spores from naturally or artificially inoculated soil in which suitable host plants have been grown. Some effort is now being made to maximize the yield of spores, not simply by efficient methods of recovery, but by discovering what factors, if any, can stimulate the fungus to produce larger numbers of spores per unit volume of soil. Sylvia and Schenck tried six approaches: (1) pruning the host shoot; (2) withholding water; (3) subjecting the symbionts to cold and darkness; (4) spraying with the herbicide. paraquat; (5) drenching the soil with P_2O_5; and (6) drenching the soil with ethazole. Unfortunately, some of these treatments sharply reduced the number of spores produced, and none increased the yield. But Ferguson found that plants grown in 15,000 cm^3 pots produced 90 times as many spores as those grown in 750 cm^3 pots, although the soil volume in the larger pots was only 20 times greater. Ferguson also noted that limited daily watering, and a temperature regime most suited to the host plant, were stimulatory to the VAM fungi. Clearly, these avenues deserve further exploration.

Starter cultures of at least 15 different named VA mycorrhizal fungi were until recently commercially available from Abbott Research Center, Long Grove, Illinois, USA. At $25 per culture, this was probably the cheapest and easiest way of initiating a research program on these fungi. Each starter culture consisted of living fungal spores, root tissue of the previous host and soil. According to the vendor, each culture weighed about 200 g, and contained enough spores to inoculate at least five pot cultures: Unfortunately, apparently because of quality control problems, these cultures are no longer available. Mixed inoculum of several VAM fungi was recently available under the trade name 'Biofert' from Biofertec Inc. Ancienne Lorette, Quebec, Canada, but quality control problems have led to its withdrawal from the market, at least for the present.

We will now discuss the two phenomena among the VAM fungi which seem to hold out the greatest promise for production of inoculum on a large scale. The first is exhibited by a recently rediscovered species, *Glomus epigaeum* which is most unusual among VAM fungi in producing macroscopic sporocarps on the surface of the soil in pot cultures. Each of these sporocarps may contain millions of spores. They develop without the usual admixture of soil found in the hypogeous sporocarps of, other species, and they can be collected without disturbing the pot culture. which can thus go on producing further sporocarps for years.

The commercial potential of *Glomus versiforme* has recently been examined by Daniels and these authors investigated several characteristics of the fungus. (1) Sporocarp production in pot culture with several hosts under different water regimes. The greatest number of sporocarps, 4.5 pot^{-1} $month^{-1}$, was produced on Sudan grass (*Sorghum vulgare*), watered manually. Sporocarps ranged from 3 to 12 mm diam. and individual sporocarps contained from 1×10^5 to 7.5×10^6 spores. (2) Germinability of spores following 3 month storage under 13 different conditions, and with or without surface sterilization spores failed to germinate, with the exception of those stored in wet loamy sand. Non-surface sterilized spores stored under 6 of the 13 conditions germinated. Germination rates were as follows. Stored in wet loamy sand. 1 per cent; in streptomycin-gentamycin solution. 13 per cent; in dry loamy sand, 33 per cent; in sterile water, 37 per cent; in sterile bentonite. 80 per cent; and as entire sporocarps, 83 per cent. Daniels and Menge also found that *Glomus versiforme* infected six of seven potential hosts tested. and produced significant increases in growth ($P = 0.05$) of four of five hosts tested.

In summary, *Glomus versiforme* can produce up to 1.8×10^7 spores $month^{-1}$ 10 cm pot^{-1} culture over an extended period. These spores can be easily harvested, stored, and used as inoculum when required. The fungus probably has a wide host range. and usually produces marked beneficial effects in its hosts under phosphorus-deficient conditions. Paradoxically, however, Daniels reported difficulty in even detecting infection caused by an isolate of *Glomus versiforme* in Sudan grass. Although the isolate number was, unfortunately, not mentioned by Daniels, it seems possible that the radically different performances reported in the twostudies were due to isolate differences. We cannot stress too strongly the need for fully documenting all isolates used in mycorrhizal research and for depositing voucher specimens in a recognized herbarium.

The second method of producing VAM inoculum we shall mention may be the one with the most potential for generating the amounts of material required for large-scale agricultural applications. We refer to infected root inoculum. Jalraji-Hare is at present working with what is believed to be a species of *Glomus* that, rarely produces extra metrical spores when grown in pot cultures. It does, however, produce very large numbers of vesicles in infected roots, and in the absence of spores she was forced to use root inoculum in her experiments. Since she obtains consistently high levels of infection with this material, it occurred to us (and to Fortin and his coworkers, from whom we had obtained our original isolate) that root inoculum was, in fact, far easier to obtain in large quantities than spores. Assuming that the dual cultures have been established using a sterile soil or soil-substitute, and that reasonably clean conditions have been maintained in the growth chambers to reduce the possibility of contamination by other unwanted VAM or non-mycorrhizal fungi, it is possible to harvest a very large amount of inoculum in a short time. All that is necessary is to pull up the infected plants, wash the root system, then cut off and use or store the infected roots, presumably refrigerated in Ringer's solution or an appropriate aqueous solution of glycerol. Of course, some kind of quality control in the form of regular staining and microscopic observations of root samples would be necessary to ensure that roots were not colonized by pathogens. But we see little to prevent the commercial development of root inoculum in the near future.

Mosse and coworkers at Rothamsted have been studying the potential for producing commercial quantities of infected root inoculum in nutrient film culture (NFT). The infected host roots are bathed in a shallow film of recirculating nutrient solution in which low levels of phosphorous are maintained by passing the solution over relatively unavailable sources of phosphorus. Although no special precautions were taken to exclude airborne spores, contamination appears to be minimal. Since tomatoes are already produced commercially by NFT, the potential exists for a crop of infected plants to be raised and the roots to be used to infect the next crop. Also Mugnier (personal communication) has developed a system for producing mycorrhizae in genetically transformed root organ tissue culture. This holds promise for large scale inoculum production and for detailed studies of spore development.

Evaluation and Selection of VAM Fungi

VA mycorrhizal fungi may be detectable as spores or sporocarps in the soil, or as hyphae, arbuscules, and in three of the four genera, vesicles, within the roots. We discussed the recovery of spores from soil in previous paragraphs. Spores are certainly one expression of the success of the symbiosis, but although some of the methods elaborated for their recovery are capable of giving almost 100 per cent retrieval, spore production is not usually employed in quantitative assessments of VA mycorrhizae. There are a number of reasons for this. (1) Spores may not be produced until the mycorrhizal relationship has been established for several months. (2) Spores may be eaten by arthropods, they may be parasitized by fungi, and they may disappear or lose their viability for a variety of reasons. (3) Spores may be of very different sizes, both among species and within them. They are sometimes too small to be readily separated from soil, as in *Glomus tenue*, or are found concentrated in occasional sporocarps, as in *Glomus fasciculatum*. (4) Some VAM fungi rarely if ever produce spores.

If it is desired to follow the development of the colonization during the course of an experiment, or to assess the level of colonization present in indigenous plants collected in the field, it is logical to study stained roots. Giovannetti compared several different methods which had been developed by various authors to estimate the amount of colonization in roots. They found that the grid line intersect method was the most accurate of those tested, so we will describe this method here.

Cleared and stained roots are spread out evenly over the bottom of a 10 cm Petri dish which is marked with a grid of lines delimiting 0.5 inch (13 mm) squares. The grid lines are systematically scanned under a dissecting microscope, and the presence or absence of colonization is recorded at each point at which a root intersects a line. If three sets of 100 intersects are recorded, and the mean value determined, the percentage of the root length colonized will be estimated with a fair degree of accuracy. This technique also allows the total length of roots in the dish to be determined. All grid intersects are counted, and the final total equals the total length of the roots in 1 cm. Newman related the length of roots in a given area to the number of times they intersect a number of straight lines placed randomly in that area. Marsh refined the method by, arranging the lines in a regular grid. and found that if the distance between lines was 14/11 of the, chosen measuring unit, then the number of intersects equalled the length of the roots, expressed in that unit. Conveniently, 14/11 cm = 0.5 in.

The responses of host plants to VAM can be measured in several ways. One of the most universally accepted parameters is dry weight production (recorded after tissues have been dried at 70°C for 72 h). It is often instructive to record dry weight of the root and shoot systems separately, since the response of the two systems to the presence of the fungus may differ. Other possible measures of response are differences in plant height, stem diameter, shoot volume and leaf 1 number and area, all of which, since they are non-destructive, can be measured repeatedly on the same plants at intervals during their growth. Crop yield is also a valid measure of mycorrhizal success. Transplant survival and disease reaction are others that may be important in particular situations. Mineral analyses or tracer studies may reflect physiological changes resulting from VAM infection. VAM inoculation of *in vitro*-produced raspberry plants (*Rubus occidentalis*) and apple trees (*Malus pumila*) facilitated their recovery from transplant shock and also accelerated their subsequent growth.

Chilvers demonstrated that onion plants (*Allium cepa*) inoculated with *Glomus caledonicum* were better able than non-mycorrhizal plants to withstand and recover from drops in temperature (in this instance to 5°C). This indicates that when host response to mycorrhization by particular VAM fungi is being evaluated, normal climatic conditions can be allowed to play their usual role in the selection of both plant and fungal survivors. Since the VAM fungus is often essentially parasitic on its host during the early phase of colonization, it is essential to allow a growth period long enough to reflect the positive effects that accrue after the fungus has developed an extensive extramatrical mycelium and is contributing to the mineral nutrition of the plant. Alternatively, multiple sequential harvests can be employed. Selection of an appropriately compatible host is also important. In previous paragraphs we listed the families of plants that are not hospitable to VAM fungi, but every host-fungus combination must be tested individually.

The extent of colonization of the root and the development of the extramatrical mycelium can; normally be directly related to the enhancement of plant growth in response to mycorrhization. However, Graham found that sometimes growth is not improved despite extensive root colonization. This can occur because the extramatrical mycelium fails to develop properly, perhaps because of the presence of some inhibitory influence(s) in the soil. Among these may be certain collembola, mites and perhaps also nematodes which can feed on the extramatrical, hyphae. Particularly in the greenhouse, arthropods can be persistent pests, and their impact on the outcome of experiments cannot be discounted.

Safir give a discussion of the interplay between a number of environmental factors (phosphorus, pH, nitrogen, water and temperature) and the host-fungus relationship. Significant differences exist among the relative efficiencies of VAM fungi as mycorrhizal partners, both among different species and among different strains of what appears to be single morphologically defined species. Although the mechanisms underlying these differentials are not well understood, there is some suggestion that much of the effect may well be due to differences in ability to infect host roots quickly and extensively. Other factors, such as the extent of the extramatrical mycelium, and the efficiency of nutrient absorption and translocation, may also be important. Extramatrical hyphae have been shown to retrieve phosphorus 27 mm from a root, phosphorus that was completely unavailable to non-mycorrhizal roots at similar distances. Owusu-Bennoah showed that the zone of depletion around mycorrhizal roots was twice that around non-mycorrhizal roots. Barrow showed that VAM fungi were equally adept at accumulating soluble and insoluble phosphorus.

Abbott have stressed the importance of knowing the response curves of mycorrhizal and non-mycorrhizal plants to application when evaluations of the impact of VAM fungi on the growth. P uptake and mycorrhiza dependency of different species of plants are being made. No stimulation normally results from mycorrhization by even the most efficient VAM fungi if available phosphorus is present at luxury levels; in fact, depression of growth can occur under these conditions. To standardize the expression of 'relative mycorrhiza dependency' for a given plant at a known level of available P, Meng used the ratio of the dry weight of the mycorrhizal plant to that of the non-mycorrhizal plant, expressing it as a percentage. This method of calculation sometimes gives a rather inappropriate 'mycorrhiza dependency' of well over 100 per cent. Plenchette proposed that the expression of mycorrhiza dependency be shifted into the normal 0–100 per cent range by application of the formula:

$$100 \times \frac{\text{Dry wt of mycorrhizal plant} - \text{Dry wt of non-mycorrhizal plant}}{\text{Dry wt of mycrorrhizal plant}}$$

to be calculated for any given level of P availability. Thus, Plenchette reports that, at 100 µg g^{-1} available P, the relative mycorrhiza dependency of carrot (*Daucus carota* var. Nantaise) is 99.2 per cent, and that of wheat (*Triticum aestivum* var. Glenlea) is 0 per cent. No value was calculated for garden beets (*Beta vulgaris* var. Dark Red), since VAM were not formed in this species. While satisfying Abbott and Robson's requirement that mycorrhiza dependency be expressed in relation to the level of available P. this calculation also permits comparisons of the responses of any species of plant to different sources of P, and to different mycorrhizal fungi.

As yet we know little about the contributions to plant growth made by VAM fungi occurring naturally in field soils (although we extrapolate a good deal from the results of our laboratory experiments). This is partly because any treatment designed to kill all spores (and thus allow comparisons to be made) is sufficiently drastic that it will also change many other factors in the chemical makeup of the soil, and probably alter its nutrient status. Plenchett found that young apple trees preinoculated with VAM fungi grew better than uninoculated trees, even when planted out in unsterilized field soil which contained an indigenous VAM flora.

It is now well established that the presence of VA mycorrhizal fungi in the roots of plants tends to reduce the incidence and the severity of soil-borne diseases in those plants. Dehne lists 56 reports of interactions between VAM and 18 soil-borne plant pathogenic fungi. 8 plant parasitic nematodes, and 3 viruses, involving a total of 21 different crop plants. The majority of these investigations showed that mycorrhizal plants had less disease, except in the case of the viruses, when symptoms were always more severe. Dehne advanced several reasons for the reduction in damage caused by fungi. (1) The mycorrhizal plants were healthier and more able to repel or resist the attacks of the pathogenic fungi. (2) As an apparently normal part of the nutrient exchange process, the cells of the mycorrhizal plants may partly digest the arbuscules of the fungus. This implies chitinolytic activity on the part of the plant, a talent that may well be turned against other invading fungi. (3) Possible infection sites on the surface of the plant roots may be preferentially occupied by the VA fungus, to which the plant is probably more susceptible anyway.

A number of factors must be taken into account when the suitability of specific VAM fungi as mycorrhizal partners for specific plants in specific areas is being considered. These include geographic distribution, frequency of occurrence, host range, soil type, pH, fertility, moisture and organic matter, persistence of inoculum in the soil, rapidity of germination and infection, and phosphorus efficiency, which is a function of the soil volume a fungus explores, the rate at which it takes up phosphorus, and the rate at which that phosphorus is translocated to the plant. Equally important are considerations of inoculum production: how easy it is to persuade the fungus to produce propagules on a large scale under controlled conditions and to store, transport and deliver that inoculum without loss of viability.

The global distribution of most VAM fungi is incompletely known. Although herbaria contain increasing numbers of collections from many parts of the world, their number is still inadequate to permit proper distributions to be extracted from them, and even that distributional data inherent in existing collections has not yet been collated. This is largely because the taxonomy of these fungi is still in a state of flux, new taxa being described every year and old ones being reassessed. Whatever the reason for the dearth of knowledge, it is most unfortunate that we have so little, because it would be very valuable in the selection of appropriate VAM partners for specific areas. In the absence of such background information we must approach each new problem by conducting extensive sampling in the field, to determine the nature of the indigenous flora, and perhaps to accumulate sufficient inoculum both to carry out an initial experimental program, and to establish pot cultures which will multiply the inoculum for future use.

Despite the apparent lack of host specificity under the artificial conditions of pot culturing in a greenhouse or growth chamber, there is some evidence that, in the field, plants select certain indigenous VAM fungi in preference to others. Schenck assessed the numbers of spores associated with monocultures of annual crop plants grown for seven years on a newly cleared woodland site. They concluded that the major influence on spore production was host plant specificity for particular VAM fungi. Nemec found that in Florida and California citrus orchards, production of spores by certain VAM fungi was favoured by high soil P, B, Ca and Mg, or salinity, while that of other species was not. As they point out, much of the apparent disagreement over the effects of local conditions on VAM fungi could probably be eliminated if species of VAM fungi were considered individually, instead of as a group. In fact, certain VAM fungi have been linked to particular kinds of soil: *Glomus mosseae* with fine-textured, fertile, high pH soils; *Acaulospora laevis* with coarse-textured, acid soils: *Gigaspora* species with sand dune soil.

For it to be worthwhile to produce and deliver VAM inoculum on a large scale, the organism involved must have been shown to be a more efficient scavenger of phosphorus than other VAM fungi present in the local soil, or to boost the soil population to a level at which it can bring about improvements in the nutrition of the desired crop. Unfortunately, phosphorus efficiency by itself is no guarantee of success for a VAM fungus. In a non-sterilized field of birds foot trefoil (*Lotus corniculatus*). introduced strains of VAM fungi improved yield, but only at the first harvest. In subsequent years the indigenous strains, presumably better adapted to local conditions, apparently replaced the introduced, more P-efficient strain, and yields returned to normal. It seems that the improvement in plant nutrition produced by the presence of VAM results mainly from the effective increase in the volume of soil that is being exploited by the plant. Therefore, a measure of the amount and extent of the extramatrical mycelium would be of considerable value. Unfortunately, such a measure is difficult to obtain with any degree of accuracy, since when the root-fungus-soil system is disturbed for sampling purposes, the fine network of extramatrical hyphae is usually extensively disrupted. Removal of a root system from soil normally results in almost complete loss of the extramatrical mycelium. A few estimates have been made, using such indicators as the degree to which the extramatrical mycelium binds together the soil around the roots. Fortunately, when Sanders carefully compared the level of root infection with the amount of extramatrical mycelium in four, VAM fungi infecting onions. they found a strong positive correlation between these two parameters. Thus the actual density of colonization of plant roots by a VAM fungus appears to be an appropriate measure of its efficiency in exploring the soil.

Nevertheless, we have little or no comparative data the respective abilities of various VAM fungi to absorb phosphorus from the soil, or to transport it to the host plant. These are two areas in which differentials may well exist. The relationship between what we may call phosphorus efficiency and speed of infection is also important. The effectiveness of a VAM fungus depends on both factors. Since plants normally have higher phosphorus requirements when young, the faster an infection can be established, the better. Thus a fungus which is phosphorus-efficient may still be an unsuitable mycorrhizal partner because it establishes infection too slowly. Once again. comparative studies are lacking.

As Abbott have pointed out, it may be possible to simplify the screening process to some extent. For example, it is obviously much easier to test fungi for the rapidity with which they can cause infection than for their ability to increase phosphorus uptake. Such simplification could be justified only if it was determined that early infection was an appropriate index of mycorrhizal suitability. Soil pH is sometimes a limiting factor in the germination of certain VAM fungal spores, so it must be noted that the prevailing pH may render some VAM fungi unsuitable for inoculation into certain soils.

Gildon isolated a strain of *Glomus mosseae* from mycorrhizae of clover growing naturally on sites heavily contaminated with Zn and Cd. This VAM strain tolerated high concentrations of these two metals. The existence of such strains suggests that when metal-tolerant plants capable of revegetating such sites are being sought, the tolerance of their endophytes should also be considered.

Fairweather recently discovered a very efficient VA mycorrhiza on tomatoes growing in an extremely phosphate-deficient soil in south-east Australia. They attempted to reconstitute the partnership experimentally, using four kinds of VAM spores recovered from the soil concerned. In view of their striking results, it is worth detailing the experimental procedures, they used.

Sand with no Olsen-extractable phosphorus was sterilized by γ-irradiation. Tomato seeds surface-sterilized in 1 per cent sodium hypochlorite for 1–2 min were sown in the sterile sand. When the first leaf emerged, the roots were washed with distilled water and 30 spores of one of the VAM fungi were placed on the roots of each seedling. Control plants were not inoculated, and remained free from infection throughout. The plants were watered as required and 50 ml of Hoagland's solution 2, containing only 25 per cent of the normal level of phosphate, and 50 per cent of the normal levels of the other nutrients, was added weekly to each 6 in pot. Plants were harvested after 49 days.

The control plants had a mean dry weight of 6.9 mg. Those infected with *Acaulospora levis, Gigaspora gigantea,* an unnamed VAM fungus and *Gigaspora margarita* had mean dry weights of 13.3 mg. 18.2 mg, 24.0 mg and 82.3 mg. respectively. *Gigaspora margarila* is obviously by far the most suitable fungal symbiont among those tested for tomatoes grown under the chosen conditions. We believe that much more of this type of research will be needed to ensure that the host-fungus combinations used in agricultural applications are optimised, as far as is possible.

Applications of VAM Inoculum

This section examines the problems related to the actual delivery or application of the inoculum, first in the laboratory, then in the greenhouse and finally in the field.

Laboratory Experiments

For complete precision in the establishment of aseptic dual cultures (host + fungus) the methods described by Hepper are appropriate, and the inoculum usually takes the form of spores.

The Agar Slant Method

A surface-sterilizedseed is germinated at the top of the slant. Pregemlinated spores are then put on the roots, and a little free water added to the bottom of the slant.

The Paper Substrate Method

A seed is germinated between a paper strip and a slide standing in a tube with its lower end immersed in a nutrient solution. Spores are then placed on the young roots.

The Fahraeus Slide Method

A seed is germinated between a slide and a coverslip held parallel to, but at a distance of 800 μm from, the slide. The lower part of the slide is again immersed in nutrient solution in a tube, and spores are added to the young roots.

It is possible to infect a root with a single spore, and Daft showed that inocula of 3 spores seed^{-1} and 250 spores seed^{-1} produced similar levels of infection in mature tomato plants. However, we would not suggest using very low spore numbers, since other workers have found that the level of inoculum used can affect the rate of infection. Higher levels of inoculum seem to shorten the lag phase in mycorrhiza development or hasten the achievement of the extensive infection desired, the plateau phase. When spores are used as inoculum, the number applied usually ranges from 50 to 500 plants^{-1}.

Greenhouse Crops

Crops that are grown in the greenhouse, or that are usually transplanted to field situations, can be infected before planting out without involving significant extra investment of time in an already labour-intensive process. This is appropriate for all container-grown seedlings, and has particular relevance to many horticultural crops. Since relatively small volumes of soil are Involved, it should not be difficult to incorporate a reasonable concentration of inoculum in this soil during its preparation. The chances for early and widespread infection are presumably Increased as the number of spores or other infective propagules increases, up to a point. The speed of infection can presumably be increased by ensuring that the propagules are in close juxta-position with the roots. Inoculum can be mixed with the soil, or placed below or to one side of the seed or the roots of the seedling, or even pipetted onto the roots. But at least one worker has suspended the spores in a viscous, non-toxic medium containing carboxymethylcellulose, then dipped the roots of seedlings in this medium before planting.

Field-sown Crops

Little experimental work has yet reached the field trial stage, but adapting to this challenge will inevitably be the next major phase of endomycorrhizal research. Inoculum for field use may be granulated, pelletized or in slurry form.

Granular Inoculum

Inoculum derived from pot cultures is roughly milled or chopped up, then dried to 5–10 per cent moisture content. It consists of a peat, perlite, vermiculite or surface substrate with an admixture of spores and infected root fragments.

Pelletized Ionculum

This has been prepared by mixing 20 parts pot culture inoculum with 1 part clay (mean particle size 16 µm) and 1 part clay (mean particle size 2–3 µm). Water was mixed with these constituents until the medium could be rolled into pellets weighing 1–2 g. Jung made pellets with a rather complex recipe involving mycorrhizal inoculum, 150 ml phosphate buffer, 50 ml bufferedacrylamide solution. 50 ml buffered N,N^1-methylenebisacrylamide solution. 1 ml freshly prepared ammonium persulfate solution, and 76 µl N,N,N,N^1-tetramethylethylenediamine, all of which form a gel within 10 min. Pellets are prepared by cutting up the gel into small pieces. The inoculum is 'polymer-entrapped', and apparently does not lose viability or infectivity after several months at room temperature.

For annual, field-sown crops, it is apparent that the inoculum must be introduced with, or at about the same time as, the seeds. The technique cited by Hayman as being most widely used is the addition of an infective soil inoculum to the furrow, below the seed, at planting time. This soil comes from around the roots of plants deliberately infected with the VA endophyte some months previously. Owusu-Bennoah and Mosse used this method of inoculation on an experimental scale for onions, lucerne and barley, but concluded that for field application, about 2.5 tons infected soil ha^{-1} would be needed. Clearly, this would be an impracticably large amount.

Ross successfully inoculated soybean with *Glomus macrocarpum,* using about 500 g inoculum m^{-2}, with an estimated spore inoculum of about 30,000 m^{-1}. Black and Tinker used a similar approach, as did Timmer, to inoculate citrus seedlings with *Glomus fasciculatum.* Five studies placed layers, pads or pellets of inoculum beneath the seeds. Kleinschmidt and Gerdemann inoculated citrus in this way, using 300 g inoculum m^{-1}. Black did the same with potatoes, applying 300 g inoculum m^{-2}, with over 30,000 spores m^{-1}. LaRue *et al.* placed pads of *Glomus fasciculatum* inoculum under peaches, at a rate of 5 g $plant^{-1}$. Owusu-Bennoah inoculated alfalfa, barley and onions in this way, using larger pads (10–20 g $plant^{-1}$). Clarke used 20 g pads of *Glomus mosseae, Glomus caledonicum* and *Glomus fascicitlarum* inoculum under barley seeds. Hall placed soil pellets containing *Glomus fasciculatum* inoculum; under seeds of *Lotus pedunculatus.* His pellets had to be placed with great accuracy, because each weighed less than 2 g and contained fewer than 30 spores. Obviously, this technique is not suited to large scale, mechanized applications.

Seeds pelleted with *Glomus fusciculatum* inoculum were used by Crush. Powell used soil inoculum pellets containing many seeds, but on a field scale this still translated into about 2 t of soil inoculum ha^{-1}. A more concentrated and less bulky inoculum composted of spores, mycelium and infected root fragments concentrated from soil by wet-sieving (or any of the other techniques described in the section on recovery of VAM inoculum from soil) has been used as a seed-coating, when formulated with a suitable binding agent. This method is applicable to crops with large seeds, but is unsuited to small-seeded crops, because the inoculum is still bulky relative to the seeds: a single spore of a VAM fungus is commonly 200–500 μm in diameter.

Witty adapted the fluid-drilling technique described by Salter 'injected' pots with a slurry containing pre-germinated seeds of white clover. *Rhizohium trifolii,* and wet-sievings of the VAM fungus, *Glomus fusciculutum,* all suspended in 4 per cent methyl cellulose. This method gave 20 per cent infection after 6 weeks, and eliminated the extra manipulations required by several other procedures.

Hayman *et al.* compared five treatments.

1. VAM inoculum and seeds broadcast over plot and raked in.
2. VAM inoculum applied in 10 g portions immediately below seeds sown in furrows.
3. VAM inoculum and pre-germinated seeds suspended in 4 per cent methyl cellulose and applied to the furrows as a slurry at the rate of 250 ml m^{-2}.
4. VAM inoculum made into 1 cm diam pellets which were rolled in seeds until 3–4 seeds adhered to each pellet. Pellets were broadcast and then thinly covered with soil.
5. Non-inoculated controls.

They found that treatments (2) and (3) above were very efficient: over 60 per cent of the plant roots were colonized after 9 weeks. Multi-seeded pellets (treatment 4) were less effective: only a third of the seedlings became extensively colonized (more than 40 per cent of root length infected). Treatment (1), the broadcasting of inoculum, was essentially ineffective, since plants in plots given this treatment had no more infection than those in control plots.

Of the treatments compared by Hayman, fluid drilling appeared to be the method 1 of choice, especially for legumes, since *Rhizobium* as well as the VA endophyte can be incorporated in the slurry. Hayman did not use chopped-up infected roots as inoculum, but mentioned that this type of inoculum, which loses much of its viability during air drying and subsequent storage, might survive better in slurry form.

Menge compared several methods of inoculation, and found that placing a layer of inoculum below citrus seed, or as a band alongside the seed, was better than pelleting seed with inoculum. It was also more efficient to have inoculum 2.5 cm below the seed than 5 cm below it.

Most of these experiments, although carried out in the field, involved hand sowing or the use of manual planters. The development work necessary to mechanize these techniques for large-scale application still lies before us, though the requisite technical expertise undoubtedly exists. Ferguson was exceptional in his use of a modified tractor-drawn fertilizer bander to apply a band of *Glomus fasciculatum* inoculum to one side of citrus seedlings, subsequently recording 48 per cent infection. Band placement of inoculum did not appear to be effective in corn and soybean. It is clear that the inoculum must be placed in the zone of root growth if it is to bring about the requisite degree of infection.

Chapter 11
Fertilizers with Organics and Biofertilizers

Plants require at least 13 mineral elements for their growth and development. All these essential nutrients are present in all soils but one or more of these are invariably present in inadequate amounts of plant usable form which makes external additions necessary (Table 11.1). Since each element has certain specific functions to perform, one cannot substitute for another. If a soil is deficient in even one nutrient, it cannot Support a good crop unless deficiency of this element is made up. This is essentially where fertilizers and other sources of nutrients step into farming.

Table 11.1: Essential Mineral Nutrients, their Forms Absorbed by Plants, Concentration in Dry Matter and Major Functions

Nutrient (Symbol)	*Forms Absorbed*	*Content in Dry Matter*	*Major Functions in the Plant*
Macronutrients			
Nitrogen (N)	NH^+_4, NO_3^-	1%–5%	Growth, protein production
Phosphorus (P)	$H_2PO^-_4$, $HPO_4^=$	0.1–0.4%	Energy compounds, root development, ripening
Potassium (K)	K^+	1%–5%	Crop quality, water regulation, stress tolerance
Sulphur (S)	$SO_4^=$	0.1%–0.4%	Protein and oil production
Calcium (Ca)	Ca^{++}	0.2%–1%	Cell structure, cell division
Magnesium (Mg)	Mg^{++}	0.1%–0.4%	Part of chlorophyll, energy transfer
Micronutrients			
Boron (B)	H_3BO_3	6–60 ppm	Cell wall development, seed setting
Chlorine (Cl)	Cl^-	0.2%–2%	Tissue hydration, photosynthesis
Copper (Cu)	Cu^{++}	5–20 ppm	Enzyme activity, chlorophyll formation
Iron (Fe)	Fe^{++}	50–250 ppm	Respiration, chlorophyll formation
Manganese (Mn)	Mn^{++}	20–500 ppm	Enzyme activation, photosynthesis
Molybdenum (Mo)	MoO^-_4	< 1 ppm	Nitrate reduction, N-fixation
Zinc (Zn)	Zn^{++}	25–150 ppm	Enzyme activity, protein synthesis

A crop is concerned more with the amounts of nutrients available for its use rather than where they have come from. Diverse Sources of nutrients can be gainfully used as long as they are able to furnish the nutrients in readily or potentially-usable forms. If a plant root is about to absorb N as ammonium or nitrate, to the best of our knowledge it neither knows nor distinguishes between it's Source be it a fertilizer, dung or green manure. Such differences are taken care of in nutrient cycles which are working round the clock in all soils. Organic and inorganic sources enter the same cycle (Figure 11.1). In soils, organic and inorganic forms are inter-convertible through mineralisation and immobilisation.

This chapter deals with two aspects (*i*) fertilizers and their role as nutrient suppliers in the context of fertility status of Indian soils and (*ii*) current status of information and issues involved in integration of fertilizers and other sources of plant nutrients towards the development of integrated nutrient supply packages. The first part is a counterpart of other chapters which deal with specific components of integrated plant nutrient supply systems (IPNS). The second part is a synthesis of available information on chemical, organic and biofertilizers.

Fertilizers

Fertilizers are mined or manufactured commercial products which contain one or more essential plant nutrients. For a material to qualify as a fertilizer, it should contain nutrients in appreciable amount and in readily or potentially usable form. In addition, a fertilizer (plant food-carrier), just like any food item should not contain or any substance which is toxic to the soils, plants or humans above permissible limits. Many countries including India have laws as to what can or cannot be labelled as a fertilizer.

Fertilizers are used with the sole purpose of improving soil fertility so that it can support larger harvests. Fertilizers represent the most common currency used by farmers to deposit plant nutrients into their soils to ensure that adequate nutrients are available to feed the crop. Plant roots do not absorb fertilizer granules as they come out of the bag or dung particles as they are in a manure heap.

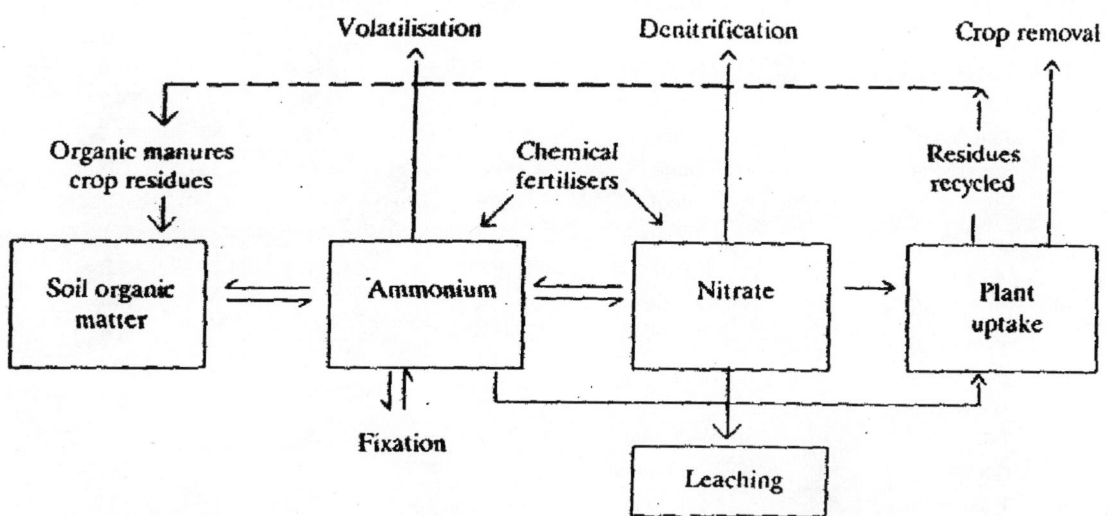

Figure 11.1: A Simplified Presentation of the Nitrogen Dynamics through Organic and Mineral Inputs in Agriculture

Plants absorb nutrients in specific ionic forms (Table 11.1) which either a fertilizer furnishes when it dissolves in soil water or various chemical and biological agents in the soil convert fertiliser currency into local currency acceptable to roots. Since fertilizer use is determined primarily by soil fertility status and nutrient requirement of crops, a brief discussion of these two aspects is considered necessary.

Nutrient Uptake and Removal by Crops

To produce each tonne or grain, a cereal crop absorbs 50–65 kg of N + P_2O_5 + K_2O plus lesser amounts or other nutrients. Absorbed nutrients are differentially distributed in different plant parts. While 70–80 per cent or absorbed N and P, and 40–60 per cent or S end up in the grains, 70–80 per cent or potash remains in the straw/stover. The destination or absorbed nutrients varies largely with the nutrient in question and crop species and to a lesser extent with growth conditions. Such distribution has important implications for nutrient gains from residue recycling which would be maximum in case or K and least for N and P.

Some estimates or nutrients absorbed by crops in relation to yield are given in Table 11.2. Nutrient uptake by intensive crop rotations can be very high, approaching 600–900 kg/ha/yr or N + P_2O_5 + K_2O. The extent or nutrient removal from a field depends on the relative composition of main and biproduce, extent or leaf fall, residues retained at harvest and those recycled later. Actual studies on these aspects in quantitative terms are few.

Table 11.2: Average Nutrient Removal/t Yield Production by Diverse Crops Under Field Conditions

Sl.No.	*Crop*	*Produce*	*Kg/tonne Produce*		
			N	*P_2O_5*	*K_2O*
1.	Rice	Paddy	20.1	11.2	30.0
2.	Wheat	Grain	24.5	8.6	32.8
3.	Sorghum	Grain	22.4	13.3	34.0
4.	Chickpea	Grain	46.3	8.4	49.6
5.	Pigeonpea	Grain	63.8	17.7	42.3
6.	Groundnut	Grain	58.1	19.6	30.1
7.	Mustard	Grain	32.8	16.4	41.8
8.	Cotton	Seed	44.5	28.3	74.7
9.	Jute	Fibre	23.5	13.0	41.7
10.	Sugarcane	Cane	1.7	0.2	2.0
11.	Potato	Tuber	3.9	1.4	4.9
12.	Cassava	Tuber	7.8	1.2	5.1
13.	Berseem	Fodder	8.6	1.4	3.8
14.	Fodder grasses	Dry fodder	9.4	1.5	14.2
15.	Grapes	Fruit	4.9	1.5	5.9
16.	Banana	Fruit	8.2	3.0	32.3
17.	Tomato	Fruit	4.1	1.5	5.9
18.	Tea	Made tea	110.0	38.0	44.0
19.	Coffee	Clean beans	37.0	6.0	47.0
20.	Tobacco	Leaf (dm)	17.0	5.0	26.0
21.	Coconut	1000 nuts	7.1	3.5	10.7
22.	*Java citronella*	Hero	0.8	0.3	1.0

Note: Data are total uptake by crop to yield one tonne of produce.

Fertility Status of Soils

The fertility status of soils indicates their nutrient supplying capability. When seen alongwith crop demand, this is the major determinant whether and to what extent external nutrient input is required. Only a fraction of the total nutrients present in soil exist in readily or potentially usable form on a short-medium term basis and therefore a knowledge of available nutrient status of soils is very valuable in planning nutrient needs of agriculture. It serves as an indicator of soil fertility and to a fair degree can be estimated by getting soil samples tested in a good laboratory.

The overall fertility status of a soil has to be assessed for each nutrient separately. Thus a composite fertility profile of a field is made up of sub-profiles, one for each nutrient. At present, deficiencies of five nutrients are of large-scale practical importance in India and therefore, much of the focus is on N, P, K, S and Zn. An overall picture of fertility status of soils is provided in Table 11.3. Fertility status of soils which are categorised as low is the poorest and they cannot support even moderate crop yields without nutrient addition.

Table 11.3: Nutrient Status of Indian Soils and the Deficiencies which Need Attention

Nutrient	*Soil Fertility Status*
Nitrogen	Low in 228 districts, medium in 118, high in 18 districts
Phosphorus	Low in 170 districts, medium in 184, high in 17 districts
Potassium	Low in 47 districts, medium in 192, high in 122 districts
Sulphur	Deficiencies scattered in 100–120 districts
Magnesium	Kerala, other sourthern states, very acid soils
Zinc	50 per cent of 150,000 soils analysed found deficient
Iron	On upland calcareous soils for rice, groundnut, sugarcane
Boron	Parts of Bihar, Karnataka, West Bengal.

In reality, available nutrient status of a soil does not have a stand alone meaning but is demand-dependent. A soil which has medium fertility for raising a 3 tonne crop will be low for growing 6 t/ha grain. A soil which is medium for local crop varieties may turn out to be low (deficient) for high yielding varieties (HYV). It is for this reason that soil test-crop response correlations assume importance and soil, crop and production potentials are to be taken into account which developing soil fertility classes.

Field experiments can be conducted to know the crop yields a soil can support on its own nutrient reserves. The rate at which soil fertility declines can be estimated from long-term experiments. Results from such experiment show that virtually no soil can support medium-high yields indefinitely without external nutrient additions. Unless this is done, available nutrient status can decline from sufficiency to deficiency levels. Even if this happens for one nutrient, the crop is not spared from its adverse effect.

All available data show that the nutrient reserves of Indian soils are being depleted and nutrient status of soils is declining. This is primarily due to negative balances caused by nutrient removals which persistently exceed nutrient additions. The root causes of nutrient depletion from soils are (*i*) increasing pressure on each piece of land to produce more to meet the needs of an expanding population (Figure 11.2), (*ii*) large gap between nutrient, removals and additions which appears to be hovering around 10 million tonnes of $N + P_2O_5 + K_2O$/year and (*iii*) crop harvesting and residue disposal methods which do not go in favour of nutrient recycling.

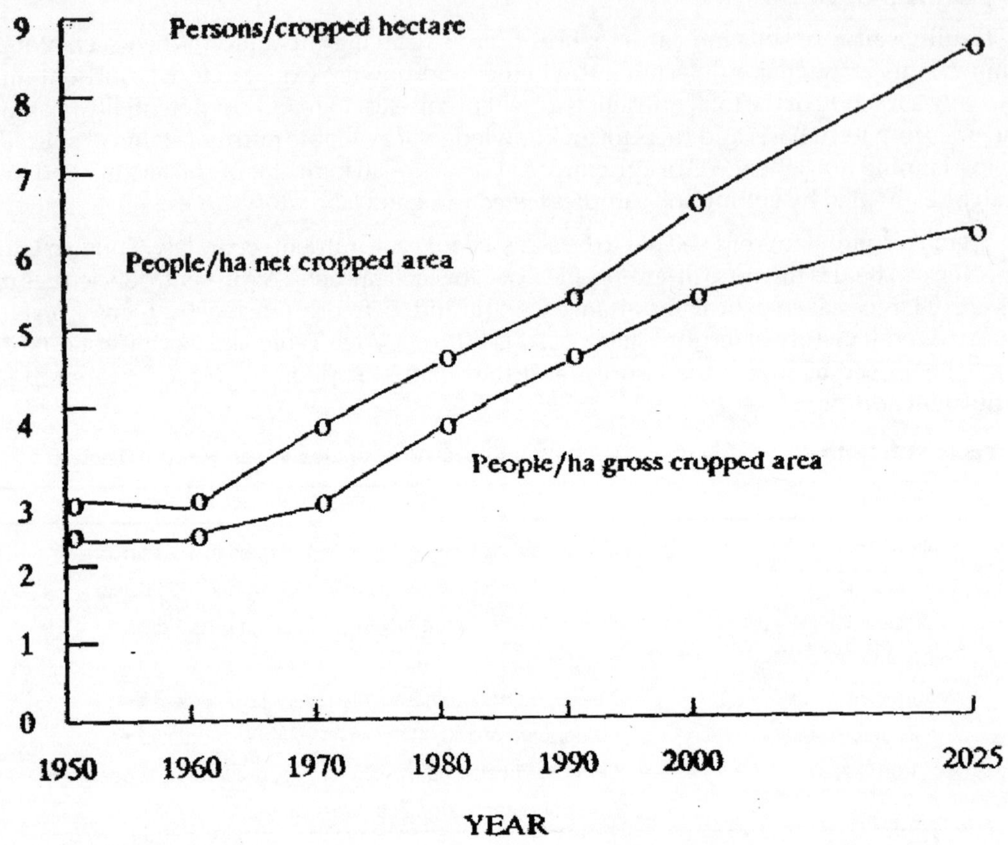

Figure 11.2: Number of People per unit of Land Already Under Cultivation

Soil nutrient reserves are still bearing the brunt of feeding the crops. Although India is the fourth largest fertilizer user in the world with an estimated consumption of 12.5 million tonnes N + P_2O_5 + K_2O during 1990–91, fertilizers account for barely one-fourth of the nutrients absorbed by crops (Table 11.4). The rest comes primarily from soil reserves, organic manures, biological N-fixation and residues as well as wastes. In an estimate made in the early 1980s, out of the total (gross) nutrient input into Indian agriculture, the share of fertilizers was 60 per cent and that of organic resources 40 per cent. Because fertilizers have been associated with about 50 per cent crop yield increases registered in recent times, their importance in Indian agriculture will grow even stronger with time.

Crop Responses to Fertilizer Application

Since fertilizers are added to upgrade the soil's nutrient supply so that high crop yields can be harvested, fertilizer application must result in an increase in crop yields. A pre-requisite for using a purchased input is that the resulting yield increase should also be sufficiently large to leave behind a net profit for the farmer. As a thumb rule, fertilizer use is considered attractive if each rupee invested in it generates extra produce worth at least Rs. 2.0–2.5 and even more in high-risk environments.

Table 11.4: Nutrient Removal by Crops and Net Contribution of Fertilizers to it in India

Nutrient	*Uptake (mmt)*	*Fertilizer Use (mmt)*	*Fertilizer Use-efficiency*	*Contribution in Uptake*	
				Fertilizers	*Others*
N	6.6	7.4	44%*	49%	51%
P_2O_5	3.0	3.0	30%	30%	70%
K_2O	11.6	1.16	50%	5%	95%
Overall	20.2	11.56		23%	77%

* Use efficiency for N, 35 per cent for rice consuming 40 per cent total N, 50 per cent for other crops. For P inclusive of some residual effects. For K, assuming low-moderate application rates, fixation, leaching etc.

Due to widespread nutrient deficiencies in soils, crops in general benefit from fertilizer application. A mass of experimental evidence on this account is available from on-station research as well as from the closer to real-life on-farm trials. A summary of response rates (kg economic produce per kg of nutrient applied) based on a large number of on-farm experiments is given in Table 11.5 for N, P and K. These results though obtained in farmers' fields had the benefit of technical supervisions and constraint-free crop husbandry. Therefore rather than representing the response rates obtained by farmers, these represent yield gains which can be obtained by them.

Table 11.5: A Summary of Fertilizer Experiments on Cultivators' Fields with High-yielding Varieties of Crops with and without Irrigation Over the 1969–1980 Period in India

Crop Details	*Number of Trials*	*Rate (kg/ha) N-P_2O_5-K_2O*	*Mean Yield (kg/ha) without Fertilizer*	*% Increase Over Unfertilized Plot*		
				N	*NP*	*NPK*
Rice (KU)	380	120–60–40	2,420	49	74	99
Rice (KI)	9,634	120–60–60	2,960	27	51	56
Rice (RI)	5,686	120–60–60	3,200	28	51	53
Wheat (RI)	10,133	120–60–60	1,550	59	95	114
Sorghum (KU)	367	90–60–60	1,100	51	80	97
Sorghum (RU)	389	90–60–60	610	56	92	120
Maize (KU)	53	90–60–60	1,230	85	107	129
Pearl millet (KU)	207	90–60–60	500	54	110	130
Finger millet (KU)	120	90–60–60	1,250	46	96	118
Chickpea (RU)	1,325	20–40–20	750	36	59	77
Pigeonpea (KU)	53	20–40–20	300	97	210	227
Groundnut (KU)	771	20–60–40	790	39	73	84
Groundnut (RI)	266	20–60–40	1,620	28	56	65
Cotton (U)	55	120–60–40	340	44	91	112
Cotton (I)	250	120–60–40	870	55	82	130

Note: K: Kharif; R: Rabi; I: Irrigated; U: Unirrigated.

Only representative and comparable results are given for sake of brevity. Sole district results are not included.

An important feature of fertilizer response rates is that these are highly management-responsive. For the same crop and in the same village, one farmer may be able to harvest 5 kg grain/kg N and another 15 kg grain/kg N. Higher figures always reflected greater degree of skill employed in fertilizer management, superior level of crop husbandry and better decision-making. The net result is efficient fertilizer use, low cost of producing a unit crop, higher net returns and minimum losses and thus minimal environmental side effects.

In addition to N, P and K, significant improvements in crop yields have been obtained from the application of sulphur and several micronutrients, of which the most important on an all-India basis is zinc. An overall summary of crop responses to S and Zn are given in Table 11.6 and 11.7. These alongwith the NPK responses discussed earlier underscore the fact that Indian agriculture is operating in an era of multinutrient deficiencies. It would not be surprising if a progressive farmer in the wheat belt has to tackle 4–6 nutrient deficiencies in order to get high crop yield and maximum profit. A tea planter in South India may have to apply 7–8 elements for obtaining more than 4500 kg made tea/ha.

Table 11.6: Crop Responses to Sulphur Application Under Field Conditions

Crop	*Studies Averaged*	*Yield Without S Application kg/ha*	*Yield Increase Due to S*			
			kg/ha		*%*	
			Range	*Mean*	*Range*	*Mean*
Wheat	32	3209	150–2120	813	4.5–109.5	25.3
Rice	27	4389	56–1720	752	0.7–39.5	17.1
Blackgram	7	787	130–200	153	17.0–26.3	19.9
Greengram	7	804	100–234	137	13.0–72.5	20.2
Pigeonpea	8	1284	195–592	282	12.4–39.9	21.7
Groundnut	23	1785	133–1480	566	8.2–106.6	31.7
Mustard	18	1122	83–839	335	10.1–92.8	30.0
Soyabean	8	1426	202–698	361	14.2–35.6	25.3
Sunflower	6	1233	70–410	249	5.8–29.7	20.2
Linseed	5	1571	47–459	246	3.2–31.3	15.7

Situations where fertilizer use will not produce a large beneficial yield increase can be (*i*) where potential as well as actual yields are low due to some environmental constraint such as inadequate sunshine (*ii*) where the potential is high but some constraint such as high degree of soil acidity or alkalinity or poor seeds are holding back good crop growth and (*iii*) where soil fertility status itself is adequate to meet crop demands. In the first case, soil nutrients are adequate to support environmentally-restricted low yields. In the second case, once the constraint is removed, by reclaiming the alkali soil/ liming the acid soil, or using good seeds, high yield potentials become realisable and responses to fertilizer are significant. In the third case, no fertilizer is required or recommended as it will not generate an economic output.

A soil which is well-supplied in nutrients cannot remain so indefinitely and becomes deficient with continuous intensive cropping if there is no nutrient input. A watch on this can be kept by fertility monitoring and on nutrients removed by harvested crops. Many soils which were deficient only in N in the early 1960s became deficient in P within a few years of growing HYVs with N-biased fertilizer practices. Heavy crop removals of potash through harvested crops have led to more extensive K-deficiencies as revealed by successive fertility maps. Major shifts in fertilizer product pattern towards S-free materials have placed an increasingly heavy burden on soil-S reserves with the result that soils in many areas are reporting S-deficiencies. As each new deficiency emerges, fertility management for that field becomes more complex and costly. Therefore balanced nutrient application assumes importance and in fact a wider role going beyond NPK. Gains from balanced fertilizer application have been adequately demonstrated, an example of which is given in Figure 11.3.

Table 11.7: Distribution of Crop Responses to Zinc in Experiments on Farmers' Fields

Crop	*Number of Trials*	*Percent Trials in the Response Range (kg/ha) of*			
		< 200	*200–500*	*500–1000*	*> 1000*
Rice	1641	32	35	22	11
Wheat	1973	47	31	15	7
Maize	188	44	25	33	8
Pearl millet	180	63	34	2	1
Sorghum	52	67	27	6	0
Finger millet	33	73	24	3	0
Groundnut	49	18	45	33	4
Cotton	25	60	36	4	0
Total	4141	42	32	17	9

It is only when correct and balanced plant nutrition is neglected that unexpeded and disappointing results can occur. A farmer who used to get good yields by applying only N and insists on doing the same after several years will be in for disappointment because in the intervening period he has been depleting the soil of other nutrients and thus deficiencies have set in which were earlier not there. He can in tact witness a declining trend in crop yields and might even conclude that fertilizer has spoiled his land. In fact what has happened is that he has been using N as a shovel to mine the soil and now his field is saying "This is what you asked for".

Very informative data are available on the dynamics of nutrient responses from a series of long-term experiments initiated during the early 1970s. Results from two locations are used here to illustrate that soil fertility management is not a one-time decision but one which needs a constant vigil if high yield are to be sustained. At Pantnagar, in the earlier years only N produced a yield increase whereas with the passage of time, all four nutrients became important in the case of wheat (Table 11.8A). For rice, N, K and Zn were important throughout even though the relative contribution of K and Zn increased with time will that of N declined. Such results contain important messages for developing balanced fertilizer schedules with a provision for periodical revision. At Palampur, N alone contributed 51 per cent to the total yield increase brought about by NPK + FYM application in the earlier years (Table 11.8B). After a decade, application-of N alone became ineffective, possibly due to severe deficiencies of P and K which had developed in the meantime. The contribution of P to total response nearly doubled and that of potash increased from *nif* in 1973–75 to 23 per cent in 1982–84.

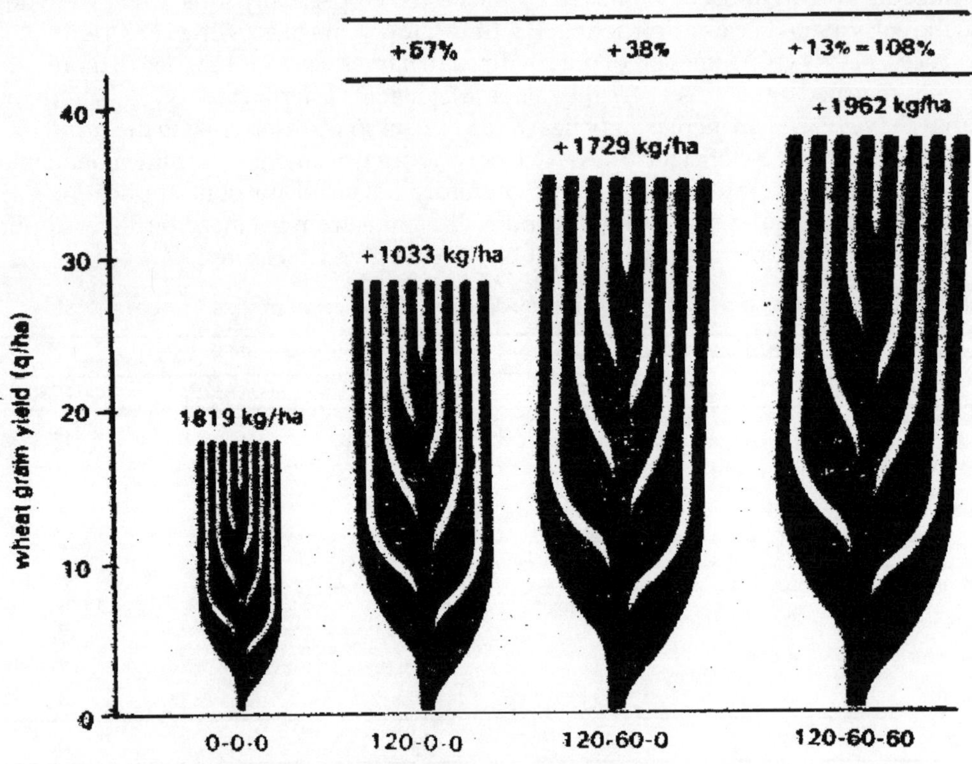

Figure 11.3: An Example of the Impact of Balanced Nutrient Applications on Wheat Productivity Based on 6869 On-farm Trials

Table 11.8: Changes in the Input Response Pattern Over Time: An Illustration

A. System; Rice and Wheat at Pantnagar	*Rice*		*Wheat*	
	1972–74	*1982–84*	*1972–74*	*1982–84*
Total response to NPK + Zn (kg/ha)	1,376	2,867	2,469	2,132
Percent contribution of N	66	43	102	75
Percent contribution or P	– 6	0	– 5	7
Percent contribution or K	21	29	0	7
Percent contribution or Zn	18	28	2	11
B. System: Maize al Palampur			*1973–75*	*1982–84*
Total grain response to NPK + FYM (kg/ha)			4,825	5,866
Percent contribution or N alone			51	0
Per cent contribution of P (over N)			22	42
Percent contribution of K (over N + P)			0	23
Percent contribution or FYM (over NPK)			27	35

Optimum Application Rates

As a result of research conducted during the past 25–30 years, optimum application rates have been worked out for various crop in different agro-ecological environments. General recommendations usually result from experiments on farmers' while site-specific recommendations can be had on the basis of soil tests.

Wherever soil testing is made use of, such recommendations are preferred over general recommendations which are middle of the road type dosages that can be just right or sub-optimal or excessive for particular fields. Optimum nutrient application rates are available separately for (*i*) local varieties and HYVs and (*ii*) for irrigated and rainfed/dryland conditions. As a thumb rule if the recommended rate for local varieties without irrigation is taken as 1, that for local varieties with irrigation or HYVs without irrigation is 2 and for HYVs + irrigation it is 3.5–4.0. These differences are due to different yield levels expected. A few examples of general fertilizer recommendations are given in Table 11.9

Table 11.9: Some Examples of General Nutrient Application Rates Recommended for Improved Crop Varieties Under Irrigation/Assured Water Supply

Crop	*State*	*kg N/ha*	*kg P_2O_5/ha*	*kg K_2O/ha*
Rice	Andhra Pradesh	100–120	30–60	0–50
Wheat	Punjab/Haryana	125	62	30–62
Maize	Bihar	60–100	40–60	25–40
Finger millet	Karnataka	100	50	50
Groundnut	Orissa	20	40	40
Mustard	Haryana	60–80	20–30	0
Soyabean	Madhya Pradesh	20	80	20
Potato	Uttar Pradesh	120	80	100
Cassava	Kerala	50–100	50–100	50–100
Sugarcane	Tamil Nadu	175–275	60	112
Cotton	Gujarat	150–240	0–120	0–120
Jute	West Bengal	44–62	23–30	23–30
Pineapple	Assam	528	88	528
Tea	Tamil Nadu	250–450	70–90	250–450
Coffee	Karnataka	120–160	90–120	120–160

One point which needs emphasis is that generally, optimum application rates also represent the cut off point, that is farmers are not advised to exceed these and application beyond the optimum will not bring in any additional net gain. Application rates below the optimum level are of course a common feature of Indian agriculture and most farmers apply according to their/resources or their notion of what is "optimum" for their conditions. It is this flexibility which distinguishes fertilizers from other agricultural chemical (pesticides) which must be applied only at the recommended rate in order to be effective.

Integration of Diverse Sources of Plant Nutrients

Major components of IPNS are mineral (fertilizers), organics (FYM, composts, green manures), industrial wastes and bio fertilizers which primarily consist of biological N-fixers. Before dealing with the integration of different nutrient sources, some basic aspects need to be discussed.

Some Basic Issues

In nature, organic and inorganic materials co-exist and constantly interact. These are parts and represent different stages of the same system. Both organic and inorganic forms of nutrients enter the same cycle and operate within its frame work (Figure 11.1). (Strictly speaking, urea is an organic compound). These do not set up different camps in the soil to furnish nutrient ions. A soil containing 1 per cent organic matter (OM) has 20,000 tonnes of it/ha in the plough layer and a soil with 5 per cent OM has 100,000 tonnes. It is this large mass which plays host to its ethnic kins which arrive in the form of a few tonnes of organic manures. A plant root about to absorb an ammonium ion cares not and knows not whether this ion has come from urea fertiliser, cowdung or has been provided by bacteria which can fix atmospheric N. Once ammonium is converted into nitrate, it is as available for crop uptake as it is vulnerable for leaching regardless of whether it originated from fertiliser, sludge or manure. Neither a fertiliser nor an organic manure will serve any purpose if it cannot release its nutrients in plant available form. Divergent view points about different nutrient sources to a large extent reveal ignorance of nutrient cycles in nature.

Second issue is that it is incorrect to club all organic manures into one category. Differences between cowdung and a succulent green manure are as large as between fertilisers and manures. In the hierarchy of organic manures, the most valuable from crop nutrition point of view are green manures while most important from utilisation and organic recycling are FYM, residues and composts. In green manures, entire nutrients assimilated by the crop are added to the soil. In crop residues, about 75 per cent of N + P and 25 per cent of K absorbed has already been taken away by the harvested grains. In case of FYM, a crop plant has already served as a source of nutrient twice before it ends up in dung, the major ingredient of FYM or rural compost (Figure 11.4). A cereal or legume straw used as fodder has first of all provided nutrients for grain production.

Then part of the nutrients left ace used for building body tissue of the animal and for milk production. If the entice urine and dung is retrieved, the N in FYM is 17 per cent of what was originally present in the plant. FYM thus represents the third stage of nutrient utilisation.

Third issue is that very little attention is paid to the nutrient content and quality of FYM and usually a casual evaluation in terms of t/ha is made. Its composition can vary by 60–70 per cent. Its performance in research station trials is superior to on-farm experience partly because FYM obtained from dairy herds maintained at university. farms is likely to be of superior quality than what would be available with small and marginal farmers. Nutrient balance studies show that the proportion of P excreted is different from N (Figure 11.5). Dung of animals in lactation is likely to have lesser, nutrients due to nutrients excreted in milk. Dung of cattle fed on cereal straw has less N and very wide CN ratios (40–50) as compared to animals fed on legumes fodder. It is acceptable if farmers overlook these aspects but many research reports make no mention of the composition and quality of manures. Some researchers study it as a source N, others for P and K while others believe that rather than nutrient sources, their major role is in improving soil physical and biological conditions. In nutrient balance studies in long-term experiments where the annual input was 12–15t FYM/ha, nutrient input of FYM was not taken into account.

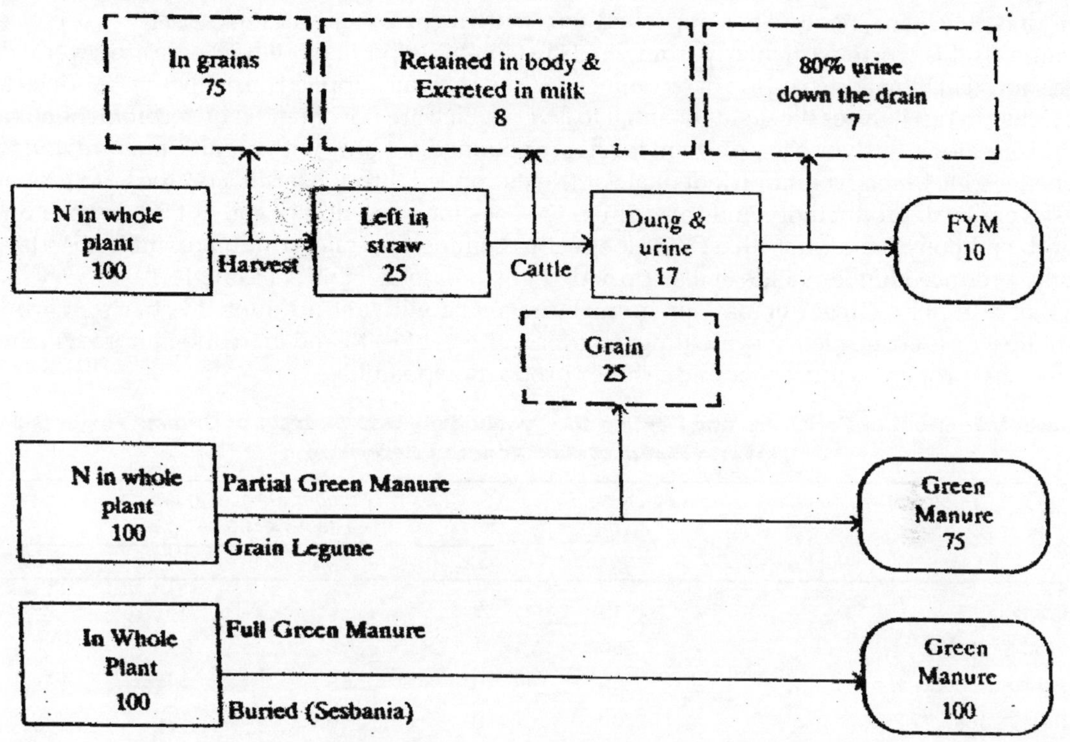

Figure 11.4: Illustration of Basic Differences Between Farmyard Manure (Biological, Residue) and Green Manure (Real Organic Manure)

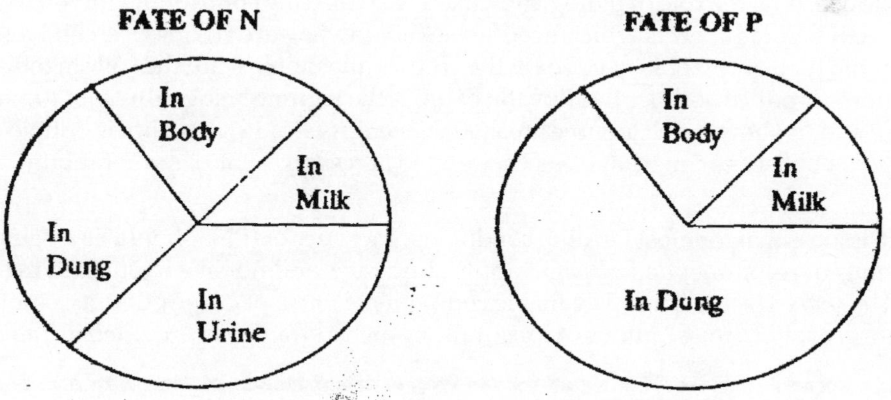

Figure 11.5: Fate of N and P Intake by Milch Buffaloes

The fourth issue is that organic manures are not just sources of nutrients, they have a profound, even sometimes the dominant, effect on soil physical properties resulting in better structure, greater water retention, more favourable environment for root and tuber growth and better infiltration of water. The beneficial effect of such properties on crop yields has rarely been given due economic importance. If organic manure has improved soil structure or water holding capacity by x per cent, how much is this worth particularly from yield sustainability point of view? There are several data which show that FYM added on top of recommended nutrient rates applied through fertilizers increases crop yields. In fact it raises the yield potential to levels which are not obtained by recommended rates of fertilizers alone. Such impacts on yield levels are obtained not only with organic manures but with mulches as well. In such condition (adequate nutrients applied through fertilizers), the FYM primarily raises overall soil productivity and in addition lessenes the depletion of soil nutrients. Data under dryland conditions demonstrate that FYM alongwith recommended rates of nutrients through fertilizers not only produced higher yields, it also provided greater yield stability (Table 11.10). It very likely served as a buffer against fluctuations in moisture availability and promoted better root growth. Therefore efforts are needed in evaluating long-term effects of FYM and in treating improvements in soil physical properties into economic terms or production potentials.

Table 11.10: Effect of Fertilizers and FYM on the Productivity and Stability of Dryland Finger Millet Over a Period of Nine Years at Bangalore

Treatment (Annual)	*Mean Yield kg/ha*	*Number of Years in Which Grain Yield (t/ha) was*			
		< 2	*2–3*	*3–4*	*4–5*
Control	1510	9	0	0	0
FYM @ 10t/ha	2550	1	6	2	0
Fertilizer 50–50–25*	2940	0	5	4	0
FYM @ 10t/ha + 25–25–12.5*	2900	0	6	3	0
FYM @ 10t/ha + 50–50–25*	3570	0	1	5	3

* Fertilizer dose is in terms of kg/ha of N-P_2O_5-K_2O

Fifth issue concerns the role of biologically fixed N. To see the contribution of Rhizobium in the correct perspective, some aspects which need to be kept in view are (*i*) does the soil have adequate population of the type of rhizobia needed by the crop in question (*ii*) are the soil rhizobia effective? Ineffective strains produce nodules but fix little N (*iii*) is the legume being grown for grain, fodder or green manure and (*iv*) how much legume area (90 per cent is rainfed) in reality is double cropped to capture carryover benefits, which have been computed at research stations where abundant water is available.

The sixth issue is that for most farming conditions, packages of IPNS should be assembled within the overall limits of recommended nutrient application rates and not in a haphazard manner. Little purpose will be served by loading all available components into a package but to assemble packages more in the form of alternative options. An example of such a strategy is provided in Table 11.11.

Table 11.11: Some Alternate Strategies for Meeting Nutrient Needs of Rice with and Without a Preceding Grain Legume

Alternate	*Nutrients (kg/ha) N-P_2O_5-K_2O*	*Fertilizer*	*Previous Legume*	*Organic Manure (4t/ha)*	*Green Manure/ Azolla/BGA*
1.	120–30–60	120–30–60	–	–	–
2.	110–27–55	70–15–35	20–0–0	20–12–20	–
3.	110–25–50	65–25–50	20–0–0	–	25–0–0
4.	110–27–55	45–15–35	20–0–0	20–12–20	25–0–0

In the following sections, attempts have been made towards presenting possible criteria for integration of diverse sources of nutrients. To the extent possible, the approach used is to compute fertilizer equivalents of different organic and biofertilizers in terms of impact on crop yield. Emphasis is not given to percent responses to inputs but rather on quantitative estimates and how much of input A and input B are required to generate comparable yield increases. It is obvious that response to an organic manure will be a combination of its effect on chemical, physical and. microbiological properties.

Farmyard Manure

Average quality FYM contains 0.5 per cent N, 0.2 per cent P_2O_5 and 0.5 per cent K_2O, that is 12 kg, nutrients/t. An assessment made several years ago showed each tonne FYM to be equivalent to 3 kg fertilizer nutrients in a single crop and 5 kg in double cropping in terms of yield responses generated. Results of field experiments with HYV rice in West Bengal revealed that 10 tonnes of bulky organic manures on dry weight basis were as effective as 40 kg fertilizer N/ha in terms of impact on paddy yields. A large number of data on response yardsticks of to FYM have become available from experiments on farmers fields. Yardsticks of responses to fertilizer application are also reported in the same publication. Because the number of trial involved is fairly large, fertilizer–FYM equivalents can be computed.

Table 11.12 provides response yardsticks of FYM applied to HYV rice in the *kharif* (monsoon) season in different geographic zones of India. On an average, 20 t FYM produced a yield increase of one tonne paddy, the figure varying from 13.5 t FYM/t paddy in the northeast to 37 t in the central zone. To calculate fertilizer equivalents of FYM, an intermediate level of fertilizer N (80 kg N/ha) is used. The table shows that 87.7 kg fertilizer N produced one tonne paddy. Therefore 20 t FYM was equivalent to 87.7 kg fertilizer N (1 t FYM = 4.4 kg fertilizer N or 230 kg FYM = 1 kg fertilizer N). The inter-zone differences reflect differences in soil properties, FYM composition and quality, temperature, moisture and application techniques.

Table 11.12: Fertiliser N Equivalent of FYM Based on Response Yardsticks for HYV Rice in Experiments on Farmers' Fields in India

Zone	FYM Applied @ 12t/ha			Fertilizer N @ 80kg/ha			Tonnes of FYM = 1 kg N
	Trials	Response kg Paddy/t	Tonnes FYM per t Paddy	Trials	Response kg Paddy/kg N	Kg N per t Paddy	
Central	11	27.0	37.0	1105	8.0	125.0	0.30
Northern	8	41.0	24.4	810	13.9	71.9	0.34
N-Eastern	236	74.1	135	1589	11.2	89.3	0.15
Southern	38	60.9	16.4	2451	10.4	96.2	0.17
All India	293	50.0	20.0	5955	11.4	87.7	0.23

A step further, since FYM is a multi-nutrient carrier, its fertilizer equivalents have been computed using composite yardsticks of response to fertilizer NPK applied in 2 : 1 : 1 ratio. Data may not be strictly comparable but an attempt can still be made. Table 11.13 provides computations based on 12t FYM/ha on one hand and 160 (80–40–40) kg/ha of N-P_2O_5-K_2O on the other. On an average, 104 kgN + P_2O_5 + K_2O or 28.8t FYM were needed to increased paddy yield by one tonne. Thus tentatively each tonne FYM can be equated with 3.6 kg N + P_2O_5 + K_2O in 2 : 1 : 1 ratio in terms of net impact on HYV rice.

Table 11.13: Fertilizer NPK Equivalents of FYM for HYV Rice Based on Experiments on Farmer's Fields

State	80 kg N + 40 P_2O_5 + 40 K_2O		12 t FYM/ha		Tonnes FYM Equivalent to 1 kg NPK (2 : 1 : 1)
	Kg Paddy per kg NPK	Kg NPK per Tonne Paddy	Kg Paddy per t FYM	Tonnes FYM per t Paddy	
Bihar	14.0	71.4	107.5	9.3	0.13
Andhra Pradesh	9.7	103.1	63.5	15.7	0.15
Orissa	11.3	88.5	49.8	20.1	0.23
Gujarat	6.1	163.9	17.9	55.9	0.34
Madhya Pradesh	11.6	86.2	13.3	75.2	0.87
Maharashtra	8.8	113.6	49.9	20.0	0.18
Karnataka	12.5	80.0	58.4	17.1	0.21
Tamil Nadu	8.2	122.0	58.3	17.2	0.14
Average	**10.3**	**103.6**	**52.3**	**28.8**	**0.28**

These computations show that fertilizer application rates can be adjusted @ 3.6 kg NPK for each tonne FYM available. However regional variations are large and there is a 6–7 fold difference in the quantities of FYM which can substitute/supplement one kg of fertilizer nutrients. It is surprising that paddy response rates/t FYM were 3.7 times higher with HYVs than with improved tall rice cultivars. Almost 74 t FYM was associated with a yield increase of one tonne in local varieties while the figure for HYVs was 20 t FYM. Are the HYVs not only more fertilizer-responsive but also FYM responsive? A potentially valuable conclusion though based only on results with rice is that since one tonne FYM (12 kg nutrients) produced as much grain as 3.6 kg fertilizer nutrients, the crop production efficiency of FYM is 30 per cent that of fertilizer nutrients on one harvest basis. Actual figure is expected to be higher than this for K and lower for N and P.

Application of FYM and fertilizers on equivalent N basis to Sehima Heteropogon grassland showed that N from FYM was 40 per cent as effective as fertilizer N in increasing crop yields, the figures with specific fertilizers being 30 per cent as compared to AS, 40 per cent in comparison to CAN and 60 per cent as effective as urea-N. In case of oats FYM-N was 50 per cent as effective as urea-N in increasing crop yield. Dry forage yield with 120 kg N + 80 kg P_2O_5/ha was 5.2 t/ha (FYM + SSP), 7.8 t/ha (urea + SSP) and 8.9 t/ha with 50 per cent N as urea, 50 per cent N as FYM and SSP. Basically similar treatment can be extended to composts if data are available.

Green Manures

Response rates to green manure are available from a large number of on-farm trials but unfortunately such data are only available for local rice varieties and not for HYVs. Therefore a comparison of green manure has been made with fertilizer responses on local rice varieties. On an average, each tonne green manure produced 39.1 kg paddy or 235 kg/ha when applied @ 6 t/ha (Table 11.14). Assuming N to be the major input through green manuring, equivalence in terms of fertilizer N has been worked out based on N-response rates. Paddy yield was increased by one tonne either through 112.4 kg fertilizer N or through 25.6 t green manure with local varieties. Thus each tonne green manure gave as much yield as 4.4 kg fertilizer N. For local varieties, the following picture emerges:

One t green manure = 28 t FYM = 4.3 kg N = 9.3 kg urea.

Table 11.14: Fertilizer N Equivalent or Green Manure on Locally Improved Tall Rice Varieties Based on Experiments on Farmers' Fields

Zone	*Green Manure, 6 t/ha*			*Fertilizer N @ 80kg/ha*			*Tonnes GM Equivalent to 1 kg N Fert.*
	Trials	*Response kg Paddy per t. GM*	*Tonnes GM/t Paddy*	*Trials*	*Response kg Paddy per kg N*	*Kg N for one t Paddy*	
Central	77	40.0	25.0	442	8.9	112.3	0.22
Northern	23	445	22.5	274	9.9	101.0	0.22
N-Eastern	158	35.6	28.1	386	8.7	114.9	0.24
Southern	325	49.0	20.4	877	8.7	114.9	0.18
Total	**583**	**39.1**	**25.6**	**1979**	**8.9**	**112.4**	**0.23**

In Tamil Nadu, one tonne of green manure is equated with 2 tonnes of FYM or compost for paddy production. On-station results show nitrogen gains of a much higher order from green manuring. A good stand of Sesbania can bring in 80–1.00 kg N/ha. In field experiments at Karnal, response to green manuring was on par willl response to 80 kg/ha of fertilizer N on an amended sodic soil (Figure 11.6). Benefits from green manuring were higher at lower level of fertilizer input but were still substantial on top of 120 kg/ha fertilizer N. The fertilizer + green manure treatment was able to raise paddy yields to a level which was not achieved with fertilizer alone. This once again underscores their favourabie impact on soil properties and productivity which goes beyond supplying nutrients. Benefits are further increased if adequate phosphate is applied to the green manure. On an average, on-station results show that green manure can be relied upon to provide 50–60 kg N and if the recommended rate is 100–120 kg N/ha a green manure can provide 50–60 per cent N. The next step is to examine the cost of raising the green manure in comparison to the fertilizer cost of 50–60 kg N which at late 1991 prices is Rs. 330–400.

In areas with an assured water supply and 2–3 months available between two main crops, green manure can compete with a pulse crop particularly where short duration pulse cultivars are available. For Indian agriculture, standard green manuring is estimated to be practiced on 6.3 m.ha, that is on 4.4 per cent of net cultivated area. Major systems using green manures are rice-based and sugarcane-based.

Rhizobium

Among biofertilizers, Rhizobium inoculants are of greatest importance. Their role is quite well defined as it is restricted to augmenting N supplies of legumes and the succeeding crop. Their importance is generally seen with reference to pulses and leguminous oilseeds but they are also responsible for much of the N provided by leguminous green manures discussed in the preceding section.

Rhizobium bacteria live in the root nodules of leguminous plants through a symbiotic relationship. The host plant gives them a place to live and energy supply and in turn gets biologically fixed N. Legumes do not get this N for free. They pay for it in advance by transferring part of the photosynthates to the nodules to meet the bacteria's energy needs. There is no free lunch as Brady puts it. Work with chickpeas shows that N-fixation starts about 15 days after sowing when nodules are small but pink and attains peak level in early stage of seed formation. Significant responses to 20–30 kg N/ha as

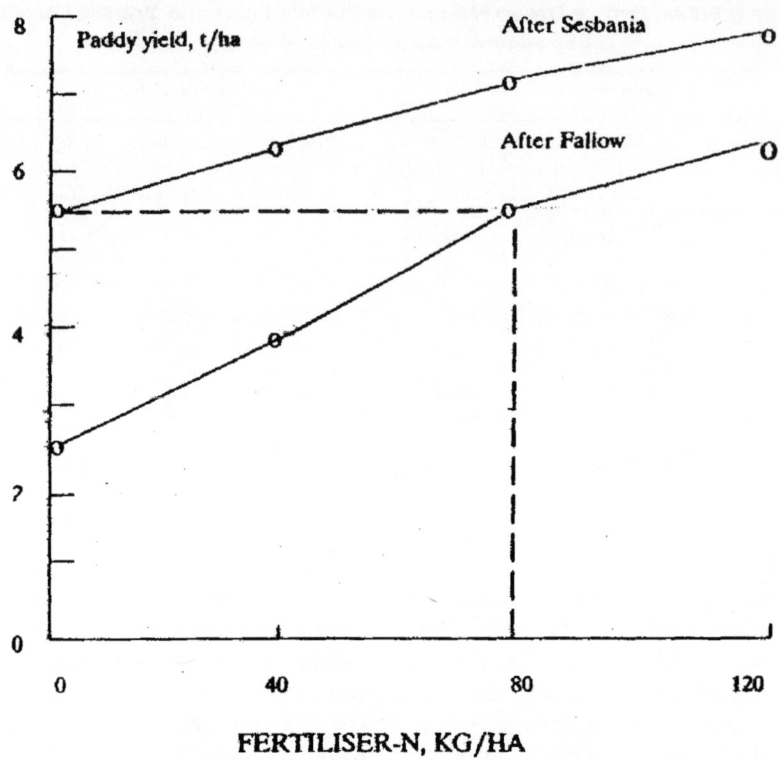

Figure 11.6: Impact at Green Manuring on Rice Yields

starter are thus to be expected under good growth conditions (Table 11.15). Since legumes occupy 34 million ha in India, the potential relevance of Rhizobia extends to almost one-fourth of net cropland area. The N-fixed by Rhizobia benefits crop production in two ways (*i*) by meeting a large part of the legume's nitrogen needs and (*ii*) by enriching the soil for the following crop's benefit.

Results of a large number of on-farm experiments show that legumes respond markedly to fertilizer N. This is understandable because legumes have high N-requirement, soils are poor in N and the N-fixation mechanism does not become functional from day one. At application rates of 20 kg N/ha, overall response rate (grain/kg N) was 14.2 in pigeonpea, 14.4 in groundnut and 15.4 in chickpea all under non-irrigated conditions (Table 11.15). Responses of such high magnitude point at the possibility that the N-fixation apparatus under natural conditions without inoculation may not be very productive.

Field surveys have shown that status of nodulation in most cases is poor to moderate (Figure 11.7). Much N-fixation thus cannot be expected unless inoculation with active and efficient strains is done and optimum pH and P supply ensured. In a survey of farmers' fields around Gwalior (M.P.), 39 per cent fields had less than 100 Rhizobia/g soil, 17 per cent had 100–1000 and 44 per cent fields had a population higher than 1000. Such studies can form the basis for delineation of areas which need inoculation on priority basis. Indian soils therefore are not only poor in chemical fertility but also poor in bio-fertility. Generally, a significant beneficial effect from using Rhizobium biofertilizer is expected if the Rhizobia population is less than 100 cells/g soil. In studies with chickpea, significant responses to inoculation were also associated with native Rhizobium population of less than 100 bacteria/g soil. At Rhizobia population of over 1000/g soil, gains from inoculation were generally small. Field surveys have also shown that several strains of native Rhizobia are ineffective. Proportion of ineffective strains was reported to be as high as 40 per cent in chickpea, 53 per cent in green gram and 63 per cent in groundnut.

Table 11.15: Response Yardsticks of Grain Legumes to Fertilizer Nitrogen in Experiments on Farmers' Fields

State	*Chickpea*		*Pigeonpea*		*Groundnut*	
	Trials	*Kg Grain per kg N*	*Trials*	*Kg Grain per kg N*	*Trials*	*Kg Grain per kg N*
Andhra Pradesh	47	16.5	56	175	258	18.0
Bihar	77	17.0			25	25.0
Gujarat			159	15.5		
Haryana	88	12.0				
Himachal Pradesh	50	11.0				
Karnataka	275	11.0	104	8.0	310	14.5
Madhya Pradesh	624	19.0	15	10.5		
Maharashtra	351	8.5			495	12.0
Orissa	71	8.5	39	19.0		
Punjab	113	10.5			62	17.0
Rajasthan	267	18.0	159	13.0	38	12.5
Tamil Nadu					384	14.0
Uttar Pradesh	408	21.5			14	12.0
Average		**15.4**		**14.2**		**14.4**

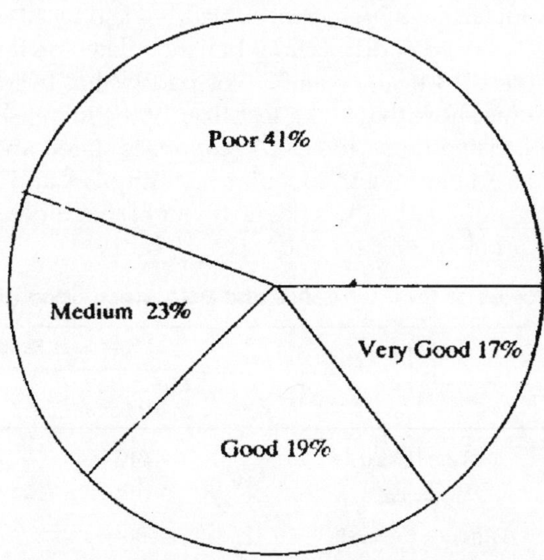

Figure 11.7: Nodulation Status of Chickpea Based on 314 Fields

Not many on-farm data are available on the impact, inoculation on grain yields. In 12 trials with chickpea, inoculated plots gave 116 kg more grain as compared to uninoculated plots. This much yield increase is associated with 8 kg fertilizer N based on available yard sticks. In another set of field demonstrations, inoculation resulted in grain yield increase of 112–227 kg/ha. In 16 results with chickpea, apparently at research stations, Rhizobium inoculation increased grain yield by 342 kg/ha (range 30–610). In 10 out of 16 cases, the inoculation response was 250–500 kg grain/ha. In terms of chickpeas response to fertilizer N, inoculation in these trials can be equated with the application of 22 kg fertilizer N/ha. Because the cost of 22 kg N is Rs. 144 and that of inoculation about Rs. 15–20/ha, the advantage of adopting Rhizobium inoculation is obvious.

Response of groundnut to inoculation comes to 273 kg/ha averaged over 29 results. Results of on-farm trials with *Kharif* groundnut gave a yield increase of 288 kg/ha due to 20 kg N/ha through fertilizer. The impact of inoculation was approximately equivalent to applying 19 kg fertilizer N/ha. Results with grass-legume pastures show that total dry fodder yield (grass + *Stylosonthes quionensis*) without fertilizer was comparable to grass yield obtained with 32 kg N/ha through fertilizer and that through pasture grass + *Centrocema pubescence* was comparable to grass yield associated with 72 kg N/ha fertilizer input.

Fertilizer N equivalents of Rhizobium biofertilizer for direct effect on legume grain yield thus come to 19–22 kg N/ha. It needs to be stated that these estimates are not based on common trial set for fertilizer and inoculation. These are tentative estimates which can be made at present. Area treated at present through Rhizobium inoculants is around 2 million ha which is 6 per cent of the area cropped annually to legumes in India.

Blue Green Algae

The BGA is an important biofertilizer for augmenting N-supplies to rice. Most frequently mentioned estimate of N fixed by BGA inoculation is 20–30 kg N/ha. Some response yardsticks to BGA inoculation at different locations are presented in Table 11.16. Data on fertilizer equivalents cannot be computed but inoculation @ 10 kg BGA/ha can be taken as worth 20–30 kg fertilizer N. In 56 on-farm trials in Madhya Pradesh, BGA (without fertilizer N) increased paddy yield by 404 kg/ha. Results from Tamil Nadu showed that BGA also increased paddy yield when inoculated on top of 100 kg fertilizer N/ha. Average impact of BGA in these 18 trials, was 373 kg paddy/ha. In Madhya Pradesh, fertilizer experiments on cultivators fields show that 40 kg fertilizer N/ha increased the yield of HYV rice in *Kharif* at 10.8 kg paddy/kg N. Using these and BGA responses stated above, BGA inoculation was equivalent to 38 kg fertilizer N. At Pantnagar, inoculation with BGA and 25 kg fertilizer N/ha gave rice yields comparable to those obtained with 50 kg fertilizer N/ha. Inoculation activity at present is estimated to cover 20,000 ha or one in 2000 ha under rice 10 India.

Table 11.16: Response Rates of Rice to *Azolla* and BGA Inoculation at Different Locations

Location	*State*	*Increase in Paddy Yield (kg/ha)*	
		Die to BGA at 10 kg/ha	*Due to Azolla at 6t/ha*
Coimbatore	Tamil Nadu	440	400
Pattambi	Kerala	140	150
Hyderabad	Andhra Pradesh	80	480
Patna	Bihar	460	780
Chinsurah	West Bengal	340	410
Titabar	Assam	790	430

Azolla

This is a potential biofertilizer, technically of proven value, for flooded rice at moderate temperature regimes. Fresh Azolla is 94 per cent water and contains 4–5 per cent N on dry basis. Azolla can be grown alongwith rice in the main field or as a green manure before planting rice or even both as a green manure and as a dual crop. Azolla in its growth habits resembles a crop in many ways. It responds markedly to phosphate application and can grow well under moderate levels of fertilizer N. Some data on response rates to Azolla have been given in Table 11.16. Generally, 6t fresh Azolla increases paddy yield comparable to 20–30 kg fertilizer N and 10–12t azolla is comparable to 50 kg N or so. In 25 experiments at 13 locations, incorporation of 6t azolla increased paddy yield by 200 to 1500 kg/ha. Inter-location variability was high but comparatively less than with BGA. Fertilizer N equivalent of azolla can be taken as 3.0–4.0 kg/tonne.

Theoretically it is possible lo raise azolla three times for the same rice crop-once as a green manure and twice as dual crop alongwith rice. In this way, total N input from azolla can go upto 75–90 kg/ha. Coupling BGA, azolla and green manuring places N-supply systems for rice is a more favourable position than other cereals. Again, much of the N needs of rice can theoretically be met from non-fertilizer sources by green manuring rice field with sesbania (50–60 kg N) followed by taking two dual crops of Azolla (another 40 kg N). At Pantnagar, green manuring with *Sesbania* followed with Azolla inoculation @ 3t/ha resulted in paddy yields similar to those obtained from 90 kg N/ha given through fertilizer. There are In fact numerous combinations possible but their critical practical feasibility from agro-economic angles under different environments has not been studied.

Conclusions

Based on discussions above, some possible fertilizer equivalents of diverse plant nutrient sources are presented in Table 11.17. These are more in the nature of an illustration. By subjecting various sources to a common parameter, integrated nutrient supply packages can be assembled. This should be done within the overall limits of recommended nutrient rates. Which should be guided by considerations of nutrient needs and impact on soil properties and not by waste disposal considerations as is sometimes done with sewage sludge to avoid soil sickness.

Table 11.17: Some Fertilizer Equivalent of Organic Manures and Biofertilizers

Component	*Input Level*	*Fertilizer Equivalent of Input in Terms on Crop Yield*
Organic manure (FYM)	Per tonne	3.6 kg N + P_2O_5 + K_2O (2 : 1 : 1)
Green manure (Sesbania)	Per tonne	4.4 kg N*
Green manure (Sesbania)	45 days crop	50–60 kg N for HYV rice
Cowpea intercropped with castor	Legume buried after 6 weeks	30 kg fertilizer N on castor
Leucaenia loppings	88 kg N in Leucaenia	25 kg fertilizer N on sorghum
Rhizobium	Inoculant	19–22 kg N
Azotobacter and Azospirillum	Inoculant	20 kg N
Blue Green Algae	10 kg/ha	20–30 kg N
Azolla	6–12 t/ha	3–4 kg/t
Sugarcane trash	5 t/ha	12 kg N/t
Rice straw + water hyacinth	5 t/ha	20 kg N/t

* For tall rice varieties. Effect expected to be higher on HYVs.

There is a need to give a quantitative and if possible an economic value to improvements caused by organic manures in soil physical properties. These quite often result in raising the yield response curve by increasing soil productivity of which fertility is one component. Fertilizers account for about 24 per cent of the nutrients removed by crops on a net bases after taking their use-efficiency into account. Estimates of animal dung used as domestic fuel have increased markedly from 30 per cent in the early 1970s to 75 per cent in the early 1990s. The fuel crisis is obviously taking its toll but routing cow dung through biogas plants (wherever possible and affordable) is a much more sound route than burning of dung cakes. A farm-level study in Ludhiana showed that farmers preferred utilisation of wet slurry with irrigation water than dry slurry.

A biogas slurry-based organo-mineral fertilizer can be formulated by using 48 kg dry grade biogas processed sludge, 11 kg urea, 31 kg single superphosphate and 15 liters water. The sun-dried manure has been suggested for growing vegetables and other cash crops. While benefits from Rhizobium inoculation to the host plant are real gains carryover benefits to the following crop under real-life conditions is not that assured. Ninety per cent area under grain legumes is rainfed/dryland where sequence (double) cropping is not a common possibility.

The realities, urgencies and pressures on the land all indicate that there will be no rest and recuperation for our soils. Nutrient removal will increase resulting ingreate depletion. Blue prints for integrated nutrient supply need to be developed in a realistic manner. Several research gap remain to be filled for all nutrient sources but even with available information, progress can be made. For example, referring to BGA, Swaminathan stated that little is known about the conditions that lead to success of inoculation or about the factors that govern survival of the organism. Much of this could apply to other organisms as well. Because nutrient application rates in most areas are well below optimum, various organic and biofertilizers can help to take these towards optimum level and to that extent supplement fertilizers. These sources will be able to substitute for part of the fertilizers, particularly N, where farmers have reached the optimum levels of nutrient application. Less than 10 per cent farmers have reached such levels.

While possibilities for integrated nutrient supply are real and attractive, most nutrient packages for high yields required to feed an expanding population from a non-expanding area will continue to be fertilizer-driven. This assessment in no way under estimates the importance of other nutrient sources most of which are gainfully integratable.

Chapter 12
Bulky Organic Manures and Crop Residues

Organic manures have been time-tested materials for improving the fertility and productivity of soils. It is however only during the last 10–15 years that these have been incorporated into integrated nutrient supply system for intensive high-yielding cropping sequences in contrast to the subsistence level production of the past.

Organic manure is a very broad term. It covers manures made from cattle dung, excreta of other animals, rural and urban composts, other animal wastes, crop residues and last but not the least green manures. This chapter is confined to bulky organic manures, composts and crop residues. Green manures and also manures made from industrial wastes and biogas plants are dealt with in next chapters.

Organic Manures and Their Composition

Materials such as farmyard manure, composts and crop residues which are bulky and supply low quantities of major plant nutrients are termed as bulky organic manures (Table 12.1). Concentrated organic manures, such as oil cakes, slaughter house wastes, fish meal, guano, shoddy and poultry manure are comparatively richer in NPK.

Farmyard Manure (FYM)

This is the most commonly used organic manure in India being readily available. It consists of a mixture of animal-shed wastes containing dung, urine and some straw. An application of 10t well rotted FYM/ha can add 50–60 kg N, 15–20 kg P_2O_5 and 50–60 kg K_2O (Table 12.1).

Composts

These are prepared through the action of microorganisms on wastes such as leaves, roots and stubbles, crop residues, straw, hedge clippings, weeds, water-hyacinth, bagasse, saw dust, kitchen wastes, and human habitation wastes. These materials undergo intensive decomposition under medium-high temperatures in heaps, windrows or pits with adequate moisture. In about 3–6 months, an amorphous, brown to dark brown humified material called compost is obtained. It is more stable in form, valuable source of plant nutrients, helps in maintenance soil organic matter and in improving soil physical condition and biological activity.

Table 12.1: Nutrient Status of Some Organic Waste Materials, FYM, Compost, Cakes and Residues

Category	*Source*	*Nutrient Content (%)*		
		N	*P_2O_5*	*K_2O*
Animal Wastes	Cattle dung	0.3–0.4	0.10–0.15	0.15–020
	Cattle urine	0.80	0.01–0.02	0.5–0.70
	Sheep and goat dung (mixed)	0.65	0.5	0.03
	Night-soil	1.2–1.5	0.8	0.5
	Human urine	1.0–1.2	0.1–0.2	0.2–0.3
	Leather waste	7.0	0.1	0.2
	Hair and wool waste	12.3	0.1	0.3
FYM Composts	Farmyard manure	0.5–1.0	0.15–0.20	0.5–0.6
	Poultry manure	2.87	2.90	2.35
	Town compost	15–2.0	1.0	1.5
	Rural compost	0.5–1.0	0.2	0.5
	Water hyacinth compost	2.0	1.0	2.3
Oil Cakes	Castor	5.5–5.8	1.8	1.0
	Coconut	3.0–3.2	1.8	1.7
	Cotton seed	3.9 (6.5)*	1.8 (2.8)*	1.6 (2.1)*
	Groundnut	4.5 (7.8)*	1.7 (1.7)*	1.5 (1.4)*
	Karanj (*Pongamia pinnata*)	3.9–4.0	0.9–1.0	1.3
	Neem (*Azadirachta indica*)	5.2	1.0	1.4
	Niger	4.8	1.8	1.3
	Mahua Butter tree (*Bassia latifolia*)	25–2.6	0.8	1.8
	Rapeseed	5.1	1.8	1.0
	Linseed	5.5	1.4	1.2
	Safflower	4.8 (7.8)*	1.4 (2.2)*	1.2 (2.0)*
	Sesame	6.2	2.0	1.2
Animal Meals	Blood	10–12	1.2	1.0
	Meat	105	25	0.5
	Horn and hoof	13.0	0.3–1.5	–
	Raw bone	3–4	20–25	–
	Steamed bone	1–2	25–30	–
	Fish	4–10	3–9	1.8

* Figures in ( )* are for decorticated material.

Composts are broadly divided into two groups such as rural compost and town or urban compost. Details regarding different methods of composting such as aerobic, anaerobic, mechanical compost plants etc. have been documented. Important factors which need to be maintained at optimum level for efficient composting are C : N ratio, C : P ratio, particle size, blending or proportioning of raw materials, moisture, aeration, temperature, destruction of pathogens and parasites and use of microbial activators. These are summarised below.

Parameter	*Optimum Value for Composting*
C : N ratio of feed	25 to 35
Particle size	10mm for agitated systems and forced aeration, 50 mm for long heaps and natural aeration.
Moisture content	50 to 60 per cent (higher values possible when using bulking agents)
Air flow	0.6 to 1.8 m^3 air/day/kg volatile solids during thermophillic stage, or maintain oxygen level at 10 to 18 per cent
Temperature	55 to 60°C held for 3 days
Agitation	No agitation to periodic turning in simple systems and short bursts of vigorous agitation in mechanized systems
pH control	Normally not necessary
Heap size	Any length, l.5m high and 2.5 m wide for heaps using natural aeration. With forced aeration, heap size depends on need to avoid overheating.
Activators	Use of efficient cellulolytic fungi and biofertilizers

Available technologies for the preparation of superior quality composts and in less time are discussed in a later section. Compost prepared from garbage, night-soil, urine, etc. contains 1.5 per cent N, 1.0 per cent P_2O_5 and 1.5 per cent K_2O (40 kg nutrients/tonne). Nutrient content of a number of organic materials is also provided in Table 12.1.

Potential and Available Supplies

Cattle and Buffalo Wastes

The cattle and buffalo population over the years has shown an increasing trend. The census conducted during 1972, 1977 and 1982 reported 224, 240, and 260 millions indicating an increase of 7.6 per cent during 1977 over 1972 and 8.3 per cent during 1982 over 1977. Population of the above animals during 1990 is estimated to be around 290 millions.

The annual excretion of wet dung and urine from cattle and buffalo thus works out to 1228 and 800 mt respectively (Table 12.2). If the entire wet dung and urine excreted by bovines are conserved for manurial purposes, its potential for supplying major plant nutrients has been worked out as 3.442 mt N, 1.307 mt P_2O_5 and 2.214 mt K_2O. It cannot be said with any certainity as how much dung is used for fuel and how much for manure though recent reports mention that almost 75 per cent of total dung may be used as domestic fuel. In contrast, during 1970s, 30–50 per cent cattle dung was reported to be used for fuel purposes.

Crop Residues

Substantial quantities of crop residues are produced in India every year (Table 12.3). Five major crops, namely rice, wheat, sorghum, pearl millet and maize alone yield approximately 236 mt straw/

stover. On an average, cereal straw and, residues on maturity contain about 0.5 per cent N, 0.6 per cent P_2O_5 and 1.5 per cent K_2O. The nutrient potential of cereal straw / residues in million tonnes from these five crops thus comes to 1.13 mt N, 1.41 mt P_2O_5 and 3.54 mt K_2O. Even if 50 per cent of these residues are used as animal feed, the rest can be recycled for their beneficial effects on soils and plants. Crop residues can be recycled either by composting, or by way of mulch or direct incorporation in the soil. It is important to recognise that the availability of crop residues in future will increase as production goes up. Its availability is directly dependent on biomass produced.

Table 12.2: Nutrient Content and Theoretical Potential of Total Dung and Urine Excretion by Cattle and Buffaloes

Source	*Annual Excretion Million Tonnes*	*Million tonnes*					
		N		*P_2O_5*		*K_2O*	
		%	*mt*	*%*	*mt*	*%*	*mt*
Wet dung	1227.8	0.15	1.842	0.10	1.227	0.05	0.614
Urine	800.0	0.20	1.600	0.01	0.080	0.20	1.600
Total NPK (mt)			3.442		1.307		2.214

Table 12.3: Potential of Farm Residues and Plant Nutrients in them

Crop	*Grain : Straw Ratio*	*Residue Production (ml)*	*% (oven dry basis)*		
			N	*P_2O_5*	*K_2O*
Rice straw	1 : 1.5	106.01	0.58	0.23	1.66
Wheat straw	1 : 1.5	80.99	0.49	0.25	1.28
Sorghum	1 : 2.0	21.04	0.40	0.23	2.17
Pearl millet	1 : 2.0	15.58	0.65	0.75	2.50
Maize	1 : 1.5	12.50	0.59	0.31	1.31
Total Pulses	1 : 1.0	13.70	1.60	0.15	2.00
Pigeonpea	1 : 2.5	6.65	1.10	0.58	1.28
Chickpea	1 : 1.0	5.05	1.19	n.a.	1.25
Sugarcane	1 : 0.2	40.92	0.35	0.04	0.50
Oil seeds	1 : 2.0	35.78	–	–	–

Technologies for Quicker and Better Compost Production

Two handicaps common to all bulky manures are their low nutrient content and large volume (bulk). Major developments to upgrade the nutrient content of composts and to hasten the composting process so that materials of better quality can be obtained in less time are briefly discussed here with particular reference to low-cost technologies.

Compost Accelerator/Inoculants

Strains of microorganisms which can hasten the process of composting of organic residues have been isolated. Such microbes are cellulolytic and lignolytic type of microorganisms. Work on the cellulolytic fungi as compost inoculant, particularly at IARI has been reviewed. Several types of wastes were chopped to 5–6 cm size and filled in the pits. Homogenised fungal culture *Trichurus spiralis, Paeciliomyces fusisporus, Trichoderma viride* and *Asperquillus* sp. was then added at 300 gm/t material. Moisture was initially maintained at 100 per cent. Rock phosphate at 1 per cent was added to narrow down the C : P ratio of substrates used as well as to hasten the composting process. After every 15 days, samples are drawn and composting mass turned upside down.

Within 8 to 10 weeks a good quality compost from paddy straw could be prepared containing around 1.7 per cent N and C : N ratio of 12.3 (Table 12.4). The beneficial effect of cellulolytic fungi in composting of dairy farm wastes has been reported. They also reported the beneficial effect of 10 per cent cattle dung as inoculant and 2 per cent rock phosphate in composting of wool waste containing 66.9 per cent organic carbon and 4.6 per cent N. The final product was obtained after 10 weeks of composting. It contained 20.6 per cent OC and 10.0 per cent N with a C : N ratio of 2 : 1.

The use of efficient cellulolytic culture not only helped in preparing compost in 8–10 weeks but also reduced the bulk by 5–10 per cent. The enrichment of total N was pronounced in the presence of the inoculant (Table 12.4A). Similar observations have been recorded with mixed crop wastes, sugarcane trash etc. at Hissar, Ranchi and, Pune. This indicates the potential of the cultures for rapid composting of dry and wide C : N ratio organic materials..

Table 12.4: Role of Various Microorganisms in Preparation of Compost from Organic Residues

Location	*Substrate*	*Total Nitrogen %***		*C : N Ratio**	
		Without Inoc.	*With Inoc.*	*Without Inoc.*	*With Inoc.*
A: Enrichment with cellulolytic fungi					
New Delhi	Sorghum stalk + wheat straw	1.41	1.65	20.0	14.3
New Delhi	Jamun leaves	1.44	1.56	20.4	15.7
New Delhi	Paddy straw	1.15	1.30	23.9	18.9
New Delhi	Paddy straw	1.52	1.76	15.5	12.3
Kanpur	Dairy farm waste****	0.57	0.63	22.6	16.6
Ranchi	Wheat straw + water hyacinth	1.00	1.37	28.4	18.6
Ranchi	Rice straw	0.90	1.20	41.0	30.0
Hissar	Mixed crop residues	1.38	1.52	24.0	20.0
Pune	Sugarcane trash	0.98	1.54	35.0	24.0
B: Enrichment with Azotobacter and P-solubiliser***					
New Delhi	Wheat straw + Sorghum stalk	1.38	1.82	20.6	12.0
New Delhi	Paddy straw	1.52	1.82	15.6	11.6
New Delhi	Paddy straw + *Leucaena* (4 : 1)	1.16	1.85	22.2	12.2
New Delhi	Chopped paddy straw	1.30	1.78	20.0	13.3
New Delhi	Unchopped paddy straw	1.26	1.72	25.2	15.5
New Delhi	Banana leaf	1.90	2.80	19.3	11.6
Kanpur	Dairy farm waste****	0.58	0.64	20.1	15.5
Pune	Sugarcane trash	1.08	1.34	26.0	33.0

* Inoculation at the start. ** Sampled after three months.

*** Inoculation one month after composting. **** Partially degraded C : N = 61.1.

Enrichment or Compost by Bio-inoculants

Enrichment of compost with N-fixing bacteria and P-solubilising fungi is one of the possible ways of improving nutrient content of the final product. This objective can be achieved by introducing efficient microbial inoculants. Details of these aspects have been described elsewhere. Inoculation with *Azotobacter* and phosphate solubilising culture in present of 1 per cent rock phosphate is a beneficial input to obtain a good quality compost which is rich in N and available P_2O_5 (Table 12.4B). The humus content was also significantly higher in material treated with microbial inoculants. The quality of compost produced from dairy farm waste was improved by its inoculation with *Azotobacter* and *Aspergillus awamori* as shown in Table 12.4B. Similarly sugarcane trash compost containing 1.34 per cent N with a C : N ratio of 26 was obtained with the bio-inoculants. Using chopped straw, compost with 1.78 per cent N and C : N ratio of 13.3 was produced with inoculants as against 1.30 per cent N and C : N ratio 20 without bio-inoculants. Banana leaf compost prepared with similar biotechnology, contained 2.8 per cent N and a C : N ratio of 11.6 against 1.90 per cent N and C : N ratio of 19.3 in the control treatment.

The effect of this technology was also tested in improvement of the quality of urban compost prepared at mechanical compost plants. Nitrogen gain of almost 100 per cent was recorded with the combined inoculation of *A. chroococcum* and *B. polymyxa* in rock phosphate amended compost. The quality of urban compost can be improved by blending it with sewage sludge in a ratio of 2 : 1. Nitrogen content of the blended urban compost with and without rock phosphate was augmented to 1.42 per cent and 1.29 per cent as compared to 0.60 per cent N in urban compost alone.

Competing Uses of Resources

Cattle Dung

It is estimated that during winter months 40 per cent of collected dung is dried into dung cakes and 58 per cent put into compost pit or heaps. However, during rainy season only about 13 per cent is dried into dung cakes and 85 per cent finds its way into the manure. About 2 per cent of cattle dung is used for other household purposes such as plastering of floor and walls. The collected dung being used for fuel purposes was estimated at 30 per cent in the early 1970s and upto 75 per cent in the early 1990s.

Crop Residues/Straw

The bulk of straw from crop residues is used as cattle feed. However, paddy straw which is produced in the largest quantity, is a poor animal feed being rich in silica and oxalic acid. In Punjab, Haryana, and Central and Western parts of Uttar Pradesh, it is not used as cattle feed but is disposed off either by burning or by any other means. Straw is used as packaging material and for making paper and building boards. It is also used as thatching material for huts and temporary dwellings in villages. Considerable amount of forest litter is burnt as fuel. Sugarcane bagasse produced is almost entirely used as fuel in boilers of sugar factories. It is also a valuable material for the production of pulp, paper and boards. Cotton gin trash is also used as fuel.

Fate of Organic Materials in Soil

The rate of decomposition of organic manures in soil depends on their C : N ratio, proportion of cellulose, hemicellulose and lignin, climatic conditions and the soil properties. Organic compounds in the manures including fresh crop residues are mineralized in the soil mainly by bacteria, fungi and actinomycetes to carbon dioxide, water and available plant nutrients and intermediate organic metabolites.

Legume residues decompose faster than cereal residues due to their higher N content and narrower C : N ratio. Addition of neem (*Azadirachta indica*) cake to cereal straw accelerated its decomposition. Addition of chemical N also speeded up decomposition of cereal straw by lowering the C : N ratio.

The addition of organic matter profoundly influences the soil microflora and different groups of microorganisms respond differently. In general, leguminous residues, followed by cereal straw supported more bacteria and actinomycetes. *Azotobacter* population was appreciably increased by the incorporation of cereal straw alone in alluvial soil, but the response in red soil was poor. Enrichment of field soil with 2, 5 and 10 t straw/ha stimulated N-fixing bacteria *Azotobacter* and anaerobic bacteria. Rice straw is particularly effective in alkaline soils, since its application decreases soil pH.

Sesbania released the maximum available N, S and P while cereal straw caused immobilisation of phosphorus and nitrogen at same intervals (Figure 12.1). In general, in all the soils studied, there was immobilisation of N during decomposition of cereal straw unless C/N ratio was lowered. However, with FYM or compost, no such negative effect was observed. The availability of N from cereal residues can be increased by the addition of N rich organic materials, *viz.*, non-edible cakes or fertilizer N. Cereal residues were more effective in increasing the organic carbon and legume residues the N content. The cation exchange capacity of soil was also improved with progressive degradation of crop residues. Low-grade rock phosphate can be used for hastening *in situ* decomposition of cereal residues. Organic N is slowly mineralised and about 30 per cent is generally available to the first crop. About

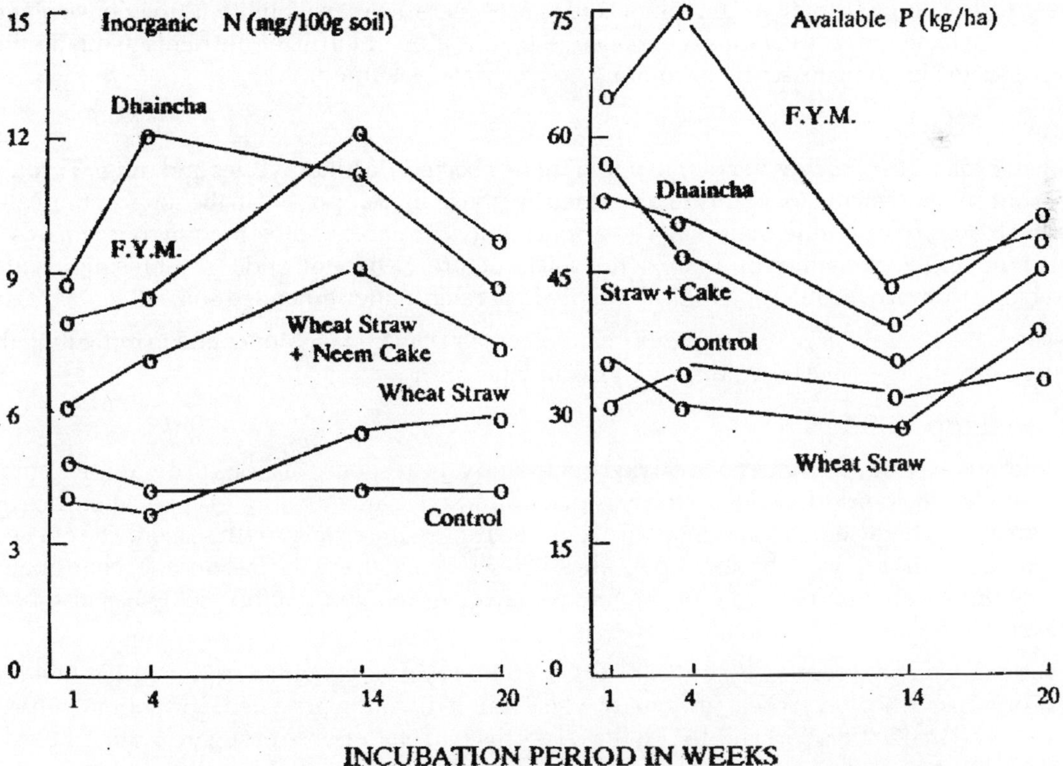

Figure 12.1: Changes in Inorganic N and Available P After Soil Incubation of Different Organic Materials

60–70 per cent of P and 75 per cent of K also is likely to become available to the first crop and the rest to subsequent crops.

Effect on Soil Properties

It is well known that organic manures influence physical, chemical and biological properties of the soil.

Physical Effects

It is the humus fraction which improves the soil physical condition. The addition of compost improves soil structure, texture and tilth. The better soil structure provides better environment for root development and aeration. The water holding capacity of the soil increases as compared to soil receiving no manure which provides protection to the crops against drought. Compost has been found to be very useful in control of soil erosion. Several studies on the impact of organic manures on improvement in soil structure were reviewed. Aggregation is quite often improved which is attributed to the action of gum compounds, polysacharides and fulvic acid component of organic matter. The type of organic material, the rate and frequency of application are important determinants in this respect.

Chemical Effects

Organic manures add nutrients to the soil thus reducing total dependence on fertilizers which involve greater energy/expense. The humic substances increase P availability as they have a very high cation exchange capacity. Humus enhances the utilisation of fertilizer nutrients by plants and helps in reducing leaching losses by promoting greater water retention.

Biological Effects

Organic manures contain a very large population of bacteria, actinomycetes and fungi. Through them not only millions of microorganisms are added but those already present in the soil are stimulated by the fresh supply of humic material. The application of organics helps the microorganisms to produce polysaccharides which build up better soil structure. N-fixation and P-solubilisation is also increased due to improved microbiological activity in organic matter-amended soil.

Besides, the above effects, role of compost in the control of plant nematodes and in mitigating the toxic effect of pesticides have been adequately documented.

Effect on Crop Yields

Numerous experiments have been carried out to study the response of crops to organic manures. The magnitude of response depends on the type of manure and its quality, time of application, dosage per unit area, soil characteristics and moisture required during the crop growth season. The effect of organic manures on crop yield has been periodically reviewed. During the last decade, coordinated multi-location research on recycling of organic resources in relation to crop yields has also been undertaken.

A summary of yield responses to 12.6 t FYM or compost/ha is provided in Table 12.5. Average response of crops such as rice, wheat (irrigated), wheat (rainfed), cotton (irrigated) and cotton (rainfed) was found to be 168, 202, 85, 56.4 and 162 kg/ha respectively. The response of sugarcane to 25t FYM/ha was found to be 8 tonnes cane/ha.

Table 12.5: Response of Different Crops to Compost/Farmyard Manure

Crop	Number of Stations	Number of Experiments	Unit	Response kg/ha to l2.6 t/ha Manure*		
				Minimum	Maximum	Average
Rice	63	34	kg paddy/ha	100	216	168
Wheat (irrigated)	31	210	kg grain/ha	82	296	202
Wheat (unirrigated)	14	71	kg grain/ha	74	140	85
Sugarcane*	19	258	t cane/ha	3.7	11.7	80
Cotton (irrigated)	10	71	kg tint/ha	48.9	96.1	56.4
Cotton (unirrigated)	25	294	kg lint/ha	11.8	22.3	16.2

* FYM to sugarcane added at 25t/ha.

The data of 340 experiments conducted on rice on cultivators' fields have been summarised. In addition, data of 35 experiments conducted on high yielding varieties at research stations were also included (Table 12.6). Most of these data belong to north eastern and southern zones. Analyses of 236 experiments, all conducted on cultivators' field in north eastern zone showed higher response yardstick (kg paddy / t FYM) of 136 in plain region of Manipur followed by northern plain of Bihar (122.8 kg). Response rates in southern plain regions of Bihar was about 30 per cent less than in northern plains of Bihar. In other agro-climatic area of this zone, *viz.*, western hilly region of Assam, coastal region of Orissa and Western region of West Bengal, the values were between 21.1 and 49.8 kg followed by southern zone with 69.0 kg paddy/t FYM. The yardstick data obtained for northern and central zone was 41.0 and 27.0 kg paddy/t FYM. As an overall average, each tonne FYM resulted in the production of 50 kg paddy/ha. Results from seven years of trials with potato showed that tuber yield was 29.9 t/ha when potato received 30t FYM/ha as compared to 28.0t/ha when no FYM was applied in the rotation. Average residual effect of FYM on the productivity of the succeeding wheat crop was 300 kg grain/ha (worth Rs. 700 at current procurement prices). Although on N basis, organic manures were less efficient than fertilizers, combined use of organic manures and fertilizers was found to be superior than the use of fertilizers alone.

Some more results on the impact of diverse organic materials on crop yields are provided in Table 12.7. Application of 5t wheat straw/ha increased the pod yield of groundnut by 95.5 per cent and grain yield of the following wheat crop by 17.1 per cent. Nitrogen uptake by ground nut crop in straw-amended plots was 100 per cent more than in the control. Straw application also increased the productivity of lentil and green gram. Impact of wheat straw incorporation on yield and protein content of soybean under pot culture conditions was reported. Incorporation of 18t wheat straw/ha in the top 7.5 cm depressed maize yield but its incorporation in 0–15 cm did not show any adverse effect. Part of this favourable effect of cereal straw on N nutrition could be due to greater utilisation of soil N as a result of better water retention in the soil profile. Significant increases in crop yield and N uptake were obtained by using cereal straw and neem cake in the proportion of 3 : 1.

Integrated use of mineral N, organic N and P-solubilising microorganisms gave higher wheat yield when the organic component was maize stubble rather than rice straw (Table 12.7). Best yields were obtained with mineral-organic-biofertilizer combination both for the direct crop of wheat and the residual crop of greengram.

Table 12.6: Yardsticks of Additional Production of HYV Rice Due to Farmyard Manure in the *Kharif* Season

Zone	*State*	*Agroclimalic Region*	*Districts/Centres*	*Number of Experiments*	*Yardsticks of Additional Production kg paddy/ t FYM at 12t FYM/ha*
Central	Gujarat	Coastal	1	2	17.9*
	Madhya Pradesh	Eastern	2	4	13.3*
	Maharashtra	Coastal	1	5	49.9*
Northern	Uttar Pradesh	Central	1	5	35.3*
	Uttar Pradesh	Eastern	1	3	46.6*
North	Assam	Western hilly	1	6	24.1
Eastern	Bihar	Northern plain	1	76	122.8
	Bihar	Southern plain	1	87	84.3
	Manipur	Plain	1	46	136.0
	Orissa	Coastal	1	16	49.8
	West Bengal	Western	1	5	27.8
Southern	Andhra Pradesh	Coastal	1	3	63.5*
	Karnataka	Central	2	22	69.5*
		Malnad	1	5	47.3*
	Kerala	Coastal	1	3	66.1*
	Tamil Nadu	Western	1	5	58.3*
	All India Average		**18**	**293**	**50.0**

Table 12.7: Effect of Some Organic Resources on Crop Yields

Crop	*Treatment*		*Control Yield (kg/ha)*	*% Increase Due to Treatment*
	Material	*t/ha*		
Groundnut	Wheat straw	5.0	1684	95.50
Wheat	Residual	–	3800	17.10
Wheat	Rice straw	120 kg N/ha*	6210**	1.00
	Rice straw + PSM	120 kg N/ha*	6210**	2.90
Wheat	Maize stubble	120 kg N/ha	4970**	26.60
	Maize stubble + PSM			35.20
Wheat	Sugarcane trash	5.0	2455	12.50
Wheat	Rice straw + water hyacinth	5.0	1599	37.87

* 120 kg N/ha input is from rice straw + fertilizer N. Rock P also added.

** Control here is 120 kg N through fertilizer and rock–P also added.

PSM = P solubilizing microorganisms.

In the black clay soil at Pune, 5t sugarcane trash/ha increased wheat yields by 307 kg/ha (12.5 per cent) as an average over N-levels and 3 years (Table 12.7). Wheat yield of 3t/ha could be obtained either by applying 120 kg N/ha through fertilizer or 60 kg N/ha through fertilizer and 2.5–5.0t of sugarcane trash. In the red loam soils, wheat yield of 3t/ha was obtained only by using 100 kg fertilizer N/ha alongwith 5t/ht of rice straw + water hyacinth. In terms of wheat productivity 100 kg fertilizer N was on par with 50 kg N + 2.5t of the above organic material and 100 kg N + 2.5t organic material was on par with 50 kg fertilizer N + 5t of rice straw plus water hyacinth.

Long-term Effects of Organic Manures

Long-term manurial experiments conducted in India during 1885–1985 showed a declining trend in productivity with application of N, P and K fertilizers alone. The decline in productivity has been associated with the onset of deficiencies of nutrients like sulphur and zinc. It could also be due to buildup of soil-borne pathogens. Integrated use of organic manures and fertilizers has been found to be promising not only in maintaining higher productivity but also for providing stability in crop production. The superiority of integrated use of organic manures and chemical fertilizers in sustaining crop productivity in intensive farming systems in comparison to chemical fertilizers alone is evident (Table 12.8). The effect has been more pronounced in case of acidic soils than in normal soils and quite similar to lime amendment.

Table 8: Effect of Fertilizer Alone and with FYM on Crop Yields in Long Term Fertilizer Experiments Initiated During Early 1970s

Soil Location	*Crop in Sequence*	*Mean Grain Yield (t/ha) Last 3 Years 1984–87)*			*Mean Grain Yield Over the Years (1970–87)*		
		100% NPK Fertilizer	*100% NPK + FYM*	*% Increase Over NPK*	*100% NPK Fertilizer*	*100% % NPK + FYM*	*% Increase Over NPK*
Alluvial	Maize*	2.2	3.3	54	2.5	3.2	28
Ludhiana	Wheat	5.2	5.3	2	4.7	4.8	2
Medium black	Finger millet*	3.3	3.7	12	2.6	3.1	19
Coimbatore	Maize	3.5	3.9	11	2.9	3.4	17
	Cowpea	0.5	0.6	20	0.5	0.6	17
Black	Soybean**	2.6		8	2.2	2.4	10
Jabalpur	Wheat	4.0	4.6	15	3.9	4.3	10
Red loam	Soybean**	1.7	2.0	18	15	1.8	20
Ranchi	Wheat	3.2	3.4	6	25	2.6	4
Literite	Rice Kharif*	2.8	3.6	29	2.9	3.4	17
Bhubaneswar	Rice Rabi	3.3	4.2	27	3.1	3.7	19
Submontane	Maize**	2.4	4.3	79	3.2	4.7	47
Palampur	Wheat	2.8	4.0	43	2.6	3.3	27
Foothill	Rice	5.8	6.8	17	6.2	7.0	13
Panthnagar	Wheat**	4.0	5.3	33	3.9	4.7	21

Average estimated N, P and K input through FYM (elemental form)

* N: 25 kg/ha; P: 5 kg/ha; and K: 20 kg/ha.

** N: 38 kg/ha, P: 8 kg/ha; and K: 30 kg/ha.

During the recent three years (1984–87) of these experiments, application of FYM on top of "optimum" NPK rates through fertilizers resulted in an extra grain yield of 200–1900 kg/ha over NPK plots for the crop which received FYM and 100–1200 kg/ha for the succeeding crops reflecting residual effect of FYM. Maximum advantage of using FYM was seen in the case of maize at Ludhiana, rice at Bhubaneswar, and both crops studied at Palampur and Pantnagar. Since FYM was added on top of optimum NPK application through fertilizers, its impact on crop yields can be attributed largely to beneficial effect on soil properties, moisture retention, better nutrient availability, more favourable conditions for root growth and supply of micronutrients.

A decline in sugarcane productivity in medium deep black soil was corrected in the composts series with mineral nutrients whereas a steady decline in cane yield was noted with the mineral nutrients in the no-compost series, emphasizing the importance of integrated use of organic and inorganic manures in maintaining stability in crop production. A combined application of organic manures and mineral fertilizers effectively checked the deterioration in the productivity of maize and wheat in acidic red loam soil. The productivity of maize declined to zero yield level after 26 crops (maize and wheat) where only mineral equivalent of N, P and K nutrients in FYM were applied for a period of 30 years (upto 1983).

Management Aspects

Organic manure can be applied to all crops and all types of soil but the rates of application will very according to the soil, crop and climate, local availability and also to the quality of manure. The subject of crop residue management and their impact on crop yields and soil properties has been recently reviewed.

Soil

The rates of application of compost will depend upon soil texture, slope and depth to water table. The addition of organic manure to sandy soils will increase the availability of moisture to the plant and thus reduce the number of irrigations necessary. In heavy textured clay soils, the addition of organics will increase premeability to water and air and increase water infiltration, thereby reducing surface run-off. One of the greatest benefits from the use of compost is to reduce the water requirement for plant growth. In rainfed and irrigated areas, this may help water conservation and thus need less irrigation.

Crops

Organic manures can be used to provide the nutrient need for growth of crops. For developing countries the use of organics on crops will not only improve the soil properties but will also cut down on the foreign exchange needs for the purchase of mineral fertilizers. The amount of compost needed for agronomic crops will depend on the nitrogen requirement of the crop as well as nitrogen content of soils. Under tropical and sub-tropical climatic condition seasonal applications are necessary to obtain good results. On-farm surveys in Haryana state indicate that farmers use organic manures preferentially on certain crops such as potato, sugarcane, cotton and least amounts (< 2t/ha) on food grains.

Rates of Application in Practice

The amount of compost to be applied to soil will be determined by (*i*) the amount of nutrients, particularly nitrogen, which can be efficiently utilized by the crops, (*ii*) the amount which will not reduce seed germination and growth rate of young plants (*iii*) the amount which can be conveniently incorporated in the soil and as stated above, the farmer's perception of its utility for specific crops.

About 25 tonnes/ha is recommended under intensive irrigated cropping conditions for sugarcane, vegetables, potatoes and rice; 12.5 tonnes/ha for irrigated or rainfed crops where potential rainfall is medium to heavy (about 125 cm) and 5–7 tonne/ha in dry areas where mean annual rainfall is about 50 cm. In dry farming areas application of 2.5 tonne/ha of compost can give significant increase in crop yield.

One cart load of FYM/compost measuring 9 m^3 weighs about 0.5 tonne. If sufficient FYM is not available, it can be applied at recommended rates to a part of the land, say 1/3 to 1/4 of the area, in rotation so that all parts of the field receive compost every three to four years. The rate of application of enriched compost will be less than the normal rates of application.

Time and Methods of Application

Most of the farmers unload manure in small piles or heaps and leave it as such for a month or so before it is spread and ploughed or disked into the soil. Some of the plant nutrients are lost during exposure to the sun and rain by volatilization or leaching. In summer this practice results in rapid drying and considerable loss of N occurs and, in rainy seasons the available N and a good portion of soil humus are washed away.

To derive maximum benefit, the manures should be applied during land preparation and incorporated into the soil having adequate moisture about two to three weeks before sowing so that nutrients are made available to the plants. The manure should be ploughed or harrowed into the soil soon after it has been carted to the field and spread over the land. An additional application can be made at the time of sowing when it is used in conjunction with mineral fertilizer in furrows. It is advisable to use compost with a phosphatic fertilizer or to prepare rock phosphate enriched compost.

Constraints in Adoption

There are several factors which effect the proper adoption of the recommendations of different types of organic manures by the farmers. There are competitive uses of organic materials such as dung cakes for domestic cooking fuel and sugarcane bagasse as fuel in sugar factories and villages.

Organic materials are a scattered resource and have to be collected. The production of organic manures involves labour for collection of materials, their processing with appropriate technology to obtain good quality compost with minimum nutrient losses, transportation to fields and incorporation into soil.

The compost prepared at the mechanical compost plants from city garbage are poor in plant nutrients. Although there is a provision to remove uncompostable materials such as metal, plastic and glass pieces before or during composting, stones, glass pieces, injection needles etc. are found in the compost which causes problems to users in its handling and application apart from injuries and health hazards.

In many parts of India, rice and wheat straw are disposed off by burning which should have been returned to soil. Proper technology either for the composting or *in situ* incorporation of residues has not been popularised amongst the farmers. The extension activities on the whole have been weak in this aspect.

The farmers do not make serious efforts in preparation and conservation of organic manures and recycling of crop and animal residues. Standard methods such as proper moisture (50–60 per cent) and turning of organic mass during composting or FYM preparation (2–3 turnings) are not followed resulting in nutrient losses and poor quality organic manures.

Future Research Needs

One of the most important needs concerns the development of suitable technology for preparation of a good quality manure which is well balanced in NPK and micronutrients. The manure should be prepared during a shorter period of about 2–3 months for use in cropping seasons. Research work in India has been conducted for hastening and enrichment of compost from certain crop residues. There is a need to develop suitable carrier materials such as sorghum seeds for preparation of cellulolytic culture inoculants for their commercial production.

Research work should be undertaken for isolation of efficient lignolytic microorganisms which can efficiently utilize organics like coir pith/dust, saw dust etc. There is a need to improve the activity of cellulolytic and lignolytic microorganisms by mutation or genetic engineering to obtain more efficient and stable isolates.

Initial studies have shown the utility of earthworms (*Eisenia foetida*) in composting of organic wastes such as dry leaves, rice straw, vegetable waste etc. Research work should be undertaken using indigenous and exotic worms in vermicomposting of organic wastes.

The effects of legume residues on succeeding crops, other than the N contribution have not been adequately studied and need to be better defined. More field information on *in situ* decomposition of crop residues in relation to soil properties and crop productivity are needed on short, medium and long term basis. There is enormous potential to convert clay wastes into organic manures. Research work should be undertaken by mechanical compost plants in collaboration with National Institutes and Agricultural Universities on enrichment of city compost and steps to increase its acceptability by the farmers initiated.

Long and medium term field trials using integrated concept of organic, mineral and biofertilizers in relation to sustainable agriculture should be conducted to work out specific recommendation for different crops and agro-climatic conditions.

Chapter 13
Green Manuring: Nutrient Potentials

The role of green manures in improving soil fertility and supplying a part of the nutrient requirement of crops is well known. Their use in crop production is recorded to have been practised in China as early as 1134 B.C. These are one of the main components of integrated nutrient supply system alongwith inorganic fertilizers and biofertilizers. In India, an estimated 6.2 million ha were green manured during 1988–89 (4 per cent of the net sown area). Andhra Pradesh (AP) and Uttar Pradesh (UP) account for 60 per cent of green manured area and 88 per cent treated area was in the six states of A.P., Karnataka, Madhya Pradesh, Orissa, Punjab and U.P.

This chapter provides an overall assessment of green manures, their significance and various management aspects in modern agriculture. Green manures can meet a part of the nutrient needs (particularly N) of crops for optimum production and to that extent can result in savings in fertilizer costs. These cannot completely replace fertilizer N if the goal is to harvest moderate-high yields on sustained basis.

Green Manures

Green manure refers to fresh plant matter which is added to the soil largely for supplying the nutrients contained in its biomass. Such biomass can either be grown *in situ* and incorporated or grown elsewhere and brought in for incorporation in the field to be manured. Just any plant cannot be used as a green manure in practical farming. Green manures may be plants of grain legumes such as pigeonpea, greengram, cowpea, soybean, or groundnut; perennial woody multipurpose legumes *viz., Leucaena leucocephala (subabul), Gliricidia sepium, Cassia siamca* or non-grain legumes like *Crotalaria, Sesbania; Centrosema, Stylosonthes* and *Desmodium.*

Leguminous plants are largely used as green manures due to their symbiotic N fixing capacity. Some non-leguminous plants are also occasionally used for the purpose due to local availability, drought tolerance, quick growth and adaptation to adverse conditions.

An ideal green manure should possess the following traits.

1. Show early establishment and high seedling vigour.
2. Be tolerant to drought, shade, flood and adverse temperature.
3. Possesses early onset of N fixation and its efficient sustenance.
4. Have an ability to accumulate large biomass and N in 4–6 weeks
5. Is easy to incorporate.
6. Is quickly decomposable.
7. Is tolerant to pests and diseases.

Leguminous Green Manures

These differ widely in nitrogen concentration and yield. Among 86 species used in India as green manures for rice their N contents ranged from 2.0 to 4.9 per cent N. Earlier results on the performance of some important green manure crops in lowland rice showed N-fixation of 74–134 kg/ha and about 200 per cent increase on paddy yield over unmanured plots.

Green manure crops suitable for various cropping situations prevailing in India are listed in Table 13.1. Abundant availability of water and sufficiently long fallow period before raising the rice crop have made the green manuring a widely adopted practice in lowland rice ecosystems. Common green manure crops in rice fields of India and their potential of biomass and N contribution in 45–60 days of growth are provided in Table 13.2.

Table 13.1: Green Manures Suitable for Some Field Crops

Field Crop	*Recommended Green Manure*
Rice	Sunnhemp, Sesbania and Wild indigo (*Indigofera tinctoria*)
Sugarcane	Sunnhemp
Finger millet	Sunnhemp
Wheat	Sunnhemp
Sorghum	Sunnhemp, Subabul (*Leucaena leucophala*)
Banana	Leaves of *Gliricidia sepium*
Potato	Lupin (*Lupinus albus*), Sunnhemp, Cowpea, Guar, Buck wheat (*Fagopyrum* sp.) horsegram

Non-grain Legumes

Recent evaluation of some of the promising green manure crops at Coimbatore indicated the potential of already popular *dhaincha* and the newly introduced stem nodulating *S. rostrata* (Table 13.3). The exotic stem nodulating *S. rostrata* of Senegalese origin has much promise for lowland rice especially with adequate irrigation. Another promising stem nodulating introduction from Madagascar is *Aeschynomene afraspera* which is capable of withstanding water stress to some extent. Its potential under Indian conditions has not been fully explored.

Though *Tephrosia purpurea* and *Phaseolus trilobus* grow rather slowly and accumulate much less N than *Sesbania,* they are more adapted to drought. *T. purpurea* has the additional advantage of self-sowing and is not browsed by cattle. Hence in deltaic areas where a single rice crop is grown due to limited water availability, *T. purpurea* is raised in the long rice fallow season without irrigation. After

3–4 months, it sheds seeds in the field which germinate after rice harvest in the next year. The green manure is incorporated 7–10 days before rice transplanting. Another drought tolerant green manure crop suitable for rainfed rice is wild indigo (Table 13.1). It can contribute 62–182 kg N/ha to the succeeding rice crop, depending on soil, climatic and cropping conditions.

Table 13.2: Some Common Leguminous Green Manure Crops for Rice Fields and their Potential N Contribution

Local Name	Botanical Name	Growing Season	Output in 45–60 Days	
			Green Matter (t/ha)	Nitrogen Contribution (kg/ha)
Sunnhemp	*Crotolaria juncea*	Wet	21.2	91
Dhaincha	*Sesbania aculeate*	Wet	20.2	86
Pillipesara	*Phaseolus trilobus*	Wet	18.3	201
Greengram	*Vigna radiata*	Wet	8.0	42
Cowpea	*Vigna sinensis*	Wet	15.0	74
Guar	*Cyamopsis tetragonoloba*	Wet	20.0	68
Senji	*Melilotus alba*	Dry	28.6	163
Khesari	*Lathyrus sativus*	Dry	12.3	66
Berseem	*Trifolium alexandrium*	Dry	15.5	67

Table 13.3: Evaluation of Green Manures for Rice Based Cropping Systems at Coimbatore

Green Manures	Output in 60 Days			Special Features
	Biomass (t/ha)	N Accumulation		
		kg/ha	kg/ha/day	
Crotolaria juncea (Sunnhemp)	16.8	159	2.65	Quick growing, easy seed production
Sesbania aculeata (Dhaincha)	26.3	185	3.08	High biomass accumulation, wide adaptability
Sesbania rostrata	24.9	219	3.65	Stem nodulating, tolerant to water logging
Tephrosia purpurea (Kolinji)	16.8	115	1.92	Drought tolerant, self seeding
Phaseolus trilobus	17.6	126	2.10	Green manure-cum-fodder-cum cover crop

Grain Legumes

Some annual grain legume crops are also used as green manure, after all or part of the grain is harvested. Greengram stover incorporated in the soil after pod harvest contributed about 60 kg N/ha to the succeeding rice crop. Evaluation of several grain legumes with *Sesbania* and *Crotolaria* showed that greengram was the best, yielding about 1t grain and 2.5t dry matter /ha equivalent to 50 kg N/ha. Greengram, blackgram (*Phaseolus mungo*) and cowpea could provide about 50–60 kg N/ha for the succeeding rice crop.

In studies at Coimbatore cowpea was the best among the grain legumes tested, contributing about 65 kg N/ha in addition to about 500 kg grain/ha. Its residues can also be used to supplement N-supplies for rice. Hence after grain harvest the legume stover grown in the pre-rice season could be incorporated as green manure to meet the partial requirement of N, provided there is no competing demand for the stover as cattle feed.

As a comparison, *S.rostrata* produced more biomass and contained 2.5–3 times more N than grain legumes in 60 days (Table 13.4). Where grains are harvested, the N-benefit to the following crop is reduced to the extent that absorbed N is taken away with the grains.

Table 13.4: Biomass Production and N Uptake of Grain Legumes and *S. rostrata* at Coimbatore in Summer (February–April)

Crop	*Grain Yield (kg/ha)*	*Biomass Production (t/ha)*	*Total N Contribution (kg/ha)*
Blackgram	409	6.11	50.7
Soybean	581	7.35	62.1
Cowpea	479 + 98 kg green pods	8.91	64.6
S. rostrata (60 days growth)		19.62	146.6

Perennial Trees and Shrubs

Loppings from such perennials are collected and used as green manure. Some of the perennials grown on bunds, avenues and waste lands and used ago green leaf manure were listed by Sanyasi. *Thespesia populnea, Cassia auriculata, C. tora, Pongamia glabra, Melia azadirachta, Calotropis gigantica, Jatropha gossypifolia, J. glandulifera, Gliricidia sepium* and *Ipomoea caruca* are some of the multipurpose perennials commonly used as sources of green leaf manure in tropical Asia. Due to the utility of multipurpose trees in Asian farming systems and large scale extension efforts on agro-social forestry, some of these trees are being widely planted for fodder, fuel, timber and green manure purposes. *Albizzia falcatoria, Calliandra calothyrsus, Erythrina* spp., *Sesbania gran diflora* several species of *Acacia* and *Prosopis, Flemingia macrophylla* and *Leucaena leucocephala* are some of these. However, not much information is available on their biomass and N accumulation capacities.

Role of Green Manuring in Cropping Systems

Green manuring is an important practice in several cropping systems though it has received relatively more attention in rice-based and sugarcane-based systems than others.

Rice-based Systems

These are the mainstay of agriculture in large areas of Asia. In single crop rice lands, farmers raise Kolinji (*Tephrosia putpurea*) as green manure in rotation with rice. This plant withstands drought and contributes substantially to the yield of succeeding rice. The summer/south west monsoon showers received prior to the main rice growing *samba* season (August–September to January–February) in the Cauvery delta of Tamil Nadu, can be utilised to raise a green manure crop. In such situations sunnhemp performed better than other green manures tested in terms of biomass production in 45 days. In double cropped rice areas, *dhaincha, pillipesara,* etc., are grown as green manure crops during summer and ploughed *in situ* for the succeeding rice crop. Growing grain legumes as catch crops in summer rice fallow is also a common practice in river command areas. After the pods are collected, plants are pulled out and incorporated into the soil as green manure.

Flood tolerance of *S. rostrata* makes it amenable for intercropping with transplanted rice. Transplanting 30 days old seedlings is better than seeding. Here one row of *S. rostrata* is planted after every 10–12 rows of rice (1.5 m interval) and incorporated 30 days later. Such a practice does not affect rice growth adversely and adds about 1.5 t of biomass/ha containing about 13 kg N/ha to rice resulting in 9 per cent increase in paddy yield as compared to sole cropping. In north India sowing green manure crop in early May is suggested for the rice-wheat system green manure crop in early May is suggested for the rice-wheat system.

Sugarcane-based System

There are two distinct zones of sugarcane cultivation in India, sub-tropical north and tropical south. Though the climate of tropical south is most suitable for sugarcane growth, 70 per cent sugarcane is grown in sub-tropical north. In south India, the response in general, is good with sunnhemp and cowpea. The response to *dhaillcha* in general, is somewhat lower and of the same order as that of guar. Raising blackgram as intercrop and incorporating the stover after removal of pods was found to improve the soil fertility in sugarcane fields. Green manuring has been a popular practice in places where feasible in Godavari delta of Andhra Pradesh and earlier estimates showed that it could provide about 50 kg N/ha in the total N requirement without loss in cane yield. Soybean grown as an intercrop is reported to contribute about 100 kg N/ha to sugarcane.

In north India, several studies have shown that continuous sugarcane cropping without green manuring depleted soil organic carbon and N content by 0.06 and 0.011 per cent respectively as compared to the soil where different crop sequences were practised. Soil pH increased from 8.0 to 8.5 after 12 years of cropping and a considerable decrease in available P and exchangeable K occurred due to continuous cropping without green manuring (Table 13.5). In multiple ratooning, the soil becomes hard due to connection with the result that the root system of successive ratoons become smaller and shallower. In such situations raising a green manure like cowpea in the inter-spaces and incorporating it helps not only in building up of soil N and release of K from non-exchangeable fraction but also in creating soil physical conditions favourable for good stand establishment and crop vigour with good root development.

Table 13.5: Soil Chemical Characteristics Twelve Years after Intensive Sugarcane Cultivation with and without Green Manuring

Soil Characteristic	*Cropping with Green Manuring*	*Continuous Sugarcane Without Green Manuring*
pH	8.0	8.5
Electrical conductivity (mmhos/cm)	0.10	0.10
Organic carbon (per cent)	0.38	0.32
Total nitrogen (per cent)	0.040	0.029
Available P (kg/ha)	17.6	8.7
Exchangeable K (kg/ha)	180.0	141.0

However, N-economy due to legume intercropping is not possible in sugarcane even if soil N and organic carbon contents are improved because the intercropped legumes cause shading on the base of sugarcane and thereby adversely affect tiller emergence. Incorporation of legumes before active tillering phase of sugarcane, however, improved cane yield. These benefits again were not because of N fixed by legumes but because of overall improvement in soil structure and efficiency of other nutrients. The time between incorporation of green manure crop and sugarcane planting should be as short as possible, otherwise the benefit of green manuring will be lost under high temperature. Further, green manuring with legumes has the limitation that it requires plenty of irrigation or assured rainfall to decompose buried plant material. Finally, legumes in the cropping system may not economise on the doses of fertilizer-N in sugarcane, but can maintain soil fertility and improve upon the efficiency of other mineral nutrients required by the crop.

Cotton-based Systems

In situ green manuring is not a common practice but green leaves from outside are brought and incorporated in the soil. In irrigated cotton in the north, Egyptian clover is sown in the standing cotton crop before final picking. The clover is partly grazed and partly ploughed in and cotton resown. Greengram and blackgram are also used as green manures after one picking in some areas. In some cases in the north, clusterbean is grown for grain and then followed by cotton. Earlier work with rainfed cotton showed either a lack of response or even yield depression as in several Maharashtra locations. At few places, sunnhemp was found to be better than *dhaincha* and blackgram.

Potato-based Systems

Research on green manuring for potato has been recently reviewed. Among the green manure crops used are sunnhemp, cowpea, horsegram, guar and lupine. Green manuring with sunnhemp was better than *dhaincha* or cowpeas. Earlier work showed lupin and buck wheat to be good green manures in the Nilgiris. Burying of material brought in from outside was as good as growing them *in situ*. Potato after sunnhemp yielded 1.5 t/ha more than potato after fallow. Contribution of *dhaincha* towards N supplies for potato range from very small, 10–12 kg N/ha to 45–52 kg N/ha. Some workers have also computed the P and K equivalents of green manuring. It is concluded that green manure should either be brought from outside or raised at a time when it does not compete with any economical crop. From many reviews of research, it is observed that sunnhemp is the best green manure crop and gives good response in many crops. *Dhaincha* is suitable where there is adequate moisture available for the growth of the crop.

Rainfed/Dryland Systems

Green manuring of dryland crops is rare but practised in small pockets in India. The scope for growing a green manure crop and ploughing it under before planting the main crops is limited. However, cowpea intercropped with castor for six weeks and incorporated as a green manure with 10 kg N/ha through fertilizer produced yields on par with 40 kg N/ha through fertilizers. Green manuring or green leaf manuring is possible in drylands under the two following situations:

1. Sowing a legume such as greengram with the earliest monsoon showers, harvesting the pods at first flush, incorporating the residues and sowing a late monsoon crop or any early post-monsoon crop, and
2. Addition of loppings of *Leucaena* as green leaf manure.

Leaves and tender twigs of *Leucaena* contain about 3 per cent N and decompose fast. While estimates of the net N contributed by such additions to the N nutrition of cereals vary considerably, the practice along with some fertilizer N is reported to hold great promise for the drylands. *Leucaena* when planted in alleys with *Rabi* sorghum and loppings used as manure, added 87.6 kg N/ha to the soil. The net effect of this input on sorghum yields was equivalent to that of 25 kg N/ha through fertilizers and incorporation of 5 t/ha of *Leucaena* gave sorghum yield which was on par with that obtained with recommended fertilizer dose. Manuring with *Leucaena* or sunnhemp made a greater contribution to sorghum yields than the addition of sorghum or safflower stubbles, as expected.

Likewise, incorporation of legume residues at 5 t/ha plus 50 per cent of the recommended nutrient rate through fertilizers outyielded the crop which received recommended rate through fertilizers in a maize-wheat rotation in the drylands of Jammu and Kashmir. The most suitable C : N ratio has been found to be 75–150. Most of the crop residues fall in this range. If not, inorganic N can be added to hasten decomposition. In drylands, such potentialities exist in several parts of India (Table 13.6). Incorporation of such residues not only improves the yield, but also improves the physical environment as well as soil fertility status.

Table 13.6: Potential Areas for Adoption of Green Manuring in Drylands

Sl.No.	*Region*	*Soil Type*	*Rainfall (mm)*	*Cropping System with Possibility of Organic Recycling*
1.	Bangalore	Alfisol	924	Cowpea-finger millet
2.	Bhubaneswar	Alfisol	1463	Greengram-finger millet
3.	Rewa	Vertisol	1168	Blackgram-wheat
4.	Agra	Inceptisol	765	Greengram-mustard
5.	Akola	Vertisol	877	Greengram-safflower
6.	Varanasi	Entisol	1113	Rice-chickpea
7.	Hoshiarpur	Inceptisol	1000	Maize-chickpea
8.	Indore	Vertisol	1053	Maize-chickpea
9.	Ranchi	Alfisol	1462	Finger millet-chickpea
10.	Dehra Dun	Inceptisol	2313	Maize-chickpea

Note: In locations 1–5, residue of rainy crop can be incorporated, and in the rest, residue of post-rainy crop can be incorporated.

Plantation Crops

Many of the plantation crops in the country are grown in hilly areas and undulating terrains and thus proper soil, water and organic matter conservation measures are of paramount importance. In most areas, interspaces between coconut palms are cultivated with tuber crops, cocoa, pepper and banana. Hence, the possibility of raising green manure in coconut plantations is rather limited even though several crops can bring in 66–153 g N/basin. The recommended radius for opening the basins in coconut is 1.8 m and such an area remains not fully utilised for each tree. Legumes like *P. phaseoloides, Mimosa invisa* and *Calapagonium mucunooides* have been recognised as suitable green manure crops for raising in coconut basins.

Fate of Green Manures on Application to Soils

Decomposition of added plant material depends on its chemical constituents and the physical and biochemical conditions in the surrounding environment. Among the organic constituents in plant material the water soluble fraction containing the least resistant components is the first to be metabolised. Cellulose and hemicelluloses do not decompose as fast as the water soluble substances, and their persistence is in the medium range. The lignins are the most resistant and consequently become abundant in residual decaying organic matter. Thus one fraction decomposes rapidly, usually during the early growth phase of the crop following incorporation and the other decomposes slowly over a longer period. The first fraction (fast N) determines the potential supply of N to the standing crop and the second fraction referred to as "slow-N" determines the residual effects. With most green manure crops, the first fraction is 50–80 per cent of the total N Residual effects (N supply to a second crop) are relatively small when green manure is applied only once, but the cumulative effects of several annual applications can be appreciable.

Most N-release studies have been conducted under laboratory and green house conditions. These indicate that, in the absence of rice plants, N release increases rapidly and reaches a plateau. In the presence of plants, it peaks and then declines. This hypothesis can be explained by the results from the field experiment presented in Figure 13.1 and 13.2.

Release patterns of C and N show that the green manure underwent rapid decomposition and mineralization upon incorporation into the soil (Figure 13.1). About 80 per cent of total N and 40 per cent of carbon present in *Sesbania* was released in about 10 days. The released N is likely to be lost by leaching, if not released in synchrony with plant (here rice) demands or held as ammonium on the exchange complex.

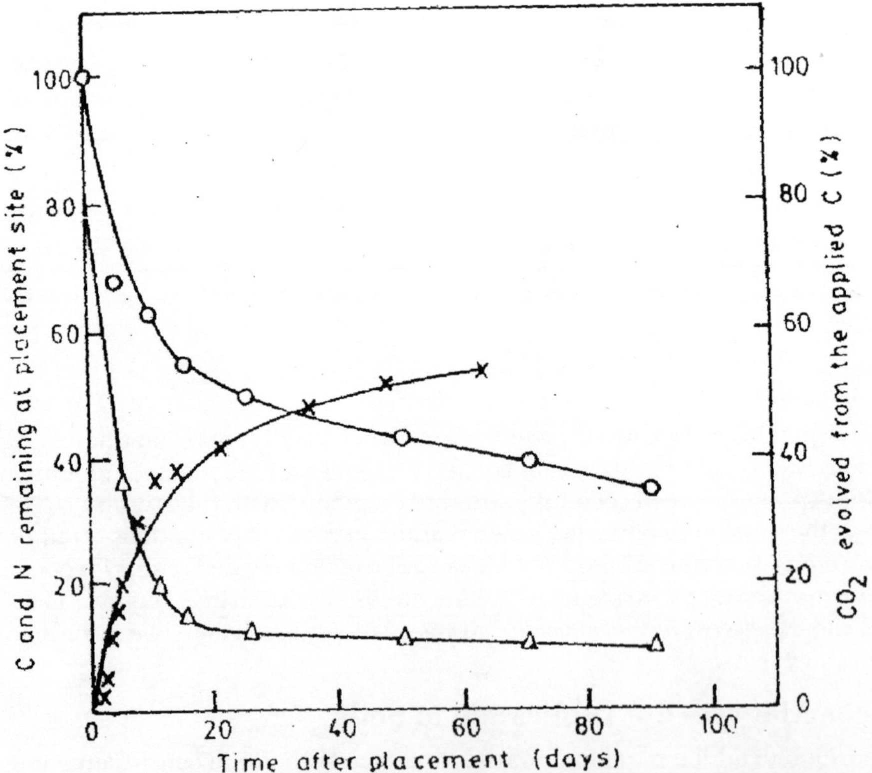

Figure 13.1: Influence of *Sesbania* Incorporation in Soil on Evolution of CO_2 in the Laboratory (x) and Changes in Carbon (O) and Nitrogen ( ) of *Sesbania* Under Field Conditions

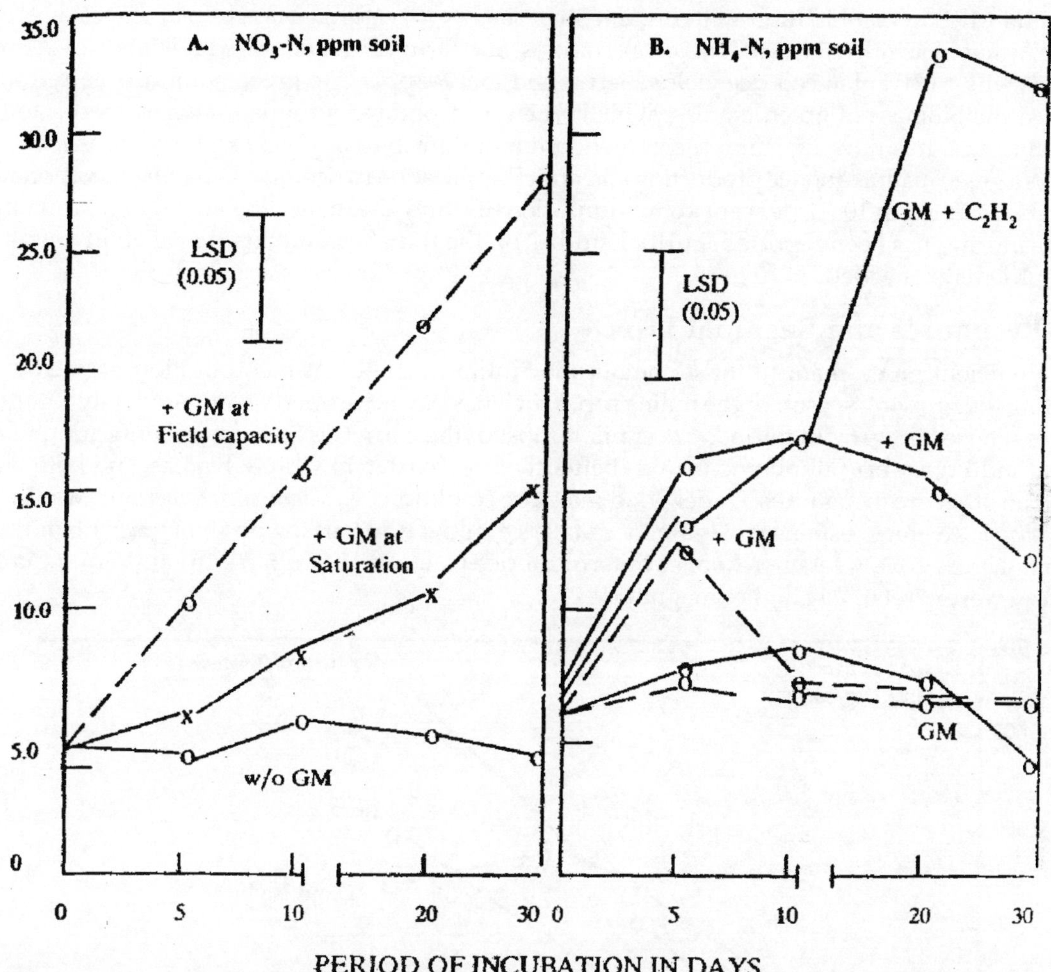

Figure 13.2: Effect of *Sesbania* Green Manure on (A) Production of NO_3 + –N at Field Capacity and Saturation Moisture Regimes and (B) on the Production of NH_4 + –N at Field Capacity and Saturation Moisture Regimes

Under field capacity moisture regime, nitrate-N increased from 5 to 28 ppm soil with time in the soils applied with *Sesbania* (Figure 13.2a). However, ammonium-N decreased abruptly after 5 days under field capacity and slowly after 10 days under saturated moisture content (Fig 13.2b). These results also showed that longer periods of decomposition allowed for *Sesbonia* may lead to the formation of NO_3^-–N which is susceptible to losses, if not absorbed by the plants.

Availability of Essential Nutrients

The benefits of green manuring are generally interpreted in terms of their capacity to provide N or substitute for fertilizer N. The role of green manures in enhancing the availability of other macro- and micronutrients has not been documented as extensively. Higher availability of P from rod phosphate (MRP) has been reported in rice due to green manuring. In studies on the response of rice to P applied as SSP and MRP in combination with organic materials (*e.g. Leucaena leucocephala*) leaves as green manure, water hyacinth (*Eichhomia crassipes*), compost and farm yard manure, combination of organic materials with MRP enhanced rice yields. Beri found that P applied to green manure in low-P soils increased the biomass of green manure which when incorporated supplied the absorbed P to the succeeding rice. It is now a general recommendation in Punjab to apply 30 kg P_2O_5/ha to *dhaincha* grown as a green manure for rice production and omit P application to rice. Green manure incorporation in rice also resulted in 10–12 per cent better utilisation of P and K enhanced availability of iron upon green manuring has been reported. Further studies on the transformation and availability of other micronutrients are needed.

Crop Responses and Residual Effects

An efficient green manure must contain maximum nutrients at incorporation, immediately preceding the growing season of the main crop. Efficiency can be properly compared only for those legumes whose best growth periods are similar. Most of the earlier trials included only sunnhemp, *Sesbania* and *Vigna* spp. Sunnhemp proved better than *Sesbania* in low rainfall areas, and both were better than other green manures. Under good moisture conditions, *Aeschynomene americana* was better than *Sesbania*. Average responses of indigenous tall rice cultivars to varying doses of green manure are shown in Figure 13.3. Average response in terms of rice yield was 240 kg/ha. In southern India the responses were of relatively higher magnitude.

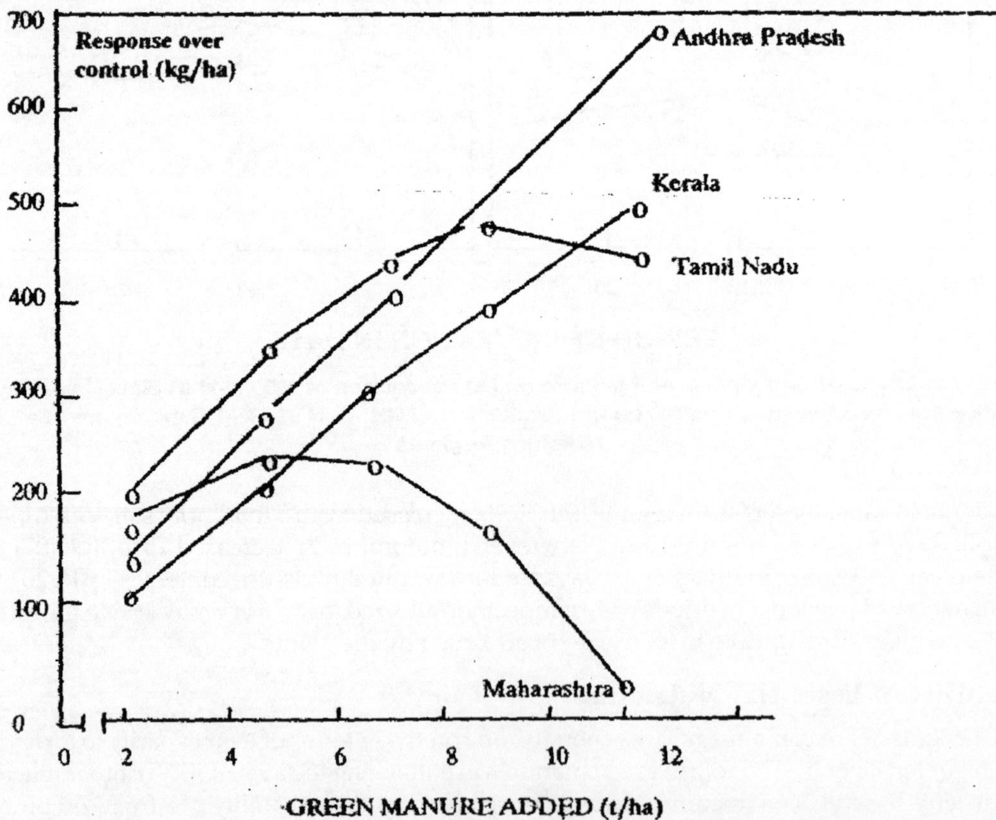

Figure 13.3: Average Response of Rice to Application of Green Manure

It is difficult to compare the relative efficiency of green manures with chemical fertilizers, even on an equivalent nutrient basis. A more rational way is to determine the extent to which green manures could substitute for nutrient elements derived from fertilizers to obtain equivalent crop yields. Based on the review of the experiments conducted in different states yardsticks of yield increases in *Kharif* rice due to green manuring @ 6 t/ha are presented in Figure 13.4.

The response of rice to green manure depends upon the time of its application. Because of succulence and narrow C–N ratio, a large proportion of the nutrients mineralised tend to leach down below the root zone even before planting rice, especially in coarse-textured soils. This leads to variations in the estimate of N substitution obtained in different experiments. However, on an average green manures can substitute chemical nitrogen to the extent of 50 to 65 kg/ha, that is 40–50 per cent if the recommended dose is 120 kg N/ha.

Green Manure Management

Management practices that influence the rate of N accumulation by a green manure crop deserve special consideration in rice-based cropping systems in order to increase the efficiency. Effective stand establishment at low cost is essential for the economic viability of the system. Treating the seeds of *S. rostrata* with concentrated sulphuric acid for 15 minutes and *Tephrosia purpurea* for 30 minutes gave the highest germination (90 per cent) and vigour index. Such treatment becomes imperative for seeds which have thick coat or which are dormant, otherwise population establishment is poor. Among the two methods of sowing tested, broadcasting was found better for *S. aculeata,* whereas *S. rostrata*

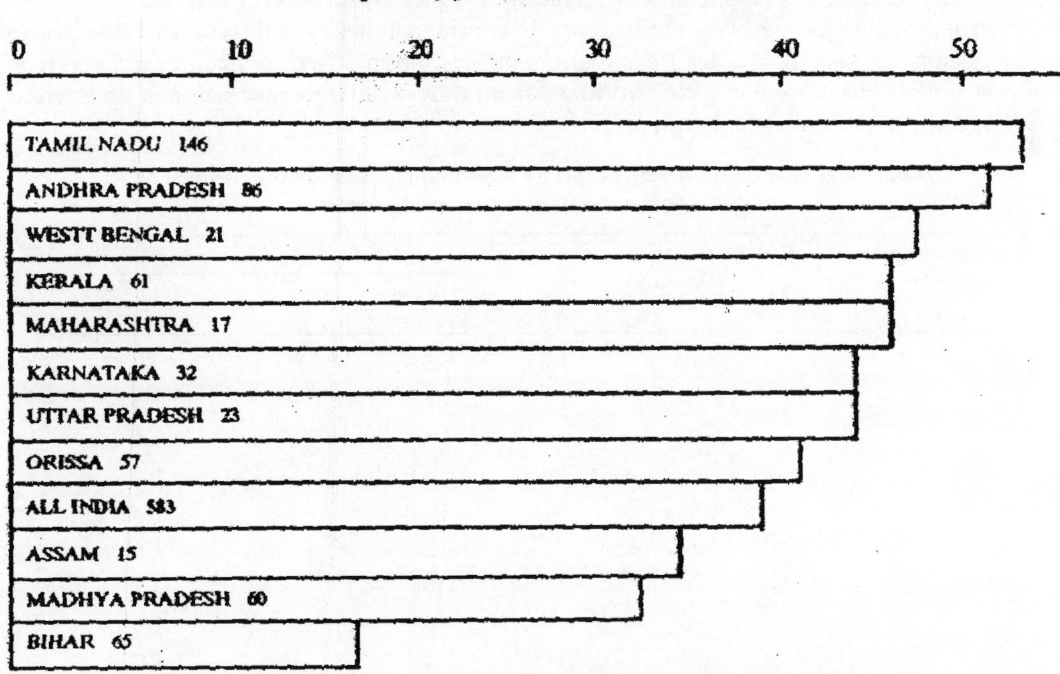

Figure 13.4: Impact of Green Manuring on Rice Yield in Different States. Number after a state represents trials averaged.

performed well when sown behind the country plough. Both the green manures produced high biomass in 60 days with a seed rate of 50 kg/ha.

When green manures are included in rice farming systems, especially during pre-rice summer season, limited irrigation facilities are needed besides rainfall. Under such situations, with 30 days irrigation frequency sunnhemp performed better, while *dhanicha* and *S. rostrata* required more frequent irrigations (once in 20 days). *S. rostrata* can also be grown successfully as an intercrop in rice in 12 : 1 ratio without affecting the growth of rice, as already discussed. At Cuttack, Orissa, interplanting *dhaincha* in a row ratio of 15 : 1 (rice spacing 20 × 15 cm) provided 360–430 kg of seeds/ha and 460–580 kg of dry stalks for fuel alongwith normal crop of rice. The establishment of 15–20 cm long hard wood cuttings of *S. rostrata* from 120 day old plants was better than soft wood cuttings, especially when dipped in 1 per cent cowdung slurry. Generally the cuttings are planted in a nursery bed for rooting. The rooted cuttings may be lifted after 30 days and planted in the field or bunds where 90 per cent establishment was observed. Using hard wood cuttings is necessary when adequate seeds are not available, especially for seed production.

Work at Cuttack during the 1950s showed that for 8 week old *Sesbania,* rice yields, decreased as the interval between burying green manure and planting rice increased from zero to 8 weeks. The reverse was true for 12 week old *Sesbania,* as the older crop became woody and needed almost 8 weeks to decompose.

Experiments conducted on soils of high permeability, where leaching of NO_3 released from mineralisation would be rapid showed that the highest rice yield was obtained when rice was transplanted almost immediately after incorporating *Sesbania* (Table 13.7). Green manures when buried just before transplanting rice, act as a slow release fertilizer and also create reducing conditions, which helps in mobilising several other nutrients. In contrast, in heavy soils where drainage is rather slow, transplanting at least 7 days after green manure incorporation is recommended since intermediary compounds, particularly organic acids, produced from the partial decomposition of green manures can adversely affect seedling establishment.

Table 13.7: Rice Yield as Affected by Age and Decomposition Period of *Dhaincha* (*Sesbania*) Green Manure

Age of Dhaincha (Days)	*Decomposition Period (Days)*	*Grain Yield (t/ha)*	
		Kamal (cv. P 2–21)	*Kanpur (cv. P 33)*
45	0	5.9	55
	10	5.6	5.1
	20	4.4	4.9
50	0	6.0	5.4
	10	5.6	5.6
	20	5.1	5.0
65	0	6.2	55
	10	5.7	5.3
	20	4.9	5.2

Residual and Long-term Effects

Estimated contribution of nitrogen by green manures sometimes exceed 100–120 kg N/ha. This is higher than the total N recommended for wet season rice in many countries. Where the N input through green manure is high occasional residual effect might be expected on second crop as this N is in organic form. Earlier as well as later work showed that the residual value of green manure applied to rice is low. These results suggest that residual effects of green manure are related to the N in the green manure and, presumably, to the quantity on N lost as well as recovered by the first crop. However, regular green manuring over a long period of time would not only improve the soil fertility but also result in noticeable residual effect in intensive cropping systems.

Results of field studies with rapeseed-mustard and the maize-mustard rotation show that green manuring rapeseed-mustard with cowpea increased seed yield by 380 kg/ha and N uptake by 31 kg/ha as an average over N levels (Figure 13.5). In the maize-mustard rotation, average direct effect on maize was a yield increase of 763 kg/ha and a further increase of 145 kg/ha in mustard yield through the residual effect. Their results showed the following input : output relations

Maize Yield (t/ha)	*Corresponding N Input*
1.0	Green manure or 42 kg fertiliser N/ha
1.5	Green manure + 24 kg N or 76 kg fertiliser N/ha
2.0	Green manure + 48 kg N or 120 kg fertilizer N/ha
2.5	Green manure + 84 kg fertilizer N/ha

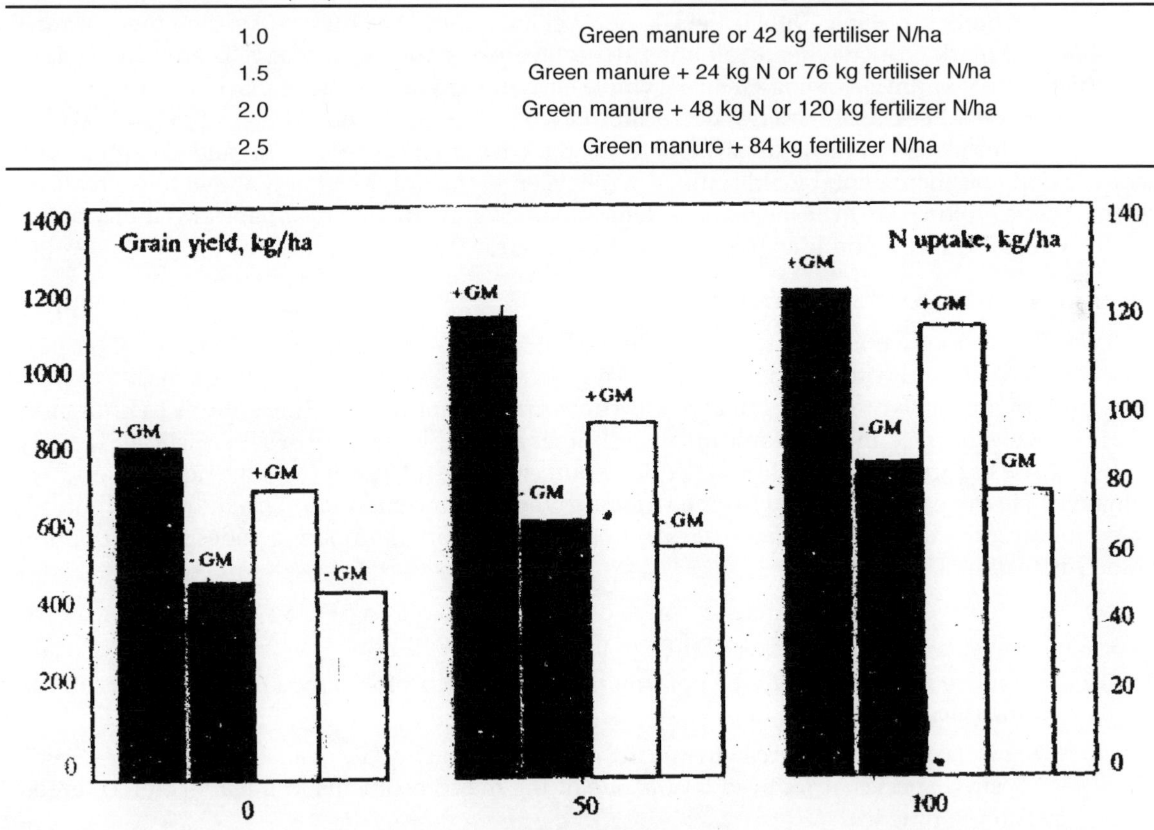

Figure 13.5: Effect of Green Manuring on Yield and N Uptake by Rapeseed Mustard at Graded Levels of Fertilizer N. Black bars are for yield, white for N uptake.

Economics of Green Manuring

Substantial information is available on the contributions of green manure towards increasing crop yields and improving soil properties resulting in conditions favourable for sustained high productivity. A critical issue for wider adoption appears to be the practical and economic feasibility of green manure technology at the farm level.

Tillage and seeding operations are major components of the labour and cash outlay by farmers who grow green manure crops. Irrigation cost is an important item in several parts of India. Where land preparation is specifically for green manure crop, this cannot be shared with another crop. High labour costs and high opportunity costs of land use are two of the major constraints to the economic feasibilily of green manure. The financial analysis seeks to compare costs and benefits, between the combined use of green manure and inorganic N, and inorganic N alone.

Overall added costs to produce a green manure crop is estimated to be about Rs. 500/ha. At current urea prices, this is equivalent to 98 kg N for small and marginal farmers and 75 kg N for larger land holders due to the recently introduced dual pricing system. In this case, it represents the opportunity cost, not necessarily the paid-out costs of producing a green manure crop. In India, the estimated ratio of benefits to costs is less than 1.0 implying that the green manure technology practised by the farmers is not profitable. But farmers do not cost their owned resources (*e.g.* their own animals and labour) at market price when evaluating the relevance of this technology. There is a merit in evaluating green manure technology within a whole farm framework to ensure that (*i*) the opportunity cost of owned resources is set with respect to alternative farmer uses, and (*ii*) the opportunity cost of land is also internalised in the analysis. However, the cost of land preparation and green manure incorporation and incremental yields gained with green manuring (over and above those realised with inorganic N alone) are to be included as determinants for evaluating the financial viability of this practice. Critical agro-economic analyses of the green manure technology are lacking.

Constraints of Green Manuring

Two observations were found to be consistent regarding the fertilizer application rates of farmers producing rice with and without green manure. They are: (*i*) fare substantially reduce N fertilizer rates when green manuring is practised, but in no case does green manure substitute entirely for inorganic N. Farmers view organic and inorganic nutrient sources as complementary or synergistic and wisely so. They do not find green manure crops adequate sources for the full nutrition of the rice crop, and, (*ii*) estimated yield increases attributed to green manure were 15 per cent in southern India and about 25 percent in eastern India. There are, however, a numbered constraints hampering widespread adoption of green manuring. There are:

1. In intensively cropped areas, farmers do not wish to set apart 6–8 weeks exclusively for growing a green manure crop with no direct cash benefit.
2. When rice is grown after wheat, farmers find it difficult to do the farm operations in the intense heat of May and June.
3. The most costly item in green manuring is the seed. Inadequate availability of quality seeds of desired species at reduced cost is one of the major problems in the adoption of green manuring practice.
4. Benefits of green manuring are not perceptible to the farmers because it is not directly visible as in the case of fertilizer N.
5. The intercropped green manures may compete for growth resources with the main crop.

Research should be focused to develop viable solutions to overcome these constrains.

Future Research Needs

With the increasing cost of fertilizers and growing awareness of ecological impact of using agrochemicals, integrated nutrient management with organic, inorganic and biofertilizer sources has become imperative for sustainable crop production. Green manuring has a special place in view of the limited availability of other organic manures. Concerted research effort is needed on the following areas to make the green manure technology feasible and viable.

1. Collection, conservation, cataloguing and evaluation of green manure species that show promise for different ecosystems.
2. Screening for N accumulation patterns under controlled conditions and under field situations.
3. Identification of the optimum combination of fast-N and slow-N release green manure species. A combination of two types of legumes may provide the best agro-ecosystem management.
4. Research to improve the establishment and stand of green-manures particularly on the aspects of seed treatments for early and uniform germination.
5. Contribution of nutrients other than N in different soils and environments.
6. Long term economic advantages including sustainability factors.
7. Multipurpose trees and short season legumes with pests and disease tolerance that can be grown on bunds or waste lands. Legumes that provide both forage and leaf green manure or fuel would be more appropriate.
8. Research on legume species with efficient rhizobium strains for symbiotic N fixation even in adverse soil environments.
9. Design of cropping systems including green manure crops without displacement of any economic crop in the system.

Conclusions

Various aspects of green manures and their management have been discussed. Diverse plant species varying from common pulses to stem nodulating legumes and trees can provide green manure. These are primarily grown and used to augment N-supplies for crop production and their biomass can deliver 25–100 kg N /ha to the soil. Most commonly observed N additions are in the range of 40–60 kg N/ha. If raising a green manure costs Rs. 500/ha, its break even in terms of urea-N at current prices is 75 kg N. The study and practice of green manuring has received relatively more attention in rice-based and sugarcane-based systems than in others. Where the green manure is succulent, there is no need to bury it well ahead of planting the main crop as its rate of decomposition and nutrient release is quite fast. Nitrogen mineralised from green manures is as susceptible to mechanisms of N-loss as is fertiliser N. Practising green manuring also results in improvement in soil physical and chemical properties and enhanced availability of other nutrients such as P and Fe (improvements on which ideally a price tag should be attached as in case of fertiliser N). Major constraints in the adoption of green manuring on a large scale are inadequate availability of good quality seed and water, competing uses of resources for raising another crops, cost: benefits in terms of N supplied and the cost of fertiliser N.

On the whole, green manuring, a 3000-year old practice is very relevant in today's context and must be made use of whereever feasible and economical in relation to the total farming system operation. Potentials, constraints and some areas requiring further research have also been discussed.

Chapter 14
Biological and Industrial Wastes: Source of Plant Nutrients

In order to meet the nutrient needs of agriculture in the coming years, the Government of India's working group on fertilizers has estimated that 16.5 million tonnes (mt) of N + P_2O_5 + K_2O will be needed by 1994–95 and 20.6 mt by 1999–2000. On no account can fertilizers, or any single input provide such large quantities of plant nutrients. All available nutrient sources have to be made use of. In addition, integrated use of mineral, organics and recyclable wastes is accepted as the most appropriate strategy for sustaining high crop yields, minimising soil depletion and value-added disposal of what are traditionally labelled as "wastes".

This chapter present a current assessment of biological and industrial wastes as sources of plant nutrients and analyses various factors which determine their usefulness. Among biological wastes, materials covered are sewage sludge, biogas slurry, waste water, fish pond effluent and some wastes of food processing industry. Pressmud and phosphogypsum (biproduct of fertilizer industry) are also dealt with. Organic resources such as farmyard manure (FYM), composts and crop residues have been dealt with other chapter.

Significance of Waste Recycling

Nutrients and water being major constraints in the development of Indian agriculture, harvesting the nutrient energy of biological and industrial wastes is of prime importance for maximising the food, feed, fodder and fuel production in the country. Further, when these wastes are recycled through land for crop production, due to the degradative and assimilative capacity of soil the pollution of streams and/or rivers receiving these wastes can be minimised to a large extent as compared their direct disposal in water resources. Role of 4 R cycles in harvesting the nutrient energy of waste and its renovation is depicted in Figure 14.1. Domestic and industrial waste waters are disposed on land (*i*) to use the nutrient potential for biomass production (*ii*) provide safe disposal to the waste water through the soil and (*iii*) prevent the pollution of streams and rivers.

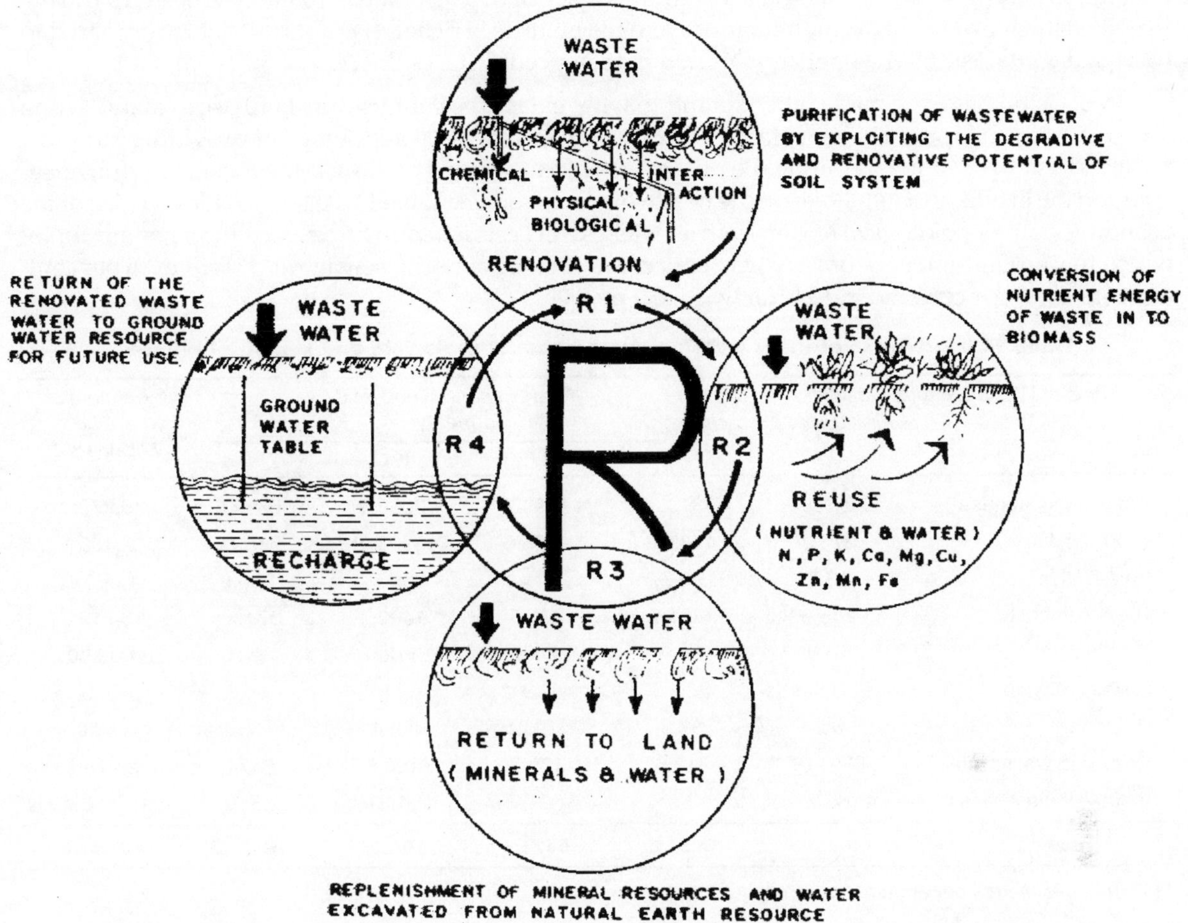

Figure 14.1: Exploitation of Nutrient Energy of Waste and Resource Conservation

Large potential for the exploitation of manurial value of biological and industrial wastes exists in India. Even at 50 per cent collection of the total agro forest waste, more than 14.5 m ha of land can be manured at the rate of 10 t/ha annually. Further, the productivity of more than 48 m ha land can be improved through biogas slurry manure at 50 per cent collection rate.

The domestic and industrial wastewaters amenable for crop production can help to increase the irrigated area by 170.4×10^3 hectares in the country. The prevention of environmental pollution through recycling of waste is difficult to quantify but it will be several times valuable as clean environment is today's need. Nutrient potential of different biological and industrial wastes has been summarised in Table 14.1. The major sources are crop residues, animal excreta, rural compost (of which dung is again a major component) domestic and industrial waste water, forest litter, and city refuse. Since dung is a major component of rural compost, nutrient potential can be over estimated if cattle dung and rural compost are taken as independent materials which they are not. If 60 per cent of the total dung available is fed into biogas plants, this can generate 470 mt of organic manure. Under the National Programme on Biogas Development about 350,000 biogas plants were set up during 1980–85 which produced 7.1 mt of manure. Biogas manure is richer in plants nutrients compared to FYM and contains 1.65–1.84 per cent N, 1.08–1.18 per cent P_2O_5 and 1.24–1.83 per cent K_2O.

A decision has to be made whether and to what extent the country wants 10 use available dung for FYM or as an input in biogas plants. Biogas route has the dual advantage of providing both fuel (gas) and fertilizer (slurry). Among the present practices, burning of dung is the most wasteful, not to mention the in effects of its smoke on the health of rural women. Table 14.1 also provides an idea of the economic value of biological and industrial wastes. It is estimated to be Rs. 86 billion per annum of which the agricultural and forestry wastes account for 94.1 per cent, sewage and sludge 3.5 per cent, city refuse 1,5 per cent and industrial waste 0.9 per cent.

Table 14.1: Nutrient Potential and Economic Value of Biological and Industrial Wastes

Type of Waste	*Total Quantity Available Million Tonnes*	*Total Nutrients Thousand Tonnes/yr*				*Economic Value Million Rs.***
		N	*P_2O_5*	*K_2O*	*Total*	
Cattle dung manuue	279.8	2813.3	1999.7	2069.3	6882.3	30,970.4
Crop residue	273.3	1283.1	1965.6	3903.9	7152.6	32,188.1
Forest litter	18.7	99.7	37.4	99.7	236.8	1,065.6
Rural compost	285.0	1431.2	861.5	1422.7	3715.4	16,719.3
City refuse	14.0	98.0	84.0	112.0	294.0	1,323.0
Sewage sludge	0.5	5.1	2.9	2.8	10.8	48.6
Pressmud	3.2	33.3	79.4	55.4	168.1	756.0
Domestic wastewater	6351.0*	317.6	139.7	190.5	647.8	2,915.1
Industrial wastewater	66.2	2.9	0.9	1.3	5.1	21.6
Total		**6084.2**	**5171.1**	**7857.6**	**19112.9**	**86,007.7**

* Million cubic meters per annum.

** Economic value computed @ Rs. 4500/t NPK.

Chemical Characteristics of Wastes and Utilisation

Chemical Characteristics

A tonne of biogas slurry provides 44 kg nutrients as compared to 19 kg by FYM and 27 kg by compost. Certain micronutrients *viz.*, Zn, Cu, and Mn are also provided through these (Table 14.2).

Nutrient potential of city refuse is estimated to be 19.5 kg/1 whereas a tonne of sewage sludge can provide 46.9 kg NPK. Pressmud, a waste from sugar mill, has a manurial value equivalent to 54.4 kg NPK/t beside micronutrient potential for Zn 237–285 mg/kg (ppm) and Cu 112–132 mg/kg. Thus the nutrient potential per tonne dry manures in the order:

Pressmud > Sewage Sludge > Biogas Slurry > Compost > FYM City refuse

Utilisation

While utilising wastes for crop production, several criteria need to be considered keeping in view their potential to create soil conditions hazardous to crop growth, quality of farm produce and its effect on consumers. These criteria may not be similar for all types of wastewater. Presence of enteric pathogens in domestic wastewater may pose several health hazards beside nitrogen limitation, while salinity, nitrogen and organics may be limiting factors for sugar mill, fruit processing and dairy waste. Distillery wastewater need special considerations for salinity, Na, Cl, N, K and colour. Chemical characteristics of wastewaters are summarised in Table 14.3.

Table 14.2: Chemical Characteristics of Biogas Slurry, Sewage Sludge, Pressmud and City Refuse

Parameters	*Biogas Slurry*	*Sewage Sludge**	*Pressmud*	*City Refuse***
Total N, %	1.40–1.84	1.08–2.34	1.12–1.19	0.58–0.61
Total P_2O_5, %	1.10–1.72	0.84–2.14	2.12–2.43	0.59–0.71
Total K_2O, %	0.84–1.34	0.53–1.73	1.98–2.03	0.67–0.73
Organic carbon, %	35.00–38.40	19.50–31.80	23.50–26.70	12.79–16.04
Zinc, mg/kg	103.00–115.60	700.4–2145.70	237.50–285.40	–
Copper, mg/kg	50.60–67.50	193.7–535.20	112.40–131.70	–
Manganese, mg/kg	231.00–294.70	176.2–465.20	–	–

Table 14.3: Chemical Characteristics of Domestic and Industrial Wastewaters

Parameters	*Industrial Wastewaters*				*Distillery Wastewater*	
	*Domestic**	*Sugar Mill*	*Fruit Proc.*	*Dairy*	*Effluent*	*Spent Wash*
pH	7.2–7.8	6.0–6.9	4.3–5.2	7.1–8.0	72–7.6	4.0–5.0
EC, uS/cm, 25°C	675–984	920–950	780–810	840–1,120	45330–40050	–
Ca, meq/L	1.64–2.72	4.50–6.50	3.37–4.42	4.30–5.75	70.60–82.40	–
Mg, meq/L	0.82–1.33	1.50–3.50	1.90–2.I0	1.30–1.70	–	800–900
Na, meq/L	4.22–6.08	2.50–2.80	1.72–2.69	2.74–3.61	11.40–120.70	–
K, meq/L	0.25–0.40	0.50–0.70	0.50–0.80	0.40–0.05	158.40–160.00	–
HCO_3, meq/L	4.92–6.45	4.50–5.85	5.50–6.10	3.60–4.40	107.60–110.00	–
Cl, meq/L	1.84–3.10	1.50–1.95	4.57–5.52	4.20–5.00	116.80–126.00	4000–7000
SO_4, meq/L	0.21–0.34	–	0.20–0.30	–	15.80–17.20	3000–7000
SAR	7.54–8.43	–	1.60–1.75	1.60–1.91	18.8–19.7	–
N, mg/L	54.92–68.35	11.60–15.30	12.10–20.60	43.1–180	1009–1200	1000–2500
P_2O_5, mg/L	18.31–23.13	6.50–7.50	2.20–4.10	31.0–132.0	62–70	120–250
K_2O, mg/L	37.52–44.24	30.00–40.00	11.25–14.50	15.5–47.0	8,000–12,000	9000–12000

Except fruit processing wastewater and distillery spent wash all the wastewaters are slightly acidic to alkaline (pH 6.0–8.0). Wastewaters are categorised under 'No problem to Increasing problem' for salinity hazard. Distillery wastewater being highly saline belongs to class 'severe problem'. Sodicity problem is confined to domestic wastewater and distillery effluent as their SAR values range from 7.5–8.4 and 18.8–19.7 respectively. These are categorised under "Increasing Problem and Severe Problem." Presence of higher concentration of N in domestic wastewater dairy and distillery waste (Table 14.3) can cause ground water contamination due to leaching of nitrates. Determination of hydraulic loading (application rates) thus need to be based on nitrogen as limiting factor so that nitrate concentration in the percolate which joins the ground water should be less than 10 mg/l.

Hydraulic loading is calculated as follows:

$$Lw(n) = \frac{(Cp)\ (P - ET) + (U)\ (4.4)}{(1 - f)\ (Cn) - Cp}$$

where,

Lw (n): Allowable hydraulic loadings based on nitrogen limits, cm/annum

Cp: Allowable nitrate concentration in percolating water, mg/1 (use 10 mg/l)

(P – ET): Normal year precipitation-evapotranspiration, cm/annum

U: nitrogen uptake by crops, kg/hr.

Cn: Nitrogen concentration in applied wastewater, mg/l

F: Fraction of applied N removed by denitrification and volatilization (use 0.20 for design)

compare the value of Lw (n) with the annual sum of Lw(i)

Lw(i) is calculated as follows:

$$Lw(i) = (ET - P)\left(1 + \frac{LR}{100}\right)\left(\frac{100}{Eu}\right)$$

where,

Lw(i): Hydraulic loading based on crop need, cm/annum

ET – P: Evapotranspiration-precipitation, cm/annum.

LR: Leaching requirements per cent (use 10–20 per cent)

Eu: Unit application efficiency for distribution system (use 60–70 per cent).

In designing the system if Lw(n) is greater than or equal to Lw(i), use annual Lw(n) for design. If Lw(n) is less than annual Lw(i), then the options are:

1. Reduce the concentration of N in wastewater.
2. Select crop with higher N uptake.

Heavy Metals and Associated Problems

Heavy metals like cadmium, lead, nickel, chromium, zinc, copper etc. are generally not present in compost, FYM and city refuse in amounts that will be toxic to the soil, plants and mankind. Some heavy metals are also essential plant nutrients. However, sewage sludge, particularly from highly industrialised areas contains heavy metals in amount that may pose toxicity problem (Table 17.4). Sludges from Ahmedabad, Delhi and Vijayawada contain more copper (400–500 mg/kg) whereas, in other cities, Cu contents range between 190-280 mg/kg. Sludge from Ahemdabad contains highest zinc (2146 mg/kg) followed by Jaipur, Delhi and Vijaywada. In Chennai, Nagpur, Kolhapur and Raipur, Zn in sludge was found to be less than 1000 mg/kg.

Table 17.4: Total and Available Heavy Metals and Plant Nutrients in Sewage Sludge

Name of City	Form	Heavy Metals, mg/kg (ppm)							Major Plant Nutrients (%)			OM (%)
		Cu*	Zn*	Mn*	Cd	Cr	Ni	Pb	N	P_2O_5	K_2O	
Ahmedabad												
(a) Pirana	Total	535.2	2145.7	215.0	3.5	60.4	32.3	76.8	1.18	1.27	1.10	33.51
	Available	50.3	201.2	11.5	0.8	–	2.3	8.3	–	–	–	–
(b) Vasana	Total	221.4	687.5	217.5	1.5	184.5	11.4	41.5	1.36	0.84	1.73	26.80
	Available	27.7	160.2	6.5	0.2	–	1.6	3.2	–	–	–	–
Delhi												
(a) Keshpur	Total	440.6	1610.4	195.2	5.5	53.5	815.2	34.5	1.67	1.27	0.53	54.72
	Available	49.5	216.2	15.0	1.4	–	52.2	6.3	–	–	–	–
(b) Okhla	Total	307.2	1260.0	190.7	1.5	82.0	19.1.5	41.7	1.11	1.50	0.57	30.15
	Available	34.6	146.3	9.2	0.5	–	7.6	6.7	–	–	–	–
Vijaywada	Total	455.6	1480.7	3025	0.8	70.8	555	34.5	1.19	1.18	0.72	44.11
	Available	41.7	200.3	26.5	0.2	–	7.4	7.4	–	–	–	–
Nagpur	Total	272.5	832.5	242.5	1.5	49.2	14.8	24.3	1.08	1.10	1.01	39.08
	Available	45.7	70.3	9.2	0.2	–	1.6	3.5	–	–	–	–
Madras	Total	210.3	935.0	465.2	8.3	38.5	60.5	16.5	0.82	1.48	0.57	40.20
	Available	17.2	77.2	245	1.3	2.2	1.9	–	–	–	–	
Kolhapur	Total	193.7	700.4	176.2	0.6	17.3	11.7	14.3	1.21	0.95	1.10	39.40
	Available	17.2	162.0	1.3	0.1	–	0.6	1.8	–	–	–	–
Jaipur	Total	265.3	1720.0	255.6	7.3	176.2	37.5	66.9	2.34	2.14	0.57	39.08
	Available	29.1	79.0	26.0	1.4	–	0.3	6.4	–	–	–	–
Raipur	Total	65.0	520.0	400.0	–	–	34.0	32.0	–	–	–	–

* These are essential micronutrients as well as but their excessive amounts are toxic.

Cadmium, a highly toxic heavy metal is observed to be in the range of 3.5–8.3 mg/ kg in sludges from highly industrialised cities like Jaipur, Chennai, Delhi (Keshavpur) and Ahmedabad (Pirana) while in other cities it varies from 0.6–1.5 mg/kg. Further highest concentration of nickel in sludge is confined to Delhi (191.5–815.2 mg/kg) while in other cities it ranges from 11.4–60.5 mg/kg. The lead is in small quantity in all the sludges. In industrialised cities it varies from 30–80 mg/kg while in others it is less than 25 mg/kg of sludge.

When heavy metal enriched wastes are applied to land they are subjected to various interactions with soil and get distributed between different fractions.

The bio availability, mobility and chemical activity of heavy metals in soil depends on nature of soil and species of heavy metals. Highly acidic conditions (pH < 5.0), low CEC and organic matter favour greater availability of heavy metals to plants, whereas, in calcareous soils and those with high CEC and organic matter the bio availability of heavy metal is appreciably low. The potential uptake of heavy metals by plants depends on species of heavy metal which plants can assimilate. Thus available metal status of soil is more important than its total content.

In general, the plants are excellent barriers to check the transloation of heavy metals from soils to consumable plant parts via root absorption except for certain accumulator species. This is specially true for nickel, copper and lead. Further, Ni and Cu have added protective mechanisms of preeminence of phytotoxicity. Threshhold phytotoxicity values for copper and nickel are 30 and 25 mg/kg respectively, thus the plants die or fail to grow much before they accumulate a metal toxic to human beings.

Zinc readily translocates from root and accumulates in foliage tissue of plants. The values for zinc in animal forage tissues with 20 per cent yield loss is estimated to be 450 mg/kg for corn, 475–570 mg/kg for sorghum, 540 mg/kg for barley, 560 mg/kg for wheat and 295 mg/kg for alfalfa. The probability of food chain hazard due to zinc, thus can not be ruled out easily as with lead, copper and nickel. However, as the zinc transfer from root and its accumulation in fruits, and grains is much less, the human beings are well protected from zinc toxicity.

Cadmium is an element of great environmental concern due to its association with several health hazards. Severely painful and crippling condition of Japanese women, known as *Itai-Itai* disease, was a result of excessive intake of Cd through rice. Its acute toxicity has been reported at 75 mg Cd/kg diet of Japanese quail. Most of the Cd intake in humans is through diet and the joint committee of WHO and FAO recommended maximum permissible weekly intake of 400 to 500 ug/Cd.

"Soil-plant barrier" does not protect the food chain hazard of Cd and it tends to accumulate with increasing soil cadmium level. Since Cd tend to be partially excluded from fruits, its content is generally higher in foliar tissues than fruits, seeds or grains and roots. Leafy vegetables like spinach is the highest cadmium accumulator. More Cd is found in soyabean and wheat as compared to corn (Figure 14.2). Accumulation of Cd in chickpea grain is restricted by physiological barrier of plants but not in case of barley. Further DTPA available cadmium is significantly correlated with Cd uptake by chickpea and barley compared to other forms of Cd in soil. Maximum permissible levels of heavy metals to be applied to soil through sewage and sewage sludge based on cation exchange capacity of soils is given in Table 14.5.

Table 14.5: Recommended Maximum Cumulative Sludge Metal Applications

Metal	*Unit*	*Soil Cation Exchange Capacity (meg/100g)*		
		< 5	*5–15*	*> 15*
Zn	kg/ha	250	500	1000
Cu	kg/ha	125	250	500
Ni	kg/ha	50	100	200
Cd	kg/ha	5	10	20
Pb	kg/ha	500	1000	2000

Pathogens and Health Hazards

The presence of pathogenic bacteria, viruses, ascaris ova and protozoan cysts in sewage and sewage sludges, and the health hazards they cause, need special consideration when these are used for agriculture. A wide variety of bacteria, viruses and protozoa are present in the wastes which can cause a large number of diseases. Work on these aspects in India has been at a rather low level but is getting more attention and several aspects are under investigation at NEERI.

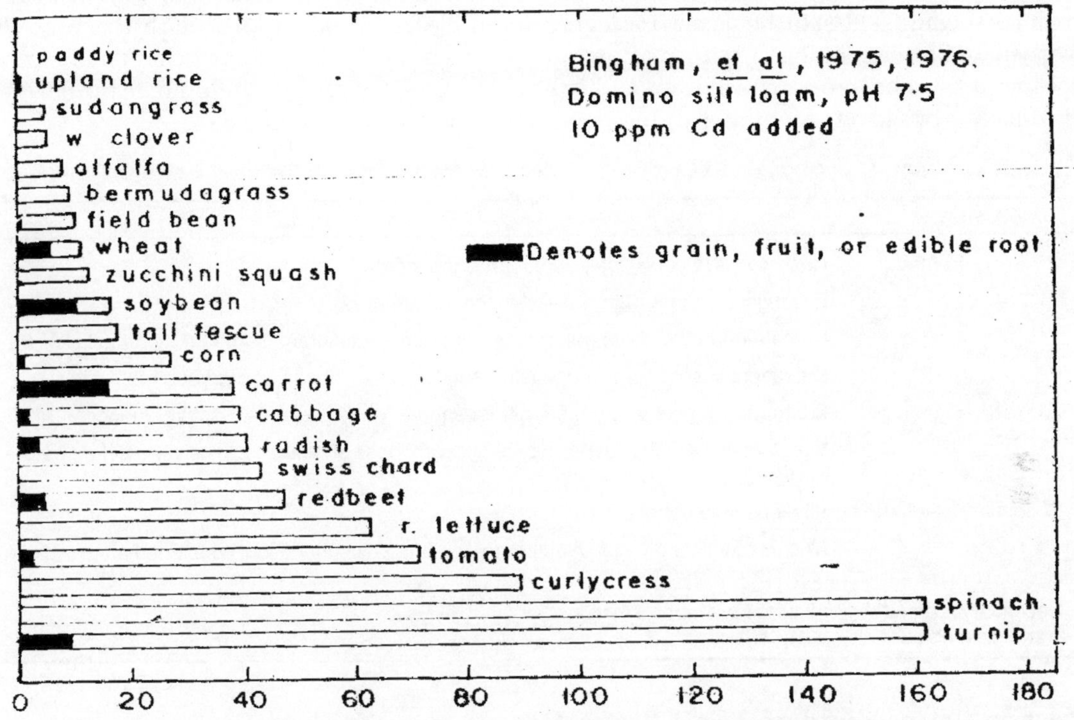

Figure 14.2: Cadmium in Plant Tissue, mg/kg Dry Weight Basis

The relative number of enteric pathogens in sewage and sludge depends on the type of sewage treatment. Primary treatment mostly consisting of settling removes 35–45 per cent pathogens while more than 95 per cent pathogen removal is confined to secondary treatment. The use of treated sewage for crop production thus minimises the health risk. The potential risk of health hazard due to bacteria and viruses to farm workers and consumers depends on persistence of pathogens in soil and on crops. Survival of bacteria and virus in soil is governed by several environmental factors *viz.*, soil pH, cations, soluble organics, moisture, organic matter, temperature and antagonism from native soil microflora.

The results of the survival study of enteric pathogens show that *Coliforms, E. coli, F. streptococci* survived in soil in significantly higher degree in untreated, 1 : 1 diluted and primary treated sewage as compared to secondary treated sewage and fish pond wastewater. Their survival extended upto 120-150 days in sufficient number to cause health hazard due to sewage irrigation except in the case of secondary treated sewage and fish pond wastewater. Thus a high degree of health hazard persists in the use of untreated and 1:1 diluted sewage and to some extent in the case of primary treated sewage.

Vegetables like *spinach, fenugreek,* cabbage and cauliflower were found to be contaminated with Salmonella where untreated sewage was used. In the case of treated sewage, its presence was negligible. However, *coliforms, F. coliforms* and *F. streptococci* were observed even with treated sewage use. Survival of virus on crops surface is shorter than in soil because they are exposed to deleterious environment. Survival of polio virus type 1 on vegetables for 14 to 36 days after irrigation on lettuce and radish was observed.

To minimise the health hazard to farm workers and consumers due to sewage use, it is desirable to treat the sewage at least to the secondary level prior to use for irrigation particularly for crops which are consumed after cooking. The primary treated sewage can be used to grow crops like cereals, pulses, seed and fiber crops. The minimum degree of treatment necessary to eliminate the health risk and suitable crops to be grown with different wastewaters are given in Table 14.6.

Table 14.6: Suggested Cropping Pattern for Irrigation with Treated Sewage

Wastewater	*Suggested Crops in Order of Preference*
Primary treated	Cash crops: Cotton, jute, sugarcane, tobacco
	Essential oil crops: Citronella, mentha, lemon grass
	Cereals and pulses: Wheat, rice, greengram, black gram, sorghum, pearl millet
	Oil seeds: linseed, sesame, castor, sunflower, soyabean and groundnut
	Fruit crops: Coconut, banana, citrus, sapota, guava, grapes, papaya, mango
	Vegetables: Brinjals, beans, lady finger etc. These should be exclusively cooked before eating
Secondary treated	All crops listed above
	All crops including vegetables borne near the soil surface, but consumed after cooking only.
Secondary treated and disinfected	All crops without restrictions.

Effect on Crops Yield and Soil Properties

Effect on Crop Yields

Biogas Slurry, Compost and FYM

Some results on the manurial value of biogas slurry (BGS), and compost/FYM for comparison on the yield of maize and finger millet were reported by Sankaran. None of the organic materials applied @ 10 t/ha could produce maize yield on par with 100 per cent nutrient needs applied through fertilizers. Taking yield increase due to 100 per cent NPK over 1 control as 100 (2184 kg/ha), the response of maize could be rated as 67 in case of biogas slurry, 53 from compost and 50 in case of FYM. Extrapolation of crop response data under different conditions reveals that BGS + 65 per cent NPK and compost + 90 per cent NPK produced yield comparable to 100 per cent NPK given through fertilizers (Figure 14.3). FYM + 75 per cent NPK was on par with 100 per cent) NPK through fertilizers. These results show that BGS has a higher rate of substitution for fertilizers nutrients as compared to FYM and compost. The residual effect of BGS also gave significantly higher yields of finger millet grain and straw over the plots receiving exclusive chemical fertilizer. These results support earlier findings.

Significant yield increases of rice and berseem with the application of wet and dry slurry have been reported. Further, BGS and FYM also showed significant residual effect for two years on yield of rice and berseem compared to yield under no manuring and use of chemical fertilizers (Table 14.7). The uptake of nitrogen by rice and berseem was appreciably higher due to use of slurry (wet and dried) and FYM and it was more so with dried slurry. Manurial potential of BGS, FYM and compost has also been reported by other workers. Biogas slurry, poulty manure, compost and pressmud have been found to be superior sources of zinc as compared to zinc sulphate particularly in Zn-deficient calcareous soil.

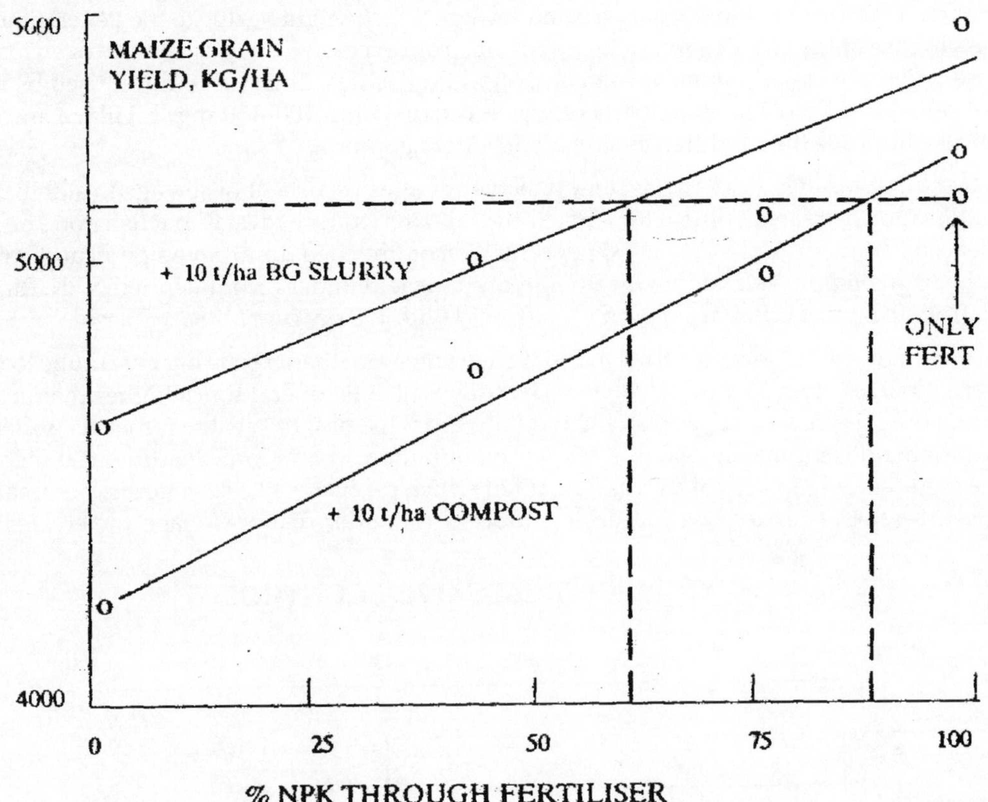

Figure 14.3: Impact of Biogas Slurry and Compost on Maize Yields at Graded Levels of Optimum NPK Input through Fertilizer

Table 14.7: Effect of Biogas Slurry, FYM and Chemical Fertilizer on Yield of Rice and Berseem

Source at 112 kg N/ha	Yield, g/pot		N Uptake mg/pot		Yield of Residual Crop, g/pot	
	Rice	Berseem	Rice	Berseem	Rice	Berseem
No manure	17.8	16.0	330	507	6.6	10.1
Wet slurry	20.4	18.4	383	590	11.8	14.4
Dried slurry	24.4	23.1	410	650	12.9	15.0
FYM	21.9	22.8	383	610	11.7	14.2
Ammonium sulphate	43.2	13.4	813	406	7.5	8.3
CD 5%	2.5	2.2	52	72	4.1	4.0

Sewage and Sewage Sludge

Field trials conducted for 5 years reveals that the primary and secondary treated sewage significantly gave more yield of wheat, cotton, paddy and potato compared to yields with recommended fertilizers (Figure 14.4). In the long run use of secondary treated sewage gave higher yields of these crops compared to untreated and primary treated sewage which might be due to the development of soil sickness in case of UT and PT. Soil sickness results from excessive organic and nitrogen loading which causes anaerobiosis and imbalance in C : N and C : P ratios. This can be minimised by using application rates based on N keeping BOD of waste water within 100–150 mg/l. Tillage and inter cultivation also improve the condition as a result of better aeration.

The two stage wastewater re-use system which envisages treatment of sewage in stabilization pond and utilisation of treated effluent for aquaculture and agriculture reveals that fish pond effluent with 100 per cent supplemental NPK (24–35 per cent of recommended dose) gave significantly higher yields of wheat, greengram and sunflower as compared to yield under recommended NPK. Rice did not benefit from fish pond effluent as compared to yield with 100 per cent NPK.

Presently there are 212 sewage farms in India covering more than 50,000 ha. Hydraulic loading on these farms ranges from 22.4 m^3 to 336.8 m^3 ha/day which is several times higher than normal loading. Thus the present system is more inclined towards disposal of sewage rather its utilization with appropriate nutrient management in view. The organic and inorganic loading also increases beyond the assimilative capacity of soil, as a result of which problems like sewage sickness, salinity and alkalinity develop. The excessive N loading induces production of more foliage and creates C : N

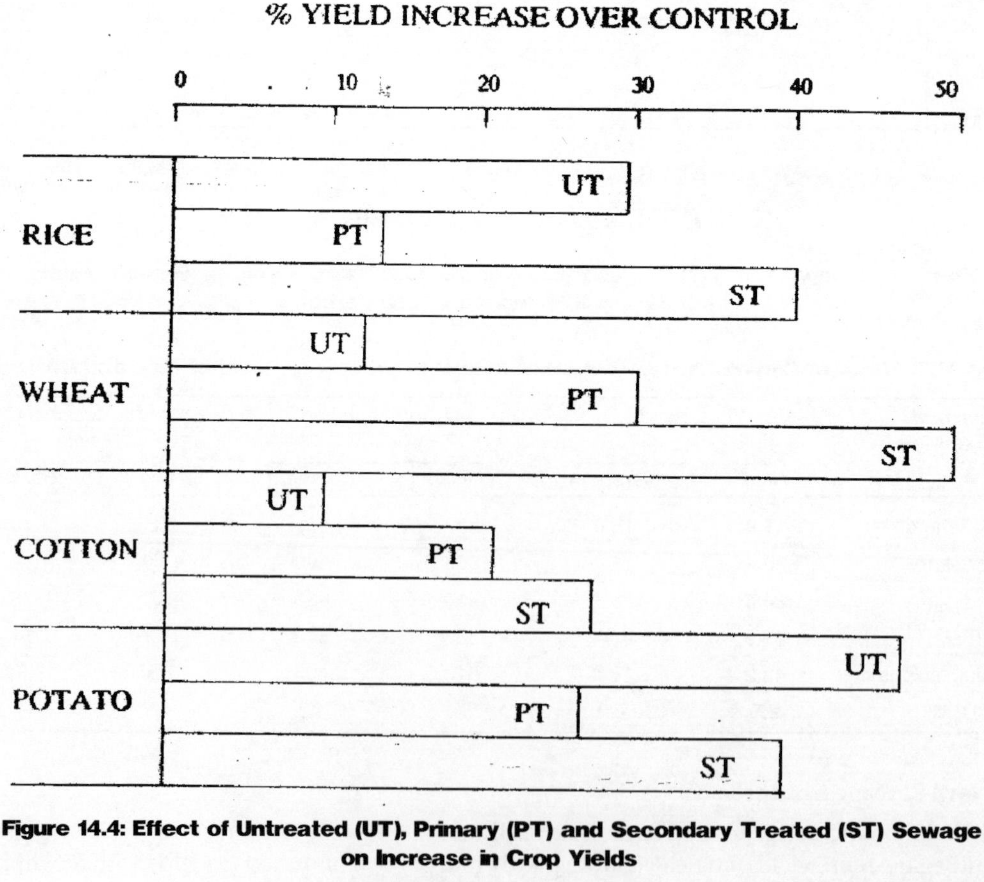

% YIELD INCREASE OVER CONTROL

0 10 20 30 40 50

RICE: UT, PT, ST

WHEAT: UT, PT, ST

COTTON: UT, PT, ST

POTATO: UT, PT, ST

Figure 14.4: Effect of Untreated (UT), Primary (PT) and Secondary Treated (ST) Sewage on Increase in Crop Yields

imbalance in plants which result in less fruiting, particularly in the case of vegetables like brinjal, tomato, chillies and flowers like chrysanthemum.

Land application of sewage has been practiced in Nagpur for 20 years covering 1,500 ha (500 with treated, 1000 with untreated) in 21 villages. In these fields cereals, pulses, oilseeds, sugarcane, vegetable and flowers are grown on a large scale. Again excessive amounts of water and nutrients are added as the loading rates vary from 3,120 m^3/ha/annum to 41,600 m^3/ha/annum, the reason being that the charges are on the basis of area instead of volume of water used. Primary treated sewage gave more yield than untreated and 1 : 1 diluted sewage (Table 14.8).

Table 14.8: Yield of Different Crops Grown with Untreated, Primary Treated and Diluted Sewage at Nagpur

Crops	*Yield, tonnes/ha*			
	Well Water with Nutrients	*Untreated Sewage*	*Primary Treated Sewage*	*Diluted Sewage (1 : 1)*
Paddy	3.78	3.30	4.25	4.10
Wheat	2.75	3.10	3.40	3.15
Soyabcan	1.60	2.10	2.34	1.90
Greengram	0.62	0.54	0.78	0.70
Chickpea	1.20	1.25	1.45	1.35
Cabbage	13.30	14.75	16.40	15.70
Cauliflower	16.40	18.15	19.70	16.90
Okra	3.10	3.40	4.75	4.02
Tomato	13.70	15.45	16.40	16.10
Brinjal	9.15	12.10	12.70	10.15
Potato	6.45	7.10	8.10	7.05
Paragrass	210.5	–	245.6	–
Sugarcane	42.70	44.40	48.45	43.25
Gaillardia	7.98	9.40	11.40	–
Marigold	5.10	7.15	7.95	–
Daizy	8.40	9.70	11.35	–
Jasmin	3.70	3.40	4.35	4.05

Sewage Sludge

Results or field trials conducted since 1983–84 at Nagpur show that sewage sludge without nutrient fortification reduced the yield of all crops studied at sludge application rates which varied from 15 to 90 t/ha. Samples of wheat, rice, sorghum and chickpea grown with 40 t/ha sewage sludge did not show any significant increase in Fe, Mn and Zn in grains compared to control treatment and Cd and Ni concentration were found to be on par with the control (Table 14.9).

Industrial Wastewater

The waters can only be used after dilution as undiluted fruit processing waste water used alone reduced crop yields. Application of wastewater diluted with ordinary water in 1 : 1 ratio increased the yield of rice by 14 per cent, but no impact was seen on the productivity of wheat, sorghum or greengram. Similarly sugarmill waste water alone or fortified with 25 per cent of NPK dose through fertilizers produced lower crop yield then the 100 per cent fertilizers treatment in long-term field experiments. Taking crop yield with 100 per cent NPK through fertilizers as 100, yield with sugar mill waste water + 25 per cent of recommended NPK through fertilizers was 93 in wheat, 91 in sunflower, 92 in sorghum and 88 in sugarcane.

Table 14.9: Nutrient and Trace Metal Content in Food Grains Grown on Sludge Amended Soil at Nagpur

Parameters	*Treatment*	*Rate*	*Rice*	*Wheat*	*Sorghum*	*Chickpea*
Protein	Control		7.82	9.46	6.98	18.72
g/l00g	Sludge	40t/ha	8.23	10.19	7.90	21.20
P	Control		0.27	0.31	0.21	0.19
g/l00g	Sludge	40t/ha	0.37	0.43	0.28	0.19
K	Control		0.41	0.37	0.30	1.67
g/l00g	Sludge	40t/ha	0.43	0.40	0.28	1.80
Fe	Control		21.42	17.10	21.40	34.72
ug/g	Sludge	40t/ha	23.47	18.70	24.30	51.40
Mn	Control		59.65	20.43	7.75	13.72
ug/g	Sludge	40t/ha	63.20	24.30	6.40	14.15
Zn	Control		10.90	19.15	11.40	17.40
ug/g	Sludge	40t/ha	13.40	23.40	14.65	19.65
Cu	Control		2.15	3.10	1.75	6.20
ug/g	Sludge	40t/ha	2.82	4.05	2.05	6.15
Cd	Control		0.01	0.02	0.01	0.02
ug/g	Sludge	40t/ha	0.02	0.01	0.01	0.01
Ni	Control		0.21	0.31	0.17	0.40
ug/g	sludge	40t/ha	0.13	0.21	0.16	0.32

Pressmud

The pressmud is a waste product from sugar mills. Sulphitation process pressmud is more useful for crop production as compared to pressmud from carbonation process. The latter being rich in calcium carbonate is more suited to acidic soils. Results of laboratory studies with sulphitation process pressmud shows that addition of 20 t/ha pressmud along with application of 75 per cent NPK dose through fertilizers resulted in 26.3 per cent and 35.2 per cent higher grain yield of sorghum and wheat respectively compared to application of 100 per cent NPK through fertilizers. Yields increases with 100 per cent NPK alongwith 20 t/ha pressmud were still higher at 31.6 per cent and 47.2 per cent (Figure 14.5). It can be tentatively concluded that 20t pressmud/ha plus 55 per cent of recommended NPK through fertilizers was as effective as 100 per cent NPK through fertilizers. Pressmud from the sugar factories using the sulphitation process has been found to be an effective source of nutrient sulphur for crops and the same is true for oxalic acid industrial waste and sulphur sludge from rayon industry. Pressmud has also been reported to be a good manure for rice.

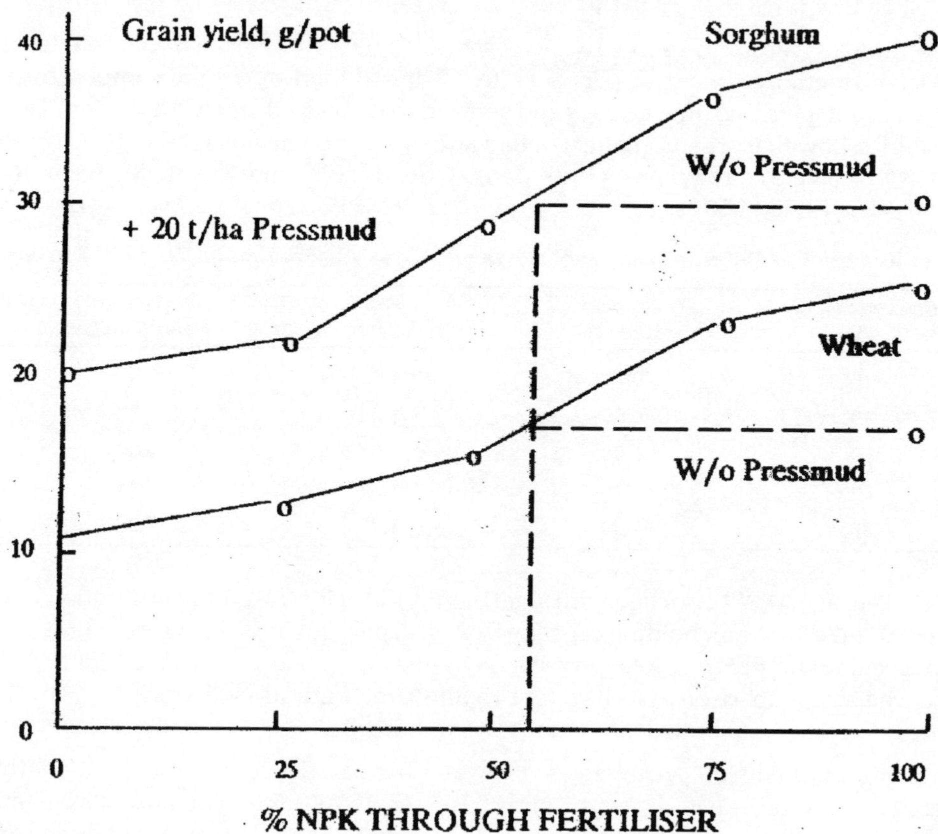

Figure 14.5: Effect of Pressmud in Combination with Graded Levels of Fertilizer on Crop Yield

Manurial potential of pressmud in restoration of manganese mine spoil productivity has also been reported. Incorporation of 100 t/ha pressmud resulted in 443 per cent, 419 per cent and 552 per cent increase in growth of teak, *shiwan* and *shishum* plants grown in mine spoil. Pressmud helps to buildup useful soil microflora and improvement in physical properties of soil. Increase in rice and wheat yield in sodic soils due to the addition of pressmud has also been reported.

Phosphogypsum

This material a biproduct of the fertilizer industry. Though not a waste in the conventional sense, its accumulated stocks of over 10 million tonnes need effective and productive disposal. It has been found to be a useful amendment for alkali soils and as also as a source of sulphur for plants. Annual production of phosphogypsum is around 1.5 million tonnes and is expected to increase. For each tonne of P_2O_5 produced in the wet process phosphoric acid, 4–5 tonne phosphogysum is discharged. At present about 50 per cent annual production is used in cement production, as a raw material for manufacturing ammonium sulphate and for soil treatment. Its current agricultural use has been estimated at 350,000 tonnes out of which 70–80 per cent is used as a soil amendment and the rest as a sulphur fertilizer.

Effect on Soil Properties

Incorporation of wet slurry markedly increased soil aggregation from 62 per cent to 70 per cent initially which further increased to 78 per cent after 60 days. Dried slurry and compost did not show immediate effect but increased after 60 days and showed that digested slurry was better than compost. Studies at NEERI show that organic matter content and infiltration rate increased from 0.78 to 1.33 per cent and 5.88 to 7.80 cm/day whereas bulk density decreased from 1.68–1.48 g/cc at the sludge loading of 90 t/ha (Table 14.10)

Table 14.10: Effect of Different Levels of Sewage Sludge Application on Some Soil Properties

Sewage Sludge t/ha	*Organic Matter (%)*	*Bulk Density, (g/cc)*	*Infiltration Rate, (cm/day)*
0	0.78	1.68	5.88
15	0.86	1.62	6.12
30	0.98	1.58	6.84
60	1.12	1.51	7.44
90	1.33	1.48	7.80

Fish pond effluent (FPE) enriched with algal biomass also increased soil infiltration rate from 4.4 cm/day to 6.00 cm/day, water holding capacity from 47.6 per cent to 59.4 per cent, aggregate stability from 58.2 per cent to 66.7 per cent, total porosity 38.2 per cent to 58.2 per cent and organic from matter 0.8 to 1.2 per cent, in comparison to well water irrigation. Incorporation of algal biomass through fish pond effluent decreased soil bulk density from 1.68 to 1.36 g/cc.

The use of undiluted fruit processing wastewater increased electrical conductivity (EC) of soil from 315 uS/cm to 980 uS/cm in clayey soil and 305 uS/cm to 780 uS/cm in sandy clay loam. Further, the rise in EC in case of 2 : 1 and 1 : 1 diluted wastewater was 1.2 to 1.6 times more than control. Soil pH did not show any appreciable change with wastewater use. Presence of lower dissolved salts in sugar mill wastewater did not cause any appreciable variation in case of EC of pre and post harvest soil during 4 years. Same was true for pH and soil organic matter.

Problems in Waste Utilisation

Today it is necessary to emphasis that the "wastes are a resource" and therefore, their management and utilization is a must. For these wastes, utilization approach rather than mere disposal needs to be adopted for conservation of resources. The information available reveals that the wastes are not properly collected, characterised, and processed. Their useful potential is also not fully appreciated in the country. Presently, less than 20 per cent of the cattle dung is actually used for biogas generation resulting in loss of energy and manure. Same is true for farm wastes.

In case of domestic and industrial wastewaters it is necessary to update the collection, treatment and minimisation of the contamination with pollutants particularly in case of industrial wastes. This will improve waste quality for crop irrigation.

Pressmud is presently dumped on ground and burned due to the lack of knowledge about its manurial potential. Similarly sewage sludge particularly from industrial cities are contaminated with heavy metals and thus may pose hazard to plants and consumers both. Segregation of toxic waste at the source will minimise the concentration of such elements in sludge. High N input through excessive hydraulic loading of sewage can lead to ground water contamination. Further, survival of pathogens in soil and on crops cause health risk. This can be minimised by proper treatment, design of hydraulic loading based on limiting factors, selection of suitable crops and adoption of modified irrigation system.

Future Research Needs

Inadequate appropriate research is still a major constraint in maximising the use of waste for crop production. Major areas which require priority attention are:

1. Optimisation of organic manuring in conjuction with inorganic fertilizer for improvement of soil productivity without creating environmental problem.
2. Fate of heavy metals in different soils amended with sewage sludge and their relation with crop uptake.
3. Identification of different forms of heavy metals in sludge amended soil and their bio availability.
4. Screening of different plant species which control the translocation of heavy metals from root to shoot.
5. Conducting field trials for land application of sewage and industrial waste water in different agro climatic zones and studies on effect of wastewaters on soil and crops to design land treatment system based on land limiting pollutants.
6. Studies on migration of pollutants from different soils and extent of ground water pollution in waste disposal areas.
7. Evaluation of pathogen survival in soils and on crops in sewage irrigated and sludge applied areas.
8. Assessment of health hazard to farm labour working sewage farms.
9. Development of modified soil and crop management practices to avoid health risk due to pathogens.
10. Extensive survey of existing waste sites to deliniate the new strategy of minimum pollution/ zero pollution.
11. Greater coordination between scientists working in agriculture, waste disposal and recycling, and human and animal health should take place to develop safe, economical and practical techniques of waste disposal.

Chapter 15
Role of Biofertilizers in Crop Production

Biofertilizers have an important role to play in improving the nutrient supplies and their crop-availability in upland crop production. Although *Rhizobium* is the most researched and well known among these, there are a number of microbial inoculants with possible practical application in upland crops where they can serve as useful components of integrated plant nutrient supply systems. Such inoculants may help in increasing crop productivity by way of increased biological nitrogen fixation (BNF), increased availability or uptake of nutrients through solubilization or increased absorption, stimulation of plant growth through hormonal action or antibiosis, or by decomposition of organic residues.

The scope of this chapter is confined to microbial inoculants of N_2-flxing bacteria, vesicular arbuscular mycorrhizae (VAM), phosphate solubilizers and plant growth promoting rhizobacteria (PGPR) and their role in production of upland crops. Aspects determining the success of inoculant technology are discussed here rather than presenting an up-to-date account of literature on the performance of biofertilizers. Rhizobial inoculants are dealt with in detail as a general example and for others, our discussion is restricted to those points which are specific to a particular group of inoculants.

Nitrogen-fixing Bacterial Inoculants

Above every hectare of land at sea level, there are 78,000 tonnes of inert nitrogen gas (N_2), Nitrogen is the most limiting nutrient for increasing crop productivity. Only a few procaryotic organisms are able to "fix" N_2 directly through a biological process. Annually BNF is estimated to be around 175 million tons of which close to 79 per cent is accounted for by terrestrial fixation (Figure 15.1). The N_2 fixers involved are metabolically diverse but the process is similar and depends upon (*i*) nitrogenase (N_2ase) enzyme complex, (*ii*) a high energy requirement (ATP) (*iii*) anaerobic conditions (for N_2ase activity) and, (*iv*) source of strong reductant. Amongst N_2-ftxing bacteria *viz.*, *Rhizobium, Azospirillum,* and *Azotobacter,* the most widely used inoculant is *Rhizobium.*

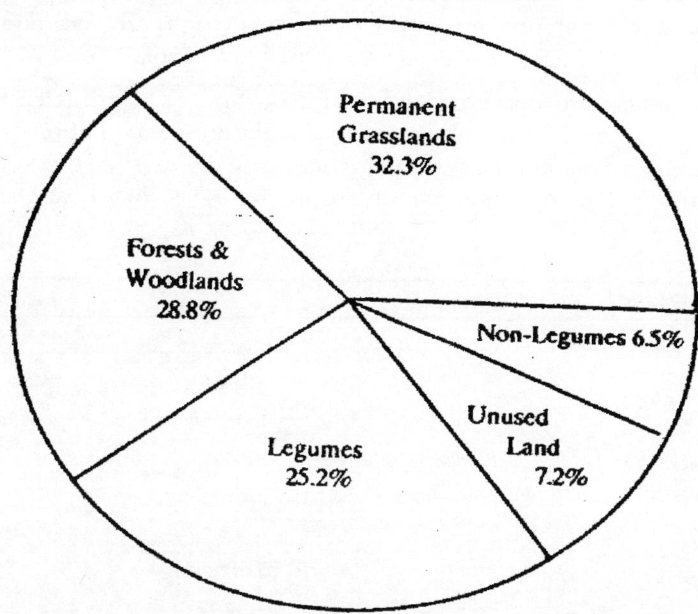

Figure 15.1: Distribution of 139 Million Tonnes of N_2 Estimated to be Biologically Fixed in Various Terrestrial Systems

Rhizobium

Symbiotic N_2-flxation by *Rhizobium* with legumes contributes substantially to total BNF (Table 15.1). *Rhizobium* inoculation is a well known agronomic practice to ensure an adequate nitrogen nutrition of legumes in place of fertilizer N.

Table 15.1: Estimates of Nitrogen Fixed by Some Legumes

Crop	*Nitrogen Fixed (kg/ha)*	*Crop*	*Nitrogen Fixed (kg/ha)*
Alfalfa	100–300	Lentil	35–100
Clover	100–150	Greengram	50–55
Chickpea	26–63	Pigonpea	68–200
Cluster bean	37–196	Soybean	49–130
Cowpea	53–85	Peas	46
Groundnut	112–152	Fenugreek	44

Classification

The genus *Rhizobium* consists of six distinct species based largely on the cross inoculation group concept. The assumption in this classification is that those leguminous plants falling within a particular infection group were nodulated by a particular species of nodule bacteria. More than twenty cross-inoculation groups have been established, but only seven have achieved prominence. The validity of the cross-inoculation groups has been questioned as many legumes are nodulated by rhizobia of other host-bacterial groups. With repeated evidence of anomalous cross-infection among the different plant groups, new classification has been proposed. The slow-growing rhizobia are grouped under the genus *Bradyrhizobium* and the fast growers under the genus *Rhizobium.*

The new classification is more complex as the same host may be nodulated by fast- and slow-growing strains. Moreover, the agricultural significance of the cross inoculation groups still remains a key feature of the established taxonomic system. In this article we have used the classification based on cross-inoculation groups. The cross-inoculation groups and *Rhizobium*-host plant associations are listed below:

Cross-inoculation Group	*Rhizobium Species*	*Host Genera*	*Legumes Included*
Alfalfa group	*R. melloti*	*Medicago*	Alfalfa
		Melilotus	Sweet clover
		Trigonella	Fenugreek
Clover group	*R. trifolii*	*Trifolium*	Clovers
Pea group	*R. leguminosarum*	*Pisum*	Pea
		Vicia	Vetch
		Lathyus	Sweet pea
		Lens	Lentil
Bean group	*R. Phaseoli*	*Phaseolus*	Beans
Lupine group	*R. lupini*	*Lupinus*	Lupines
		Ornithopus	Serradella
Soybean group	*R. joponicun,*	*Glycine*	Soybean
Cowpea group	*Rhiizobium* sp.	*Vigna*	Cowpea
		Cojanus	Pigeon pea
		Cicer	Chickpea
		Lespedeza	Lespedza
		Crotolaria	Sunnhemp
		Arachis	Groundnut
		Phaseolus	Lima bean
		Pueraria	Kudz

Need for Inoculation

The most important point is do we need inoculation of legumes in a region where these crops have been grown over long periods? Development of an inoculant industry in many countries has been largely motivated by the desire to introduce legume species to new areas. Most cultivated tropical soils are assumed to have relatively large populations (> 100/g dry soil) of rhizobia capable of nodulating the legumes grown in such soils. The need to inoculate the legumes grown on cultivated soils must be assessed by considering the interacting factors between the soil, the host plant and *Rhizobium.*

Presence of nodules on plant roots does not necessarily mean that sufficient N_2 is being fixed for maximum benefit to the host plant. In groundnut or pigeonpea nodulation occurs naturally at most locations due to the cross-species promiscuity of the cowpea rhizobia. However, the ability to fix high amounts of N (efficiency) is governed by the symbiotic capability between *Rhizobium* and the host plant. Hence, it maybe necessary to introduce superior (more competitive and efficient) strains of *Rhizobium* to ensure adequate N_2 fixation for maximum growth and yield of the host plant. In a survey of groundnut crops grown in farmers' fields in southern India, 52 out of 95 fields showed inadequate nodulation with less than 1.0 per cent N_2-fixing (acetylene reducing) activity of what can be obtained under reasonable field conditions. The results of surveys of farmers' grain and fodder legume crops also revealed poor nodulation in large areas and good noduation only in a few pockets. Poor nodulation in farmers' fields could be due to several factors *e.g.* lack of appropriate rhizobia in soil, deficiency or toxicity of a nutrient, unfavourable conditions like prolonged water logging, unsuitable pH, abundance of bacterial predators, pests and disease attack, etc. Although adequate nodulation was observed in some parts, ineffective nodules exceeded the number of effective nodules. Out of 87 ground nut rhizobial strains isolated from different parts of India, only 5 were found to be effective.

It should be realized however, that poor nodulation can be due to poor plant growth resulting from pests, diseases, and nutrient deficiencies. In Karnataka, trials on farmer's fields with pigeonpea showed dramatic increase in nodulation due to application of diammonium phosphate (DAP) alone than to inoculation with *Rhizobium* alone. The plots receiving DAP and *Rhizobium* yielded 100 per cent more than the control plots. Similarly in groundnut, fertilisation with B, Co, Mo and Zn in a medium calcareous soil, with and without *Rhizobium* inoculation significantly increased nodulation, percentage of effective nodules and plant dry matter. Insect damage can seriously affect the N_2-flxing ability of nodules. Extensive nodule damage to pigeonpea by a Dipteran larva, *Rivellia angulata* in farmers' fields reduced yields significantly. The extent of nodule damage is greater in pigeonpea grown in Vertisols (up to 86 per cent) as compared to 20 per cent in Alfisols. In brief, entire area under grain and forage legumes needs attention in order to improve BNF. However, the solution may not be in using biofertilizers alone. We need to follow a holistic system's approach to relieve the other constraints like nutrient deficiencies, pests and diseases for harnessing maximum benefits from use of biofertilizers.

Competitiveness and Effectiveness of Strains

In soils lacking rhizobia nodulating a particular legume, inoculation with efficient strains increased yields. In soils which contain established native *Rhizobium* populations, the introduced strains should be competitive and efficient. The degree of establishment and persistence of an inoculant strain generally decreased with increase in population density of the native rhizobia. However, some inoculant strains have succeeded in forming more nodules even in the presence of active indigenous competing rhizobia *e.g.* NC92 on groundnut. Little is known of the factors controlling competitiveness but host cultivar, soil properties, soil microl1ora, environmental factors and the nature of the competing strains influence the success of inoculant strains in nodule formation. The success of the strain NC 92 in terms of nodule formation increased with repeated inoculation (Table 15.2). Higher inoculum rate of 10^6–10^6 cells per seed (10^6 = 1 million) at the initial inoculation helped in early establishment. Competition between *Rhizobium* strains and inoculution response was less pronounced in the presence of soil mineral N than under conditions where such N was immobilised and unavailable.

In many rice-growing areas, legumes are grown after paddy, using residual moisture. In such fields, less than 100 cowpea group rhizobia/g soil were observed and continuous cultivation of paddy had an adverse effect on *Rhizobium* survival. Under such conditions inoculation with effective strains showed significant responses in chickpea and pigeonpea. It is generally difficult to displace indigenous rhizobia with inoculant strains. Rupela successfully displaced inefficient strains with an efficient inoculant strain through soil solarization and inoculatum.

Table 15.2: Persistence of Inoculum Dtrain NC 92 Over Two Seasons on Groundnut

Season		*% Nodules Formed on Groundnut Plants*	
1st	*2nd*	*72 days*	*116 days*
After Sowing	After Sowing		
Uninoculated	Uninoculated	9 (5)*	11 (8)
Uninoculated	Inoculated	31 (27)	27 (25)
Inoculated	Uninoculated	28 (25)	42 (32)
Inoculated	Inoculated	39 (41)	75 (54)
SE		± 2.5	± 5.4

* Data analysed after arcsine transformation: original means in parenthesis.

Factors Affecting Performance of Inoculant Strains

Crop responses to inoculation with biofertilizers are not as dramatic as those with fertilizer N. Being biological agents, these are subjected to a range of hostile environments and their survival and efficiency is governed by several factors. Generally, there is a decline in the rhizobial population on seeds but conventional wisdom is that multiplication should occur as the rhizosphere forms, so t-hat accelerated germination can also assist in ensuring an adequate population. The seed coat of dicots is often carried on the top of the cotyledons into the open air, so that only a part of the inoculum maybe left to multiply within the rhizosphere. In case of crops grown on residual moisture, such as chickpea, the inoculated rhizobia cannot move downwards with the growing root from the top soil where inoculated resulting in poor nodulation. Secondly, deep sowing results in good crop stand but affects nodulation adversely.

Carrier-based inoculants are usually coated on seeds for the introduction of bacterial strains into the soil. However, alternative inoculation methods are necessary where seed treatment with fungicides and insecticides is needed or where seed of crops such as groundnut and soybean can be damaged when inoculated with an adhesive. In addition, use of superphosphate as the P source can be harmful for *Rhizobium* because of contact with the acidic fertiliser. Often the soils themselves are acidic and lime coaling of seed has been a popular measure for additional protection. The normal carrier-based inocula can be successfully applied separately from the seed. While all methods of inoculation were successful under favourable conditions, "liquid" and "solid" methods were superior to seed inoculation under adverse conditions. Increased groundnut yields were obtained when incoulation was done by applying a slurry of peal-based incoulum in the seed furrow (Table 15.3).

At ICRISAT, a bullock-drawn seed drill commonly used by farmers has been modified for simultaneous *Rhizobium* application in the seed furrow. Soil properties can also affect the survival inoculated rhizobia. For example, out of 11 locations tested for response of groundnut CV Robut 33-1 inoculation with strain NC 92 failed to increase yields at two locations; namely Tirupathi and Kadiri. Subsequent analysis of soil samples from Tirupathi revealed a high (150 ppm) available manganese content. Manganese and aluminum can be toxic to symbiotic N_2 fixation even if they are not at a level high enough to affect plant growth. Soil acidity and alkalinity can also pose problems for symbiotic N_2 fixation. For such problem areas, specific strains with the ability to overcome such adverse conditions need to be selected as inoculants. Significant differences were observed among pigeonpea rhizobiam strains for their ability to nodulate and fix N_2 under saline conditions.

Table 15.3: Effect of Fungicide and Method of Rhizobium Inoculation on Nodulation by Strain NC92* on Groundnut

Seeds Treatment	*Method of Rhizobium Inoculation and % Nodules Formed by Strain NC 92**		
	Liquid	*Seed*	*Unioculated*
Untreated	3 (27)	22 (20)	4 (2)
Captan	28 (23)	7 (4)	3 (1)
Thiram	25 (18)	6 (4)	7 (2)
Dithane	19 (10)	14 (9)	7 (3)
Bavistin	24 (16)	14 (8)	10 (3)
Mean	25 (19)	13 (9)	6 (2)

SE mean for comparing Rhizobium inoculation mean within a fungicide treatment is ± 5.5.

* Nodules typed by ELISA 60 days after sowing. Data analysed after arcsin transformation: original means in parenthesis.

Yield Response to Inoculation

The field performance of inoculation is variable. For example, with chickpea significant improvement in grain yield was reported from 7 out of 16 and 6 out of 12 locations, predominantly in central and northern India with yield increases varying from 14 to 30 per cent over the control plots. In pigeonpea significant increases in early nodulation due to inoculation were not always well correlated with the final grain yields. Increase in grain yield of the pigonpea inoculated with effective *Rhizobium* ranged from 19 to 68 per cent over uninoculated controls. In groundnut, inoculation responses varied from decreased yields to significant increased yields over unioculated controls.

Sometimes, legumes yields are not increased by inoculation but N concentration in grains or plant parts is increased over the uninoculated control. In cases where both types of responses are not observed, it might simply result in a saving of soil N which might be useful for the succeeding crop. Nodulating legumes benefit the succeeding cereal crops in terms of increased yields in comparison with the yield of cereal grown after cereal or non-nodulating lines of the legumes. Such increased yields are attributed to enrichment of soil N due to N_2 fixation and the N conservating effect of legumes.

However, all such benefits cannot be explained by N effect alone, and several factors *e.g.* improved soil biological, physical and chemical properties, reduced weed and disease severity and insect infestation, etc. are responsible. The benefits of preceding legume crop on a cereal crop yield in terms of fertilizer N equivalents range from 20 to 123 kg N/ha (Table 15.4). The benefits of inoculating legumes in terms of increased economic yields are also not consistent. The possible explanation could be any one of the above mentioned factors for example in chickpea, in the presence of adequate available N in soil or through fertilizer application, the non-nodulating lines of chickpea produce similar/same yields as those of well nodulated plants. Further, such non-nodulating plants do not show any apparent N-deficiency symptoms.

Azotobacter and *Azospirillum*

Although many genera and species of N_2-fixing bacteria have been isolated from the rhizosphere soil of various cereals, mainly members of *Azolobacter* and *Azospirillum* genera have been widely tested under field conditions. *Azospirilla* and *Azotobacters* are active N_2 fixers under laboratory conditions, ubiquitous in geographic distribution and cause a variety of carbon and energy sources for their growth on combined N or N_2. A survey of 200 fields in the traditional sorghum- and millet-growing areas in north-western India showed that the most probable number (MPN) of N_2 fixers varied 1000 fold, from 100 to 100,000/g soil. Out of 3760 isolates obtained from these soils, 42 per cent showed N_2ase activity *in vitro*. In another study, out of 546 isolates obtained from the rhizosphere of pearl millet grown at ICRISAT Center only 17 per cent isolates showed N_2ase activity *in vitro*. Pearl millet rhizosphere was dominated by azospirilla (72 per cent of total N_2-fixing isolates) followed by enterobacters (12 per cent), azotobacters (11 per cent) and 5 per cent pseudomonads. Similar information is lacking for other crops.

Table 15.4: Residual Effect of Preceding Legume on Cereal Yield in Terms of Fertilizer N Equivalents

Preceding Legume	*Following Cereal*	*Fertilizer N Equivalents (kg N/ha)*
Berseen	Maize	123
Sweet clover	Maize	83
Chickpea	Maize	60–70
Groundnut	Pearl millet	60
Cowpea	Pearl millet	60
Chickpea	Pearl millet	40
Lentil	Pearl millet	40
Peas	Pearl millet	40
Pigeonpea	Wheat	40
Lathyrus	Maize	36–48
Pigeonpea	Pearl millet	30
Greengram	Pearl millet	30
Pigeonpea	Maize	20–49
Peas	Maize	20–32
Lentil	Maize	18–30

The mechanisms by which the plants inoculated with these bacteria derive positive benefits in terms of increased grain, plant biomass and N uptake are attributed to small increase in N input from BNF development and branching of roots, production of plant growth hormones, enhancement in uptake of NO_3^-, NH_4^+, $H_2PO_4^-$, K^+, Rb^+, and Fe^{2+}, improved water status of the plants, increased nitrate reductase activity in plants, production of antibacterial and antifungal compounds. Inoculation of sorghum with *Azotobacter* and *Azospirillum* resulted in a marked decline of shootfly (*Atherigona soccata* Rond.) damage as compared to uninoculated control through increased levels of phenol content in the shoots of *Azospirillum* inoculated plants.

Yield Responses to Inoculation

Plant responses to inoculations with azotobacters and azospirilla in cereals and non-cereals are often reported in terms of increased grain yield, plant biomass yield, nutrient uptake, grain and tissue N contents, nitrogenase activity, early flowering, tiller numbers, greater plant height, leaf size, increased enzyme levels in plant parts, increased number of spikes and grains per spike, thousand grain weight, increased root length and volume, reduced insect and disease infestation. Recent reviews have evaluated the worldwide crop responses to inoculation with azotobactors and azospirilla. The results indicated that in many cases inoculations increase plant yields but such increases are variable (statistically significant or otherwise and also sometime negative). The responses varied with crops, cultivars locations, seasons, agronomic practices, bacterial strains, level of soil fertility, and interaction with native soil microflora. In an evaluation of the reported worldwide success of *Azotobacter* and *Azospirillum* inoculation, it was concluded that statistically significant yield increases were obtained in approximately 60 per cent of the trials in USSR, Israel and India.

Multi-locational trials in India showed that seed inoculation with *Azosp. brasilense* increased the mean grain yields of pearl millet significantly at 6 and with sorghum at 4 out of 9 locations tested. The yield increase with pearl millet varied from –10 to 17 per cent and with sorghum from 7–31 per cent.

In inoculation trials with pearl millet in India average increases in grain yield with inoculation were higher (11 per cent) in case of *Azosp. lipoferum* and 8 per cent with *Aztb. chroococcum* (ICM 2001) over the uninoculated controls (Table 15.5). Furthermore, experiments with *Azosp. lipoferum* have shown significant increases as compared with *Azotb. chroococcum* inoculation. Increased yields with *Azotobacter* inculation ranged from 2 to 45 per cent in vegetables, 7 to 28 per cent in cotton and 9 to 24 per cent in sugarcane.

Table 15.5: Summary of Pearl Millet Inoculation Experiments with Two Cultures Conducted During 1982–86 at Different Locations in India

Item	*A. lipoferum (ICM 1001)*	*A. chroococum (ICM 2001)*
Field experiments conducted	24	24
No. of experiments which showed significant increase in grain yield	11	8
	with average increase of	
	18.7%	13.6%
No. of experiments which showed non-significant increase in grain yields	10	12
	with average increase of	
	9.3%	8.3%
No. of experiments which showed no response	1	2
No. of experiments which showed reduction in grain yield	2	2
	with average reduction of	
	2.7%	4.5%
Average increase in grain yield due to inoculation	11%	8%

Effect of Soil Nutrients

Soil and fertilizer N affect the response to inoculation, as expected. Generally, good responses to inoculation are obtained at intermediate levels of N fertiliser. Largest differences in yield were obtained when the soil was adequately but not excessively fertilised. In multi-location experiments with pearl millet, higher increases in grain, plant biomass, and total N uptake were observed with zero N + inoculation and the extent of response declined with the increasing levels of applied N (Table 15.6). Grain yields obtained from zero N treatments inoculated with N_2-ftxing bacteria were similar to the yields from the noninoculated plots receiving 20 kg N/ha. It is not uncommon to observe yield increases equivalent to < 20 kg N/ha depending on locations, soil fertility and other factors. Such benefits in terms of increased grain yields and N uptake could not be explained in terms of BNF based on ARA and ^{15}N based studies.

Table 15.6: Mean Grain Yield and Total N Uptake by Pearl Millet Inoculaled with N_2-fixing Bacteria at Different N Levels

N Levels (kg/ha)	Bacterial Culture		Non-inoculaled Control	Mean	S.E. ±
	Azosp. lipoferum	*Aztb. chroococcum*			
		Grain yield (t/ha), mean of 7 locations			
0	1.8 (16)*	1.8 (16)	1.5	1.7	
20	2.0 (10)	1.9 (4)	1.8	1.9	0.059^{NS}
40	2.0 (6)	2.0 (3)	1.9	2.0	
Mean	1.93	1.88	1.76		
SE±		0.033**			0.036**
CV (%)		20			
		Total plant N uptake (kg/ha), mean of 3 locations			
0	32.2 (27)*	29.9 (18)	25.3	29.1	
20	37.0 (13)	36.6 (12)	32.6	35.4	
40	39.2 (8)	37.3 (3)	36.2	37.6	1.065**
Mean	36.1	34.6	31.4		
SE±		1.72^{NS}			0.831**
CV(%)		24			

* Figures in parentheses indicate percentage increase over respective control.

** P: < 0.01; NS: Non significant.

Frequency of Inoculation

Most trials have measured the effect of one time inoculation, but in a few cases the benefits of continued inoculations were studied. Three years of continued inoculation enabled 4 millet crops (3 main crops and one succeeding crop) to assimilate 26 kg extra N/ha over the uninoculated plots. These increases were observed along with a 2–3 fold increase in the Most Probable Number (MPN) and enzyme-linked immunosorbant assay (EUSA) counts of azospirilla and azotobacters. Such results suggest that every year the crop needs to be inoculated. Use of FYM, green manures or other organic amendments enhanced the benefits from inoculation. These bacteria establish in the rhizosphere in low numbers after inoculation. There is a need to study the reasons for decline in their numbers and to find the agronomic practices that may help the inoculated bacteria to establish in large numbers in the rhizosphere.

Phosphate Solubilizing Microorganisms

A group of heterotrophic microorganisms are known to have the ability to solubilize inroganic P from insoluble sources. There are:

Bacteria: *Bacillus megaterium, B. circulans, B. subtilis, Pseudomonas straita, P. rathonis.*

Fungi: *Aspergillus awamori, Penicillium digitatum, Trichodemia* sp.

Yeast: *Schwanniomyces occidentails.*

Mechanism of Action

The solubilization of P by these microorganisms is attributed to excretion of organic acids like citric, glutamic, succinic, lactic, oxalic, glyoxalic, malcic, fumaric, tartaric and α-ketobutyric. These microorganisms weather rock phosphate and tricalcium phosphate by decreasing the particles size reducing it to nearly amorphous forms. The action of organic acids has been attributed to their ability to form stable complexes with Al^{++}, Ca^{++}, Fe^{++}, and Mg^{++}. In addition to P-solubilization, these microorganisms can mineralize organic P into a soluble form. These reactions take place in the rhizosphere and because the microorganisms render more P into solution than is required for their own growth and metabolism, the surplus is available for plants to absorb. The P-solubilizers also produce fungistatic and growth-promoting substances which influence plant growth.

The efficiency/performance of these microorganisms is affected by availability of a carbon source, P concentration, particle size of rock phosphate and other factors temperature and moisture. Significantly more P was taken up by wheat inoculated with *B. circulans* from apatite and SSP when ammonium sulphate (an acid forming fertilizer) was applied as a basal dressing. Similarly, P uptake by berseem and rice was significantly increased by inoculation in the presence of FYM over the no FYM controls.

Yield Responses to Inoculation

Responses of grain crops, vegetables, and forages, to seed inoculation with P solubilizers along with organic amendments and rock phosphate, bone meal or SSP have been studied. In earlier work carried out in USSR 5–10 per cent yield increase due to inoculation with *B. megaterium* var. phospbaticum popularly known as 'phosphobacterin' was observed in about one third trials. During the 1970s, out of 37 field trials conducted in India, 10 trials showed significant increases in yields in the case of wheat, rice, maize, chickpea, pigeonpea, soybean, groundnut and berseem. Significant increase in soybean yield was obtained due to inoculation with *B. polymyxa* or *P. striata* along with rock phosphate application over the control, whereas application of 80 kg P 20s/ha through SSP did not result in similar increase. In wheat and rice no significant increases were observed due to inoculation. In multi-locational trials conducted with different crops, increased yields (0–50 per cent) were obtained due to inoculation with a culture ('Microphos') with or without rock phosphate addition. Potato tuber yields were increased dramatically (50–60 per cent) due to inoculation with *Aspergillus awamori, P. striata* and *B. polymyxa* cutlures. The beneficial effect of inoculation was also observed on the crops grown after the inoculated crop.

Vesicular-Arbuscular Mycorrhizae (VAM)

The symbiotic association between plant roots and fungal mycelia is termed as mycorrhiza (fungus root, plural mycorrhizae). As per the revised classification the arbuscular mycorrhizae are formed by non-septatephycomycetous fungi belonging to the following orders.

These fungi are found associated with majority of agricultural crops. They are ubiquitous in geographic distribution occurring with plants growing in arctic, temperate and tropical regions alike. VAM occur over a broad ecological range from aquatic to desert environments. These fungi are obligate symbionts and have not been cultured on nutrient media. However, there is a report that VAM fungus *Glomus aggregatum* was cultured axenically and maintained on a synthetic medium. However, until other laboratories confirm such a success, this report should be viewed cautiously.

Order	Sub-order	Family	Genus
1. Endogonales		Endogonaceae	*Endogone*
2. Glomales	*i) Glominae*	*i) Glomaceae*	*Glomus Sclerocystis*
		ii) Acaulosporatede	*Entrophosphora* Acaulospora
	ii) Gigasporinae	*i) Gigasporaceae*	*Gigaspora Scutellispora*

Mechanism of Action

VAM fungi infect and spread inside the root. They possess special structures known as vesicles and arbuscules. The arbuscules help in the transfer of nutrients from the fungus to the root system and the vesicles, which are 'saclike' structures store P as phospholipids. There is little host specificity for VAM but the competitive ability of a given species with native strains may influence the dominance of a certain endomycorrhizal fungus in a root system. VAM have been associated with increased plant growth, and with enhanced accumulation of plant nutrients, mainly P, Zn, Cu and S mainly through greater soil exploration by mycorrhizal hyphae. It has also been suggested that VAM stimulate plant growth by physiological effects other than by enhancement of nutrient uptake or by reducing the severity of diseases caused by soil pathogens.

Root Colonisation

The survival and performance of VAM fungi is affected by the host plant, soil fertility, cropping practices, biological and environmental factors. Maximum root colonization and sporulation occurs in low fertility soils. However, positive yield increases in barley grown in soil with 40 ppm available P ($NaHCO_3$-extractable) were observed due to VAM inoculation. Internal P concentration of roots rather than external P concentration in soil controls root colonization by VAM fungi. Root P concentration controlled root colonization through root exudation: roots with high P concentration reduce root exudation, thereby affecting VAM colonization.

Application of FYM stimulated VAM while long fallows reduced mycorrhizal colonization of crops grown later. Cereals in rotation with legumes showed higher root colonization and more number of VAM propagules than cereals grown in monocropping. In mixed cropping, infection in the host plant wheat was reduced by the nun-host mustard. Higher temperatures generally result in greater root colonization in temperate zone. However, the reverse may be true in the tropics. Fungicides such as benumyl applied to soil depressed VAM populations while nematicides increased it. Drastic practices such as soil fumigation, soil solarization or prolonged waterlogging can reduce number of VAM propagules in soil.

Yield Responses to Inoculation

Crop response to VAM inoculation is governed by soil type, host variety, and VAM strains in addition to the biotic and abiotic factors mentioned earlier. In general, field experiments with VAM inoculation are fewer than those reported for *Rhizobium, Azospirillum,* or *Azotobacter*. The major constraint for field trials with VAM has been the inability to produce 'clean pure' inoculum on a large scale as the fungi are obligate symbionts and have to be maintained and multiplied on living host plants.

The results of field trials conducted in India indicate that VAM inoculations increased yields significantly in around 50 per cent trials and the response varied with soil type, soil fertility, and YAM cultures (Table 15.7). This scenario is similar to that for *Rhizobium* or *Azospirillum*. In such a situation, until suitable methods are evolved to multiply VAM on a large scale for field inoculation of crops directly sown in the field, the best strategy to utilize VA mycorrhizal fungi for crop production is to concentrate on crops normally grown in nursery beds where they can be easily inoculated with selected strains and then transplanted.

Preparation of Inoculum

Being an obligate symbiont, inoculum can be supplied in the form of infective soil, infected roots and soil sievings. Infective roots and growth medium from pot cultures open to the atmosphere could become contaminated with pathogens (fungi, bacteria, nematodes). To produce clean VAM inoculants, the approach of using root organ cultures infected with sterilized single spores of VAM fungi has been proposed and standardised to produce 10–100 kg of inoculum in a phased manner. Such cultures of produced and used even only for nursery crops and forest trees would go a long way in improving soil productivity particularly of the wastelands and reclaimed soils.

Table 15.7: Effect of Mycorrhizal Inoculation on Crop Yields Under Field Conditions

Crop	*Mycorrhizal Over Control %*	*Increase*	*Remarks*
Soyabean	*G. fasciculatum* *Rhizobium + G. fasciculatum*	26.4^{NS} 18.5^{NS}	Increase in shoot and grain mass and P content not significant
Fingermillet	*G. fasciculatum* Isolate M6 Isolate M14	–0.6^{NS}–15* 8*–26* 1^{NS}–24*	Response to VAM inoculation was reduced with increasing rates of P application
Cowpea	*G. fasciculatum* *Rhizobium + G. fasciculatum*	8.6^{NS}–18.5^{NS} 10^{NS}	With increasing P levels response to VAM inoculation decreased
Pigeonpea	*G. fasciculatum* *Rhizobium + G. fasciculatum*	36.8^{NS} 21.2*	Mean of 3 yrs data. With VAM alone significant yield increase observed only during first year
Chickpea	*G. vessiforme*	11*	Shoot dry matter and P contents significantly increased over control
Groundnut	*G. fasciculatum* *Rhizobium + G. fasciculatum*	20.9* 10*	Combined inoculation with VAM and Rhizobium significantly increased pod yield over Rhizobium inoculation
Groundnut	9 cultures of VAM	NS	No response in terms of pod yield was observed in 2 trials. Certain VAM cultures increased nodulation by Rhizobium strains
Tomato	*G. fasciculatum* *Aztb. vinelandii + G. fasciculatum*	16.6^{NS} 7.9^{NS}	Inoculation with Azotobacter and VAM did not increase yield significantly over Azotobacter inoculation alone. N and P contents also increased
Tomato	*G. fasciculatum* *Rhizobium + G. fusiculatum*	7.4* 8*	Shoot dry mass and N contents were not influenced
Chillies	*G. fasciculatum* *G. macrocarpum* *G. albidum* Isolate 14	8.1^{NS}–20* 0–0 37*–43* 17^{NS}–24*	In red sandy loam soil VAM cultures showed differential response with the same host cv. Significant increases were observed mainly in the presence of 37.5 kg P/ha application.
Chillies	*G. fasciculatum* *G. macrocarpum* *G. albidum* Isolate 14	41*–48* 8^{NS}–19^{NS} 13.5^{NS}–41* 11^{NS}–37*	In black clayey soil higher increases in yield were observed without P and application

*: Significant at P: < 0.05; NS: Non significant.

Plant Growth Promoting Rhizobacteria

A group of rhizosphere bacteria (rhizobacteria) that exert a beneficial effect on plant growth is referred as plant growth promoting rhizobacteria or PGPR. Although biological control of soil pathogens by microorganisms has been studied since long, interest and research in this area has increased steadily since 1965. PGPR belong to several genera, *e.g., Actinoplanes, Agrobacterium, Alcaligenes, Amarphosporangium, Althrobacter, Azotobacter, Bacillus, Cellulomonas, Enterobacter, Erwinia, Flavobacterium, Pseudomonas, Rhizobium* and *Bradyrhizobium, Streptomyces* and *Xanthomonas*. The first commercially used bacterial biocontrol agent was *Agrobacterium radiobacter* strain 84 which controls crown gall caused by *A. tumefaciens*. *Bacillus* spp. are appealing candidates for biocontrol because their endospores are tolerant to heat and desiccation. Currently *Pseudomonas* spp. are receiving much attention as PGPR. Fluorescent *Pseudomons* strains also suppress major plant pathogens. Take all, an important root disease of wheat caused by *Gaeumannomyces graminis* var. tritici is controlled by fluorescent *Pseudomonas.*

Mode of Action

PGPR are thought to improve plant growth by colonizing the root system and pre empting the establishment of or suppressing deleterious rhizosphere microorganisms (DRMO) on the roots. The PGPR improved potato growth and yield in short but not long-rotation soils, primarily by suppressing cyanide-producing DRMO. The mechanisms of pathogen suppression by PGPR include substrate competition and niche exclusion, production of siderophores and antibiotics and induced resistance. Large populations of bacteria established on planting material and roots become a partide sink for nutrients in the rhizosphere thus reducing the amount of C and N available to stimulate spores of fungal pathogens or for subsequent colonization of the root.

Fluorescent Pseudomonads are especially suited for rapid uptake or scavenging of nutrients, since they are nutritionally versatile and grow rapidly in the rhizosphere. Certain areas on the root, such as cell junctions and points of emergence of lateral roots, appear to be favoured for colonization by many microorganisms including pathogens, because root exudates are abundant there. Inoculating planting material with PGPR presumably prevents or reduces the establishment of pathogens at these sites. Siderophores are low molecular weight, high affinity Fe^{+++} chelators that transport iron into bacterial cells. When grown under Fe-deficient conditions, fluorescent pseudomonads produce yellow-green fluorescent siderophore-iron complex. This highly efficient iron scavenging mechanism is thought to compete with that of fungal pathogens, thereby creating an iron-deficient environment deleterious to fungal growth. The most convincing evidence has been the fact that siderophore-minus mutants are less suppressive to pathogens in the rhizosphere than parental strains.

Yield Response to Inoculation

In field trials with wheat, potato, sugarbeet and zinnia conducted on experimental and commercial scale, 40 out of 63 trials (63 per cent) showed significant results with yield increases varying from –7 to 136 per cent with an average increase of 7–35 per cent in different crops over the control. Seed treatment with *B. subtilis* strain A13 increased yield of carrot by 48 per cent, oats by 33 per cent and groundnut up to 37 per cent and it has been marketed as a treatment for groundnut in USA. A multitude of factors could account for inconsistent results, given the complex interactions among host, inoculated organism, other rhizosphere organisms and the environment which are discussed earlier. Research with PGPR in India is at a low level.

Future Research Needs

Most important characteristic common to most biofertilizers is the unpredictability of their performance. Available data show that the success rate with any biofertilizer in terms of significant impact on yield is up to 60–65 per cent. On the whole increased yields due to inoculation with N_2-flxing bacteria would contribute significantly to the economy of the subsistence farming. Taking example of pearl millet grown in India, and assuming that inoculation would increase grain yield by 10 per cent on 50 per cent of 11.3 million ha with an average yield of 600 kg/ha, the increased yield would be 0.35 million tons, worth Rs. 718 million at 1991–92 grain procurement prices.

In order to harness potential benefits of biofertilizers in commercial agriculture, the consistency of their performance must be improved. This requires research in many diverse areas as these biological systems involve complex interactions among the host, other rhizosphere microflora and fauna and the environment. The survival and competitive ability of the strains to be introduced must be improved. Very little is known about the competitiveness of the microorganisms and factors governing. Research is needed on soil physical and chemical factors that influence establishment of the inoculated strain in the rhizosphere and full expression of its traits for benefitting the crop. Once these factors are identified, it may be possible to manuipulate them in the field to enhance the consistency of their performance.

Research to understand the mechanisms by which the introduced microorganisms benefit the crops is critically important. Although organisms are known for their specific functions, *e.g.*, rhizobia, azospirilia and azotobacters are known for their N_2-fixing ability and VAM for increasing P uptake, research has shown that other properties like production of siderophores, hormones or antibiotics help the host plants to increase productivity. Identifying such important traits will allow more efficient selection of new strains.

Efficient strains with selective traits to perform well under adverse conditions like soil acidity, salinity, moisture stress and for insect resistance need to be selected or developed and tested for their performance. This applies more to azospirilla, VAM and PGPR in whose case the action mechanisms are complex and not fully understood. Once the traits/factors governing competitive ability and efficiency of the strains of interest are revealed, screening for more efficient and competitive strains would be rewarding.

In case of VAM concerted efforts are needed to culture it in the laboratory. Unless we find a way to multiply these fungi on large scale, it is difficult to test their performance under field-scale conditions.

More research on formulation and efficient delivery of the biofertilizers is needed. Search for newer synthetic carrier materials which can be uniform, nontoxic, simple to use and can sustain large population of introduced microorganisms for a longer time must be pursued vigorously.

Strategy for Successful Use of Biofertilizers

For increasing crop yields through biofertilizers, the following strategy is suggested using the case of *Rhizobium* inoculants as an example. Most important constraints to effective exploitation of BNF technology in India are the quality of the inoculants:

1. Lack of knowledge about inoculation technology for the extension personnel and the farmers.
2. Effective inoculant delivery system.
3. Formulation of the policy dictating the desire to exploite BNF successfully.

The history of inoculant manufacture and of many strain collections is full of examples of organisms which look like rhizobia but are not! Contaminated cultures contribute to the problems which placed the inoculant industry of Austraiia in peril in the early 1950s. Many inocula of poor quality were sold and the losses at sowings of new legumes into poor soils were enormous. This was repeated in India during the late 1970s and early 1980s when BNF attracted attention because of escalating oil prices resulting in fertilizer shortages. Many small scale industries sprang up to manufacture biofertilizers. They sold poor quality inoculants to the farmers who in due course lost faith in biofertilizers. Several rhizobial inocula from the Indian manufacturers were examined at ICRISAT for their infectivity tests. Irrespective of private or public institution origin, the majority failed to pass the published standards. There must be strict quality control mandatory on all biofertilizer producers irrespective of their status as private/public or government organization.

For success of biofertilizers in India, concerted efforts right from production, demonstration to distribution will be required. There is some hope of success with a mission-oriented approach under which at least production of mother cultures in lyophillised form must be centralised. Freeze dried rhizobial cultures multiplied in culture broth and immediately used for impregnating the carrier was found as approach extended shelf life for the inoculants as well as adequate numbers of viable rhizobia on seed are assured under tropical conditions. The scientists involved in biofertilizer projects in the Universities and ICAR institutes can identify the target crops, areas to be covered and recommend strains for preparing biofertilizers. People involved in biofertilizer production should be trained microbiologists who are aware of the pitfalls in the processes involved. The NGO organisations like Bharatiya Agro-Industries Foundation (BAIF) have their network upto village level, have experience in handling microorganisms and also have facilities to produce inoculants. Such agencies along with the National Biofertilizer Development Centre can play an important role in popularising biofertilizers in India. The next step is convicting and educating the farmers regarding the benefits of these inoculants. The pricing of the biofertilizers must be controlled if private agencies are involved, otherwise if farmers don't see the significant effects in term of economic yields, they may be scared away from using biofertilizers.

At this stage the policy issue arises that biofertilizers should be used or considered as an insurance for harnessing BNF to its maximum potential taking systems approach. As discussed earlier in chickpea, the non-nodulating or low nodulated plants look similar in appearance to well nodulated plants but this is at the cost of soil or fertilizer N. We must take the view that in the end we may derive benefit in terms of maintaining or improving the productivity of our soils. We should not be disappointed by not seeing the direct benefits in some cases. As evident from the Karnataka experience where nodulation in pigeonpea improved substantially by addition of DAP rather than by inoculating with *Rhizobium* alone. A holistic approach to improve production of legumes (or other host plants) is needed and we must ensure that all the constraints for good plant growth other than N nutrition are alleviated for better performance of BNF technology.

Chapter 16

Biofertilizers for Flooded Rice Ecosystem

Rice is grown in about 42 m ha in India. It contributes 75 mt. (44 per cent) to the total food grain production of 170 mt. Apart from water management, supply of nitrogen is a key factor in the realisation of potential yields from modern high yielding varieties. The rice plant absorbs about 20 kg N/t paddy produced. Due to the poor N-status of soils, N application is a must for harvesting moderate-high yields. Rice is estimated to receive 40 per cent of the total fertilizer N applied in India.

Rice, however, is not exclusively dependent on fertilizers or FYM for external N supplies because the crop can receive sizable N input from green manures and certain biofertilizers. The most important biofertilizers for flooded rice are Azolla and Blue-Green Algae (BGA). Both can grow along side rice. In addition, Azolla can also be used as a green manure.

This chapter takes stock of available information on Azolla and BGA for flooded rice ecosystem. Any integrated N-supply system for flooded rice must feature one of these biofertilizers which have the potential to contribute 20–68 kg N/ha depending on the intensity of multiplication. Sustainability of rice production will depend on how well these inputs have been integrated with other sources of N. Both these biofertilizers are members of the plant kingdom and they themselves require certain inputs at optimum level for growth and N-fixation.

Azolla

The aquatic fern Azolla is distributed in both temperate and tropical rice growing regions, and fixes atmospheric nitrogen (N_2) in symbiotic association with a heterocystous BGA, *Anabaena azollae*. Azolla contains 0.2–0.3 per cent N on fresh weight basis and 3-5 per cent N on dry weight basis. Azolla has been used as a fertilizer for rice in Vietnam and China for centuries. However, its use as a biofertilizer for rice in other countries is a relatively recent development. About 90 strains belonging to seven species are being maintained at CRRI, Cuttack which is the premier centre for Azolla research in India. Of the seven species, *Azolla pinnata* is most widely distributed in India.

Growth and N-fixation

The growth rate, total biomass and N content of Azolla provide the estimate of its potential for agricultural use. The environmental conditions and nutrient availability greatly influence fern growth. In India, the fresh weight of *A. pinnata* increases 2–6 fold in a week. The BGA in the Azolla cavities is very efficient in N-fixation and can meet the fern's total N requirement. However, the fern is able to utilize both fixed N and soil N simultaneously and maintains a high rate of N-fixation in presence of combined N. Under ideal conditions, it has a potential of fixing more than 10 kg N/ ha/day. At Cuttack, *A. pinnata* fixed 75 mg N/g dry wt./day and produced a biomass of 347 t fresh wt/ha in a year which contained 868 kg N, as much as in 1900 kg urea. About 30–100 kg N/ha can be fixed in a month.

Wide variability is observed among Azolla strains with regard to growth and N fixation. Among different *A. pinnata* strains, the Vietnam green and Bangkok strains performed better than the India (Cuttack) strain (Figure 16.1). One strain of *A. caroliniana* is found to fix more N than many *A. pinnata strains*. This *A. caroliniana* strain can grow round the year in rice fields and has better tolerance to snails, other pests and diseases. One crop of Azolla provides 20–40 kg N/ha to the rice in about 20–25 days. Even though the estimates of N input vary considerably, the N-fixing potential of Azolla is fully established.

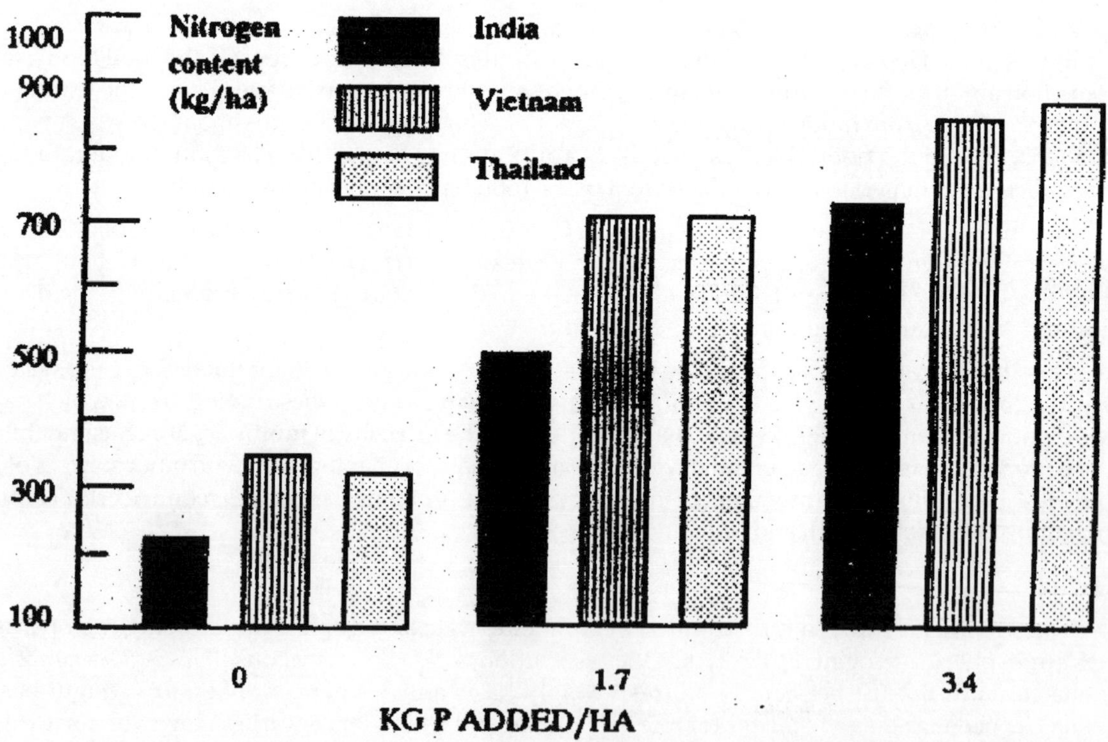

Figure 16.1: Effect of P Application on the N Content of *Azolla pinnata* Isolates Grown for One Year

Factors Affecting Growth and N-fixation

Water

Azolla prefers to grow in a free-floating state. Good water control is desirable for its successful multiplication. The multiplication rate is drastically reduced if the soil is just moist and the fern dies upon complete drying. However, one strain of *A. caroliniana* is found to survive on the moist soil for a longer period as compared to the strains of other species. A water depth of 5–10 cm is recommended for good growth but depths upto 30 cm do not have any adverse effect.

Mineral Nutrients

Azolla requires all the essential plant nutrients for normal growth. Because of its aquatic nature, these elements must be available in the water. The deficiency of anyone element adversely affects its growth and N-fixation. In these respects Azolla behaves as an agricultural crop. Phosphorus is a key element and its deficiency results in poor growth, pink or red colouration, curling of roots and reduced N content. The effect of P and Ca deficiency on the growth and N-fixation is more intense than that of K or Mg deficiency. Threshold level of P in Azolla is about. 0.2 per cent–0.3 per cent P on dry weight basis and its uptake by the plants increases with increasing levels of P in the growth medium as described later. Under favourable conditions, addition of one kg P results in fixation of about 5–10 kg N.

Azolla is considered to be an efficient scavenger of potassium and may serve as a K-source for rice in K-deficient soils. Iron is also important and deficient ferns turn yellowish, due to decreased chlorophyll content. Cobalt and molybdenum are required for efficient functioning of the N-fixing system. The deficiency of one element affects the uptake of others for example, P-deficiency results in increased uptake of Fe and Zn whereas Mg deficiency reduces the uptake of K but increases the uptake of Fe, Co and Mn.

Light

Both the partners of the fern-alga association carry out photosynthesis which helps it to maintain a high rate of N-fixation. Growth and N-fixation of Azolla are influenced by both quality and intensity of light. Azolla prefers a certain degree of shading, particularly during summer, and reaches its full at 25–50 per cent of full sun light. When the fern and rice are grown together in a dual cropping system, Azolla growth is below its potential due to shading by rice canopy which increases with the advancement of rice growth. The shading effect is more in the wet season than the dry season. *A. caroliniana* is more tolerant to shading than other species. The growth of Azolla dual crop thus depends on leaf area index of rice, weather conditions and fertility status of flood water. The day length also affects the growth of Azolla. The growth is better at higher latitudes than in the tropics as a result of longer days during the Azolla growing season.

Temperature

Temperature is perhaps the most important environmental factor that limits the growth of Azolla and 25–30°C is optimum for most species. *A. pinnata* is successfully grown in the rice fields at Cuttack round the year (14–35°C) but its growth is better during July to December. Azolla growth and N accumulation in relation to water temperature and solar radiation are presented in Figure 16.2. The diurnal variation in temperature is important in determining the temperature response of Azolla. The temperature response varies among Azolla species/strains. *A. filiculoides* and *A. rubra* are cold tolerant whereas *A. mexicana, A. microphylla, A. caroliniana, A.nilotica* and some varieties of *A. pinnata* have better tolerance to high temperature. *A. filiculoides* does not grow well at high temperature but it can tolerate low temperature upto 5°C.

There appears to be a synergistic relation between tolerance to high temperature and light intensity. At high light intensities the optimum temperature for its growth shifts upwards. Temperature response is also influenced by population density. Tolerance to high temperature is low when Azolla attains a stationary phase. The high temperature causes a severe damage to Azolla indirectly through its stimulatory effects on population of Azolla pests. The extreme temperatures result in reddish brown colouration in Azolla but change in colour usually does not affect its N-fixing activity. The high temperature damage on Azolla can be reduced by shading, draining the field or flushing.

pH and Salinity

A pH of 5–8 is optimum for Azolla growth although it can survive in the pH range of 3.5–10.0. The influence of pH on growth of Azolla is related to its effects on nutrient availability. For example, saline soils which usually have a pH of more than 7 may be deficient in phosphorus, zinc and copper. Flooding results in a shift in the pH of many acidic and alkaline soils toward neutral. However, such a change in pH is not noticed in those acid soils which are poor in organic matter. Such soils can have toxic levels of Al, Fe and a low level of P. The optimum salt concentration for growth of Azolla is 90–150 mg/liter. The salinity and alkalinity problems occur mostly in coastal regions and in poorly drained soils.

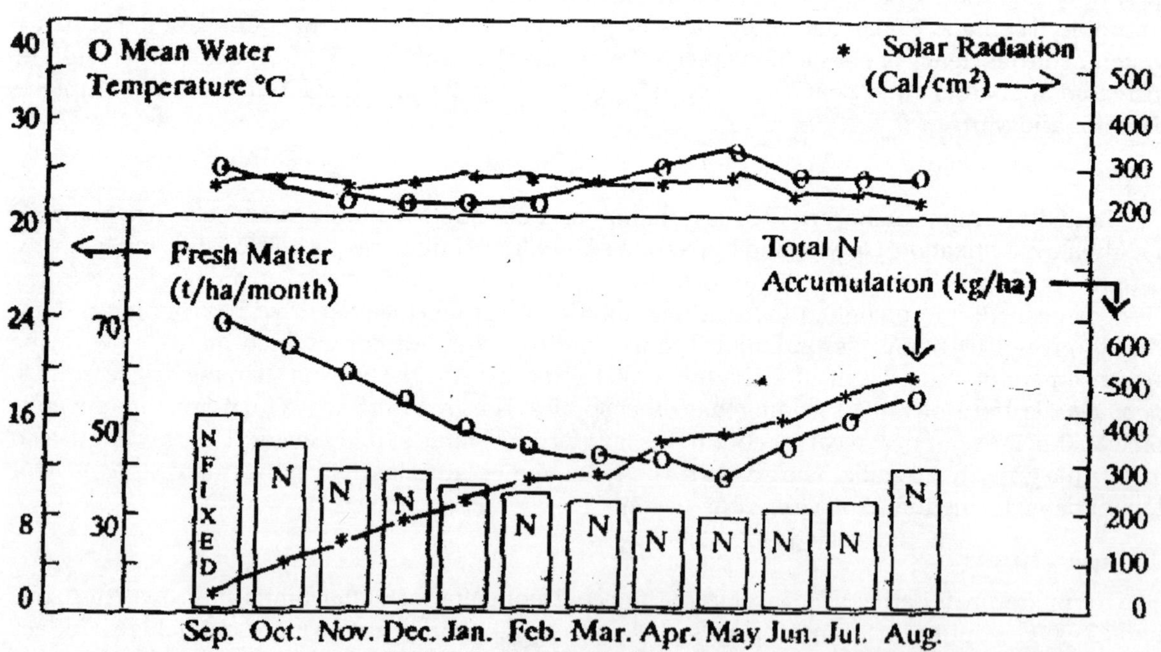

Figure 16.2: Fresh Matter and N Increase per Month and Cumulative N Increase in Relation to Water Temperature and Solar Radiation in a Continuous Cropping of Azolla (average of two years and mean of nine treatments)

Management Practices

Azolla is grown both as a green manure before transplanting and as a dual crop after transplanting of rice. In rice-Azolla dual cropping system, the benefits from *Azolla* can be maximised by proper management. The field is ploughed, leveled and divided into small plots by raising bunds. Water depth of 5–10 cm is maintained during Azolla multiplication.

Rate and Time of Inoculation

The amount of inoculum significantly affects the growth rate of Azolla. The rate of inoculation is usually higher for multiplication plots than for dual cropping. The inoculum of 0.1–0.3 kg/sq m is recommended for multiplication plots and 0.05–0.1 kg/sq m (0.5–1.0 t/ha) for dual cropping. Under poor growing conditions or in areas with excessive algal growth, higher inoculum rate should be used. Azolla inoculation in early stages of rice growth is preferable because a delay reduces the fern growth due to increased shading by rice. However, formation of an Azolla mat during panting might cause damage to rice seedlings, particularly if it covers the seedlings due to rise of water level. Thus, it is recommended to inoculate the fern after 1 week of planting. In case of Azolla raised as a green manure it is inoculated in the main field 15–20 days before transplanting rice.

Fertilizer Application

Effect of fertilizer N on Azolla growth and N-fixation depends on the N rate as well as on Azolla strain. Application of small quantity of N is beneficial for its growth under stress conditions. In studies at CRRI 90 kg N/ha did not have any significant effect on Azolla grown as a dual crop with rice whereas higher N doses reduced its biomass and N content in the fallow fields. A strain of *A. caroliniana* is found to grow better and fix more N than India (Cuttack) strain of *A. pinnata,* when grown alongwith N fertilizers. It seems that N-fixing in Azolla is not strongly inhibited by fertilizer N and its poor growth in presence of N is mainly due to the stimulatory effects of N on growth of other organisms which compete with Azolla. Application of FYM and urea super granules causes less reduction in Azolla growth than urea or ammonium sulphate. The inhibitory effect of N fertilizers on, Azolla can be lowered by their deep placement in furrows, which also improves fertilizer use efficiency. Azolla grows better when basal N dose of rice is supplied through Azolla or other green manures rather than through chemical fertilizers.

The recommended dose of P in inoculum production plots is 4–8 kg P_2O_5/ha at, weekly intervals. The highest Azolla biomass in Azolla-rice dual cropping is obtained at 10–15 kg P_2O_5/ha applied in 3 equal splits at 7-day intervals (Table 16.1). Split application of fertilizer P helps in producing higher *Azolla* biomass through maintenance of a continuous supply of P in the flood water. Water soluble P-sources are found to be more effective than rock phosphates. Animal dung at 1.5 t/ha was found to enhance the Azolla biomass on par with 10 kg P_2O_5 ha as superphosphate. The application of potassium at 5–8 kg K_2O/ha and 50 kg ash/ha at 7-day intervals improves the growth of Azolla. Its growth and N-fixing activity are also increased by application of small quantities of iron (1.0 kg Fe/ha) and molybdenum (0.15 kg Mg/ha).

Method of Rice Planting

The growth of Azolla dual crop is significantly influenced by density of rice which can be manipulated by changing the method of planting. Higher Azolla biomass and rice yield are produced by using the double narrow-row method of rice planting, as against the conventional line planting. In double narrow-row method, pairs of narrow rows are spaced at 40 cm with 7 cm gap between two narrow rows and bills are 15 cm. apart. The fern is grown in the broader spare between two pairs of narrow rows. The benefits of this method are particularly higher when Azolla inoculation is delayed. Azolla can also be successfully grown with randomly planted or direct seeded rice but its biomass is relatively lower than that under line planting. Rectangular planting is better than square planting and the fern growth is higher when rice rows are planted in east-west direction.

Table 16.1: Effect of Varying P Levels on Growth and N-fixation of Azolla in Planted Rice Field (mean of two seasons)

Phosphorus kg P/ha	*Azolla Caroliniana*			*Azolla pinnata*		
	Fresh wt. (t/ha)	*ARA (n mol Ethylene/g Fresh wt./hr)*	*N Yield (kg/ho)*	*Fresh wt. (t/ha)*	*ARA (n mol Ethylene/g Fresh wt./hr)*	*N Yield (kg/ha)*
0	6.1	446	9.5	4.9	301	8.5
4.4	12.4	745	22.3	8.1	468	17.8
6.6	11.8	870	22,3	9.0	532	19.5
8.8	12.2	885	22.4	8.7	570	19.9

* P applied in 3 equal splits at 7-days intervals after Azolla inoculation.

Insects, Diseases and Weeds

Azolla is often attacked-by insects and the entire crop is damaged within a few days if they are not timely controlled. Damage due to insects is higher in the hot and humid weather because of their increased multiplication rates. The species of *Chironomus, Pyralis* and *Nymphula* are the common insects of Azolla in India. Several species of snails are also known to damage the Azolla plants. However, these insects are so far not reported to cause damage to the rice crop. Insecticides commonly used to control rice insects are also effective against major insects of Azolla and their application increases fern growth. The application of natural pesticides like *neem* oil, *karanj* oil or *neem* cake also stimulates fern growth. Commonly used herbicides at rates recommended for rice usually have a detrimental effect on growth and N-fixation of Azolla. Toxicity is observed in the order butachlor > benthiocarb > 2,4-DEE > 2,4-DNa. Azolla should be applied about 3 weeks after application of butachlor or benthiocarb, whereas Azolla and 2,4-DNa can be used together. The other herbicides like oxadiazen and perdimethalen also inhibit Azolla growth.

The fungal and bacterial infections on Azolla are usually observed when its mat is left unharvested for a longer period, particularly in hot weather. Their attack is believed to follow the damage by insects or dessication. Fungal diseases of Azolla can be controlled by spraying of 0.5–1.5 per cent benlate solution. One isolate of *A. caroliniana* is observed to grow well in the Indian rice fields and possesses greater tolerance to many diseases and insects, including snails. The abundance of algal growth is harmful for the fern growth and algae should be removed manually, by drying the field for a few days or by application of 10 kg copper sulphate/ha.

Method of Utilization

Azolla can be used both as a green manure before transplanting and as a dual crop after transplanting of rice. Both practices are feasible in India, but dual cropping is more practicable, although growth and N-fixation of Azolla as well as yield of rice are generally higher in green manuring than the dual cropping. Adoption of Azolla green manuring is restricted to areas having water available before transplanting.

For the purpose of green manuring after, the fern forms a mat, field is drained (if possible), ploughed and rice seedlings transplanted. The plants which do not get buried into the soil during ploughing grow further and contribute to N supply from Azolla. When Azolla is raised as a dual crop, after mat formation (3-4 weeks), water is drained out and Azolla is incorporated into the soil with the help of a weeder or rake. Unincorporated Azolla mat also decomposes due to excessive growth and benefits the rice crop.

Application of basal dose of N through legume green manuring or fertilizer is recommended along with Azolla dual cropping for better tillering of rice. Presence of a thick Azolla mat reduces availability of light and nutrients to weeds and suppresses their growth.

Impact on Rice Yield and Soil Fertility

Availability of Azolla-N to Rice

Azolla provides N to rice after its decomposition just like any organic matter. Decomposition of unincorporated Azolla is slow as compared to incorporated Azolla (Figure 16.3). Incorporated Azolla takes about 8–10 days to decompose and releases about 67 per cent of its N within 35 days. The decomposition and N release vary among Azolla species/strains. Fungi and bacteria play a major role in Azolla decomposition and their population substantially increases during this period. The high temperature stimulates decomposition perhaps due to increased microbial population in the field. The N from Azolla is released slowly in comparison to fertilizer N and its availability to the first rice crop is about 70 per cent of ammonium sulphate N. However, Azolla releases its N faster than BGA. Fresh Azolla is superior to dry Azolla as regards N release.

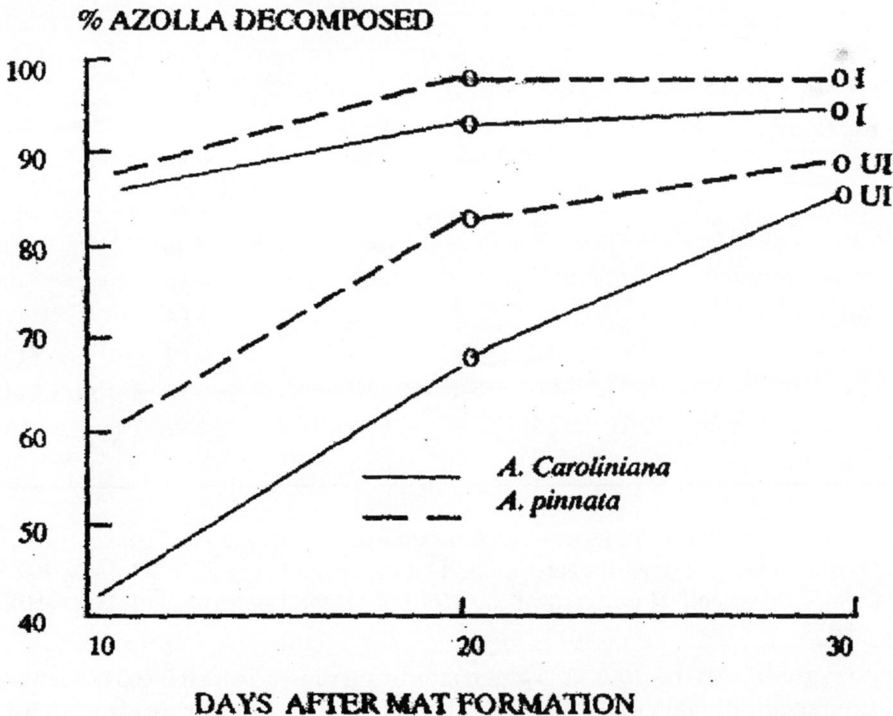

Figure 16.3: Rate of Decomposition % of Incorporated (I) or Unincorporated (UI) Azolla

Effects on Rice Yield and Soil Fertility

Studies carried out for 15 years at CRRI have shown that Azolla green manuring produces 10–20 t/ha of green matter containing 20–40 kg N, whereas its dual cropping provides 20–30 kg N/ha. The supply of N can be further increased by raising more than one crop of Azolla, first as a green manure and then as a dual crop alongside rice. Azolla green manuring and two dual crops are reported to fix about 90 kg N/ha whereas green manuring plus one dual crop or two dual crops fix about 60 kg N/ha. The N-input possibilities are as follows:

Azolla as green manure	20–40 kg N/ha
Azolla as dual crop once	20–30 kg N/ha
Azolla as dual crop twice	40–60 kg N/ha
Azolla as green manure + one dual crops	50–60 kg N/ha
Azolla as green manure + 2 dual crops	70–90 kg N/ha

The crop of Azolla green manuring or dual cropping increases the grain yield of rice on par with application of 30 kg N/ha through fertilizer (Table 16.2). Experiments carried out by the ICAR's Model Agronomy Project at different locations (Table 16.3) as also the results of international trials conducted at 37 sites in 10 countries including India produced similar results. Ten tonne Azolla/ha gave 95 per cent of the yield increase which resulted from using 60 kg fertilizer N.

Table 16.2: Effects of Different Methods of Azolla Utilization on Rice Yield (mean of two seasons)

Treatments	*Azolla Fresh wt. (t/ha)*	*Azolla-N (kg/ha)*	*Paddy Yield (t/ha)*	*N-efficiency kg Paddy/kg N*
No nitrogen	–	–	2.71	–
Fertiliser (30 kg N/ha)	–	–	3.16	15.0
Fertilizer (60 kg N/ha)	–	–	3.57	14.3
Fertiliser (90 kg N/ha)	–	–	3.90	13.2
Azolla basal (AB)	21.9	31.9	3.25	16.9
30 kg N/ha + AB	22.5	33.7	3.62	14.3
Azolla dual cropping once (AD)	12.9	20.3	3.10	19.2
30 kg N/ha +AD	11.6	18.3	3.54	17.2
AB + AD	34.2	51.9	3.49	15.0
30 kg N/ha + AB + AD	33.0	48.7	3.87	14.7
AB + Azolla dual cropping twice (2AD)	43.8	67.8	3.72	14.9
30 kg N/ha + AB + 2AD	41.3	63.9	4.14	15.2

In multi-location trials by the All India Coordinated Rice Improvement Project, incorporation of 6t Azolla/ha was more effective than dual cropping. Data averaged over 55 trials show that 6t Azolla/ha plus 25 kg urea N produced 94 per cent of the yield obtained by applying 50 kg fertilizer N/ha (Table 16.4).

The increase in grain yield due to Azolla green manuring ranges from 0.5–2.0 t/ha, whereas dual cropping produces an additional yield of 0.5–1.5 t/ha over the control. Paddy is worth Rs. 2300/t at 1991–92 procurement prices. The rice yield produced by Azolla green manuring followed by dual cropping is equivalent to that from 60 kg N /ha as fertilizer. Two Azolla dual crops in addition to its green manuring produce higher yields than the application of 60 kg N/ha, but two dual crops alone are not as effective as 60 kg N/ha. Combined use of Azolla and N fertilizer gives best results. The response of Azolla is better in the dry season than the wet season. Azolla green manuring performs better with short duration rice varieties whereas dual cropping is more suited for medium or long duration varieties. Grain yield response also varies with Azolla species/strains.

Table 16.3: Impact of Azolla and Fertilizer N on Rice Yields in Model Agronomy Trials at 4 Locations

Treatment	*Paddy Yield (t/ha)*				
	Cuttack	*Bhubaneswar*	*Kharagpur*	*Tilabar*	*Mean*
Control	2.75	2.94	2.48	2.92	2.77
Azolla (10 t/ha) basal	3.29	4.00	3.75	3.63	3.67
Azolla dual cropping*	3.08	3.60	3.05	3.12	3.21
30 kg fertilizer N/ha	3.05	3.38	3.74	3.00	3.29
60 kg fertilizer N/ha	3.25	4.20	4.19	3.25	3.72

* One t/ha fresh Azolla inoculated and incroporated 20 days later.

Table 16.4: Impact or Azolla Green Manuring on Rice Productivity in Multi-location Trials by AICRIP (mean or 4 seasons)

State	*Trials*	*Paddy yield, t/ha*				
		No-N	*25 kg N*	*50 kg N*	*Azolla (6t/ha)*	*Azolla (6t/ha) + 25 kg N/ha*
Andhra Pradesh	3	3.60	4.58	5.01	4.15	4.79
Assam	10	2.89	3.16	3.65	3.15	3.21
Bihar	2	3.58	4.18	4.33	4.49	4.41
Karnataka	2	3.88	4.35	4.55	4.39	4.47
Kerala	3	2.89	3.36	3.55	3.00	3.28
Madhya Pradesh	4	2.18	2.65	3.10	1.16	1.39
Maharashtra	3	2.81	3.69	4.05	3.50	4.24
Manipur	3	3.01	3.16	3.45	3.34	3.80
Punjab	2	2.34	3.50	3.91	2.61	3.66
Tamil Nadu	10	3.89	4.62	4.81	4.50	4.69
Uttar Pradesh	10	2.06	2.77	3.29	2.52	2.92
West Bengal	3	4.10	4.96	5.15	4.45	4.93
Average	55	3.10	3.75	4.07	3.44	3.82

Besides supply of N and other nutrients to the rice, application of Azolla is believed to bring about an improvement in soil fertility level. Its effect on physical properties is less studied as compared to the chemical properties. A few studies have shown that bulk density and soil resistance are reduced whereas aggregate stability and available moisture content are increased due to application of Azolla. Increase in the contents of organic carbon, N, P, and K in the soil treated with Azolla is known. There are also reports on the increased availability of micronutrients like Fe and Mn in the soils as a result of Azolla incorporation. The release of oxidizable organic compounds during Azolla decomposition results in a lowering of redox potential of the soil. All the Azolla-N is not utilized by the same rice crop and subsequent crops are also benefited.

Economic Aspects

Azolla biofertilizer technology is labour intensive. It is economically feasible in India, which is a manpower resourceful country. The major inputs for Azolla cultivation are labour, irrigation water, phosphate fertilizer and pesticide. The costs of these inputs are location specific. At CRRI, current cost of Azolla dual cropping, using recommended inoculum of 500 kg/ha has been estimated at Rs. 50–54/ha. This estimate does not include the cost of fertilizer P and pesticide because they are also used for the rice crop. The P accumulated by Azolla is released into the soil after its decomposition. Moreover, SSP can be replaced by animal dung or cattle slurry and Furadan by some cheaper biopesticides. Cost of labour, superphosphate and pesticide for production of 10 t Azolla which was taken as Rs. 95 in 1982 can now be taken as around Rs. 130. This incidently is equal to the cost of 20 kg N as urea at late 1991 prices. The net return and benefit cost ratio are generally higher with Azolla as compared to the N fertilizers (Figure 16.4). The investment of one rupee in Azolla dual cropping gave a net return of Rs. 21–24 as against a net return of Rs. 5 with fertilizer N. Apart from supply of N, other benefits of

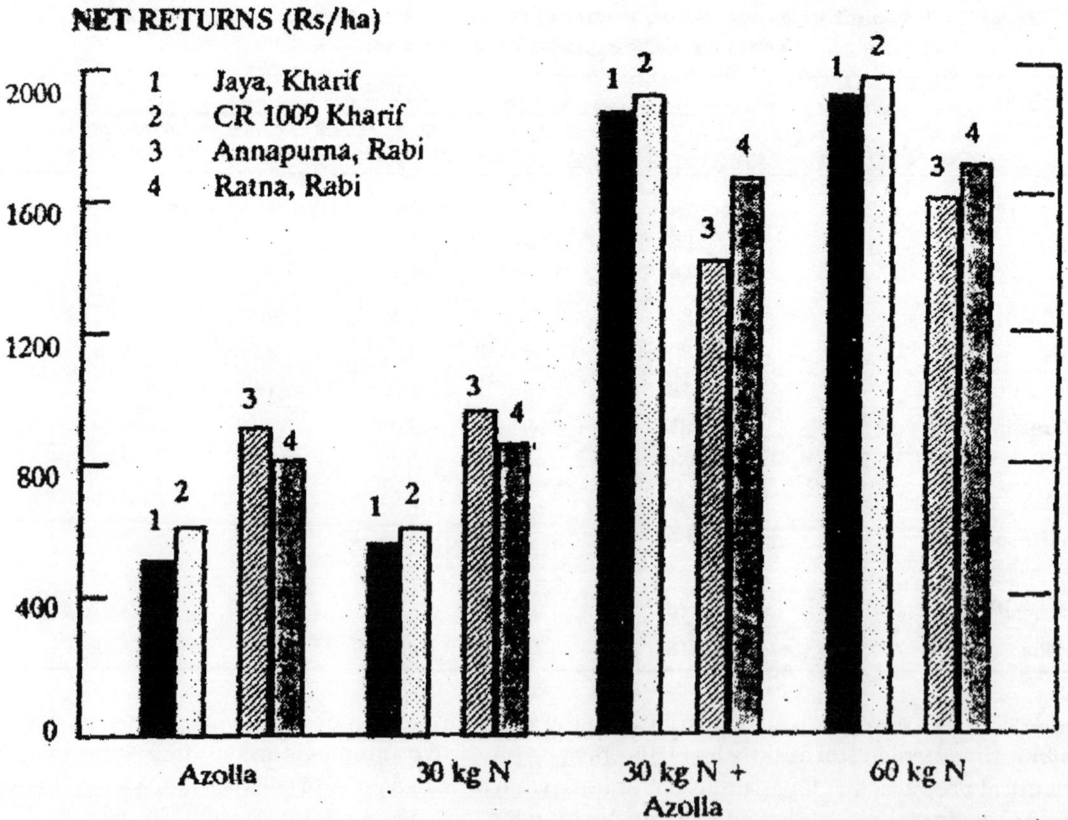

Figure 16.4: Net Returns Using Azolla and Urea in Rice Production (Rs. 26 = I US $, late 1991)

using Azolla such as increased availability of other elements, improvement of soil fertility and reduction of weed growth should also be given due weightage while comparing it with N fertilizer.

Suitable Agroclimatic Conditions

Exploitation of Azolla as a biofertilizer is a practical possibility in flooded soil conditions. It can survive for only a few days if soil dries during intermittent rains. Azolla forms sporocarps which can survive drying, but sporulation is restricted to certain part of the year and factors responsible for sporulation are not clearly understood. The optimum temperature for most Azolla species is within 20–30°C, although some strains can grow at 40°C or more. Concern about high temperature is due to its stimulation effect on growth of insects that attack Azolla. One *A. carolilliana* strain is being successfully grown year-round at CRRI and is found to be comparatively tolerant to many insects and diseases. The soil pH is all important factor that determines suitability of Azolla in a given area. It is useful in neutral to slightly acidic soils but can also be used under adverse soil pH conditions after some economically feasible amendments. Frequent rains are favourable for Azolla, until low light due to cloudy conditions inhibits the growth. The soils having high available P in the flood water are good for Azolla, but it can also be grown in P-deficient soils with application of phosphate fertilizer. Thus, the most favourable conditions for utilising Azolla are availability of plenty of water, temperature legumes of 20–30°C, near nautral soils and adequate P in the growth medium (soil).

Adoption Constraints and Future Research Needs

For adoption of Azolla technology, it is necessary to develop production and distribution system as is being followed in Vietnam and China. Fresh Azolla is being used in these countries for inoculation in rice fields. However, development of suitable technology for using sporocarps for field inoculation can help in overcoming the difficulty of growing fresh material throughout the year, its transportation to distant places and facilitate the maintenance of Azolla germplasm. At present, fresh Azolla plants are to be inoculated in paddy fields for rapid multiplication. The susceptibility of Azolla to temperature extremes poses a serious problem, particularly in its maintenance beyond rice cropping season. Damage from high water temperature during summer can be reduced by shading, draining of flushing water. However, identification of heat or cold tolerant Azolla species/strains is also important. The heat tolerant strains should also possess tolerance to major insects and diseases. The selection of Azolla species/strains with high efficiency of P absorption would be useful, especially where flood water P content is low.

Blue-Green Algae (BGA)

The BGA are distributed world-wide and contribute to the soil fertility in many agricultural ecosystems. The first account of agronomic potential of BGA in rice was presented by De, who attributed the natural fertility of tropical paddy fields to these N-fixing organisms.

Nitrogen Fixing Potential and N-input

The N-fixing potential of BGA can be estimated by evaluating its biomass, N-content and N-fixing activity. More than 125 strains of N-fixing free-living BGA are common in flooded paddy fields. Quantitative studies during the last decade showed presence of a high density (2×10^3 to 4×10^6 CFU/g dry soil) of N-fixing BGA in the Indian paddy fields where *Aulosira, Anobaeno, Cylindrospermum, Calothrix, Nostoc* and *Tolypothrix* occur predominantly. The National Facility for BGA, located at the IARI campus in New Delhi houses a collection of 563 BGA strains.

Although much work has been done in India on utilization of BGA as a source of N for rice and their potentiality has been assessed in terms of grain yield response, estimates of BGA biomass are very much limited. Studies conducted at CRRI indicated biomass variations of 4.3 to 27.6 t fresh wt/ ha, 99 to 822 kg dry wt/ha and N-yield of 14 to 32 kg/h. The predominant species found at CRRI were *Aulosira, Cylindrospermum, Nostoc, Anabaena, Aphanothece* and *Gloeotrichia.* The BGA biomass varies widely in dry matter (0.2–5 per cent), ash (5–70 per cent) and N-content (2–13 per cent). The N-fixing activity of BGA has been frequently studied by acetylene reduction activity (ARA) which also varies largely. At CRRI, the estimate of ARA varied from 189 to 259 for *Aphanothece* and from 249–357 n mol ethylene/ mg chl/h for *Aulosira.* In a study on the N-fixation potential of 62 BGA strains, the N-fixing capacity (mg N/50 ml/28 days) ranged from 0.84–2.45 in *Anabaena,* 0.77-5.83 in *Calothrix,* 0.31–1.15 in *Hapalosiphon,* 0.44–4.04 in *Nostoc,* 0.30–3.41 in *Scytonema,* 2.55–5.67 in *Tolypothrix* and 0.6–2.34 in *Westiellopsis.* One tonne of fresh BGA contains about 1 to 2 kg N and 500 kg dry BGA contains about 15–20 kg N/ha. On an average, a BGA mat corresponds to 20–25 kg N/ha. In contrast to Azolla, growth and N-fixation of BGA is higher in planted fields than in fallow fields. The various contributions of BGA in paddy fields are noteworthy.

Factors Affecting Growth and N-fixation

Water

BGA also prefer standing water for their normal growth. Unicellular BGA form balls or scum and filamentous BGA form mat on the water surface. Most BGA prefer 5–10 cm standing water. However, abundant growth occurs on moist soil also. Most BGA prefer clean standing water than turbid water. Water management directly affects growth and N-fixation. They possess a strong adaptibility to dessication. Alternate drying and rewetting of paddy fields encourages BGA growth, which supresses weeds and controls grazer populations as well. *Calothrix, Tolypothrix* and *Aulosira* perform better in areas where rice is grown rainfed or where irrigation is not assured.

Mineral Nutrients

BGA need all the plant nutrients for their growth and N-fixation. Phosphorus is the primary growth limiting nutrient which should be present in available form in the flood water. Application of fertilizer P enhances growth and N-fixation. Phosphorus deficiency causes drastic reduction in BGA growth and N-fixation. BGA growth is closely related to available P of the soil. The growth is poor at 0.2–5 ppm available P whereas vigorous growth is observed above 6 ppm P. The P content of BGA is 0.05 to 0.18 per cent on dry weight basis. Another essential nutrient for BGA growth is molybdenum. The minimum Mo level for BGA growth is 0.2 ppm which is often available in paddy soils. However, during drying period when BGA are active in N-fixation, Mo might become limiting. Iron deficiency sometimes reduce BGA growth. Magnesium is important to ensure adequate amounts of divalent cations which are required for proper functioning of nitrogenase and glutamine synthetase. Calcium requirement for maximum BGA growth is also known. Addition of calcium carbonate enhances their growth and N-fixation. Other nutrients required by BGA are adequately present in the flooded soil.

Degree of susceptibility of BGA to ammonium-N seems to be strain-dependent. Although a sizeable number of strains appear to remain unaffected upto 75 ppm of ammonium N, the safe limit appears to be around 50 ppm, equivalent to 100 kg N/ha. Such high concentrations are not common under field conditions due to sub-optimal N application rates and the widespread practice of split application.

Light

BGA are phototrophic microorganisms which occur in the light zone as well as on soil surface in the flood water. Light availability which influences BGA growth and N-fixation, depends on the season, latitude, cloud cover, plant canopy and turbidity of water. The growth of BGA is better at higher light intensity. However, the rice crop canopy is reported to decrease light availability to BGA by 50 per cent after 15 days, 85 per cent after one month and 95 per cent after 2 months of transplanting. Growth and N-fixation of BGA are positively correlated with solar radiation.

Temperature

Optimum temperature for BGA growth is about 30–35°C and low temperature decreased their growth. In submerged soils, daily variations in temperature are moderated by flood water. Temperature is rarely a limiting factor for BGA in tropical paddy fields.

pH and Salinity

pH is the most important factor which determines the BGA growth and N-fixation. The optimum pH for BGA growth in culture media ranges from 7.5 to 10 and its lower limit is about 6.5 to 7. Under natural conditions, BGA growth is better in neutral to alkaline soils, although their occurrence is reported in a wide range of acidic and alkaline pH. *Aulosira fertilissima* and *Calothrix brenissima* have been found to be ubiquitous in Kerala rice having pH from 3.5 to 6.5. The relationship between soil pH and number of N-fixing BGA species has been established. BGA are relatively salt tolerant. Role of BGA in the reclamation of alkali soils was reported 30 years ago.

Management Practices

Agronomic practices for rice cultivation influence the growth of BGA. Generally, tillage has a detrimental effect because of incorporation of BGA in the soil and decreased availability of light. Mid season tillage increases availability of P and Fe in paddy fields which favours BGA growth.

Rate and Time of Inoculation

BGA is inoculated in fresh form, dry form or as soil-based inoculum. Inoculation of fresh BGA is better than dry BGA or soil-based inoculum, since fresh BGA establishes early in paddy fields and grows faster. Fresh BGA at 30–60 kg/ha and dry BGA or soil based inoculum at 3–10 kg/ha is recommended for multiplication plots and transplanted paddy fields. However, use of high inoculum helps to cover the paddy fields early. BGA are inoculated one week after transplanting paddy. It is advised to inoculate BGA in morning hours so that it gets longer duration of sun light. BGA is inoculated on the clean standing water under clear sky conditions.

Fertilizer Application

The type and quantity of fertilizers and their application techniques influence the growth of BGA. Nitrogen fertilizers generally inhibit BGA growth and N-fixation. In field studies at Cuttack, growth of *Aulosira* sp. was reduced to 25 per cent at 120 kg N/ha. The N-fixing activity was reduced by 14 per cent at 40 kg N/ha and 23 per cent at 80 kg N/ha (Table 16.5). Inoculation of BGA in paddy fields at various levels of N-fertilizer showed a consistent decrease in BGA growth with increasing N. However, other workers did not observe such effect and reported that ARA was not depressed in soil-BGA system with 40 ppm NH_4–N in the stagnant water. Deep placement of N is recommended in fields where BGA is used for obtaining dual benefit of higher N-use efficiency and increased N-fixation.

Table 16.5: Effect of N-fertilizer on N-fixing Activity (ARA) of *Aulosira* sp. Under Flooded Conditions (mean of three seasons)

N-levels (kg N/ha) and splits	*ARA (n mol. ethylene/mg protein/h)*		
	30 Days After Inoculation	*60 Days After Inoculation*	*Mean*
0	334.0	357.5	345.8
40 (50 : 25 : 25)	287.3	307.5	297.4
80 (50 : 25 : 25)	249.5	283.5	266.5
Mean	290.3	316.2	

Phosphorus is often a limiting factor for growth and N-fixation of BGA. A strong positive correlation between available P and abundance of BGA in rice fields is reported. The P requirement of BGA is 40 to 80 kg P_2O_5/ha in Indian paddy fields. The ARA measurements have also indicated a significant effect of applied P on N-fixation. Application of 40 kg P_2O_5/ha increased yield of *Aulosira, Aphanothece* and *Gloeotrichia* by 2.5, 3.5 and 5 times respectively. Split application of P is better than one basal application and P-use efficiency is generally low in acid soils. Application of lime and Mo has also been reported to increase BGA growth and N-fixation. Potassium does not have a stimulatory effect on their growth and N-fixation. This could be due to adequate K-status of the soil.

Method of Rice Planting

Transplanting is better for BGA growth as compared to broadcasting rice. BGA growth is more in line planting than randomly planted or direct seeded rice. The growth of BGA is significantly influenced by planting density of rice and wider spacing favours better growth than the close spacing.

Insects, Diseases and Weeds

Predators and pathogens limit BGA growth. The cyanophages, parasitic bacteria and fungi cause rapid death of large BGA blooms, but their occurrence in paddy fields is rare. Generally, zooplankton, mosquito larvae, cladocerans, copepods, ostracods and snails damage BGA establishments. In CRRI paddy fields, *Canthydrus lactabilis, Idiopoma dissimilis* and snails were found to be major BGA predators. Further, mucilage forming BGA are less susceptible to grazing than low-mucilage forming BGA. Although pesticides used against rice pests also check BGA pests, such studies under field conditions are limited. However, their effects could be stimulatory or inhibitory depending on nature and concentration of the chemicals. Field experiments at CRRI revealed that Cytrolane and Furadan at 2.5 kg/ha enhanced growth and N yield of *Aulosira* sp. On the other hand, diazinon, gammaxene and BHC were inhibitory to its growth. However, toxicity of insecticides decreases with the increase of incubation period. BGA growth is stimulated at low doses of herbicides but their higher doses are inhibitory. BGA could tolerate high concentration of fungicides. Testing of 27 strains of N-fixing BGA revealed that most BGA could tolerate high levels of pesticides. Field doses of pesticides generally do not harm BGA in flooded fields.

Method of Inoculum Production

Under laboratory conditions, BGA inoculum can be produced in nutrient medium in glass containers or small tanks, which is efficient but energy intensive and expensive. Laboratory grown inoculum often fails to establish in the paddy fields. On the contrary, open air soil culture method is simple, less expensive and easily adaptable by farmers. It is based on the use of a starter culture with multi-strain inocula of *Aulosira, Tolypothrix, Scytonema, Nostoc, Anabaena* and *Plectonema* which is multiplied in shallow trays or tanks with 5–15 cm standing water in 4 kg soil/sq.m and addition of 100 gm superphosphate. A thick BGA mat is formed on the soil surface in about 15 days and tray is allowed to dry in the sun. BGA flakes are collected and stored for use. A simple method of large scale inoculum production in the paddy fields has been developed at CRRI using indigenous mixed inocula of *Aulosira* sp., *Aphanothece* sp. and *Gloeotrichia* sp. with addition of 40 kg P_2O_5/ha. BGA production in trays varies from 0.4 to 1 kg dry wt./m^2 in 15 days whereas in the paddy fields it varies from 3.3 to 366.5 kg dry wt/ha/month.

Method of Utilization

In paddy fields where native N-fixing BGA populations are high, inoculation of outside strains might not be useful. Studies conducted at CRRI have indicated that growth of native N-fixing BGA in paddy fields could be encouraged by split application of 40 kg P_2O_5/ha. Both inoculated and native BGA form a mal in about 20–25 days. After mat formation or appearance of thick BGA bloom, the water is drained out and BGA incorporated into the soil with the help of a weeder or rake. The unincorporated BGA also decompose due to reduced availability of light resulting from increased rice canopy. However, incorporated BGA produces higher rice yield as compared to unincorporated mat.

Impact on Rice Yield and Soil Fertility

Availability of BGA-N to Rice

Nitrogen fixed by BGA becomes available to the rice crop after their decomposition as is the case with Azolla or any organic source. Field experiments for 4 consecutive years indicated that only one-third of BGA-N was absorbed by rice plants in the first year and rest of it remained in the soil as residual N. The ^{15}N-tracer studies have indicated that about 50 per cent of N fixed by BGA is released into the surroundings. The recovery of N from incorporated BGA is higher than from unincorporated BGA. Fresh BGA is superior to dry BGA as regards the release of N.

Effect on Rice Yield

Field experiments conducted in India have reported about 14 per cent yield increase over control, which corresponds to 450 kg grain/ha. Experiments at CRRI indicated that BGA increased grain yield by 6 to 35 per cent. Increased rice yield due to BGA was confirmed in all India trials (Table 16.6). The response to BGA application is better in dry season than the wet season and long duration rice varieties show higher response as compared to short duration varieties. In deep water rice, BGA occur as epiphyte and increase rice yield. Use of mixed BGA inocula is superior to unialgal inoculum. Among different BGA strains, agronomic potentially of *Aulosira* sp. is established in Indian paddy fields. Grain yield measurements suggest that algalization produces a cumulative residual effect. This is attributed to build up of organic N and BGA populations in soil which facilitates better establishment of BGA. Comparison of N fertilizer and BGA indicates that BGA inoculation is equivalent to application of 20–30 kg N/ha.

Many workers have reported increased soil-N due to BGA amendments. Studies at CRRI have indicated that inoculation of *Aulosira* in the rice fields at 60 kg fresh wt./ha registered a significant increase in soil-N whereas indigenous BGA populations increased soil-N by 4 to 6 per cent. Application of BGA also increases the organic carbon and available P content of the soil.

Table 16.6: Impact of Blue-Green Algal (BGA) Inoculation on Grain Yield of Rice in Different States*

Treatments	Grain Yield (t/ha)			
	Madhya Pradesh (56)	Tamil Nadu (111)	Kerala (4)	Andhra Pradesh (5)
0 kg N/ha	2.42			3.64
0 kg N/ha + BGA	2.82			4.43
60 kg N/ha	3.50			
40 kg N/ha + BGA	3.63			
75 kg N/ha		5.24		
50 kg N/ha + BGA		5.11		
90 kg N/ha			3.56	
60 kg N/ha + BGA			3.84	
100 kg N/ha		4.70		
75 kg N/ha + BGA		5.21		
100 kg N/ha + BGA		5.48		
150 kg N/ha				5.84
100 kgN/ha + BGA				6.04

Number of field trials in ( ).

* States with only 1–2 trials results not included.

Economic Aspects

BGA biofertilizer usage in rice cultivation is economically rational. Inoculum is given to farmer at Rs 1/kg at present. Net gains of Rs. 483 and Rs. 630 were reported in the early 1970s. It has been estimated at a state farm of Tamil Nadu that BGA production units could make a net profit of 625 US dollars from a threshing floor of 8.5 × 7.1 m size. A study in farmers field at Nasaratpet village in Chingleput district of Tamil Nadu has reported a net income of Rs. 2000 from 0.2 ha land by using BGA biofertilizer technology.

Suitable Agroclimatic Conditions

BGA generally prefer to grow in flooded conditions, but most species are tolerant to drying and intermittent rain favours their growth. BGA require relatively high temperature and light intensity for their optimum growth. Cloudy conditions are detrimental for their growth. BGA growth is better in alkaline soils but they can be successfully grown in acidic soils after suitable amendments. Soils having high available P content are good for growth of BGA. However, application of fertilizer P can enable them to grow in low P soils.

Adoption Constraints and Future Research Needs

Lack of suitable technological recommendation to the farmers is a major constraint for large scale adoption of BGA. The inoculated BGA often fail to establish and their establishment is sporadic. Factors responsible for colonization and growth of inoculated BGA in different environmental conditions are not clearly known. However, low pH, low temperature, P deficiency, heavy rainfall, cloudy weather and grazing by invertebrates are some of the possible reasons of poor BGA establishment. The quality and composition of BGA inoculum is also important. The antagonism and competition among various BGA strains needs to be studied. The basic research on ecology of inoculated and indigenous BGA and physiology of N-fixation under anaerobic and microaerobic conditions is necessary. Many studies on algalization are empirical and confined to grain yield measurement. Such studies should also include qualitative and quantitative analysis of BGA population in soil and inoculum, chemical analysis of soil, record of major climatic parameters, estimation of BGA biomass and N-fixing activity and residual effects.

Conclusions

The importance of Azolla and BGA as biofertilizers for rice and some other crops has been established. These are essentially organic N-fertilizers specific for rice cultivation. Both need adequate P. BGA can withstand higher temperature and variable moisture regimes than Azolla but the latter is more versatile as it can be grown as a green manure and/or as dual crop. Large scale use of Azolla is presently limited due to lack of production and distribution system for supply of fresh material to farmers for field inoculation. BGA have advantage over Azolla in this respect because its inoculum distribution is easy since dry BGA can also be used for field inoculation. The selection of Azolla or BGA should be primarily based on soil and climatic conditions prevailing in a given locality.

The Government of India has established over 60 centres in various parts of the country for development and use of biofertilizers, but Azolla is yet to get proper place in this programme. It is necessary to set up *Azolla* production centres at the block level for ensuring timely supply of fresh inoculum to farmers so that large sections of farming community can be benefited from this technology. Integrated crop nutrition can only be practiced if its various components and know how for their use is available at the farm level.

Chapter 17
Production, Distribution and Promotion of Biofertilizers

Biological routes of improving soil fertility for optimum crop production are vital components of integrated nutrient supply systems. These routes are operated by microorganisms who either synthesise plant-usable forms of nutrients (N_2 to NH_4) or increase the availability and root accessibility of nutrients already present in the soil, as in case of P. Though most of these organisms are present in the soil and have been on the job for centuries, as manageable agricultural inputs they have received attention only during the 20th century. Due to several reasons, their importance is on the increase and therefore their production and distribution aspects assume practical significance.

Such microorganisms have somehow come to be called as "biofertilizers" a term which many consider to be arbitrary, a misnomer and even slangish but nevertheless widely used. In the strictest sense, real biofertilizers are the green manures and organics (materials of biological origin which are added to deliver the nutrients contained in them). We believe that what are commonly referred to as biofertilizers should be referred to as inoculants after the name of microorganism they contain *viz.*, *Rhizobium* inoculant or *Azospirillum* inoculant.

Definition and Classification

Microbial inoculants are biologically active products containing active strains of specific bacteria, algae, fungi, alone or in combination, which may help in increasing crop productivity by way of helping in the biological nitrogen fixation, solubilization of insoluble fertilizer materials, stimulating plant growth or in decomposition of plant residues. A number of biofertilizers are now available in India. Depending upon the nutrients provided, these can be broadly classified on next page.

In India, systematic study on biofertilizers started 70 year ago with the first report of the isolation and identification of *Rhizobium* from different cultivated legumes by Joshi in 1920. This was followed by extensive research by Gangulee, Sarkaria and Madhok on the physiology of the nodule bacteria and its inoculation for better crop production. Important milestones in production, development and promotion of biofertilizers in India are presented in Table 17.1.

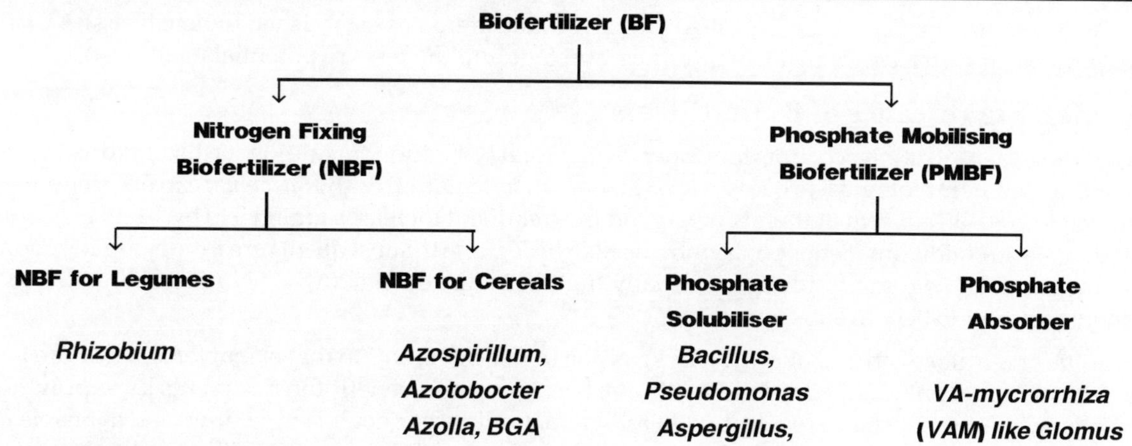

Table 17.1: Some Milestones in Research, Production, and Promotion of Biofertilizers in India

Year	Events
1920	First study on Legume-Rhizobium symbiosis by N.V. Joshi
1934	Earliest documented production of Rhizobium inoculant by M.R Madhok
1939	Discovery of N-fixation by Blue Green Algae (BGA) in rice field by P.K. Dey
1939	Report on the performance of *Azotobactor* in rice soil by B.N. Uppal
1956	First commercial production of Biofertilizer
1957	Study on solubilisation of phosphate by microorganisms by Sen and Pal
1958	First attempt to standardise quality of legume inoculant by A. Sankaran
1960	First isolation of new non symbiotic N-fixing organism *Derxia gummosa* in the world by P.K. Dey and Miss R. Bhattachayya
1964	Spurt in demand of Biofertilizer for soybean particularly in Madhya Pradesh
1968	All India Pulse improvement Project and Soybean Project set up by ICAR where Rhizobium study got priority
1969	Use of Indian peat as carrier reported by V. Iswaran
1970	Scope for use of charcoal, lignite and FYM as alternate carriers to peat reported by V. Iswaran
1975	Coal as alternate carrier to peat reported by J.N. Dube
1976	Indian Standard Specification for Rhizobium
1977	Use of ISI mark for Rhizobium
1979	Initiation of All India Coordinated Project on BNF
1979	ISI standard for *Azotobactor* inoculant
1983	Setting up of National Project on Development and Use of Biofertilizer by Ministry of Agriculture, Government of India
1985	First National Productivity award on Biofertilizer
1988	Setting up of National Facility Centre for BGA at IARI

Today, *Rhizobium* and Blue-Green Algae can be considered as established biofertilizers; *Azolla, Azospirillum* and *Azotobacter* are at an intermediate stage and the rest are potential materials.

Practical Significance of Biofertilizers

Biological N-fixation accounts for 69 per cent of total N-fixation (including fertilizer industry) in the world and non-biological processes for 31 per cent. Inoculation with *Rhizobium* can help legumes to meet upto 80–90 per cent of their N needs and the treatment increases grain yield by 10–15 per cent under on-farm conditions. Benefits of symbiotic BNF by legume to subsequent cereal crops are common and measurable. Legume residues are usually high in N, have a narrow C : N ratio and generally mineralise laster to benefit subsequent crops.

Blue green algae can add about 20–25 kg N/ha to rice fields and to that extent fertilizer N can be saved or supplemented. In addition, BGA have been shown to benefit the rice plants by supplying growth promoting substances such as gibberellic and indole acetic acid. Temperature and available P are two most important factors which determine the success of *Azolla. Azospirillum* and *Azotobacter* have also shown promise as biofertilizers. At present there is a great demand for *Azospirillum* from farmers of Tamil Nadu as they have obtained significant responses in rice and sugarcane due to its inoculation.

Results with P-solubilising microorganisms invariably show that inoculation of seeds with these increases grain yield. These also have the potentiality of being used as co-inoculants for legumes which have high P requirement. VAM-fungi facilitate the accumulation of P by plants through mycelial network, but the multiplication of VAM on commercial scale is yet to arrive.

One tonne Rhizobium inoculant is equivalent to 100 t of fertiliser N (considering minimum N-fixation of 50 kg/ha and 0.5 kg/ha application dose) and 1 tonne of BGA is equivalent to 2 t of fertilizer N (considering minimum N-fixation of 20 kg/ha from 10 kg/ha application dose). Monetary benefits derived from the use of biofertilizers are presented in Table 17.2.

Table 17.2: Some Example of Benefits Derived from the Use of Biofertilizers Based on Extensive Field Trials in India

Crop/Biofertilizer	*Grain Yield (kg/ha)*		*Increase in Yield*		*Monetary Benefits from Inoculation (Rs/ha)*
	Control	*Inoculated*	*(kg/ha)*	*%*	
Chickpea/*Rhizobium*	1956	2228	272	13.9	1204
Pigeonpea/*Rhizobium*	1985	2182	197	9.9	1054
Rice/Blue green Algae	4175	4650	475	11.4	1015
Rice/*Azolla*	2800	3480	680	24.3	1514
Sorghum/*Azospirillum*	3130	3700	570	18.2	1149
Pearl Millet/ *Azospirillum*	1430	1789	359	25.1	720
Colton/*Azotobacter*	1254	1339	85	6.8	343

Calculated on the basis of 1991–92 official procurement/support prices. Cost of biofertilizer: Rs 20/ha for bacteria, Rs 50/ha for Azolla, Rs 10/ha for BGA.

Requirement of Biofertilizers

Estimated annual requirement of *Rhizobium* innoculum vaies from 1,250 to 15,000 t. Highest requirements are apparently based on an over-simplified approach multiplying the total legume area by dosage/ha. If 25 per cent of area is annually treated, 3750 t inoculum is needed for 30 m ha,. Present production is about one-fourth of this.

Estimated requirement of BGA varies from 168,000 t for treating 16.8 m ha under wetland rice to 230,000 t for treating 23.6 m ha of rice in eight major states to as high as 400,000 t apparently for inoculating the entire rice area in India with 10 kg/ha. The authors have also arrived at an estimate closer to that of Venkataraman. Present production is less than 0.1 per cent of this.

Good estimates should actually be based on the native bacterial population per unit soil, nodulation map for Rhizobium, natural distribution of microorganisms, environments offering greatest potential and presence of any major constraints such as nutrient deficiencies. Requirements based on multiplication of total area a by dosage/ha do not serve any purpose and provide exaggerated estimates which cannot be used as a basis for planning or setting up production facilities.

Production Technology of Biofertilizers

Rhizobium

Based on the physical nature and carrier materials used, various types of biofertilizers are manufactured by different producers. These are carrier based inoculant, agar based inoculant, broth culture and dried culture. New developments in biofertilizer productions like (*i*) freeze-dried inoculants (*e.g.* BAIF, IARI, India) (*ii*) Rhizobium-paste (*e.g.* KALO Inc. USA), (*iii*) granular inoculant (*e.g.* Soil implant of Nitragin, USA) (*iv*) pelleting (*e.g.* Pelinoc of Nitragin), (*v*) Polyacrylamide entrapped rhizobia (*e.g.* Agrosoke) and (*vi*) Pre-coated seeds (*e.g.* Prillcote of New Zealand) appear to be more promising for inoculation success in tropical legumes. Advantages and disadvantages of most of the types mentioned above have been critically examined and carrier based inoculant has been regarded as most suitable for commercial purposes. Since different species of Rhizobium bacteria infect different legumes, as discussed eariler, a whole range of inoculants are needed. Their production technology is however the same.

In India, septic methods of production of *Rhizobium* biofertilizer is followed comprising mixing of broth with unsterilised carrier. Sterilisation of carrier in bulk is done in autoclave by most of the producers, where complete sterilisation is not achieved. The fool proof method of sterilisation of carrier material is through gama radiation but such facility is lacking. To achieve production of quality biofertilizer, carrier should be sterilised in sterilisable poly propylene pouches, inoculum introduced by injecting and cured for at least two weeks at 27–30°C before despatch.

Sources of Mother Cultures

A key input for the production of inoculum is the availability of mother cultures which can form the basis for mass production. Some promising strains of Rhizobium are listed in Table 17.3. Well known institutions like Indian Agricultural Research Institute, Jawaharlal Nehru Krishi Vishwa Vidyalaya, Mahatma Phule Agricultural University, Tamil Nadu Agricultural University, International Crops Research Institutes for Semi-Arid Tropics (ICRISAT), National Biofertilizer Development Centre, CSIRO (Australia) and Niftal (Bangkok) supply proven strains mostly on request.

Particularly for proven strains of BGA, the national BGA facility at IARI, the BGA project at Madurai Kamraj University, and 60 BGA sub centers can be contacted. National Facility has a collection of 398 strains of BGA. The IARI as well Tamil Nadu Agricultural University can be contacted for *Azospirillum* and *Azotobacter* and the major institution engaged in research on Azolla is the Central Rice Research Institute at Cuttack. University of Agricultural Sciences, Bangalore and ICRISAT are important centers of work on Mycorrhiza. The Asian Mycorrhiza Network is located at the Tata Energy Research Institute, New Delhi.

Table 17.3: Some Promising Rhizobium Strains

No.	*Host Legumes*	*Promising Rhizobium Strains*	
		*Indian**	*Foreign***
1.	Groundnut	NC-92	Niftal 1000
		IGR-6	Niflal 968
		IGR40	Nitragin 8A11
2.	Pigeonpea	CC-1	P-241
		F-4	Niftal-S69
		BDN-A2	Nitragin-21A16
3.	Chickpea	F-6	Nitragin-27A3
		F-76	Nitragin-27A8
		H-45	Ca-142
4.	Soybean	SB-103	Niftal-102
		SB-119	Nitragin-61A148
		UASB-29	11TA-2l27
5.	Lentil	L-6	Niftal-640
		L-1	C 4201
6.	Pea	RL(P) 114	Nitragin 128A13
7.	Greengram	M-10	UPLB-M6
		GMBSI	THA 301
		MO-5	Nitragin 176A22
8.	Blackgram	DU-2	THA-301
		Ku-1	Niftal-442
		BDN-F	UPLB M-6
9.	Cowpea	DC-28	Niftal-174
		DC-2	Niftal-I89
		DC-6	Nitragin-176A22

* Principal Investigator (Microbiology), AICPIP (except Groundnut and Soybean).

** USDA Rhizobium Culture Collection Catalog, Beltsville, MD.

Carriers

Following are the most commonly used carrier in preparation of biofertilizers

Rhizobium: Peat/lignite/charcoal

Blue-Green Algae: Soil based

Azotobacter: Peaf/lignite/charcoal

Azospirillum: Peat/lignite/charcoal/mixture of soil and FYM (3 : 1).

There are many other carrier materials being tried for *Rhizobium*, like burnt paddy husk, burnt coconut shells, and vermiculite.

Among the various methods available for production of BGA in India, *e.g.* trough method, pit method and field scale production, the last one is important from farmers' point of view. Its important features are summarised in Table 17.4.

Table 17.4: Main Items in the Production Technique of BGA and Azolla

Particulars	*BGA/cent (40 sq.m.)*	*Azolla/cent (40 sq.m.)*
Technique	Field plot method	Azolla nursery plot technique
Fresh cattle dung	–	10 kg
Single Superphosphate	2 kg	300 gm
Furadan	200 gm	100 gm
Culture	5 kg	8 kg
Period	15–20 days	15–20 days
Season	March–August	September–February
Yield	30–40 kg BGA	60 kg Azolla biomass

Production of Biofertilizers

There are nearly 30 biofertilizer producers in India. These are from Government (Central and State), Institutional Agencies, Agricultural Universities and private companies. The main production is of *Rhizobium*. Recently some organisations have started the production of *Azospirillum*, *Azotobacter*, Phosphate solubiliser and mycorrhiza. Production is demand-oriented.

Rhizobium

The current average production of rhizobium innoculum in India is around 1000 tonnes. During (1985–86 to 1988–89, production declined in 1987–88, otherwise the level of production remains unchanged. State governments and institutional agencies account for bulk of the production (Figure 17.1). The present trend of production of some leading organisations is given in Table 17.5. Other organisation mainly Agro Industries as in Assam, Orissa, Punjab, Rajasthan, and M.P Oilseed Growers Federation are also producing Rhizobium biofertilizer which has not been included in the Table because of non-availability of production figures consistently for the last five years.

Table 17.5: Current Status or Rhizobium Production by Some Important Organisations in India

Sl.No.	*Organisation*	*Starting Year*	*Capacity (tonnes)*	*Production (Tonnes)*				
				84–85	*85–86*	*86–87*	*87–88*	*88–89*
State Government								
1.	Department of Agriculture Govt. of Tamil Nadu	1976	200	143.60	134.6	159.14	170.99	157.58
2.	Department or Agriculture Govt. of U.P. (10 labs.)	1977-78	220	na	196.20	166.40	124.80	131.20
3.	Department of Agriculture Govt. Himachal Pradesh	1979	10	1.14	0.99	0.77	1.10	0.89

Contd...

Table 17.5–Contd...

Sl.No.	Organisation	Starting Year	Capacity (tonnes)	Production (Tonnes)				
				84–85	85–86	86–87	87–88	88–89
Agricultural Universities								
4.	Haryana Agricultural University, Hisar	1974	35	23.09	12.04	12.10	10.01	10.00
5.	Tamil Nadu Agricultural University, Coimbatore	1971	20	4.40	5.51	5.53	3.55	1.61
6.	Mahatma Phule Agricultural University, Pune	1974	10	3.12	3.01	3.91	2.48	na
Industries								
7.	M.P. Agro Industries Corporation, Bhopal	1983	150	9.76	13.08	58.06	92.25	98.52
8.	Maharashtra Agro Ind. Corp., Pune	1981-82	250	197.31	177.84	222.80	90.78	16.13
9.	Gujarat State Fertilizer Company, Vadodara	1983–84	100	0.85	9.20	12.10	5.40	26.60
10.	Gujarat Slate Cooperative Mktg. Fed.	1976	80	69.43	32.64	26.72	20.04	30.26
11.	NAFED, Indore	1984–85	150	30.14	80.64	79.88	69.98	32.49
12.	BAIF, Pune	1983	50	40.00	50.00	125.00	na	na
13.	Rajasthan State Agro Ind. Corp. Jaipur	1987	100	na	na	na	35.94	58.54
14.	Micro Bac India, 24 Parganas, W.B.	1977	100	na	75.00	61.86	41.54	70.45

na: Data not available.

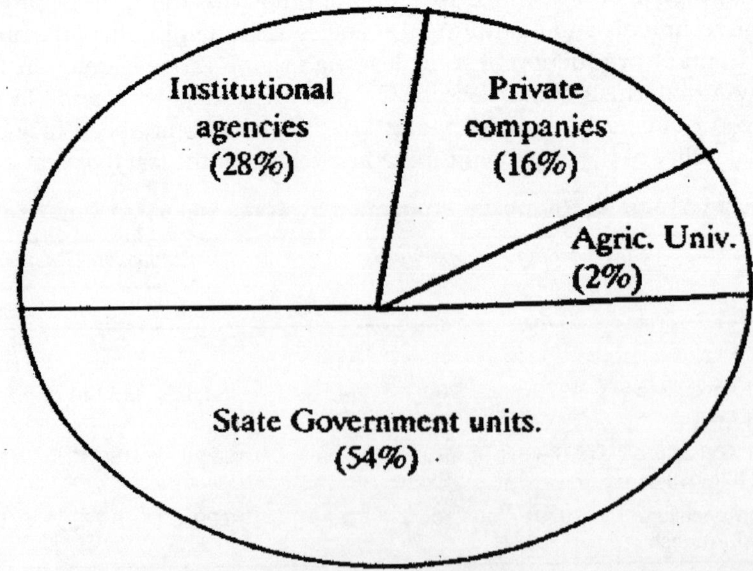

Figure 17.1: Contribution of Different Organisations in *Rhizobium* Production

Azospirillum and *Azotobacter*

Out of a few organisations engaged in their production, bacterial cultural scheme of Tamil Nadu at Cuddalore and Gujarat State Fertilizer Company Ltd. (GSFC) at Baroda have made considerable headway as shown below:

Year	*Tamil Nadu*		*GSFC*
	Azospirillum (t)	*Azotobacter (t)*	*Azospirillum + Azotobacter (t)*
1984–85	10.8	2.4	0.85
1985–86	36.1	8.9	14.50
1986–87	56.2	9.7	29.00
1987–88	103.1	7.3	30.23
1988–89	126.3	0.9	68.48
1989–90	136.3	0.0028	not available

Blue Green Algae

Sixty sub-centres in different parts of the country under National Project on Biofertilizer produce BGA. The current annual production from these centres is approximately 200 tonnes. The locations of these sub-centres are:

State	*BGA Sub-centres*
Andhra Pradesh	Krishna, E. Godavari, Nizamabad (2), Guntur, Warangal Nalgonda (2), Karimnagar, Anantapur
Assam	Guwahati, Jorhat
Bihar	Bhagalpur, Patna
Gujarat	Nawagam
Himachal Pradesh	Una
Jammu and Kashmir	Jammu
Karnataka	Bangalore, Bellary, Chitradurga, Gangavathi, Mandya, Mysore, Shimoga
Kerala	Pattambi
Madhya Pradesh	Bastar, Raipur
Maharashtra	Akola, Bhandara, Pune, Thane
Manipur	Imphal
Meghalaya	Garo Hills
Orissa	Balasore, Berhampur, Chiplima, Kendrapara, Keonjhar, Mahisapurt, Puri, Semiliguda
Punjab	Ludhiana
Tamil Nadu	Attiyandal, Coimbatore, Salem, Thanjavur (2), Tirur, Trichy, Vandarayanpet
Uttar Pradesh	Baharaich, Kanpur, Kumarganj (Faizabad)
West Bengal	Baruipur, Berhampore, East Midnapur, Kalyani, Singur, West Dinajpur, West Midnapur

Thus, on the whole, present production pattern is in the order:

Rhizobium > *Azospirillum* > BGA > *Azotobacter* and is depicted in Figure 17.2.

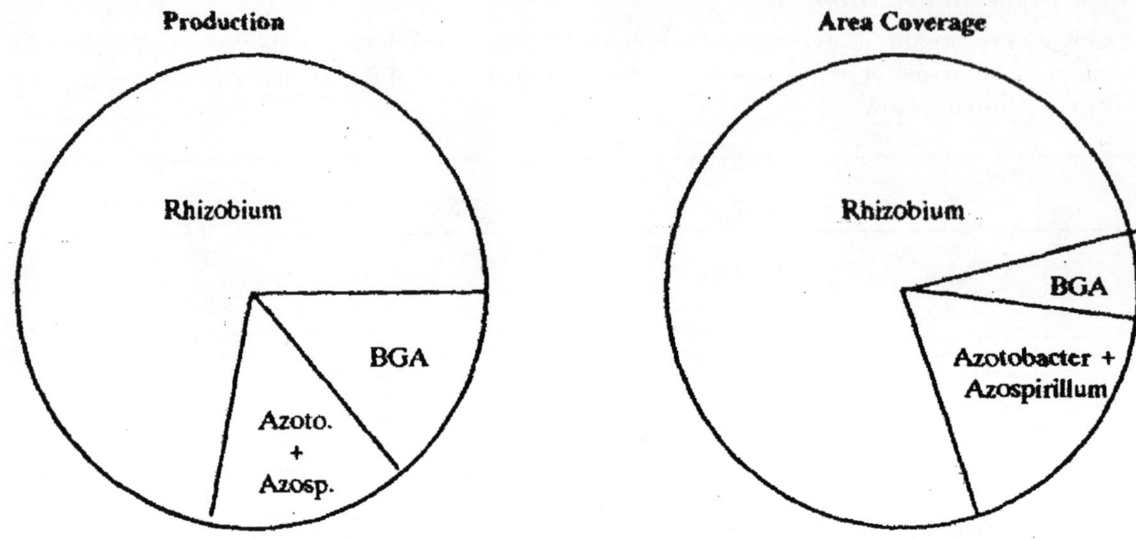

Figure 17.2: Share of Biofertilizer in the Present Level of Production and Area Coverage

Standards and Quality Control

Quality control of biofertilizers is a very important aspect. In the initial years (1965–75), soybean *Rhizobium* was in great demand, and many private companies mushroomed and distributed substandard inoculants, not only for soybean but for other pulses also. The effect or inoculation was not promising. This eroded the belief of farmers in biofertilizers and it is a challenge to regenerate their faith. This can only be achieved by supplying standard quality material at field level. Quality standards have been set up for *Rhizobium* and *Azotobacter* by the ISI (now BIS, Bureau of Indian Standards).

Rhizobium (Standard IS: 8268-1976)

The inoculant should be carrier based and pH of the carrier should be between 6.0–7.5. It shall contain a minimum of 10^8 viable cells/g. of carrier on dry mass basis within 15 days of manufacture and 10^7 within 15 days before expiry date marked on the packet when stored at 25–30°C. It should have maximum of six months expiry period from the date of manufacture. Each packet shall be marked legibly to give the following informations–name of product, crop for which intended, name and address of manufacturer, type of carrier, batch or code number, date of manufacture, date of expiry, net quantity of the area meant for and storage instructions. Directions for use shall be printed on the packet and each packet may also be marked with ISI certification mark. *Rhizobium* inoculant should result in 50 per cent or more dry matter compared to control. Another edition has been published in 1986 with the same IS number.

Azotobacter (IS: 9138-1979)

ISI guideline is more or less same as for *Rhizobium.* Regarding efficiency of the inoculum, the minimum amount of nitrogen fixed shall be not less than 10 mg per g. of sucrose utilised. Details are available in the ISI bulletin.

Bacterial cultures are usually sold in polythene bags (14 × 22 cm) and usually contain 200 g culture for treating 12–15 kg seeds. Packet sizes vary from 100–250 g. Cost per packet of culture produced under the National Project is Rs. 4/200 g or Rs. 20/kg. Bacterial cultures sold commercially are available at Rs. 8–20 per packet (Rs. 55–100 per kg). Price also varies with the strain. For example, the price of ordinary *Azotobacter* inoculum in one case is Rs. 55/kg and that of improved *Azotobacter* is Rs. (80/kg or Rs. 20 per 250 g packet. Cost of bacterial culture required per hectare varies from Rs. 20–100 depending upon the crop. It is lowest in the case of pulses and highest in the case of vegetables where 10 packets of 100 g each may have to be used.

Government Support and Programmes

Under the National Project on Development and Use of Biofertilizers, the Government of India, Ministry of Agriculture established a National Centre at Ghaziabad and six regional centres at Bangalore, Bhubaneswar, Hisar, Imphal, Jabalpur and Nagpur. These have been established to accomplish the following objectives:

1. To produce, distribute and promote use of the effective strains of *Rhizobium*, Blue Green Algae, *Azospirillum* and *Azotobacter* to supplement and improve the efficiency of chemical fertilizers.
2. To organise field demonstration and training programmes on effective use of biofertilizers.
3. To create an agency for quality control of biofertilizers.
4. To provide technical support to institutions other than Government Departments who are engaged in production/distribution of biofertilizers.

Production of biofertilizers under the national project is summarised in Table 17.6. The project is intended more to serve as a center for encouraging, catalysing and coordinatory all-India developments rather than as a commercial production facility.

Table 17.6: Production Capacity with the National Project on Biofertilizers

Centre	*Location*	*Production Capacities (t/yr)*		*Area to be Covered*
		Rhizobium	*BGA**	
National (HO)	Ghaziabad, U.P.	75	2.5	All India
Regional Centres				
Northern	Hisar	50	3	J&K, Punjab, Haryana, H.P., U.P.
Eastern	Bhubaneswar	50	3	Orissa, Bihar, West Bengal
Southern	Bangalore	50	3	Karnataka, Kerala, Tamil Nadu
Central	Jabalpur	50	3	M.P., Rajasthan, Gujarat
North Eastern	Imphal	50	3	Assam, North-Eastern Region
Western	Nagpur	50	3	Maharashtra, Goa, A.P.
	Total (t/yr)	**375**	**20.5**	

* For distribution of mother culture only.

Under the project it is envisaged to extend one time financial support upto Rs. 150,000 per BGA centre already established under administrative control of State Governments and Agricultural Universities to produce 10 mt soil base BGA per year per center. There is also a provision of one time financial support to the maximum of Rs. 1.3 million per unit for *Rhizobium* production. Twelve units belonging to State Governments, agro industries, fertilizer and seed companies have already received such Grant in Aid (Table 17.7). It is expected that under the project, with 1 National Centre, 6 regional centres 60 BGA sub-centres and 20 *Rhizobium* production units, the production can be raised to 1875 mt of *Rhizobium, Azospirillum, Azotobacter* biofertilizer and 600 t of BGA.

Table 17.7: Statement of *Rhizobium* Product Units Sanctioned Under the National Project on Development and Use of Biofertilizer 1990–91

Sl.No.	*State/Agency*	*Location of the Project*	*Capacity t/Year*	*Million Rs. Grant Sanctioned*
1.	Andhra Pradesh			
(*i*)	A.P. Agril. University, Hyderabad	Amravati (Guntur)	75	Rs. 1.30 (1989–90)
(*ii*)	Godavari Ferti. and Chem. Ltd., Secundrabad	Secundrabad	75	Rs. 1.30 (1990–91)
2.	Gujarat			
(*iii*)	Gujarat State Co-op. Mktg. Fed., Ahmedabad	Nawagon	75	Rs. 1.05 (1989–90)
(*iv*)	Gujarat State Ferti. Co., Vadodara	Vadodara	75	Rs. 1.30 (1989–90)
3.	Haryana			
(*v*)	Haryana Agril. University, Hisar	Hisar	75	Rs. 1.30 (1989–90)
4.	Maharastra			
(*vi*)	Bhartiya Agro Inds. Foundation, Pune	Waghoti (Pune)	75	Rs. 1.26 (1989–90)
5.	Orissa			
(*vii*)	Orissa Agro Ind. Corpn., Bhubaneshwar	Bhubaneswar	75	Rs. 1.30 (1989–90)
6.	Punjab			
(*viii*)	Punjab Agro Ind. Corpn., Chandigarh	Ludhiana	75	Rs. 1.18 (1989–90)
7.	Rajasthan			
(*ix*)	Rajasthan Agro-Ind. Corp., Jaipur	Jaipur	90	Rs. 1.30 (1988–89)
8.	Tamil Nadu			
(*x*)	Madras Fertilizer Ltd., Madras	Madras	100	Rs. 1.30 (1990–91)
(*xi*)	Deptt. of Agriculture	Salem	200	Rs. 1.30 (1990–91)
(*xii*)	Deptt. or Agriculture	Kudumiamalai	2	Rs. 1.40 (1990–91)

Department of Biotechnology, Ministry of Science and Technology is also supporting two projects on Technology Development and Demonstration of Bio-fertilizers for *Rhizobium* and BGA. The project of *Rhizobium* has collaborating centres at Delhi (IIT), Kolkata, Pune, Jammu Tawi, Burdwan, Pantnagar with headquarter at IARI, New Delhi. The BGA project has collaborating centres at Coimbatore, Pune, Luck now, Kolkata and New Delhi with the headquarter at Madurai Kamaraj University, Madurai. The objective is to carry out research and development, product formulation, demonstrations, quality assurance, training and extension.

To enable scientific research and development in the field of biofertilizer other infrastructural facilities in the premier institutions in India are:

1. National facility on Microbial Type Culture Collection (MTCC) and Gene Bank at the Institute of Microbial Technology (IMTECH) Chandigarh–for providing a good and well-maintained collection of authentic strains of microbes for basic and applied research in the areas of food, fuel and health.
2. National facility on BGA collection at the IARI, New Delhi.
3. National Facility for Plant Tissue Culture Repository (NFPTCR) at the National Bureau of Plant Genetic Resources (NBPGR), New Delhi.

To encourage the producers and recognise outstanding performance the Government has instituted National Productivity Award on Biofertilizer, only on *Rhizobium.* This award was given for the first time in 1985–86. Details oft his award are available from the NPC and the National Biofertilizer Development Center.

Constraints

There exist several constraints in the field of biofertilizer. These are mainly: (*i*) at production and distribution level (*ii*) at the field level and (*iii*) at marketing level.

Production and Distribution Level Constraints

These essentially consist of the (*i*) unavailability of appropriate and efficient strains (*ii*) unavailability of a good carrier or (*iii*) poor storage facilities.

The strains selected for production should have as far as possible qualities like competitive ability over other strains for nodulation of the host, N-fixing ability over a range of environmental conditions, nodulation and N-fixing abilities in the presence of soil N, ability to multiply in broth and survive in inoculant carriers, ability to survive when incorporated in seed pellets, persistence in the soil, ability to migrate from the initial site of inoculation, ability to colonize soil away from the influence of host roots, ability to survive adverse physical conditions such as dessication, heat or freezing and strain stability during storage.

It is very difficult to get strains which have all these qualities. Attempts have been initiated to set up culture collection bank at IARI and at the centres under the National Project on Biofertilizer. Some Agricultural Universities have also a good collection of strains. Some efficient strains have been listed in Table 17.3.

For the carrier based inoculum, the carrier should have high water holding capacity, no heat of wetting, nearly sterile, chemical and physical uniformity, be non-toxic, be biodegradable and non-polluting, have near-neutral or easily adjustable pH, supports rhizobial growth, be amenable to nutrient supplements, rapid release of rhizobia in the soil, manageable in the mixing, curing, and packaging operations, suitable for all rhizobia, available in powder and granular forms, adequately available and at reasonable cost.

Peat has been universally recognised as the most suitable and standard carrier. But non-availability of good quality peat in India has led to the development of alternative carriers. Although many solid materials have been evaluated but till today only lignite and charcoal are considered as alternate though not as good as peat. Out of these, charcoal is more commonly used because it is readily available though in terms of quality it is not very effective. Water extending organic polymers have also been tried. However, unless a good alternative carrier is decisively identified, the efficiency of biofertilizer may not be achieved to the desirable level.

Storage and Distribution

Successful introduction of biofertilizers on farmers' field is restricted to certain crops and locations, either because of (micro)-environmental and biological reasons (soil-*Rhizobium*-plant relationship), or because of the "human factor". With regard to the last category, poor inoculant quality is not necessarily the result of small-scale production activity but can be caused by poor standards, transport, inadequate distribution and storage facilities. Inoculants can be stored for 6 months or longer at 20°C but lose their effectiveness within a few hours at 40°C or higher. No established sale network for biofertilizers exists except in a few cases. In practice, State Departments of Agriculture place supply order mostly with their own production units, from where packets are transported to district headquarters. Before it reaches the field, the packets pass through VLWS and farmers. During delivery from production units to sowing fields, high temperature to which these tiny microorganism are exposed, result in their death resulting in poor quality biofertilizer. Such situations occur frequently in most states where packets are transported during summer (+40°C) for use in *kharif* crops. Such high temperatures do not prevail in South Indian states specially in Tamil Nadu and consumption of biofertilizer has increased many folds there during last two year. Encouraged with this, four biofertilizer units are being launched in Tamil Nadu (2 by State government and 2 by fertilizer companies).

There is a need for awareness of proper inoculant storage conditions at each point in the distribution chain. Unfortunately, this is not available at the distribution outlets. It needs immediate improvement.

Constraints at Field Level

There is low level of acceptance of biofertilizer at farmers level. Because of slow release of nitrogen, the response is not immediate as is visible in case of chemical fertilizers. The constraints for incomplete expression of *Rhizobium* can be a combination of several factors such as:

1. Soil and climate problems such as unfavourable pH, high temperature, drought, high nitrate level, presence of certain elements in toxic level, or deficiency of P, Cu, Co and Mo.
2. Competition from native strains and compatibility with the host legume cultivar, supply of energy (carbohydrates) to the nodules, prevalence of plant diseases etc.
3. Use of sub-standard inoculants or faulty inoculation techniques, any adverse effect of agro chemicals and unfavorable conditions such as water logging.

Detailed discussion on field level constraints is available elsewhere.

Regarding BGA, despite years of research, actual use of algal inoculation is still minimal. Present production of 200 t BGA is sufficient to treat only 20,000 ha (one in 2000) at 10 kg/ha. Its potentiality shown at Nagpur, Maharashtra is not reflected in the fields of Uttar Pradesh, Madhya Pradesh and Bihar. The reason for this, may be the imbalance between test tube and field studies, the utilisation of grain yields as the only criterion in inoculation experiments (without collecting information on the environmental characteristics, indigenous N_2-fixing BGA, fate of inoculated BGA etc.), under estimation of the potential of indigenous strains and possibly sub-optimal dose of inoculant.

It is possible that the generally recommended dose of 10 kg/ha inoculum adds BGA to the soil at a density far below the average density of indigenous N_2-fixing BGA in rice soils.

Market Level Constraints

Marketing of biofertilizers is not easy as the product contains living organisms. It involves the quality of product, its usefulness, packaging consumer demand, faith and also promotional aggressiveness of the producer. Though use of biofertilizer in India started in early 1970s consumption has not picked up basically due to unawareness of farmers and substandard material in the market. Since farmers have not widely accepted the biofertilizer, its of ftake depends on the indent given by different State Governments. Production share of State Government is 54 per cent followed by Institutional agencies (28 per cent) and 16 per cent by private companies (Figure 17.1). This clearly indicates that when State Governments have their own production units and the placing of indent depends on them, it is very difficult for other producers to market their product in those states. Because of these most of the private companies who distributed biofertilizers during 1970's have closed their units because of not having any indent (orders) from State Governments.

Considering the constraints in marketing of biofertilizers, a financial support is being extended from National Project on biofertilizer also to fertilizer and seed producing companies with established marketing network. Gujarat State Fertilizers Company and Madras Fertilizer Ltd. have launched their production units. It is expected that consumption of biofertilizer will increase to a desirable level. Intensive extension activities through compact demonstration programme, wide publicity through mass media can help in creating awareness among farmers on biofertilizer which may also help marketing.

Areas for Future Development

During the last 30 years considerable efforts have been made to popularise biofertilizers but the success has not yet been achieved to desirable extent. India is one of the largest *Rhizobium* producing countries with a production capacity of 2750 t/year. The consumption of *Rhizobium* and BGA is not enough to utilise the installed capacity. Consumption is by and large through various projects operating under Central/State Governments through which the packets of biofertilizers are distributed free or at subsidised rates to the farmers with the intention to popularise this low-cost environment friendly input among farmers. It has not yet gained a foot hold in the package of practices followed by farmers. The consumption of *Rhizobium* biofertilizer to the tune of 2 million packets each in Madhya Pradesh and Tamil Nadu is basically through effective extension. Among the various steps needed for successful production, promotion and distribution of biofertilizers, ones requiring immediate consideration are as the following:

Training

Intensive education of extension officer, dealers, farmers and exposition of policy makers through training, field demonstrations and audio-visual aids to generate awareness about the importance and practical value of biofertilizers in modern farming.

Improvement in Production Technology

Work on suitable substitutes for peat as carrier needs to be intensified. At present the production of biofertilizer is being done by using septic conditions introducing acid producing contaminants which are not inducive to the growth of *Rhizobium.* An improvement in mixing and packaging system is required.

Need for Preparation of Biofertilizer Map

The quantification of native soil microbial status will be helpful in determining inoculation requirement and therefore a survey to prepare such map should be conducted and priority areas for introduction of different biofertilizers should be identified.

Region-specific Effective Strains

Effective strains should be selected and deposited in the culture collection bank with proper cataloging.

Necessary Quality Control Acts

Quality control is essential and should be ensured right in the production units. Appropriate legislation should be enacted to prescribe the required standards for good quality inoculants.

Proper Storage Facilities

Biofertilizers have to be kept under conditions favourable for their growth and multiplication of microorganisms. A good quality *Rhizobium* inoculant can be stored for more than one year without effecting its quality under 5°C. Proper storage facilities at the production unit, sale points and during transportation will help to maintain required level of viable cells.

Conclusions

Significant developments in the production and distribution of biofertilizers are taking place in India. At present, the major emphasis is on inoculants of *Rhizobium* followed by *Azospirillum,* BGA and *Azotobacter.* The National Biofertilizer Development Center established by the Government of India is the nodal agency for planning, coordination, monitoring and quality control of biofertilizers. It has six regional centers. Though bulk of the present production activity is from units set up by state governments, cooperatives, NGOs and the industry including fertilizer companies are also on the scene and their role is expected to grow.

Biofertilizers are vital components of the integrated plant nutrient supply systems and inspite of variability in performance, low cost makes their use attractive but not in a haphazard manner. Current production of biofertilizers is around 1,400 tonnes out of which 1000 t is of *Rhizobium.* In terms of coverage, present production can treat 2.4 m ha consisting of 2 m ha under legumes, 0.4 m ha of other upland crops and about 20,000 ha of rice.

A great deal of work is needed for working out realistic demand estimates for biofertilizers, delineation of areas which require inoculation on priority basis and environments which can derive full benefit from this input. Development of suitable carriers, streamlining of marketing and distribution channels and strengthening of extension and training activities also are sectors which deserve greater attention.

Chapter 18

Effect of Biofertilizers on Growth

Pungam is extensively used for afforestation of watersheds in the drier parts of the country. It is drought resistant, moderately frost hardy and highly tolerant to salinity. It is being propagated by direct sowing or by transplanting one year old seedlings raised in nursery.

Microorganisms that are used as biofertilizers stimulate plant growth by providing necessary nutrients as a result of their colonization at the rhizosphere (*Azotobacter, Azospirillum, Pseudomonas,* phosphate-solubilizing bacteria, and *Cyanobacteria*) or by symbiotic association (*Rhizobium,* mycorrhizae and *Frankia*). The role of biofertilizers has already been proved extensively in annual crops, but its exploitation in perennial trees in India is scanty. In forestry, few research reports are available to demonstrate that biofertilizers stimulate the growth, biomass, nodulation and YAM-colonization. Keeping this in view, the present, study was designed to elicit information on the effect of biofertilizers on the growth and development of pungam (*Pongamia pinnata*) (L.) Pierre.

Material and Methods

The study was carried out at Forest College and Research Institute (11°19'N, 76°56'E, 300 above MSL). Six month old uniform sized seedlings were chosen for biofertilizer inoculation. The seedlings were transplanted in 30 × 45 cm polybags filled with nursery mixtures of red soil, sand and FYM (2 : 1 : 1). The experiment was set up in a completely randomized design and replicated thrice. The nursery soil mixture of the polybags were inoculated with biofertilizers, *viz.,* (*i*) *Rhizobium,* (*ii*) phosphobacteria, (*iii*) Vesicular-Arbuscular Mycorrhizae (VAM), (*iv*) *Rhizobium* + phosphobacteria, (*v*) phosphobacteria + VAM, (*vi*) *Rhizobium* + VAM, (*vii*) *Rhizobium* + phosphobacteria + VAM. Uninoculated seedlings were maintained as control.

Biofertilizer inoculation was prepared with a base of peat soil. Two hundred grams of *Rhizobium* and phosphobacteria were weighed and mixed with 3 kg of well decomposed and powdered FYM separately. Fifty grams of this inoculum mixture and twenty grams of VAM inoculum were applied to each polybag at 5 cm depth near the root zone.

Six seedlings in each treatment were selected at random and observed initially, 2 and 4 months after inoculation for number of leaves, leaf area (LICOR Model LI 3000 Leaf Area Meter), root volume, total dry matter, DGR, NAR, number of nodules and VAM colonization. The results were subjected to analysis of variance and tested for significant differences ($P < 0.05$) after Panse.

Results and Discussion

A significant increase in number of leaves (47.7 per cent) (Table 18.1), total leaf area (44.4 per cent) (Table 18.1) and root volume (47.3 per cent) (Table 18.2) was achieved over control by inoculating the seedlings with *Rhizobium*, phosphobacteria and VAM conjointly. The results corroborated the findings of Sekar in Shola species. The microorganisms that are used as biofertilizers colonize in the rhizosphere and stimulate the plant growth by providing necessary nutrients, thanks to their symbiotic association. The association may also regulate the physiological process in the ecosystems by involving in the decomposition of organic matter, fixation of atmospheric nitrogen, secretion of growth promoting substances, increased availability of mineral nutrients and protection of plants from pathogens besides increasing the availability of nutrient elements at the root zone.

Table 18.1: Effect of Biofertilizers on Number of Leaves, Total Leaf Area and Total Dry Weight of Pungam Seedlings

Biofertilizers (B)	*Number of Leaves Months After (M) Inoculation*				*Total Leaf Area ($cm^2.plant^{-1}$) Months After Inoculation*				*Total Dry Weight ($g.plant^{-1}$) Months After Inoculation*			
	0	*2*	*4*	*Mean*	*0*	*2*	*4*	*Mean*	*0*	*2*	*4*	*Mean*
Rhizobium (R)	18.7	43.7	48.7	37.0	312.6	431.1	569.0	437.5	9.14	27.56	36.13	24.27
Phosphobacteria (P)	18.3	45.3	47.7	37.1	305.4	437.9	569.0	464.0	9.45	32.35	41.01	27.60
VAM	19.0	46.3	53.3	39.6	302.0	462.5	648.6	491.6	9.08	27.61	39.80	25.50
R + P	20.0	48.7	54.3	41.0	305.0	498.9	710.2	504.3	9.25	31.15	40.12	26.84
P + VAM	19.0	47.7	57.3	41.3	286.4	499.6	651.3	479.1	9.44	29.88	40.60	26.64
R + VAM	19.3	52.3	62.0	44.6	297.6	556.1	739.0	531.0	9.11	32.22	42.39	27.91
R + P + VAM	19.7	58.7	68.3	48.9	303.0	641.2	850.0	589.0	0.01	43.18	46.94	33.04
Control	19.7	37.3	42.3	33.1	291.8	408.0	524.1	408.0	9.11	23.15	31.88	21.38
Mean	19.2	47.5	54.3		300.4	488.6	675.1		9.20	30.89	39.86	

	SEd	CD (P = 0.05)	SEd	CD (P = 0.05)	SEd	CD (P = 0.05)
B	1.04	2.10	24.75	49.76	0.839	1.686
M	0.64	1.29	15.15	30.47	0.514	1.033
B × M	1.81	3.64	42.86	86.19	1.453	2.921

An increase of 54.5 per cent in total dry weight over the control was evident in seedlings inoculated combined with *Rhizobium*, phosphobacteria and VAM (Table 18.1). Similar increase in biomass production due to VA-mycorrhizal inoculation was documented in *Acacia* spp.; and *Albizia* spp. In the present investigation, individual inoculation VAM also increased biomass relative to the control though not to the tune of combined biofertilizer inoculation. The increase in seedling biomass production may be strongly correlated with increased accumulation of N due to *Rhizobium* and P due to VAM and phosphobacteria.

Tripartite inoculation of *Rhizobium*, phosphobacteria and VAM registered higher number of nodule production (69.9 per cent) than uninoculated control (Table 18.2). Dual inoculation with *Rhizobium* and *G. fasciculatum* improved the nodulation status in *L. leucocephala* compared to the single inoculation.

In the present experiment, the combined inoculation of *Rhizobium*, phosphobacteria and VAM registered higher VAM-colonization (65.6 per cent) than the uninoculated seedlings (56.1 per cent) (Table 18.2). Such an increase in root colonization due to VAM and *Rhizobium* inoculation was reported in *L. leucocephala*.

Table 18.2: Effect of Biofertilizers on Root Volume, Number of Nodules and Vesicular-arbuscular Mycorrhizae Colonization of Pungam Seedlings

Biofertilizer (B)	*Root Value (mm³) Months After (M) Inoculation*				*Number of Nodules Months After Inoculation*				*VAM Colonization % Months After Inoculation*			
	0	*2*	*4*	*Mean*	*0*	*2*	*4*	*Mean*	*0*	*2*	*4*	*Mean*
Rhizobium (R)	13.0	25.0	29.0	22.3	9.0	19.0	21.3	16.4	51.0 (45.6)	58.3 (49.8)	66.0 (54.3)	58.4 (50.0)
Phosphobacteria (P)	13.3	26.3	29.0	22.9	9.3	12.3	23.3	15.0	51.3 (45.8)	57.7 (49.4)	65.7 (54.1)	58.2 (49.8)
VAM	13.5	25.0	30.0	22.8	10.0	11.3	17.3	12.9	51.7 (45.9)	61.7 (51.8)	70.0 (56.8)	61.1 (51.5)
R + P	12.8	26.0	30.7	23.2	10.0	15.7	18.3	14.7	51.0 (4.56)	59.0 (50.2)	66.0 (54.3)	58.7 (50.0)
P + VAM	13.4	31.3	34.7	26.5	9.0	13.0	17.0	13.0	51.3 (45.8)	63.3 (52.7)	94.0 (57.4)	61.9 (52.0)
R + VAM	12.9	36.7	42.0	30.5	9.0	18.0	20.0	15.7	51.3 (45.8)	64.0 (53.1)	72.7 (58.5)	62.7 (62.7)
R + P + VAM	13.6	43.3	49.7	35.5	8.3	20.0	29.3	19.2	52.0 (46.1)	66.7 (54.7)	78.0 (62.0)	65.6 (54.3)
Control	13.3	21.7	27.0	20.7	9.0	10.3	14.7	11.3	51.3 (45.8)	55.0 (47.9)	62.0 (52.0)	56.1 (48.5)
Mean	13.2	29.4	34.0		9.2	14.9	20.2		51.4 (45.8)	60.7 (51.2)	69.0 (56.2)	

	SEd	CD (P = 0.05)	SEd	CD (P = 0.05)	SEd	CD (P = 0.05)
B	1.19	2.40	0.88	1.77	0.44	0.89
M	0.73	1.47	0.54	1.09	0.27	0.55
B × M	2.07	4.16	1.53	3.07	0.77	1.55

The relative growth rate (RGR) of *P. pinnata* seedlings inoculated with *Rhizobium*, phosphobacteria and VAM was higher (0.0071 $g.g^{-1}.day^{-1}$) compared to the Control seedlings (0.0045 $g.g^{-1}.day^{-1}$) (Table 18.3). Similarly, increase was also evident in crop growth rate (CGR), the quantum of increase being 141.2 per cent over the control. However, the net assimilation rate (NAR) was not significantly influenced due to the treatments. The CGR is a function of dry matter production hence, the increase in CGR and RGR in the present investigation might be due to the increased dry matter production in the seedlings inoculated with *Rhizubium* + phosphobacteria + VAM.

The results of the present study are in strong support of combined inoculation of the *P. pinnata* seedlings with *Rhizobium*, *phosphobavteria* and VAM. It is hence recommended that seedlings produced for large scale afforestation should be subjected to such an inoculation treatment to ensure better plant growth and survival.

Table 18.3: Effect of Biofertilizers Growth Rate, Crop Growth Rate and Net Assimilation Rate of Pungam Seedlings

Biofertilizer (B)	*RGR ($g.g^{-1}.day^{-1}$) Months After Inoculation (M)*			*CGR ($g.m^{-2}.day^{-1}$) Months After Inoculation*			*NAR ($g.m^{-2}.day^{-1}$) Months After Inoculation*		
	0–2	*2–4*	*Mean*	*0–2*	*2–4*	*Mean*	*0–2*	*2–4*	*Mean*
Rhizobium (R)	0.0080	0.0020	0.0050	0.307	0.144	0.225	3.60	1.48	2.54
Phosphobacteria (P)	0.0089	0.0017	0.0053	0.382	0.145	0.263	4.54	1.44	2.99
VAM	0.0080	0.0025	0.0052	0.309	0.203	0.256	3.56	1.47	2.52
R + P	0.0089	0.0018	0.0053	0.365	0.149	0.257	4.04	1.49	2.77
P + VAM	0.0083	0.0023	0.0053	0.341	0.179	0.260	3.85	1.43	2.64
R + VAM	0.0091	0.0020	0.0056	0.385	0.169	0.277	4.05	1.53	2.79
R + P + VAM	0.011	0.0028	0.0071	0.569	0.145	0.357	5.62	1.83	3.73
Control	0.0067	0.0024	0.0045	0.234	0.062	0.148	2.92	1.38	2.15
Mean	0.0086	0.0022		0.361	0.149		4.02	1.51	
	SEd	CD (P = 0.05)	SEd	CD (P = 0.05)	SEd	CD (P = 0.05)			
B	0.00049	0.00100	0.0287	0.0584	2.734	N8			
M	0.00024	0.00050	0.0143	0.0292	1.146	2.341			
B × M	0.00069	NS	0.0406	0.0826	1.596	3.257			

Chapter 19

Biofertilizer: A Supplementary Nutrient

Biofertilizer have an important role to play in improving nutrient supplies and their crop availability in the years to come. They are of environment friendly non-bulky and low cost agricultural inputs. A biofertilizer is an organic product containing a specific micro-organisms in concentrated form which is derived either from the plant roots or from the soil of root zone (Rhizosphere). High productivity in irrigated areas like sugarcane leads to higher nutrient turnover and a larger gap between nutrient supply and removal. Therefore, there is strong need to have complementary use of available source to plant nutrient including biofertilizer along with mineral fertilizers for maintenance of soil productivity.

The microorganisms responsible for fixing atmospheric nitrogen has been recognised since time immemorial, however, commercialization of their use is a relatively new innovation. In view of their apparent low-cost and risk-free application leading to yield advance both for current as well as for subsequent crop growth or residual fertility. Their large scale application in intensive farming call for systematic efforts. Their use especially important in view the fact that in the country to be self sufficient in the production of nitrogenous and phosphatic fertilizers. Among the biofertilizers *Azotobacter, Azospirillum, Acetobacter* are the important for nitrogen fixation, *Bacillus* sp. and *Aspergillus* sp. are important for phosphate solubilization and other soil mineral nutrients.

Experiment and Results

Azotobacter

This is a group of bacteria which are free living nitrogen fixer. The mechanism by which the plants, inoculated with *Azotobacter,* derive possible benefits in terms of increased grain, plant biomass and nitrogen uptake which in turn are attributed to small increase in nitrogen input from biological nitrogen fixation, development and branching of roots, production of plant growth hormones, vitamins, enhancement in uptake as NO_3, NH_4, H_2PO_4K and Fe, improved water status of the plant, increased nitrate reductase activity and antifungal compounds.

Table 19.1: Effect of *Azotobacter* Culture on Sugarcane Yield (T/ha)

Nitrogen Levels	Incoulated	Uninoculated	Mean
250 kg/ha	146.43	132.97	139.70
300 kg/ha	152.77	145.70	149.23
350 kg/ha	156.73	149.07	152.90
400 kg/ha (control)	154.67	149.98	152.32
Mean	152.65	144.43	

SE ± 1.83; C.D. (5 per cent) 5.52.

Fixation of nitrogen to the soil by use of *Azotobacter* culture is a natural phenomenon and sufficient research work has been carried out on the role of *Azotobacter* culture in sugarcane cultivation at Sugarcane Research Station, Padegaon. The results, in general, indicates that application of *Azotobacter* at the rate of 5 kg/ha helps in reducing nitrogen dose by 50 kg/ha with increase in cane yield by 5 to 10 per cent. However, it can help in reducing nitrogen dose even up to the tune of 100 kg/ha without loss in yield.

The *Azotobacter* plays an important role in improving cane germination, establishing of seedlings, survival of tillers, increase the population of millable cane and improve the soil fertility.

Azospirillum

It is an associative micro-aerophilic nitrogen fixer. It colonizes the root mass and fixes nitrogen in close association with plant. It fixes nitrogen in an environment of low oxygen tension. These bacteria induce the plant roots to secrete a mucilage which creates low oxygen environment and helps to fix atmospheric nitrogen. High nitrogen fixation capacity, low energy requirement and abundant establishment in the roots of sugarcane and tolerance to high soil temperature makes these most suitable for tropical conditions. Use of *Azospirillum* inoculum under saline-alkali conditions is also possible because their strains are known to maintain high nitrogen activities under such stress conditions.

The field experiment revealed that *Azospirillum* inoculation increase the cane yield up to 4.47 tonnes/hectare and could save 25 per cent nitrogen dose in medium black soil under Adsali planting.

Table 19.2: Effect of *Azospirillum* Culture on Sugracane Yield

Treatment	Yield (t/ha)	Available N in Soils (kg/ha)
100% N (control)	122.19	179.55
75% N + *Azospirillum*	126.44	169.95
50% N + *Azospirillum*	117.10	150.60

SE± 1.37; C.D. (5%) 4.10.

Phosphate Solubilizing Microorganisms

Phosphorus is the most important micro-nutrient required by sugarcane and micro-organisms for their normal growth and development. However, Indian soil test, generally low to medium in available phosphate and not mote than 30 per cent of applied phosphate is available to the current crop, remaining part gets converted into relatively unavailable forms.

Table 19.3: Effect of *Azotobacter* and *Azospirillum* Inoculation Under Graded Levels of Nitrogen on Yield and CCS of Surgacane (Average of 3 years)

Cultures	*Nitrogen Levels kg/ha*				
	175	*200*	*225*	*250*	*Mean*
Control	103.50	105.73	107.03	110.47	106.68
	(12.25)	(12.71)	(13.07)	(13.23)	(12.81)
Azotobacter	106.70	108.20	11.97	112.67	109.88
	(12.58)	(12.83)	(13.42)	(13.50)	(13.08)
Azospirillum	108.23	109.87	112.63	113.43	111.04
	(12.59)	(12.93)	(13.55)	(13.69)	(13.19)
AZT + AZP	108.53	110.73	114.33	115.00	112.14
	(12.59)	(13.11)	(13.85)	(13.95)	(13.37)
Mean	106.74	108.63	111.49	112.89	
	(12.50)	(12.89)	(13.47)	(13.59)	

AZT: *Azotobacter*, AZP: *Azospirillum*.

Figures in parantheses denote the CCS in MT/ha. (commercial cane sugar)

1. Yield	C.D. (P-0.05)	CCS C.D. (P-0.5)
2. Culture	0.32	0.032
3. N levels	1.78	0.102
4. Interaction	3.79	0.33

Many common heterotrophic bacteria like *Bacillus megaterium* and *B. circulans* and fungi like *Aspergillus awanori* and *Penicilium striata* possess the ability to bring sparingly soluble/insoluble in organic and organic phosphorus into soluble forms by secreting organic acids. These organic acids lower soil pH and in turn brings about dissolution of unavailable forms of soil phosphorus. Some of the hydroxy acids may chelate Ca, Al, Fe and Mg resulting in effective availability of soil and hence its higher utilization by sugarcane plants. The field experiments showed that the inoculation of phosphorus solubilizing cultures to sugarcane sets at the time of planting increased cane and sugar yield significantly over uninoculated control. It could be reduced the phosphate dose by 50 per cent and could be applied in the form of rock phosphate which is the cheaper source of phosphorus.

The other groups of micro-organisms like *Acetobacter dizotrophicus* a nitrogen fixer, *Thiobacillus thiooxidans* sulphur and iron oxidizers and Vasicular Arbuscular Mycorrhiza (VAM) a phosphate solubilizers are found to be effective but needs more experimentations at multilocations and soil types for confirmation and recommendation.

Table 19.4: Effect of Phosphate Solubilizing Culture on Cane and Sugar Yield

Treatment	*Cane Yield*	*CCS (T/ha)*
Control (N + K)	145.28	21.71
N + K_2O+ P.S. Culture	152.48	22.63
Rock phosphate + N + K_2O	164.07	25.65
Rock phosphate + N + K_2O + P.S. Culture	166.81	22.49
Super phosphate + N + K_2O + culture	194.81	28.67
Rock phosphate + 50% + N + K_2O	149.35	20.95
Rock phosphate 50% + N + K_2O + P.S. Culture	158.70	23.61
Superphosphate 50% + N + K_2O	171.48	24.29
Superphosphate 50% + N + K_2O + P.S. Culture	174.86	25.68
Rock-P + Super-P 50% + N + K_2O	182.79	26.00
Rock-P + Super-P 50% + N + K_2O + P.S. Culture	191.87	27.43

P.S. Culture: Phosphate Solubilizing Culture.

SE. ± 3.20.

C.D. 5 per cent: 9.40.

Chapter 20

Bioinoculations and Biofertilizer on Growth

India, due to increase in population, may face the problems of food. So contribution towards the agricultural productivity improvement is necessary. *Sorghum* is the staple food crop of Maharashtra State. The main principle of obtaining maximum yield from *Sorghum* hybrid is judicious use of chemical fertilizers which at present is proving a set back in increasing the yield as the chemical fertilizers are not only in short supply but also are expensive. Use of chemical fertilizers is limited to rich farmers and they show certain drawbacks like their manufacture depends on energy derived from fossil fuels which are getting depleted at faster rate. Many crops suffer from potassium deficiency because of excessive use of nitrogen based fertilizers, excessive potassium treatment decrease available nutritive food such as ascorbic acid etc., continuous use of ammonium fertilizers increase soil acidity and radioactivity, lastly chemical fertilizers have less effect on succeeding crops. On the other hand biofertilizers are advantageous over chemical fertilizers due to low cost, simple methodology of production, no any hazard to agroecosystem and considerable residual effect on succeeding crop. It is eported that microbial inoculants application gives equivalent output as derived from the application of 30 to 40 kg/ha. N. Biofertilizer makes a significant contribution towards the development of strategies for productivity improvement and may helps for economizing the production cost. Since long effect of nitrogen fixers, phosphate solublizers singly has been studied by many workers on many crops. But little work on combined effect of inoculant has been reported.

Keeping this view in mind an attempt has made to see the effect of bioinoculants on growth parameters of *Sorghum* using *Azotobacter, Azospirillum,* VAM and their different combinations. The same experiment is compared with recommended dose of fertilizer treatment and with no any treatment as a control.

Material and Methods

Pure cultures of *Azotobacter chrocooccum, Azospirillum brasilense* and VAM fungus, *Glomous fasiculatum,* were obtained from Sorghum Research Unit, Akola and ICRISAT, Hyderabad. Keeping all the recommended packages of practice the experiment was laid out in split plot design with 3 replication, using 24 treatment combinations. The main plot treatments were recommended full dose of fertilizer 80N : 40P : 40K kg/ha, three forth dose of fertilizer 60N : 30P : 30K and no fertilizer. While the subplot treatment were seed treatment with *Azotobacter, Azospirillum,* VAM, Nitrogen Fixers + VAM and no seed treatment.

Seed treatment was carried out using jaggery solution as an adhesive which was prepared by boiling 100 gms of jaggery in 1 litre of water. The seeds were treated in a pan by sprinkling the adhesive on seed cultures 10 gms of bioinoculant was use for treating 250 gms of seed. The seed material was eventually coated with cultures, seeds were treated two hours prior to sowing. Sowing was done by dibbling seeds. Germnation studies were carried out in pot cultures. 10 seeds were treated with bioinoculants singly and combination. Germination count was taken 10 days after sowing. Simultaneously germination studies in laboratory was also carried out using Rad dolls paper towel method 100 seeds were plated on two well moistened blotters (48 × 48 cms) and then covered with another well moistened blotters of the same size and then roll was prepared. For each treatment 100 seeds were tested. The rolls were then kept in upright position in a germination chamber at 30°C. To provide continuous moisture to the germinating seed, one end of the roll is kept in small amount of water. Germination count was recorded on 10th day after seeding. Simultaneously field trials was undertaken at various fertilizer levels and bioinoculant seed treatment. Observation regarding growth parameters like germination count, root and shoot length, biomass production and seedling vigour index were recorded by selecting 4 plants randomly at an interval of 15, 30, 45, 60 days. The data recorded for the growth parameters was analysed by standard statistical methods and all the treatments were tested for their significance at 5 per cent probability levels.

Results and Discussion

Effect of different treatments on germination count in pot cultures (Table 20.1) indicates that with increase dose of fertilizers germination percentage also increased but the increase was not significant, seed germination using single and composite culture inoculation in the pot studies revealed that maximum percentage of seed germination was observed when seeds were treated with *Azotobacter* + *Azospirillum,* followed by *Azotobacter+ Azospirillum* + VAM. Germination percentage with treatment VAM culture was better than in combined treatment with *Azotobacter* + VAM and *Azospirillum* + VAM and single inoculation of *Azotobacter* and *Azosporillum.* Seed germination percentage observed in paper towel test in laboratory conditions also showed similar effects those in pot studies. From the literature referred there are few reports on the effect of the combined culture inoculation especially on the germination in *Sorghum.* However, present studies revealed that seed inoculation with *Azotobacter* and *Azospirillum* alongwith VAM increased the seed germination of *Sorghum* compared to single culture inoculation. Seed bacterization with either *Azotobacter* or *Azospirillum* had improved the seed germination in *Sorghum* and the present results are in acquiescence with those reported by Lazarov *et al.*

Lazarov *et al.* in 1964, Mallikarjunaiah *et al.* in 1982 and Gaskins *et al.,* 1979 reported increase in root and shoot length of sorghum when treated with *Azotobacter* and *Azospirillum* respectively. From the literature referred no citation on root and shoot length in *Sorghum* with combined culture inoculation could be traced. As well as no reports with effect of VAM on root and shoot length in *Sorghum* in pot studies are available. In present studies it was observed that under laboratory conditions *Azotobacter* and *Azospirillum* seed treatment increases root and shoot length, *Azospirillum* + VAM had significant effect on root and shoot length and hence confirms the findings of above workers. Further it was also noticed that full dose of recommended fertilizers contributed in increasing root and shoot length in pot (Table 20.2).

Table 20.1: Effect of Different Treatments on Germination Count, Root Length and Shoot Length (cm) in Pot After 10 Days

No.	Particulars				Germination %		Root Length (cm)		Shoot Length (cm)		SVI Seeding Vigor Index
1.	Fertilizer Treatments										
	T_1–full dose of fertilizer				84.16		12.16		10.7		1940.72
	T_2–three-fourth dose of fertilizer				76.66		9.87		9.6		1492.57
	T_3–no dose of fertilizer				75.41		8.70		8.6		1304.59
	'F' test				NS		Sig		Sig		
	SE (M) ±				2.01		0.16		0.12		
	CD at 5%				–		0.66		0.49		
2.	Seed treatments										
	S_1–Seed treatment with *Azotobacter*				77.77		9.6		8.9		1438.74
	S_2–Seed treatment with *Azospirillum*				77.77		9.8		9.2		1477.63
	S_3–Seed treatment with VAM				78.88		10.0		9.3		1522.38
	S_4–Seed treatment with *Azotobacter* + *Azospirillum*				84.44		10.5		10.2		1747.90
	S_5–Seed treatment *Azotobacler* + VAM				75.55		10.4		10.1		1548.77
	S_6–Seed treatment *Azospirillum* + VAM				76.66		11.4		10.2		1655.85
	S_7–Seed treatment *Azotobacter* + *Azospirillum* + VAM				82.22		10.9		10.0		1718.39
	S_8–No seed treatment				76.66		9.1		8.3		1333.88
	'F' test				NS		Sig		Sig		
	SE (M) ±				3.81		0.18		0.41		
	CD. at 5%				–		0.51		1.17		
	Bar notation	T_1	T_2	T_3							
	Root length (cm)	12.36	9.87	8.70							
	Shoot length (cm)	10.7	9.6	8.6							
	Bar notation	S_6	S_7	S_4		S_5	S_3	S_2		S_1	S_8
	Root length	11.4	10.9	10.5		10.4	10	9.8		9.6	9.1
	Bar notation	S_6	S_4	S_5		S_7	S_3	S_2		S_1	S_8
	Shoot length	10.2	10.2	10.1		10	9.3	9.2		8.9	8.3

SE (M) ±: Standard error of mean; Cd: Critical difference; Sig: Significant.

Lazarov *et al.* reported more root and shoot length of *Sorghum* due to *Azotobacter* inoculation where as Umali reported increased root and shoot development by *Azospirillum* inoculation under field conditions, present results of mean root and shoot length was found to be maximum in even full dose of fertilizers where as minimum in the treatment T-3 control. In case of seed treatment it was found that treatment with nitrogen fixers + VAM shows maximum root and shoot length which was significantly superior over control. (Table 20.2) Hence, present studies confirms the findings of above worker. Result of pot studies regarding increase in root and shoot length were also confirmed under field conditions.

Table 20.1(a): Effect of Seed Treatment on *Sorghum* with Biofertilizer in Paper Towel Test

Particulars	*Germination %*
Azotobacter	80 per cent
Azospirillum	80 per cent
VAM	81 per cent
Azotobacter+ Azospirillum	87 per cent
Azotobacter+ VAM	80 per cent
Azospirillum + VAM	82 per cent
Azotobacter + *Azospirillum* + VAM	85 per cent
No seed treatment	78 per cent

Kapulnik *et al.* reported increase in root and shoot weight of *Sorghum* with *Azospirillum* treatment, Nagarjun reported that combined effect of VAM + *Azospirillum* increases biomass of *Sorghum* whereas Sarig noticed more dry matter produced with *Azospirillum* inoculation as compared to control. Pocovasky and when seeds were treated with *Azotobacter* + *Azospirillum,* followed by *Azotobacter* + *Azospirillum* + VAM. Germination percentage with treatment VAM culture was better than in combined treatment with *Azotobacter* + VAM and *Azospirillum* + VAM and single inoculation of *Azotobacter* and *Azospirillum.* Seed germination percentage observed in paper towel best in laboratory conditions also showed similar effects those in pot studies. From the literature referred there are few reports on the effect of the combined culture inoculation especially on the germination in *Sorghum.* However, present studies revealed that seed inoculation with *Azotobacter* and *Azospirillum* alongwith VAM increased the seed germination of *Sorghum* compared to single culture inoculation. Seed bacterization with either *Azotobacter* or *Azospirillum* had improved the seed germination in *Sorghum* and the present results are in acquiescence with those reported by Lazarov.

Lazarov reported increase in root and shoot length of sorghum when treated with *Azotobacter* and *Azospirillum* respectively. From the literature referred no citation on root and shoot length in *Sorghum* with combined culture inoculation could be traced. As well as no reports with effect of VAM on root and shoot length in *Sorghum* in pot studies are available. In present studies it was observed that under laboratory conditions *Azotobacter* and *Azospirillum* seed treatment increases root and shoot length, *Azospirillum* + VAM had significant effect on root and shoot length and hence confirms the findings of above workers. Further it was also noticed that full dose of recommended fertilizers contributed in increasing root and shoot length in pot (Table 20.2).

Lazarov *et al.* reported more root and shoot length of *Sorghum* due to *Azotobacter* inoculation where as Umali reported increased root and shoot development by *Azospirillum* inoculation under field conditions, present results of mean root and shoot length was found to be maximum in even full dose of fertilizers where as minimum in the treatment T-3 control. In case of seed treatment it was found that treatment with nitrogen fixers + VAM shows maximum root and shoot length which was significantly superior over control. (Table 20.2) Hence, present studies confirms the findings of above worker. Result of pot studies regarding increase in root and shoot length are also confirmed under field conditions.

Kapulnik *et al.* reported increase in root and shoot weight of *Sorghum* with *Azospirillum* treatment, Nagarjun reported that combined effect of VAM + *Azospirillum* increases biomass of *Sorghum* whereas Sarig noticed more dry matter produced with Azospirillum inoculation as compared to control. Pocovasky and Fuller reported that *Azospirillum* + VAM incombination results more dry matter in *Sorghum* as compared to uninoculated control.

Table 20.2: Effect of Different Treatments on Root and Shoot Length of *Sorghum* (CSH-14) in Field Trails

Sl.No.	Particulars	Mean Root and Shoot Length (cm) at Various Intervals in Days after Sowing									
		15		30		45		60		Mean	
1.	Fertilizer Treatments	RL	SL	RL	SL	RL	SL	RL	SL	RL	SL
	T_1–full dose of fertilizer	21.19	7.58	100.41	28.29	103.95	68.74	106.16	97.40	82.92	50.50
	T_2–three-forth dose of fertilizer	17.36	7.24	97.12	24.80	98.66	63.78	100.29	90.43	78.35	46.56
	T_3–no dose of fertilizer	15.47	6.39	90.41	22.11	92.29	60.11	97.08	83.05	73.81	42.92
	'F' test	Sig	Sig	Sig	Sig	Sig	Sig	Sig	Sig	Sig	Sig
	SE (M) ±	00.689	0.12	00.602	00.34	1.45	00.63	1.45	0.36	1.04	0.36
	CD at 5%	02.706	0.48	02.362	01.34	5.60	2.49	5.72	1.41	4.07	1.43
2.	Seed treatments										
	S_1 (*Azotobacter*)	17.52	7.33	94.11	24.76	96.66	61.26	98.55	91.07	76.71	46.10
	S_2 (*Azospirillum*)	18.20	6.96	95.11	26.15	95.66	64.07	101.11	90.26	77.52	46.86
	S_3 (VAM)	19.11	7.24	98.00	26.10	99.22	63.95	101.66	91.06	79.49	47.08
	S_4 ($S_1 + S_2$)	17.18	7.18	95.55	25.48	96.88	68.02	102.21	91.44	77.95	48.28
	S_5 ($S_1 + S_3$)	17.76	7.05	97.22	24.06	96.77	65.28	101.22	90.20	78.74	46.65
	S_6 ($S_2 + S_3$)	18.20	6.87	98.90	24.91	99.22	64.78	102.66	93.17	79.74	45.72
	S_7 ($S_1 + S_2 + S_3$)	19.40	7.41	101.44	26.24	105.11	69.51	106.44	94.24	83.09	49.35
	S_8 (no treatment)	16.62	6.50	89.33	22.84	95.55	56.80	96.00	81.88	74.37	42.01
	'F' test	Sig	NS	Sig	Sig	Sig	Sig	Sig	Sig	Sig	Sig
	SE (M) ±	0.576	0.21	1.36	0.66	2.01	0.96	1.35	1.02	1.32	0.71
	CD at 5%	1.644	–	3.90	1.88	5.71	2.76	3.85	2.90	3.75	2.51

Bar notation	T_1	T_2	T_3					
Root length (cm)	82.92	78.35	73.81					
Shoot length (cm)	50.50	46.56	42.92					
Bar notation	S_6	S_7	S_4	S_5	S_3	S_2	S_1	S_8
Root length	83.09	79.74	79.49	78.74	77.95	71.52	76.71	74.37
Bar notation	S_6	S_4	S_5	S_7	S_3	S_1	S_1	
Shoot length	49.35	48.28	47.08	46.86	46.65	46.10	45.72	42.35

SE (M) ±: Standard error of mean; Cd: Critical difference; Sig: Significant.

Biomass production in *Sorghum* as affected by different treatments in field (Table 20.3) clearly indicates that increase in the level of fertilizers biomass of the *Sorghum* also increases linearly. Maximum green weight was recorded at full dose of fertilizer followed by ¾ dose and was significantly superior over control. Treatment with *Azotobacter* + *Azospirillum* + VAM gave significantly highest biomass over all the treatments followed by *Azotobacter* + *Azospirillum* and *Azotobacter* + VAM. Lowest fresh biomass was recorded in control treatment.

Table 20.3: Effect of Different Treatments on Biomass of *Sorghum* in gm at Various Intervals

Sl.No.	Particulars	Biomass of Plant in Days									
		15		30		45		60		Mean	
1.	Fertilizer treatments	F	D	F	D	F	D	F	D	F	D
	T_1 full dose of fertilizer	5.2	1.61	129.80	26.32	247.16	64.45	461.45	122	210.90	53.60
	T_2 three fourth dose of fertilizer	3.4	1.42	124.00	22.81	215.00	58.95	392.58	113.16	183.74	49.09
	T_3 no dose of fertilizer	2.8	1.38	101.50	20.22	179.91	52.37	333.16	104.37	154.34	44.59
	'F' test	Sig	Sig	Sig	Sig	Sig	Sig	Sig	Sig	Sig	Sig
	SE (M) ±	0.137	0.07	3.60	0.57	6.61	0.59	7.3	0.85	4.41	0.52
	CD at 5%	0.53	0.61	14.13	2.23	25.96	2.32	28.70	3.30	17.33	2.12
2.	Seed treatments										
	S_1 (*Azotobacter*)	3.8	1.36	116.1	22.80	219.22	55.77	339	108.77	169.95	47.17
	S_2 (*Azospirillum*)	3.8	1.34	122.3	20.85	222.55	58.22	391.5	108.88	185.00	47.32
	S_3 (VAM)	4.0	1.38	21.4	23.80	239.77	61.00	411.4	115.66	194.14	50.46
	S_4 ($S_1 + S_2$)	3.9	1.63	121.2	23.52	233.88	58.33	426.11	121.22	196.27	51.18
	S_5 ($S_1 + S_3$)	3.7	1.65	111.2	23.13	212.55	55.22	414.00	115.00	185.3	48.75
	S_6 ($S_1 + S_3$)	3.8	1.70	122.00	22.30	232.00	64.22	424.40	118.88	195.50	51.78
	S_7 ($S_1 + S_2 + S_3$)	4.3	1.80	129.30	26.18	241.11	66.77	489.11	137.77	215.95	51.78
	S_8 (no treatment)	3.3	0.90	104.00	19.30	183.11	50.22	270.22	95.11	140.15	41.38
	'F' test	Sig	Sig	Sig	Sig	Sig	Sig	Sig	Sig	Sig	Sig
	SE (M) ±	0.25	0.09	3.90	0.83	7.6	0.92	12.58	1.86	6.08	0.925
	CD at 5 per cent	0.72	0.28	11.13	2.37	21.84	2.62	35.90	5.33	17.39	2.65
	Bar notation	T_1	T_2	T_3							
		210.90	183.74	154.34							
	Bar notation (D)	53.60	49.09	44.59							
	Bar notation (F)	S_7	S_4	S_6	S_3	S_5	S_2	S_1	S_6		
		215.95	196.14	195.50	194.14	185.3	185.00	169.95	140.15		
	Bar notation (D)	S_7	S_6	S_4	S_3	S_5	S_2	S_1	S_8		
		58.13	51.78	51.18	50.46	48.75	47.32	47.17	41.38		

F: Fresh biomass; D: Dry Biomass; SE (M) ±: Standard error of mean; CD: Critical Difference; Sig: Significant.

Similar trend was noticed with dry matter production as in fresh biomass with various treatments.

From the studies, it was concluded that *Azospirillum* was more effective as compared to *Azotobacter* specially in some characters like root and shoot length and biomass production etc. The results are encouraging in most parameters studied with combined application of *Azotobacter* + *Azospirillum* + VAM followed by other combinations of biofertilizers over single culture inoculation.

Chapter 21

Significance and *Azospirillum brassilense* and *Pseudomonas* on Growth

Under subtropical climatic condition with low N and P in alfisol will never be provided the effective crop productivity and its establishment. To achieve with sustainable crop productivity the use of certain beneficial microbes can be considered. Biofertilizer have recently gained with momentum for effecting the substantial increase in crop yield under various ago climatic condition. Role of Diazotrophs on the crop yield was documented. To mobilize the impounded P in alfisol the use of phosphobacteria is useful. The present investigation has been taken up to assess the significance of nitrogen fixer and P mobilizes on the improvement of growth and yield of finger millet in alfisol.

Material and Methods

A field experiment was conducted during Kharif 1995 season under alfisol (pH 5.6; EC: 0.34 mmh/cm^2) to record the performance of two cultivars of fingermillet (*Elucine crocana*) *viz.*, CO-11 and CO-12 with bioinoculation of nitrogen fixing *Azospirillum brasilense* and P solubilising *Pseudomonas striata* and were inoculated alone and in combination. These inoculants were applied as seed, seedling and soil application at the rate of 2, 3 and 10 packets in which each packet consisting of 200 g of active inoculants at 10^7 cells/g in a peat based carrier system. To compare the efficacy of these microbial inoculants N and P controls were also maintained. Each treatment was replicated five times and conducted in RED. Seedlings were transplanted at 20×10 cm spacing in 3×5 m^2 plots. During vegetative phase of the crop vigour index (20 days after planting), plant biomass, root and shoot growth at 30 and 45 days and tillers initiation at 45 days were recorded at harvest the yield attributes *viz.*, finger counts earhead length, 1000 grain weight, haulms and grain yield were recorded and harvest index was calibrated.

Results and Discussion

The results on the effect of nitrogen fixer and P mobilizes on the growth and yield attributes of two cultivars of finger millet (CO-11 and CO-12) in alfisol is presented in Tables 21.1 to 21.4.

Table 21.1: Influence of *Azospirillum* and *Pseudomonas* on Growth Attributes of CO-11 Ragi

Treatments	At 20 Days Vigour Index	At 30 and 45 Days After Sowing						
		Plant Biomass (g/Plant)		Root Growth (cm/plant)		Shoot Growth (cm/plant)		Tillers Production (number/plant)
		30	45	30	45	30	45	45
Control	585	2.61	4.02	8.1	12.3	15.1	32.2	2.9
N-control	790	4.92	7.39	13.2	17.2	26.1	48.2	4.3
P-control	620	3.72	5.93	11.2	15.2	22.4	42.3	3.8
Azospirillum	720	4.26	6.12	14.5	19.6	30.7	52.1	4.6
Pseudomonas	660	3.98	5.87	12.4	16.4	25.1	48.3	3.9
Azos. + Pseudo.	740	5.23	8.41	16.6	21.6	33.7	54.6	4.8
CD (P = 0.05)		0.32	0.61	0.08	1.2	3.3	4.1	NS

Table 21.2: Influence of *Azospirillum* and *Pseudomonas* in Yield Attributes of CO-11 Ragi

Treatments	Finger (no./plant)	Ear Head Length (cm/plant)	1000 Grain (g)	Haulms Yield (t/ha)	Grain Yield (t/ha)	Per cent Increase Over Control	Harvest Index
Control	4.0	5.7	2.3	5.214	3.160	–	60.6
N-control	6.6	7.9	2.7	5.880	3.765	18.6	63.2
P-control	6.0	6.9	2.6	5.440	3.634	15.0	66.8
Azospirillum	7.2	7.5	2.7	5.885	4.075	29.0	69.5
Pseudomonas	6.6	6.9	2.6	5.695	3.730	18.0	66.0
Azos. + Pseudo.	7.3	7.9	2.9	6.290	4.315	36.6	68.5
CD (P = 0.05)	NS	0.23	NS	0.34	0.21		

Growth Attributes

Under alfisol bioinoculation of nitrogen fixing *Azospirillum* and P mobilising *Pseudomonas* as seed, seedling and soil broadcasting effected enhanced establishment and vigour indexes in two cultivars of finger millet. Similarly the various growth biometrics *viz.*, plant biomass, root and shoot growth at 30 and 60 days after transplanting in main field was found significantly increased over untreated control. The effect being registered with the dual inoculation of both the bio inoculants which might be due to the provision of nitrogen and growth promoting substances (IAA, GA) by *Azospirillum* and the possible solubilisation of fixed P (as alumina and iron phosphate) by *Pseudomonas* created with sustainable growth of the crop in alfisol. Apparao reported seeding of *Bacillus subtilis* found to increases the yield of pigeonpea. Similar to the present investigation, Upadhyaya *et al.* evinced the role of nitrogen fixing microbes in the rhizosphere of finger millet found to record high nitrogenase activity and it was altered by the varieties, age and nature of the soil. Gopal critically reviewed the role of biofertilizers on the sustainable crop productivity.

Table 21.3: Influence of *Azospirillum* and *Psudomonas* on Growth Attributes of CO-12 Ragi

Treatments	At 20 Days Vigour Index	At 30 and 45 Days After Sowing						
		Plant Biomass (g/Plant)		Root Growth (cm/plant)		Shoot Growth (cm/plant)		Tillers Production (number/plant)
		30	45	30	45	30	45	45
Control	560	3.01	5.11	7.3	1.9	16.2	31.1	3.6
N-Control	820	4.32	9.36	11.2	16.3	23.2	43.1	6.2
P-Control	670	4.29	7.93	10.1	13.8	22.1	40.2	4.6
Azospirillum	790	4.96	7.20	12.1	16.1	28.2	46.1	5.8
Pseudomonas	660	4.72	7.21	10.2	17.3	23.6	42.2	4.8
Azos. + Pseudo.	820	6.21	9.95	15.2	21.4	31.4	32.6	4.7
CD (P = 0.05)		0.31	0.43	1.1	1.7	2.3	2.9	NS

Yields Attributes

At harvest the co-inoculation of *Azospirillum* and *Pseudomonas* significantly enhanced the straw yield and grain yield in two cultivars of finger millet with reference to grain productivity bioinoculants cumulatively recorded 38.0 and 36.6 per cent increased yield over their respective control. Kundu reported the yield of wheat was increased due to the inoculation of *Azotobacter* and phosphobacteria. Saxena critically analysed the role of biofertilizers on the crop productivity.

Table 21.4: Influence of *Azospirillum* and *Pseudomonas* on Yield Attributes of CO-12 Ragi

Treatments	Finger Length (cm/plant)	Ear Head Weight (g/plant)	Haulms Yield (t/ha)	Grain Yield (t/ha)	Per cent Increase Over Control	1000 Grain (g)	Harvest Index
Control	4.0	6.1	5.31	3.26	–	2.2	60.7
N-Control	6.6	8.0	5.90	3.77	16.8	2.8	63.2
P-Control	6.0	6.9	5.43	3.64	15.0	2.6	66.8
Azospirillum	7.0	7.2	5.86	4.08	29.0	2.7	69.6
Pseudomonas	6.6	6.9	5.69	3.73	18.0	2.6	66.1
Azn. + Pseudo.	7.4	7.9	6.29	4.32	36.6	2.9	68.6
CD (P = 0.05)	NS	0.12	0.3	NS			

Among the two cultivars of finger millet CO-12 performed better than CO-11 in alfisol tilak and Singh reported the response of pearl millet to be inoculation of P mobilizes and nitrogen fixer. Prabakaran *et al.,* also indicated the significance of *Azospirillum* and P solubiliser on growth and establishment of *Tectona grandis* in alfisol.

Conclusion

A field study was conducted to assess the significance of nitrogen fixing *Azospirillum brassilense* and P mobilizing *Pseudomonas striata* on the growth and yield of two cultivars of finger millet (*Elusine crocana*) in alfisol. Both the inoculants were applied at rate of 15 packets (2 for seed treatment; 3 packet for seedling dipping and 10 packets for soil application). These inoculants tried at as a single and dual inoculation and comprised with N and P controls. The results revealed that the single inoculation of *Azospirillum* and *Pseudomonas* recorded 34 and 21 per cent increased grain yield in CO-11 ragi whereas their combination yielded 38.9 percent over control. In CO-12 cultivar the single inoculation had effected 27 and 18 per cent increased yield over their control and in combination the effect was at 36.6 per cent.

The possible synergistic effect of enhancing the biogrowth as well as the yield productivity in finger millet by the seed, seedling and soil inoculation of nitrogen fixing *Azospirillum* brasilense and P-solublising *Pseudomonas sriata* in alfisol might be due to the following reasons:

1. Inherent production of auxins by both microbes (IAA, GA) will bring out with better germinability of seedling and fast establishment after transplantation in the main field can bring out required no of plant population and its better stand in alfisol.
2. By the associative symbiosis the *Azospirillum* can able to supply the need based nitrogen for the crop during its various growth phases will bring out with better plant growth and yield enhancement.
3. The possible role of *Pseudomonas* in solublising impounded P (iron and alumina P) to available form by ways of organic acid production (acetic formic, propionic acids) oxidation of Sulpher to sulphur dioxide and the convert as sulphuric acid might solubilise the fixed P in soil or by means of chelation effect the P mobility can be improved for crop used.

Over all influences by these two microbes played a vital role in supplying N and P to the finger millet and found enhanced the growth yield over the untreated control.

Chapter 22

Application of *Mycorrhizae* and *Rhizobium* on Biomass Production

The depletion of forest cover leading to the scarcity of fuelwood and animal coupled with increasing wasteland formation is a major concern of the developing nations. Leguminous trees and shrubs have been suggested as a renewable source of fuel and wood products and are known to have symbiotic association with VAM Fungi and *Rhizobium.* VAM Fungi which constitute a group of important soil microorganisms are ubiquitous throughout the world are known to improve the plant growth through better uptake of nutrients. They also improve the activity of N fixing organisms in the root zone. VAM Fungi can increase the drought resistance. Their extra-radical hyphae can influence rhizosphere architecture and improve host water dynamics. Shortage in the availability of the woody biomass do not only to cover exploitation but also to poor productivity as a result of growth limiting factors in the environment. VAM Fungi have been reported to occur naturally to Saline/Alkaline environments and have been linked with increased plant biomass and development in saline soils. Application of biofertilizers as VAM Fungi and *Rhizobium* can facilitate higher establishment and increase biomass of trees species in saline soil sites. The present chapter deals with the impact of interaction between VAM Fungi and *Rhizobium* on the biomass of *Acacia nilotica* (Highly promising multipurpose tree species) in saline and forest soils during different phases of plant growth.

Material and Methods

The pots (40 × 30 × 30 cm) filled with thoroughly sieved sterilized saline and forest soil from Muzaffarpur and forest range Bettiah respectively. The saline soil was amended with forest soil in 1 : 1, 2 : 1 and 3 : 1 ratios. Surface sterilize seeds of *Acacia nilotica* were soaked in water for 48 hours at room temperature to break the dormancy and also to soften hard seed coat. Seed were sown about I. 5 cm deep in the soil and regularly watered.

Rhizobium inoculum (3×10^9 cells/seed) was mixed in a cooled solution of sugar and arabic gum. Such mixing forms a slurry to which the experimental seeds were added. Inoculated seeds were placed on the paper and dried in shade. These seeds were sown to demonstrate the impact of *Rhizobium* in relation to production of biomass yield. Inoculum of *Glomus fasciculatum* (Thaxter sensum Geredemann) were obtained from the Bhartiya Agro Industry Foundation, Pune. Inoculation (300 VAM Spores/seed) was done by placing the inoculum 2 cm below the seed in pot filled with soil. The experiment was laid out as completely randomised designed and comprised of 20 treatments. They were provided normal condition for growth. 5 replicates were taken for each set of experiments. The plants were harvested after third, sixth and ninth month of growth. Spores were isolated by wet sieving and decanting method. Root infection was evoluted by a microscopic examination of cleared and stained root samples following Phillips and Haymna. The percentage of root infection was recorded by the Grid line intersection method of Giovanetti. Data on physical growth parameters shoot dry weight, number of leaves, number of nodules and survival percentage was recorded for each plant and statistically analysed by ANOVA.

Results and Discussion

The results of the study are presented in Tables 22.1 and 22.2. The plant biomass, *i.e.* shoot height, root length, collar diameter, shoot and root dry weight, number of leaves and number of nodules increased with the increase of plant age in different treated plants of *Acacia nilotica.* However, the increase was more prominent in VAM and *Rhizobium* inoculated plant. Inoculation of saline soil with VAM fungi significantly increase plant height. The maximum plant height (226.5 cm) with on increase of 30.5 per cent over control was observed in plant growing in saline soil mixed with forest soil (1 : 1) and 24.5 per cent increase in case of forest soil. Inoculation with *Rhizobium* did not significantly improve growth in any soil treatment in respect of VAM treatment.

Table 22.1: Effect of VAM and *Rhizobium* on Growth Response, Nodulation and Biomass Production of *Acacia nilotica* after 9 Months

Treatments	Shoot Height (cm)	Root Length (cm)	Collar Diameter (cm)	Shoot Dry Wt. (g)	Root Dry Wt. (g)	No. of Leaves Per Plant	No. of Nodules Per Plant
T_1	130.5	15.2	12.5	29.96	4.5	220	5
T_2	182.0	16.2	12.7	30.50	4.6	226	16
T_3	173.5	19.9	12.9	31.60	4.4	326	10
T_4	139.6	17.3	12.7	30.41	4.7	270	7
T_5	142.3	17.6	12.8	30.49	4.4	280	20
T_6	149.4	18.9	12.8	31.21	4.1	390	22
T_7	150.4	19.4	12.9	31.49	4.2	440	18
T_8	162.6	18.9	31.1	32.00	4.1	445	24
T_9	161.6	19.6	13.6	32.09	4.6	490	25
T_{10}	162.8	19.9	31.7	32.99	4.7	501	21
T_{11}	165.9	20.5	13.8	33.00	4.8	530	29
T_{12}	174.6	22.4	14.9	33.15	5.1	930	23
T_{13}	195.8	23.3	15.1	34.10	5.6	1060	31
T_{14}	226.5	24.0	16.2	35.40	7.7	1155	65
T_{15}	171.3	21.3	14.3	40.60	7.9	835	49
T_{16}	193.2	22.1	14.5	33.01	8.5	890	34

Contd...

Table 22.1–Contd...

Treatments	*Shoot Height (cm)*	*Root Length (cm)*	*Collar Diameter (cm)*	*Shoot Dry Wt. (g)*	*Root Dry Wt. (g)*	*No. of Leaves Per Plant*	*No. of Nodules Per Plant*
T_{17}	204.1	23.1	14.6	34.11	6.9	970	50
T_{18}	175.2	21.4	14.4	34.91	7.0	844	51
T_{19}	195.5	22.4	14.6	33.90	7.5	895	41
T_{20}	206.4	23.1	14.7	34.35	7.9	901	52

T_1: Saline soil; T_2: Forest soil; T_3: Saline soil + forest soil (1 : 1); T_3: Saline soil + forest soil (2 : 1); T_5: Saline soil + forest soil (3 : 1); T_6: Saline soil + *Rhizobium;* T_7: Saline soil + VAM; T_8: Saline soil + *Rhizobium* + VAM; T_9: Forest soil + *Rhizobium;* T_{10}: Forest soil + VAM; T_{11}: Forest soil + *Rhizobium* + VAM; T_{12}: Saline soil + forest soil (1 : 1) + *Rhizobium;* T_{13}: Saline soil + forest soil (1 : 1) + VAM; T_{14}: Saline soil + forest soil (1 : 1) + *Rhizobium* + VAM; T_{15}: Saline soil + forest soil (2 : 1) + *Rhizobium;* T_{16}: Saline soil + forest soil (2 : 1) + VAM; T_{17}: Saline soil + forest soil (2 : 1) + *Rhizobium* + VAM; T_{18}: Saline soil + forest soil (3 : 1) + *Rhizobium;* T_{19}: Saline soil + forest soil (3 : 1) + VAM; T_{20}: Saline soil + forest soil (3 : 1) + *Rhizobium* + VAM.

Table 22.2: Effect of VAM and *Rhizobium* on Percentage VAM Colonization, Spore Population and Phosphorus Content of *Acacia nilotica* after Nine Months of Plant Growth

Treatments	*Root Infection (%)*	*VAM Spores per 100 g Soil*	*P Content in Shoot (%)*	*Seedling Viability*
T_1	–	–	0.03	30.7
T_2	13.2	65	0.04	95.0
T_3	12.6	61	0.05	65.4
T_4	14.0	76	0.06	63.3
T_5	15.2	86	0.05	64.4
T_6	–	–	0.04	35.5
T_7	39.1	110	0.05	35.4
T_8	46.3	115	0.07	46.5
T_9	–	–	0.08	95.1
T_{10}	56.4	156	0.07	94.1
T_{11}	65.4	186	0.08	95.4
T_{12}	–	–	0.08	66.3
T_{13}	66.4	196	0.09	65.4
T_{14}	85.6	210	0.11	66.9
T_{15}	29.1	76	0.07	96.7
T_{16}	64.1	176	0.08	63.4
T_{17}	69.4	186	0.09	64.2
T_{18}	28.4	105	0.06	65.1
T_{19}	65.3	176	0.08	63.6
T_{20}	71.2	187	0.09	64.3

Treatments T_1 to T_{20} as per Table 22.1.

Maximum dry matter yield was observed in plant growing in saline soil and mixed with forest soil (1 : 1) and inoculated with VAM Fungi in different Plant growth *i.e.* 3 months, 6 months and 9 months.

Collar diameter of plants growing in saline soil amended with forest soil (1 : 1) and colonised with VAM fungi and *Rhizobium* was highest.

The maximum number of leaves (13-1155) was observed in plants grown in saline soil mixed with forest soil (1 : 1) and inoculated with VAM fungi and *Rhizobium* and minimum in control (8-226). Maximum number of nodules were observed in plants growth in forest soil treated with VAM fungi with an increase of 95.4 per cent over control. Similar trend was also recorded in all soil groups treated with either VAM or with VAM fungi and *Rhizobium.* In general plants grown in forest soil had more number of nodules than that of those grown in salinc soil. Maximum number of VAM spore in soil and percentage infection in root were recorded in saline and forest soil (1 : 1) treated with VAM and *Rhizobium.*

The pH value varied from 7.5 to 9.5 in different combination but maximum biomass of plant and VAM spore were recorded in pH 8.5 amended with saline soil and forest soil (1 : 1) treated with VAM and *Rhizobium.*

Seedlings survivability percentage was 95 per cent in forest soil and 30.7 per cent in saline soil. Inoculation of soil with VAM fungi and *Rhizobium* individually improved seedling survival percentage in saline soil than that of either forest soil or saline soil amended with different proportion of forest soil.

This study indicate that none of without VAM seedling grew as fast as those with VAM fungi in saline soil. The impact of VAM fungi on biomass of *A. nilotica* was positive but this trend was not noticed with *Rhizobium* alone or with dual inoculation. This could perhaps be due to incompatibility by *Rhizobium* strain with this plant species or its ineffectiveness in soils. The biomass of plant was significantly more in combination of VAM and *Rhizobium* as compare to VAM and *Rhizobium* alone and also with control treatment.

In other studies also conducted by authors, VAMF and Rhizobium have been found effective in plant establishment, nutrient uptake, leaf area, photosynthesis efficiency and biomass in drought effected plants. However, there was a tendency towards drought evidence rather than drought tolerance. Soil salinity may influence the growth and activity of VAM fungi and *Rhizobium* through several mechanisms, either discceetly or interactively. The exact mechanism by which VAM fungi tolerate unbalanced, ionic conditions as in salty soils is not known. Since it has not so far been possible to maintain VAM fungi in auxanic culture, it is extremely difficult to distinguish between direct and plant mediated effects on their biomass. It may be concluded that a suitable combination of VAM fungi and *Rhizobium* may be useful in ensuring higher biomass produced establishment of *A. nilotica* in saline soil.

Chapter 23
Effect of VAM Fungi on Banana Plants

Banana is one of the important commercial fruit crops which is commonly grown through out the humid tropics and subtropics. The potential value of plant tissue culture (PTC) technology is being commercially exploited by various organizations all over the world. This technique provides an existing tool to multiply banana plants vegetatively in large numbers with in a short period of time.

The potential of arbuscular mycorrhizal fungi and phosphate solubilizing microorganisms as biofertilizers and bioprotectors to enhance micropropagated plantlets is recognized, but not well exploited with because of the current agronomic practices with their implications for environment.

In the present study an attempt has been made to discuss the response of micropropagated banana plants to dual inoculation with PSB and VAM.

Material and Methods

The experiments were conducted in the department of Botany, Bangalore University Bangalore. The methodology followed for the experiments is detailed below.

Phosphate Solubilizing Microorganism

Bacillus megatherium was procured from the Department of Agricultural Microbiology U.A.S. Bangalore.

Mycorrhizal Inoculum

Glomus mosseae and *Glomus fasciculatum* inoculum were procured from U.A.S. Bangalore.

Plant Material

Micropropagated banana plantlets variety Nanjanagudu Rasbale were procured from Tissue Culture Laboratory, Lalbagh, Bangalore.

Phosphate solubilizing microorganism and vesicular arbuscular mycorrhizal fungi were added to 4 weeks old micro propagated banana plants. In addition to 4 week (acclimatization stage) a further 13 and 16 weeks of growth in green house was allowed before harvest.

Treatment

T_1: Un-inoculated control.

T_2: Phosphate solubilizing bacteria, (PSB), (*Bacillus megatherium*) (*B.m*)

T_3: *Glomus mosseae* alone (*G.m*).

T_4: *Glomus fasciculatum* alone (*G.f*).

T_5: *G.m* + PSB.

T_6: *G.f*+ PSB.

The following observations were made.

Mycorrhizal Percent Infection

The percent infection of banana plant roots were estimated according to Philips and Hayman.

Estimation of Mycorrhizal Spores

Extra matrical chlamydospores produced by the mycorrhizal fungus was estimated by wet sieving and decanting methods outlined by Gerdman and Nicolson.

Plant Growth Parameters

Root length and shoot length were recorded for each treatment.

P-Content

Phosphorus content was estimated by Vanadomolybdate phosphoric yellow colour method.

Ca and Mg Content

Ca and Mg contents were estimated by the Versanate titration method.

Na and K Content

Na and K contents were estimated by flame photometric method.

Results and Discussion

Banana plant growth, biochemical composition and nutrient uptake of mineral elements were greatly improved by simultaneous inoculation with VAM and PSB.

The development variation found in inoculated and uninoculated plants resulted in significant increase in shoot length and root length.

Combined inoculation with *G. mosseae* and *B. megatherium* substantially increased the survival of banana plants (*Musa paradisiaca* L.). The bacterium *Bacillus megatherium* in combination with *G. mosseae* gave a significant increase in plant height and highest frequency of root infection. Dual inoculation considerably increased mineral elements than single inoculated plants (Table 23.1 and Figures 23.1 and 23.2).

Dual inoculation with *G. mosseae* and *B. megatherium* resulted in significant increase in Ca concentration than *Glomus fasciculatum* treated with *B. megatherium* alone.

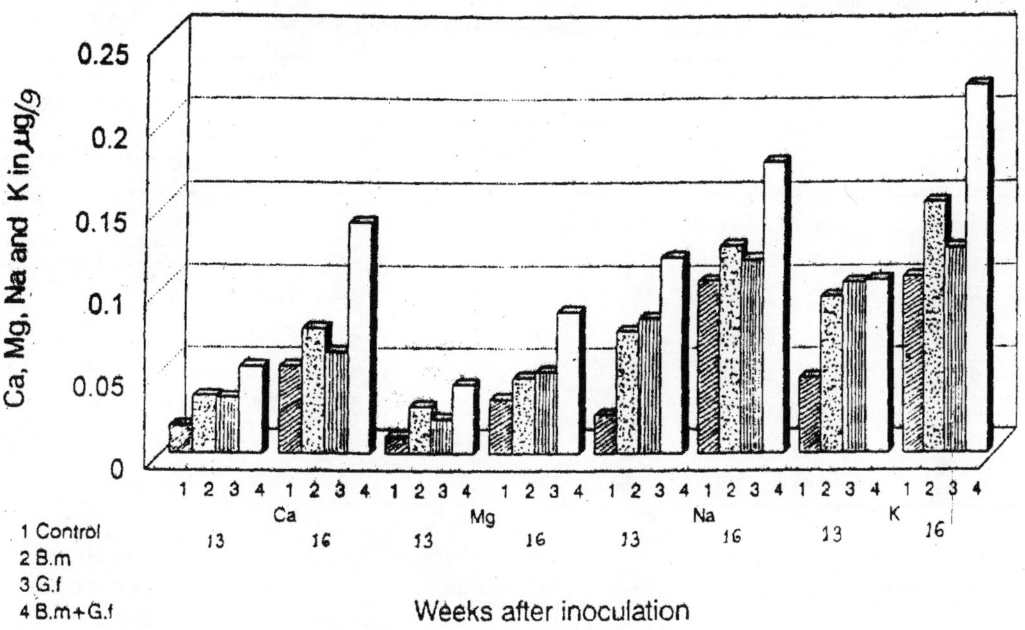

Figure 23.1: Effect of Dual Inoculation on Mineral Nutrients (G.f. + B.m.)

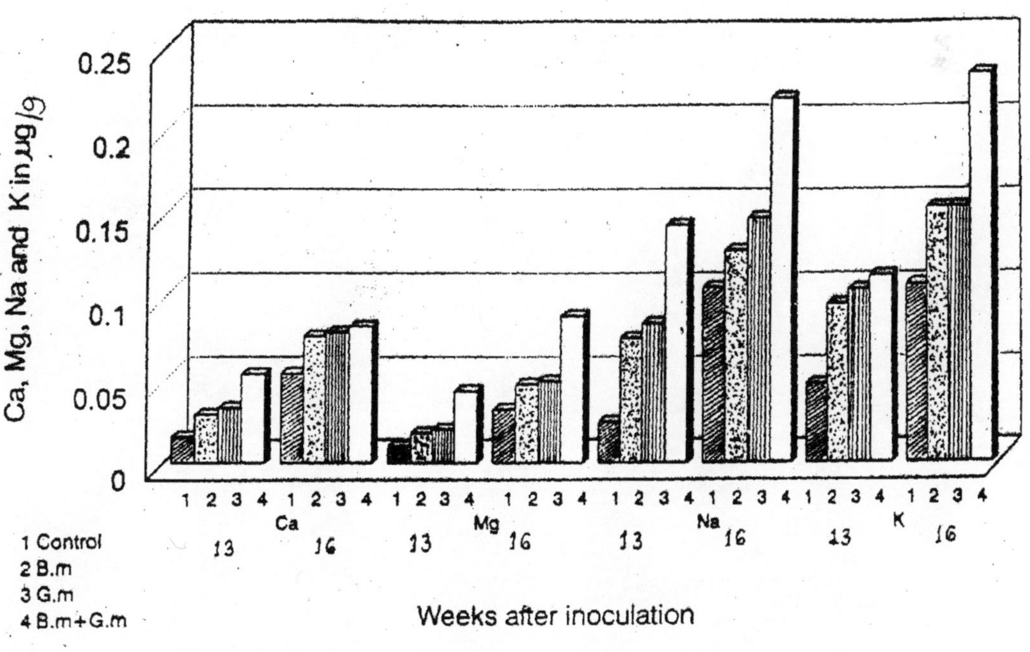

Figure 23.2: Effect of Dual Inoculation on Mineral Nutrients (G.m. + B.m.)

Table 23.1: Response of Micropropagated Banana Plants (*Musa paradisiacal* L.) to Dual Inoculation with PSB and VAM

Treatment	*% Infection*		*Spores/100 ml*		*Shoot Length (cm)*		*Root Length (cm)*		*Total-P (μg/g)*	
	Weeks After Incubation									
	13	*16*	*13*	*16*	*13*	*16*	*13*	*16*	*13*	*16*
Uninoculated control	30	45	69	112	10.23	11.43	7.31	9.58	7	10
B. megatherium alone	45	54	128	180	12.45	16.12	9.68	12.58	13	21
G. fasciculatum alone	64	81	212	228	11.89	15.31	7.69	11.81	12	22.5
B.m + G. f.	73	91	224	348	13.28	16.28	9.25	12.92	15	26
Uninoculated control	36	40	68	109	10.31	11.38	7.1	9.72	8	10
B. megatherium alone	40	50	132	179	12.68	16.00	9.82	12.70	12	21
G. mosseae alone	64	91	262	336	12.47	15.23	9.01	11.38	16	25
B.m. + G.m.	81	100	268	338	13.40	16.40	9.93	13.34	18	27.5

The response due to mixed culture inoculation was more than single inoculation showing the synergistic effect of the two types of organisms. The response may be due to supply of two major plant nutrients and other factors. The results indicated that the mixed culture inoculation can be safely used for better growth and survival of micropropagated banana plants in green house.

The present research showed that the mycorrhizal fungi colonizing the micropropagated banana plants showed differences in growth rate. This is in confirmation with results respond on seven banana cultivars.

The exact mechanism by which VAM and PSB contribute to enhanced growth and survival of micropropagated plants were not fully understood.

Chapter 24
Mungbean with Solubilizing Bacteria

The introduction of high yielding varieties, better irrigation potential, plant protection measures and judicious use of chemical fertilizer have contributed to a break through in Agriculture. But the fast depleting fossil fuel resources, escalating cost of chemical fertilizers and environmental hazards caused by their use are some of the constraints for better crop yield. These constraints have forced researchers to think of better and economic alternatives. The use of microbes, have however, drawn the attention which are not only economic source of fertilizers, yet increase the availability of nutrients to the plants.

Among the major plant nutrients, phosphorus (P) is next to nitrogen required for better crop yields. It is estimated that 98 per cent of Indian soil contain insufficient amounts of available phosphate to support maximum plant growth. Application of phosphatic fertilizers is therefore essential for optimum crop yield. But the main problem concerning phosphatic fertilizer is its fixation with soil complexes within a very short period of application rendering more than two-thirds unavailable. Thus, the non-availability of soluble forms of phosphate limits the growth of plants. However, some microbes solubilize insoluble (Fe-P, Al-P etc.) phosphate and make it available to the plants resulting in higher crop yields. The beneficial effect of phosphate dissolving microorganisms (PDMOS) is mainly due to the production of organic acids, enzymes and growth promoting substances.

Significant increase in crop yield and N_2-fixation was observed when plants were inoculated with mixed cultures of associative N_2-fixers, mixtures of *Azospirillum* and VAM fungi and *Azotobacter* spp. and phosphate solubilizing bacteria.

Furthermore, favorable effect of phosphate solubilizing bacteria on survival of N_2-fixers have also been found encouraging. But sufficient work has not been conducted on interactive effect of *Bradyrhizobium* sp. (vigna) with phosphate solubilizing bacteria in Indian conditions. As both sets of microorganism has a specific role in plant growth, the present study was aimed to investigate the interaction of *Bradyrhizobium* sp. (vigna) with phosphate solubilizing bacteria in *in vitro* condition and their interactive effect on the performance of mungbean crop under pot conditions.

Material and Methods

Bacterial Strains

Bradyrhizobium sp. (vigna) strain A-4 was obtained from the Department of Soil Science, G.B. Pant University of Agriculture and Technology, Pantnagar, whereas phosphate solubilizers, *Pseudomonas striata* and *Bacillus polymyxa* were obtained from the culture collection centre, Division of Microbiology, IARI, New Delhi. *Bradyrhizobium* sp. (vigna) strain A-4 was grown and maintained on yeast extract mannitol agar medium whereas phosphate solubilizers on Pikovskaya's medium.

In vitro interaction study: The interaction study between N_2 fixing and phosphate solubilizing bacteria was made using yeast extract mannitol and Pikovskaya's broth. Solubilization of tri-calcium phosphate (TCP) in Pikovskaya's broth with single and mixed culture was studied. In each case, one ml thick suspension of bacteria (OD 1.0) was inoculated. The two organisms added in equal proportion in the combined treatment. The flasks were incubated at 30 ± 2°C and observation on growth of the organisms were made at regular intervals. The amount of phosphate solubilized by the organism was determined colorimetrically by stannous blue colour method as outlined by Jackson, 1967.

Pot Experiment

A pot experiment on mungbean [*Vigna radiata* (L.) Wilczek] var. T-44 was conducted on unsterile soil of sandy clay loam texture with pH 8.5 and WHC, 40 per cent. The soil was uniformly packed in pots (20 cm high and 20 cm internal diameter) at the rate of 4 kg per pot. The treatment consisted of phosphorus application at the rate of 198.36 mg/pot through Mussoorie rock phosphate together with 20.5 mg $N.kg^{-1}$ soil as urea in the presence and absence (control) of *Bradyrhizobium* sp. (vigna) strain A-4 and phosphate solubilizers. A total of eight treatments were maintained and each treatment was replicated three times. Ten surface sterilized seeds soaked in liquid culture (Approx. 10^9 Cells/ml as determined by the plate technique method) were sown in each pot. The plants were thinned to 4 uniform plants after ten days of germination. The pots were watered regularly to maintain optimum moisture level. The plants were harvested after 45 days and data was analysed statistically.

Nodulation Pattern and Nodule Rating

The nodulation pattern was determined as outlined by International Network of legume inoculation trials under *nif* TAL project of University of Hawaii, College of Tropical Agriculture and Human Resources, Honolulu, Hawaii, USA. The nodules were rated using the formula of Burton:

$$\text{Nodule rating} = \frac{10 \times \text{no. of plants with tap root} + 5 \times \text{no. of plants with nodulation close to tap root} + 1 \times \text{no. of plants with scattered nodulation}}{\text{Total no. of plants}}$$

Results and Discussion

The results on interaction between *Bradyrhizobium* sp. (vigna) strain A-4 and phosphate solubilizers (*P. striata* and *B. polymyxa*) in liquid culture medium on their growth pattern indicated that N_2-fixer and phosphate solubilizers can grow together and no antagonistic behaviour of one organism towards another was noticed. This finding suggested the feasibility of using them together as microbial inoculants. The population of *Bradyrhizobium* sp. (vigna) A-4, however was low when grown alone in the modified Pikovskaya's liquid medium as compared to their growth in YEM broth (Table 24.1). The decrease in N_2-fixer population may be due to non-solubilizing properties of insoluble phosphorus by the organism. Similar findings have been reported by Epsito with *Azotobacter* sp. which showed a decreased growth and N_2 fixation in culture medium devoid of phosphorus.

Both bacteria (*P. striata* and *B. polymyxa*) were found to solubilize tri-calcium phosphate (TCP) to varying degree (Table 24.2). Maximum solubilization of TCP was observed at 6th day of incubation than 13th day and 20th day. The solubilization of TCP by *P. striata* and *B. polymyxa* was however greater when they were grown in combination with *Bradyrhizobium* sp. (vigna) (37 per cent increase over *P. striata* alone in *P. striata + Bradyrhizobium,* 32.1 per cent increase over *B. polymyxa* alone in *B. polymyxa + Bradyrhizobium*). Increase in solubilization could be attributed to the organic acids secreted extracellularly by *Bradyrhizobium* in the medium (Halder *et al.,* 1990). The result also indicated an inverse correlation between phosphate solubilization and pH of the medium (Table 24.2). Phosphate solubilization, however, decreased with increase in pH. In all probability, the decrease in pH may possibly be due to production of organic acid and hence increased phosphate solubilization.

Table 24.1: Population of Phosphate Solubilizer and Nitrogen Fixer (Ч 10^6/ml) in Modified Liquid Medium and YEM Broth

Treatment	*Period of Incubation (hours)*						
	0	*24*	*48*	*72*	*96*	*120*	*144*
In YEM broth							
Bradyrhizobium sp. A-4	0.020	0.182	8.4	350	62	226	10
In modified medium							
Bradyrhizobium sp. A-4	0.020	0.011	0.029	0.53	0.55	0.78	0.12
B. polymyxa	0.04	4.9	425	1800	3800	1100	250.20
P. striata	0.030	2.89	230	992.0	2410	420	1.5
P. striata+	0.30	3.5	41	560	220	4.1	0.97
Bradyrhizobium sp. A-4	0.020	0.041	0.211	16.8	27.6	44.0	41.0
B. polymyxa +	0.06	6.2	600.0	1300	3600	490	140.0
Bradyrhizobium sp. A-4	0.02	0.031	0.91	10.9	4.0	245.0	30.0

Table 24.2: Solubilization of Tricalcium Phosphate by Phosphate Solubilizing Bacteria in Modified Liquid Medium as Affected by *Bradyrhizobium* sp.

Treatment	*Period of Incubation (Days)*					
	6		*13*		*20*	
	pH of Medium	*Soluble P (mg/50 ml)*	*pH of Medium*	*Soluble P (mg/50 ml)*	*pH of Medium*	*Soluble P (mg/50 ml)*
P. striata	5.00	13.00	5.59	11.40	5.88	7.40
B. polymyxa	5.10	12.80	5.49	10.91	5.89	9.82
P. striata +	5.02	17.81	5.23	12.92	5.76	10.52
Bradyrhizobium		(37)*				
sp.–(vigna)		(53)+				
B. polymyxa +	5.01	16.91	5.30	13.20	6.01	10.10
Bradyrhizobium sp.		(32.10)**				

*: % increase over *P. striata*

**: % increase over *B. polymyxa*

+: % increase over *B. polymyxa + Rhizobium* sp.

The results obtained on the associative effect of N_2-fixer and phosphate solubilizers in culture medium promoted us to assess their performance on mungbean crop in natural soil under pot culture conditions. The combination treatment of *B. polymyxa* + *P. striata* + *Bradyrhizobium* sp. (vigna) A-4 was found to be statistically superior of all the treatments (Table 24.3). The increase in dry matter of plants, grain yield and nodule number and nodule weight over uninoculated plant or plant receiving only rhizobial inoculation can be due to the synergistic effect of plant genotype and the microorganisms application, as all other factors remained identical for both sets of plant (*i.e.* inoculated and uninoculated plants). Among the dual inoculation, maximum number of nodules (30/plant) was however, produced on plant inoculated with *P. striata* + *Bradyrhizobium* sp. (vigna) A-4 as compared to *B. polymyxa* + *Bradyrhizobium* sp. (vigna) treatment. The increase in nodulation can be due to increased supply of phosphorus to the host plant or due to the production of plant hormones by the organisms. But the nodules were smaller in size (5.88/plant) as compared to the treatment receiving *B. polymyxa* + *Bradyrhizobium* sp. A-4 and hence, low dry weight. The production of low number of nodules in general, on mungbean in all the treatments under identical set of condition could be due to inefficient release of bacteria from infection threads or to limitation in bacterial proliferation within the infected cells.

Table 24.3: Effect of N_2-fixer and Phosphate Solubilizers on Dry Matter, Yield and Nodulation of Mungbean (T-44)

Treatment	*Root Dry Wt. per Plant (g)*	*Shoot Dry Wt. per Plant (g) (g)*	*Grain Yield per Plant (g)*	*No. of Nodules per Plant*	*Nodules Dry Wt. per Plant (g)*	*Nodulation Pattern*	*Colour of Pigment*	*Nodulating Rating*
Control	0.16	1.00	0.35	15.00	3.20	8	White	2.25
Bradyrhizobium sp. A-4	0.31	1.56	0.75	26.00	5.46	6	P R	8.75
	(93.75)*	(56.00)*	(114.30)*	(73.30)*	(70.62)*			
Pseudomonas striata	0.41	150	0.52	18.00	3.78	8	White	2.75
Bacillus polymyxa	0.34	1.50	0.57	16.00	3.36	9	White	2.00
B. polymyxa + *P. striata*	0.30	1.52	0.58	17.00	3.57	8	White	2.75
B. polymyxa + *Bradyrhizobium* A-4	0.366	1.57	1.06	28.00	6.30	6	P R	7.75
B. striata + *B. Bradyrhizobium* A-4	0.40	1.85	1.11	38.0 86.6	5.88	6	P R	7.75
B. polymyxa + *P. striata* + *Bradyrhizobium* sp.	0.48	1.91	1.50	36.00	7.56	5	P R	7.75
	(200.0)*	(91.0)*	(228.50)*	(140)*	(136.25)*			
			(53.3)+	(38.40)+	(38.46)+			
CD at 5%	0.004	0.030	0.005	0.154	0.003			

*: % increase over control.

+: % increase over *Rhizobium* inolculation alone.

The presence of indigenous rhizobia that might be delaying the nodulation cannot be ignored, as the soil taken for study was unsterile. Similarly root dry weight, shoot dry weight and grain yield was recorded maximum (200 per cent, 228.6 per cent, and 91 per cent respectively increase over control) with triple inoculation (Table 24.3). However, the effect of dual inoculation *P. striata* + *Bradyrhizobium* sp. (vigna) resulted in higher dry matter production and yield than *B. polymyxa* + *Bradyrhizobium* sp. (vigna). Substantially increased yield in soybean, pigeonpea and French bean due to inoculation with *Rhizobium* spp. has also been reported by Khan *et al.*, in mungbean, Algawadi using N_2-fixing and PSB in barley and Mallik and Sanoria in lentil crop.

Colour of the uninoculated nodule pigments as well as the treatment devoid of *Bradyrhizobium* sp. (vigna) inoculation were found to be white suggesting the presence of ineffective species of rhizobia. On the other hand, inoculated legumes showed pink to red colour pigmentation indicating an effective association between legumes and rhizobia. The nodulation pattern of mungbean varied between 5–9. The nodulation pattern of *Bradyrhizobium* inoculation indicated a root system with larger nodules concentrated on tap root whereas interaction of *Bradyrhizobium* sp. (vigna) with *B. polymyxa* and *p striata* exhibited a nodulation located mostly on secondary roots. As compared to nodulation pattern, nodulation rating seemed to be a better qualitative procedure for differentiating *Rhizobium* sp. and phosphate solubilizers varying in degree of effectiveness. Single inoculation with *Bradyrhizobium* as well as *Bradyrhizobium* inoculated with phosphate solubilizers exhibited a high degree of nodule rating (Table 24.3). Inoculation with *Bradyrhizobium* sp. (vigna) A-4 (8.75), *Bradyrhizobium* sp. A-4 + *B. polymyxa* (7.75) and *Bradyrhizobium* sp. A-4 + *P. striata* + *B. polymyxa* (7.75) were found to be highly efficient in pot trials. Since the nodulation rating is based on the classification of plants on the basis of nodule in tap root, close to tap root and scattered nodulation, a high nodulation rating indicated tap root nodulation in most of the plants. It is the tap root that develop earliest and hence supply N_2 to the planet for a longer period of time resulting in better crop yield. The synergistic interaction been *P. striata* or *B. polymyxa* with *Bradyrhizobium* and increased phosphate solubilization thus, may prove beneficial for developing a mixed inoculant for increasing crop productivity.

Chapter 25

Performance of Asymbiotic Biofertilizers

Now-a-days advance agriculture as application of chemical nitrogenous fertilizer is not economical because the applied fertilizer, only 50 per cent of the nitrogen is utilized by the plants and the remaining is being lost through denitrification or leaching. In contrast, biological system fix atmospheric nitrogen at no cost and at constant rate high permit immediate incorporation into plant proteins. Recent findings revealed the association of tropical grasses with dinitrogen fixing bacteria which under favourable conditions, contribute significance to the nitrogen economy of plant.

The increasing demand for nitrogen in agriculture could therefore, be solved by the exploitation of biologically fixed nitrogen, *Azotobacter* and *Azospirillum* are known to be responsible for dinitrogen fixation but also secret growth promoting substance like auxins, gibberalins, cytokinins and some fungistatic substances in grasses and grain crop, like sorghum, bajra, maize, rice and wheat etc. The VAM are known for solubilization of insoluble phosphorus and they play significant role in phosphorous uptake and translocation, beside uptake of other nutrient elements, such as zinc, sulphur and copper.

The present chapter therefore deals with the study of individual and combined effect of VA-mycorriza (*Glomus fusiculatum*), *Azotobacter chroococcum* and *Azospirillum brasilense* at different doses of chemical fertilizer on yield, dry matter weight, plant uptake of N and P and residual of N and P of sorghum hybrid CSH-14, so as to minimise the dose of chemical fertilizer.

Material and Methods

A field trial was conducted at Department of plant pathology field, Dr. Panjabrao Deshmukh Krishi Vidyapeeth, Akola, during kharif 1992–93. The soil of the experiment site was medium black, having pH 7.55, total nitrogen 150.64 kg/ha organic carbon 0.42 per cent available P_2O_5 20.75 kg/ha and available K_2O 416.00 kg/ha.

Soil of experimental plot was low in nitrogen and very poor in phosphorus content and rich in potassium. Split plot design block design with three main treatment and eight subtreatments, replicated thrice. The culture of VAM (*Glomus fasiculatum*) was obtained from Dr. Jalali, Haryana Agricultural University, Hissar and those *Azotobacter chroococcum* and *Azospirillum hrasilense* from ICRISAT Hyderabad.

The three doses of fertilizer *viz.*, full dose of fertilizer (80 : 40 : 40) T_1, 3/4 dose of fertilizer (60 : 30 : 30) T_2 and no fertilizer application T_3. The dose of fertilizers were applied of which half dose of nitrogen and full dose of phosphorus and potash at the time of sowing and remaining nitrogen dose was applied month after sowing as top dose. The biofertilizer seed treatment were *Azotobacter* (S_1), *Azospirillum* (S_2), VAM (S_3), *Azotobacter* + *Azospirillum* (S_4), *Azotobacter* + VAM (S_5) *Azospirillum* + VAM (S_7) and No seed treatment (S_8).

The healthy seeds of *Sorghum* cv. CSH-14 were inoculated with the culture at the rate of 25 g/kg seed by following slurry method. The inoculated seeds were then dried in shade and dibbled (at the rate of 8 kg/ha) at spacing of 45 cm between row and 15 cm between plants. The observations on root length, shoot length, and green matter weight were recorded at 15-day interval taking randomly selected plants.

Table 25.1: Effect of Different Interaction on Grain Yield (q/ha), Fodder Yield (q/ha) Plant Uptake of N and P (q/ha) and Residual N and P in Soil (kg/ha)

Observations	*Treatment*	S_1	S_2	S_3	S_4	S_5	S_6	S_7	S_8	
Grain yield	T_1	43.22	43.67	46.37	46.08	45.58	46.19	47.86	42.51	N.S.
	T_2	36.24	36.24	37.37	38.60	39.44	38.31	41.66	35.83	
	T_3	25.91	25.83	26.13	26.58	27.88	27.90	29.98	24.10	
Fodder yield	T_1	96.08	96.32	98.23	97.70	99.89	99.88	106.03	94.19	Sig.
	T_2	88.88	90.10	90.30	92.55	93.40	93.60	95.97	82.42	SE(m) ± 0.69
	T_3	73.30	76.62	77.33	77.91	78.22	78.42	81.07	74.93	CD% 1.89
Plant uptake of N	T_1	218.11	218.52	219.00	220.81	223.28	222.65	225.5	216.15	Sig.
	T_2	209.99	211.19	213.45	215.33	217.51	216.35	218.80	204.82	SE(m) ± 0.68
	T_3	182.99	184.44	185.68	188.36	190.18	189.43	192.34	176.21	CD% 1.87
Plant uptake of P	T_1	28.57	29.09	29.66	30.1	31.46	30.81	34.04	28.25	Sig.
	T_2	24.43	25.64	26.39	26.81	28.40	27.81	31.58	24.18	SE(m) ± 0.26
	T_3	23.85	23.92	24.65	24.98	25.94	25.77	26.79	20.61	CD% 0.71
Residual N in soil	T_1	153.89	154.94	156.09	160.12	162.04	161.14	163.14	151.90	Sig.
	T_2	145.46	147.32	148.33	152.05	154.22	153.32	156.22	141.9	SE(m) ± 0.30
	T_3	136.47	137.88	138.85	143.27	146.81	145.54	148.42	131.76	CD % 0.82
Residual P in soil	T_1	28.57	29.09	29.66	30.10	31.46	30.81	34.4	28.25	Sig.
	T_2	25.43	25.64	26.39	26.81	28.40	27.88	31.58	24.18	SE (m) ± 0.26
	T_3	23.85	23.92	24.65	24.98	25.94	25.77	26.79	20.61	CD % 0.71

Residual of N and P (20.75 kg/ha) Before sowing (150.64 kg/ha)

Seed treatment = S_1: *Azotobacter*, S_2: *Azospirillum*, S_3: VAM

S_4: *Azotobacter + Azospirillum*, S_5: Azotobacter + VAM

S_6: *Azospirillum* + VAM, S_7: *Azotobacter + Azorpirillum* + VAM, S_8: No treatment

T_1: Full dose of NPK (80 : 40 : 40), T_2: 3/4 dose of NPK (60 : 30 : 30), T_3: No fertilizer.

For estimation of the dry matter weight, five plants were randomly selected from each plot at an interval of 15 days and were air dried followed by oven drying at 60°C till constant weight were obtained. Grain yield data and fodder yield data were recorded after harvest of the crops.

For the soil analysis, the soil samples were collected from 20 cm soil depth from each subplot before rowing and after harvesting. The samples were ground to fine powder to estimate residual nitrogen and phosphorus by Kjeldahl's Method and Olsen method respectively.

For the plant analysis, the plant samples were cut into pieces, dried and converted into fine powder which was used to estimate uptake of nitrogen by Kjeldahl method and phosphorus by Olsen method.

Table 25.2: Effect of Main and Subtreatments on Grain Yield (q/ha), Fodder Yield (q/ha), N, P Uptake and N, P (kg/ha) Residue in Soil

Treatments	*Grain Yield*	*Fodder Yield*	*Plant Uptake*		*Residue*	
			N	*P*	*N*	*P*
T_1 Full dose of fertilizer	46.18	98.52	220.50	61.11	167.91	30.84
T_2 ¾ dose of fertilizer	38.05	90.93	213.44	54.52	158.85	27.68
T_3 No fertilizer	26.79	77.55	186.26	50.82	151.12	24.06
SE (m) ±	0.18	0.47	0.25	0.24	0.12	0.41
CD 5%	0.73	2.35	2.5	2.35	1.22	3.6
Sub Treatments						
S_1 *Azotobacter*	35.32	86.75	203.70	53.50	155.27	25.95
S_2 *Azospirillum*	35.69	87.68	204.75	53.79	156.71	26.21
S_3 VAM	36.62	88.88	206.04	55.57	151.76	26.90
S_4 *Azotobacter* + *Azospirillum*	37.08	89.38	208.16	54.62	161.81	27.29
S_5 *Azotobacter* + VAM	37.26	90.40	209.48	56.30	163.33	28.15
S_6 *Azospirillum* + VAM	39.83	90.69	210.32	57.29	164.35	28.60
S_7 *Azotobacter* + *Azospirillum*	39.83	94.35	212.21	59.43	165.93	30.80
S_8 No seed treatment	34.15	83.86	199.06	51.23	151.85	24.34
SE (m) ±	0.39	0.40	0.39	0.24	0.17	0.15
CD 5%	1.13	1.14	1.25	0.63	0.56	0.47

Results and Discussion

It is revealed from the data (Table 25.1) that the application of recommended dose of fertilizer along with *Azotobacter* + *Azospirillum* + VAM gave more grain yield (47.86 q/ha) followed by VAM (46.37 q/ha) and *Azospirillum* + VAM (46.19 q/ha). Fodder yield also obtained same trained. In the present investigation significant result were obtained for grain yield and fodder yield over uninoculated control, positive result were obtained in combined inoculation rather than single inoculation (Table 25.2), there were seen to be synergistic effect in imposing the grain and fodder yield. The effect of inoculant and fertilizer showed significant results, however interaction effect were found non significant.

Among the two non symbiotic nitrogen fixer *Azospirillum* was noticed to be superior over *Azotobacter.*

In present investigation it was observed that grain and fodder yield was more due to VAM inoculation as compared to *Azotobacter* or *Azospirillum* inoculation. This increase can be attributed to antimicrobial activity, secretion of growth promoting substances, fixing of atmospheric nitrogen and increased P availability to plant. Maximum uptake of P_2O_5 was noticed in combined inoculation and inoculation of VAM compared to *Azotobacter* + *Azospirillum.* Raju *et al.* had reported that along N and P_2O_5 is also increased in VAM inoculated plant compared to non inoculated plant.

The present chapter showed that soil poor in N and P but application of biofertilizer inoculation with fertilizer minimize the cost of production also increased the grain yield but also save nitrogen and phosphorous some extent without affecting normal yield of sorghum.

Chapter 26

Effect of *Azospirillium* on Quality of Sugarcane

The use of organic and inorganic fertilizers have got prime role in maintaining the soil fertility and also to increase the productivity. Organic manures being slow in releasing nutrients and available in insufficient quantities, can not meet the total nutrient requirements in crop production. Hence the use of chemical fertilizers became imperative to exploit the further crop potential. However, use of inorganic fertilizers are becoming scare and costly in recent years. In addition to this, excessive use of inorganic fertilizers results in environmental pollution and increases salinity and alkalinity of soil. With this view the use of biofertilizers in the form of carrier based inoculants became more relevant and of economic importance.

Azospirillum is a associative symbiotic nitrogen fixing bacterium. Anderson found that *Azospirillum lipoferum* was able to fix in or on the roots of sugarcane. According to him apart from atmospheric nitrogen and making it available to plants as well as promoting uptake of minerals ions such as nitrate, potassium, phosphate in addition to growth hormones.

Next to nitrogen, phosphorus is vital and non-renewal nutrient for plants and microorganisms. About 98 per cent Indian soils are poor in available phosphorus and efficiency of phosphate fertilizers utilization fixation of applied soluble phosphate. Phosphate solubilizing microorganisms posses the ability to dissolution of insoluble form of phosphate to soluble form of phosphate.

Material and Methods

The experiment was conducted during 1992–93 to 1994–95 at three different co-operative sugar factories of Maharashtra *viz.* Vasantdada Shetkari S.S.K. Ltd., Sangli, Terna Shetkari S.S.K. Ltd., Osmanabad and Satara S.S. K. Ltd., Bhuinj. The objective of the present investigation was to study the interaction effect of Nitrogen fixing bacterium *Azospirillum* and phosphate solubilizing culture on yield and quality of sugarcane var. Co 7219. The trial was conducted on medium black soil in randomised block design with eleven treatments and three replications. For present study the

Azospirillum and P-solubilizing culture developed by Vasantdada Sugar Institute, Pune is used at the rate of 7.5 kg/ha each at the time of planting by soil application method.

Treatment Details

T_1: 50 per cent N and 100 per cent P and K

T_2: 75 per cent N and 100 per cent P and K

T_3: 50 per cent P and 100 per cent N and K

T_4: 75 per cent P and 100 per cent N and K

T_5: 50 per cent N and *Azospirillum* + 100 per cent P and K

T_6: 75 per cent N and *Azospirillum* + 100 per cent P and K

T_7: 50 per cent P and P-solubilizing culture + 100 per cent N and K

T_8: 75 per cent P + P-solubilizing culture + 100 per cent N and K

T_9: 50 per cent N and P + *Azospirillum* + P-solubilizing culture + 100 per cent K

T_{10}: 75 per cent N and P + *Azospirillum* + P-solubilizing culture + 100 per cent K

T_{11}: Control (Only 100 per cent NPK)

Results and Discussion

Growth Parameters % (Table 26.1)

In case of germination all the treatments remained statistically not significant over control. The significantly highest tillering ratio at 120 days was recorded in the treatment having 75 per cent N and P *Azospirillum* + P-solubilizing culture *i.e.* 2.28 over control (1.97). The same treatment was found to be significantly superior in case of plant population and it was 88.25 × 1000/ha over control (81.45 × 1000/ha) while all other treatments remained statistically non significant.

Table 26.1: Mean Growth Parameters as Affected by *Azospirillum* and P-solubilising Culture

Trial No.	*Germination % (60 days)*	*Tillering Ratio (120 days)*	*Millable Cane/ ha ('000)*	*Girth of Cane (cm)*	*Millable Cane Heights (m)*
T_1	62.54	01.77	75.59	08.81	02.18
T_2	61.83	01.88	78.94	08.76	02.33
T_3	62.15	01.73	80.66	08.72	02.24
T_4	62.96	02.05	83.21	09.04	02.35
T_5	60.52	02.03	80.55	08.92	02.42
T_6	65.23	02.10	85.13	08.99	02.51
T_7	66.60	02.05	81.78	09.28	02.39
T_8	64.85	02.15	83.81	09.55	02.57
T_9	65.27	02.19	83.36	09.54	02.63
T_{10}	66.08	02.28*	88.25	09.57*	02.68*
T_{11}	63.07	01.97	81.45	08.90	02.35
C.D.	N.S.	00.28	05.47	00.44	00.17

As far as girth of cane is concerned the treatments having 75 per cent P + P-solubilizing culture, 50 per cent N and P + *Azospirillum* + P-solubilizing culture and 75 per cent N and P + *Azospirillum* + P-solubilizing culture showed significantly higher cane girth, *i.e.* 9.55, 9.54 and 9.57cm respectively over control (8.90 cm). The same treatment was found significantly superior in case of millable cane height.

Caner and Sugar Yield (Table 26.2)

Results of cane and sugar yield of three different locations are pooled and presented in Table 26.2. Data revealed that the treatment having 50 per cent N and P *Azospirillum* + P-solubilizing culture and 75 per cent N and P + *Azospirillum* + P-solubilizing culture recorded significantly higher cane yield *i.e.* 105.31 and 109.74 MT/ha over Control (90.29 MT/ha).

Table 26.2: Mean Cane Yield and CCS (MT/ha) as Affected by *Azospirillum* and P-solubilizing Culture

Treatment No.	Cane Yield (MT/ha)			Pooled Mean	CCS (MT/ha)			Pooled Mean
	Terna SSK	Vasantdada SSK	Satara SSK		Terna SSK	Vasantdada SSK	Satara SSK	
1	76.63	91.00	84.60	81.68	09.69	10.97	11.53	10.73
2	78.40	87.40	83.88	81.72	10.22	11.20	11.76	11.06
3	74.80	95.33	86.45	82.10	08.75	10.76	12.08	10.53
4	80.34	99.33	89.60	85.58	10.57	11.50	12.97	11.68
5	90.36	77.00	81.24	85.09	12.33	11.45	11.14	11.64
6	92.70	101.40	91.60	93.78	13.17	13.95	12.54	13.22
7	98.42*	96.67	89.45	95.14	13.66*	13.73	12.90	13.43
8	103.33*	100.13	91.90	99.18	14.64*	14.18	12.88	13.90
9	109.05*	95.20	104.70	105.39*	15.75*	14.54	15.10	15.13*
10	113.38*	101.67	108.33	109.74*	16.45*	15.16	15.64	15.75*
11	89.20	92.73	90.70	90.21*	12.41	14.27	12.56	13.08
C.D.	6.60			9.06	1.07			1.66

This indicates that by use of *Azospirillum* and P-solubilizing culture increased the cane yield by 19.45 MI/ha with saving of 25 per cent chemical nitrogen and phosphorus.

In case of CCS, the significantly highest CCS was recorded in the treatment having 75 per cent N and P + *Azospirillum* + P-solubilizing culture at all 3 locations *viz.*, Terba Shetkari S.S.K., Vasantdada Shetkari S.S.K., Sangh and Satara S.S.K. and it was 16.45, 15.94 and 15.64 MT/ha respectively over control. The pooled data of 3 locations revealed that the treatment having 75 per cent N and P + *Azospirillum* + P-solubilizing culture showed significantly highest CCS *i.e.* 15.75 MT/ha over control (13.08 MT/ha).

Summary and Conclusion

An experiment was conducted at three different locations of Maharashtra State *viz.* Vasantdada Shetkari S.S.K. Ltd., Sangli, Terna Shetkari S.S.K. Ltd., Osmanabad and Satara S.S.K. Ltd., Bhuinj during 1992–93 to 1994–95 with an object to find out the interaction effect of *Azospirillum* and phosphate solubilizing culture on yield and quality of sugarcane. It was observed that the application of

Azospirillum and P-solubilzing culture at the rate of 7.5 kg/ha each at the time of planting by soil application method gave significantly maximum cane and sugar yield (109.74 and 15.75 MT/ha) with saving of 25 per cent inorganic nitrogen and phosphorus while in control the cane yield was 90.29 MT/ha and sugar yield was 13.08 MT/ha.

Chapter 27

Bio-inoculants for Recycling Banana Wastes

Banana is grown over an area of 1,64,000 hectares in India. This crop is mainly grown in Tamil Nadu, West Bengal, Kerala, Maharashtra etc., on a commercial basis, though it is grown in other areas too.

Tiruvaigundam taluk of Tuticorin District is a place wherein Banana is extensively grown on a commercial basis in the Tamiravaruni basin, where there is a heavy crop of Banana. The Banana wastes after harvest are found to lie across the water courses causing overflow of channels, leakage due to seepage and loss in water leading to secondary problems like mosquito menace. Moreover the mosquitoes in the presence of high moisture ravage through the deteriorating Banana wastes causing disease as well environmental pollution.

Further the hard peduncles after removal of hands mostly of lignin and hemicellulose content are left thrown in the market place causing bruises to the customers.

Material and Methods

Thus, realising the vulnerability of the situation and to facilitate recycling and to produce a good manure with the available wastes, experiments were conducted in the Department of Soil Science and Agricultural Chemistry, Agricultural College and Research Institute, Killikulam of Tamil Nadu Agricultural University in the year 1995–96 with the locally existing Biowastes left in the market place *viz.* Banana wastes.

Fungal Bio-inoculants *viz. Agaricus, Pleurotus* and *Penicilium notatum* were inoculated under different levels of moisture *viz.* 40 per cent and 60 per cent with two different processes of decomposition *viz.,* Aerobic and Anaerobic in a miniature form using Pot-culture experiments with Randomised Block design comprising of twelve treatment replicated thrice *viz.,* Trash, Peduncles, Pseudostem.

The waste were spread in layers and they were detected of the exact moisture content gravimetrically so as to maintain the moisture level, periodically.

The samples were aerated once in 3 days by frequent stirring in aerobic method and in anaerobic method a sandy paste was made over the surface of the pot and water was sprinkled over the sandy lauer once in 3 days. The decomposition process was allowed to continue for 2 months until the manure attained a black colour with friable nature.

In anaerobic method, the sandy paste spread over the pot was disturbed once in month to aerate the pot.

After the decomposition period of 2 months, the pots were overturned and the manure was dug from the pot and shade dried and analysed for their nutrient contents.

The bio-inoculants *viz., Agaricus, Pleurotus and Penicilium* were inoculated at the rate of 3 g/pot and 3 ml/pot (1 ml = 10^4 cfu, colony forming units) respectively with a capacity of 6 kg wastes per pot.

Initial analysis of the waste materials as well as the nutrient content of the manure at different intervals *viz.,* 1 month after decomposition and 2 month after decomposition were analysed using Piper, 1966. The wastes were incorporated initially on a moisture free basis. The treatment included were as detailed below:

1. Control (Aerobic)
2. *Pleurotus* + 40 per cent moisture (Aerobic)
3. *Pleurotus* + 60 per cent moisture (Aerobic)
4. *Penicilium* + 40 per cent moisture (Aerobic)
5. *Penicilium* + 60 per cent moisture (Aerobic)
6. Control (Anaerobic)
7. *Pleurotus* + 40 per cent moisture (Anaerobic)
8. *Pleurotus* + 60 per cent moisture (Anaerobic)
9. *Penicilium* + 40 per cent moisture (Anaerobic)
10. *Penicilium* + 60 per cent moisture (Anaerobic)

Results and Discussion

The nutrient contents of the Banana wastes are depicted in Table 27.1

Table 27.1: Banana Wastes and their Nutrient Content (%) on Over Dry Basis

Name of the Part	*Nutrient Content (%)*						
	N	*P*	*K*	*Ca*	*Mg*	*Total Carbon*	*Crude Fibre*
Pseudo stem	0.18	0.50	2.20	1.48	1.32	54	20
Leaf	0.68	0.86	1.72	0.80	0.65	24	5
Pendruncle	0.15	0.30	1.60	0.58	0.47	70	30

The results also revealed that (Table 27.2) there was a drastic reduction in dry matter after 2 months of decomposition *viz.,* 0.93 kg.

Total carbon content in percentage worked out to be on an average 46.9 per cent. Further the moisture content of the manure also got reduced with advent of time, *viz.,* 2 months after decomposition, *i.e.* 37 per cent.

Table 27.2: Effect of Different Levels of Moisture and Bio-inoculants on Dry Matter Reduction (kg), Moisture Content (%), Total Carbon (%)

Treatments	Dry Matter Reduction (kg)		Total C (%)		Moisture (%)		Total Nitrogen (%)		Total P (%)		Total K (%)	
	1 Month	2 Months	1 Month	2 Months	1 Month	2 Months	1 Month	2 Months	1 Month	2 Months	1 Month	2 Months
Control Aerobic (A)	198	1.80	62.54	52.30	15.78	8.89	0.27	080	0.45	0.48	0.52	0.48
P + 40% M-A	1.00	0.70	48.60	35.70	36.40	30.40	0.68	2.50	1.07	1.45	1.12	0.52
P + 60% M-A	1.00	0.64	52.30	43.90	45.20	35.90	0.59	1.95	1.03	1.27	1.06	1.75
Pe + 40% M-A	1.37	0.98	52.30	38.80	45.40	28.70	0.58	1.95	0.82	1.34	0.80	1.19
Pe + 60% M-A	1.00	0.67	57.80	45.30	48.30	33.30	0.54	1.85	0.81	1.26	0.76	1.00
Control Anaerobic (AA)	2.03	1.50	80.68	56.30	40.28	26.30	0.25	0.90	0.57	0.45	0.54	0.52
P + 40% M-AA	1.54	0.79	54.60	47.60	57.30	48.40	0.56	1.85	0.95	1.27	1.00	1.61
P + 60% M-AA	0.74	0.32	58.80	49.30	62.40	57.10	0.52	1.75	0.91	1.08	0.90	1.19
Pe + 40% M-AA	1.89	1.01	58.40	49.70	59.70	46.50	0.56	1.75	0.74	0.95	0.74	1.28
Pe + 60% M-AA	1.93	0.86	59.60	50.80	65.30	55.30	0.50	1.55	0.72	0.91	0.68	1.15
Mean	1.45	0.93	58.56	46.92	47.59	37.07	0.51	1.69	0.81	1.05	0.81	1.22
S.D.	0.47	0.39	8.01	5.89	14.05	14.11	0.13	0.50	0.19	0.33	0.20	0.42

A: Aerobic; AA: Anaerobic; P: *Pleurotus*; Pe: *Penicilium*; M: Moisture.

Regarding the total nutrient contents *viz.*, Total N, Total P, Total K, it averaged out to 1.69 per cent N, 1.05 per cent P, 1.22 per cent K after 2 months of decomposition as compared to one month decomposition period which recorded lower values.

Among the inoculants, the dry matter reduction was higher in *Pleurotus, viz.*, 0.61 kg as compared to that of *Penicilium* which recorded 0.88 kg after 2 months decomposition process.

Similarly the total carbon content in (per cent) was found to be lower in *Pleurotus* 44.1 per cent after 2 months of decomposition.

The total nutrient content *viz.*, N, P, K were found to be higher in *Pleurotus* than *Penicilium, viz.*, 2.01 per cent, 1.27 per cent and 1.64 per cent respectively.

Thus among the inoculants *Pleurotus* was found to enrich the manure in a better way and proved to be an efficient convertor of waste with higher order of degradation.

Among the moisture regimes, 40 per cent level of moisture was found to be congenial as compared to that of 60 per cent moisture level which could aid faster and better decomposition.

Among the decomposition processes the anaerobic method showed higher nutrient contents especially K after 2 months of decomposition as well the total carbon content remained low *i.e.* 50.6 per cent after 2 months of decomposition process. Whereas, the nitrogen and phosphorus remained high in aerobic process, though there was not much variation between both the processes (Table 27.3).

Thus die use of Bio-softwares will definitely serve as "Bio-Golds" to harness the manurial turnover in India and help a great deal in tackling the fertilizer demand realised nowadays.

Table 27.3: Effect of Method of Decomposition and Time Interval on the Manurial Content

Parameters	*Methods*			
	Aerobic		*Anaerobic*	
	1 month	*2 months*	*1 month*	*2 months*
Dry matter reduction (kg)	1.27	0.96	1.63	0.90
Total carbon (%)	54.71	54.50	62.40	50.60
Moisture (%)	38.20	27.40	56.99	46.70
Total N (%)	0.53	1.81	0.48	1.56
Total P (%)	0.84	1.16	0.78	0.93
Total K (%)	0.85	0.99	0.77	1.15

Chapter 28

Pressmud as Plant Growth Promoter

Maharashtra and Uttar Pradesh (U.P.) are the major sucrose producing States of India. It, therefore, obviously follows that along with sugar, they generated many byproducts of sugar and distillery wastes which need closer attention for their utilisation on merit. The chief byproducts are bagasse (a potential source of fibre for pulp and paper industry), Pith (an amorphous parenchymatous core tissue, rich in sugar, constitutes 90 per cent calorific value of bagasse, a source of fuel for sugar industry), molasses (a vital raw material for distilleries), tillage from distilleries (as a source of biogas) and pressmud, a multi-ingredient and vital source of macro-micro-nutrients, matrix for microbes, source of phytosterols, PGRs and plant protecting steroids.

On a global scenario, India being the largest producer of sucrose, it is estimated that it generated about 3.2 million tonnes of pressmud (filter mud/scum). Although, sugar industry looked at it as a waste, with the emphasis on recycling this type of waste through imaginative biotech applications, we feel that pressmud could be analyzed for its macro- and micro-nutrients and assigned at least a few value-added applications so that application of waste generated wealth. The present review article examines its chemical characteristics, physical nature and explores the feasibility of assigning it futuristic applications, hopefully for sustained agricultural production.

Material and Methods

Pressmud was obtained from Madhukar Sahakari Sakhar Karkhana, Faizpur and Purushottamnagar Sakhar Karkhana, Shahada during the crushing season December–May.

AR grade chemicals and demineralized water used for analysis. Total and volatile solids were estimated by gravimetry.

Ash contents were determined by burning at 600°C for 30 minutes until constant weight.

Cellulose was determined as per Updegraff. Hemicellulose was determined as per Deschatelets. Lignin was estimated as described in Methods of Wood Chemistry. Organic carbon was analyzed by

Walkley-Black method. Wax was estimated by extraction in benzene and noting down the difference in the weight of pressmud before and after extraction. Protein was estimated as per Lowry. Total carbohydrates were estimated by phenol sulphuric acid method as per Dubois *et al.* Reducing sugar was estimated by DNSA method as per Miller. The above methods were routinely used for the analysis of pressmud. Its representative analysis on dry weight basis is summarized in Table 28.1.

Table 28.1: Constituents-wise Profile of Representative Pressmud

Constituents	%
Total solids	100
Volatile solids	70
Ash (Non volatile solids)	30
C/N ratio	30 : 1
Cellulose	25
Hemicellulose	27
Lignin	4
Organic carbon	24–27
Solubles	38
Wax (Benzene Extract)	9
Proteins	7
Total Carbohydrates	2
Reducing sugars	0.4

About 1.2 per cent nitrogen, 2.5 per cent phosphorous, 2 per cent potassium, 260 mg/kg zinc and 120 mg/kg copper are present.

Results and Discussion

Physically, pressmud is soft, spongy, bulky (light weight) and amorphous blackish brown material.

Literature has reported uses of pressmud in (*i*) distemper and paint industry, (*ii*) cement and filter aid industry, (*iii*) metal polish and animal feed supplement and (*iv*) its sterols/steroids in pharmaceutical industry. However, these are only laboratory scale ideas, without estimates of their viability and pollution potential. A lot of work on techno-economic viability of such applications is desired before they mature into a concrete application. Therefore, farm application of pressmud in the near future appears worth exploring, crop-wise, in experimental plots, so that its suitability on scientific basis is established in this context, its application for the ecofriendly restoration of manganese and coal mine overheads or mine spill soil by Juwarkar is worth mentioning.

From Table 28.1, the following facts have clearly emerged upon analysis of pressmud ingredients:

1. The C : N ratio of 30 : 1 is optimal for the growth of a number of crops. These includes flowers, vegetables, cereals, pulses, oil seeds, sugarcane, bamboo and eucalyptus.
2. Its cellulose and hemicellulose content in relation to lignin has shown that allelopathic effect, if any, on plant growth is already neutralized.
3. Its meager amount of reducing sugars are hoped to catalyze rapid microbial growth and the same would be promoted over a long period by total carbohydrate and subsequently sustained

by composting of both, cellulose and hemicellulose. Thus, a sustained enrichment of organic carbon in the soil appears assured for comparable productivity.

4. The presence of 7 per cent proteins in the pressmud would be a source of sustained release of amino acids by continuous proteolysis through rhizobial microflora. It appears now well established that amino acids facilitate overcoming the physiological constraints and thereby promote increased chlorophyll synthesis, availability of more photo-assimilates and increased growth rate.
5. Tandon have found pressmud superior to sewage sludge, biogas slurry, compost, farm yard manure and soil conditioners from municipal wastes. This would not have been possible without pressmud affording physical, chemical, biological and possibly pharmaceutical benefits through porous texture, more aerobicity for the roots, more water holding capacity, possibly sustained release of moisture, sustained release of tricontanol (PGR), harbour of microflora and phytosterols/steroids in wax affording plant protection against plant pathogens.

The above extrapolation of different benefits is merely a reflection of its micronutrients and as yet unidentified ingredients.

Basically, it would afford a porous texture to the soil and thereby make more air available for the respiration of the root system, in turn rendering more energy available for absorption of nutrients from the rhizosphere. Incidentally, due to its pH 5.0, it has the potential to amend alkaline soils which otherwise are barren. Such amended soils have shown excellent productivity in forestry. Being insoluble in nature, it could serve as a matrix for the growth and sustenance of beneficial microbes. Department of Biotechnology, New Delhi has been examining various matrices for serving as a vehicle for the application of biofertilizers. We propose to examine if pressmud could enhance shelf life of nitrogen fixing and/or phosphate solubilizing bacteria/fungi.

Its various inorganic constituents such as nitrogen, phosphorus and potassium are anticipated to supplement fertility of soil and thereby reduce dependence on capital-, energy-, pollution-, and cost-intensive chemical fertilizers. Further more it micronutrients, notably zinc and copper, would supplement micronutrient fertility without additional cost. In fact, Tandon *et al.* have converted these nutrient values into economic values to Rs. 75 crores/annum.

About 10 per cent wax content of pressmud has the dual potential to serve as a source for plant growth regulators such as tricontanol and phytosterols which have the potential for neutralizing the virulence of plant pathogens. Finally, the unidentified trace nutrients of press mud would render invaluable fertility value by complementing several physiological processes in plant and rhizobial microflora, providing sustained growth rate. In the light of these facts, potential multiplier effect of pressmud, renders it a most attractive candidate for biotech applications. Its use in agriculture/forestry appears to be least polluting and most cost-effective, rewarding and sustainable.

Thus, cumulatively, pressmud appears to be a fountainhead of a number of vital attributes, directly or indirectly, immediately or on sustained basis. Potentially or otherwise and therefore, its value-addition appears within the realm of human efforts, without additional processing/efforts/cost.

In fact, ignorance of pressmud composition attracted least attention of the business community as yet and hence the putrifield piles on the premises of sugar mills have become a source of aerial pollution. Its utilisation per se would not only serve as an avenue for additional income to the sugar industry but its application would control aerial pollution also and simultaneously supplement fertilizer value in turn affording higher agricultural productivity.

In the absence of (*a*) alternative plans by the sugar industry (*b*) competing technology in the offing, and (*c*) a competing product emerging from or for pressmud, it remains a wealth for farmers. This simple, efficient and economic way of pressmud management could place it rightfully as the biofertilizer of 21st century.

Chapter 29
Biofertilizer for Multipurpose

Importance of ectomycorrhizal (EM) fungi in reforestation of eroded lands have been recognized for a long time now. The inoculation of forestry plant seedlings with these fungi has become a standard protocol. Early stage EM fungi are the choice as they have a broad host range. *Laccaria laccata* (Scop. Fr.) Berk. and Br. is one of the widely distributed early stage fungus with a broad host range forming ectomycorrhiza with *Eucalyptus, Larix, Pines* and others. Early stage fungi like *Laccaria laccata* are very suitable for inoculation of seedlings which have to be placed in skeletal or mineral soils. With the help of this fungus, time required for the generation of plantable seedling can be reduced to half. Tolerance to the extreme of soil conditions evoked the interest in applying this fungus to develop wastelands into green forest covers. This fungus exhibited antagonistic behaviour against some potential root pathogens.

Various forms of fungal inoculum can be used for inoculating tree seedlings, but the most recommended form is vegetative mycelium. Vegetative mycelium can be produced by using either solid substrate cultivation or submerged cultivation. Sorghum grain and sphagnum-vermiculite carrier have been used to produce inoculum through solid substrate cultivation. Inocula produced through these processes are not physiologically consistent and are difficult to maintain. Furthermore, substrate that remains at the end of the process is difficult to remove and invites contamination during application and finally leads to the loss of inoculum. Most of these problems can be overcome by producing physiologically consistent mycelium through submerged cultivation. Sturdy inoculum then can be produced by physical entrapment of this mycelium in alginate gel. The objective of this investigation was to examine the possibility of using granulated inoculum of *Laccaria laccata* for inoculation of seedlings of *Eucalyptus* hybrid and *Acacia nilotica* in nursery.

Material and Methods

Species

Laccaria laccata strain ITCC 3334 was obtained from the Indian Type Culture Collection Centre, Indian Agriculture Research Institute (IARI), Delhi. Seeds of *Acacia nilotica* and *Eucalyptus* hybrid were obtained from Tata Energy Research Institute (TERI), New Delhi.

Inoculum Preparation

Laccaria laccata was maintained on agar slants of Modified Melin-Norkan's (MMN) medium. This strain was cultivated under submerged conditions in a fermenter Biostat C-22L in chemically defined medium. Mycelium obtained was washed thoroughly with sterile distilled water to remove the traces of remaining nutrients. Rinsed mycelium was homogenized in a presterilized Warring blender type homogeniser. Slurry of mycelial fragments was mixed with equal volume of sterile 4 per cent sodium alginate solution and added drop by drop to sterile 0.1 M calcium chloride solution with continuous stirring. After hardening, the beads obtained were rinsed with sterile distilled water. Each bead contained approximately 100 colony forming units. The alginate beads were stored at 4°C in sterile distilled water before using it for inoculation of plant seedlings.

Treatment

Preparation of Soil-vermiculite Mixture

One volume of vermiculite was mixed with one volume of soil. Either sterile or non-sterile soil-vermiculite mixture was used in experiment. Sterilization of soil-vermiculite mixture was carried out by keeping the mixture for 2 hrs at 121°C under 15 psi and allowing it to cool down for 24 hrs. Polyethylene bags were then filled with the sterile or non-sterile mixture.

Inoculation of *Acacia nilotica*

Seeds of *Acacia nilotica* were graded on the basis or their weight. Seeds of weight ranging from 0.145 to 0.165 g were soaked in 9 N sulphuric acid for 15 minutes and washed thoroughly with sterile distilled water. Acid treated seeds were then soaked for 24 hrs in distilled water. Inoculum beads at the rate of 10 beads/bag were placed at a depth of six to eight centimetre. Acid treated seeds at the rate of one seed/bag were placed at a depth of one to two centimetre. Control seedlings did not receive inoculum beads. Three repeats of twenty seedlings per treatment were placed in nursery.

Inoculation of *Eucalyptus* Hybrid

Eucalyptus hybrid seeds were sowed in a presterilized soil-vermiculite mixture. Seedlings at the four leaved stage were thinned down to one seedlings per bag, either containing presterilised or non-sterile soil-vermiculite mixture. Inoculum beads at the rate of 10 beads/seedling were placed near the root zone while the control seedlings did not receive any inoculum bead. Three repeats of twenty seedlings per treatment were placed in nursery.

Results

Inoculation with the *Laccaria laccata* significantly affected the growth of seedlings. *Acacia nilotica* showed increment of 185 per cent (Figure 29.1) with respect to uninoculated seedlings in sterilised soil-vermiculite mixture. Growth of *Eucalyptus* hybrid was significantly higher when inoculated with *Laccaria [accata* than the uninoculated plant seedlings (Figure 29.2). In sterile soil-vermiculite mixture, *Eucalyptus* hybrid showed increase up to 147 per cent (Figure 29.2, treatment A, B) to inoculation. Inoculation of *Eucalyptus* hybrid with *Laccaria laccata* showed improvement up to 112 per cent in comparison to its non-inoculated seedling in non-sterilized soil-vermiculite mixture (Figure 29.2, treatment C, D). Contribution of the microorganism in improvement of plant growth is also apparent (Figure 29.2, treatment A, C). Contribution in growth due to inoculation in non-sterilized condition is quite significant but not as high as in sterile soil-vermiculite mixture. Analysis of variance showed that there is no significant difference between the inoculated *Eucalyptus* hybrid seedlings in sterilised and non-sterilised soil-vermiculite mixture (Figure 29.2, treatment B, D).

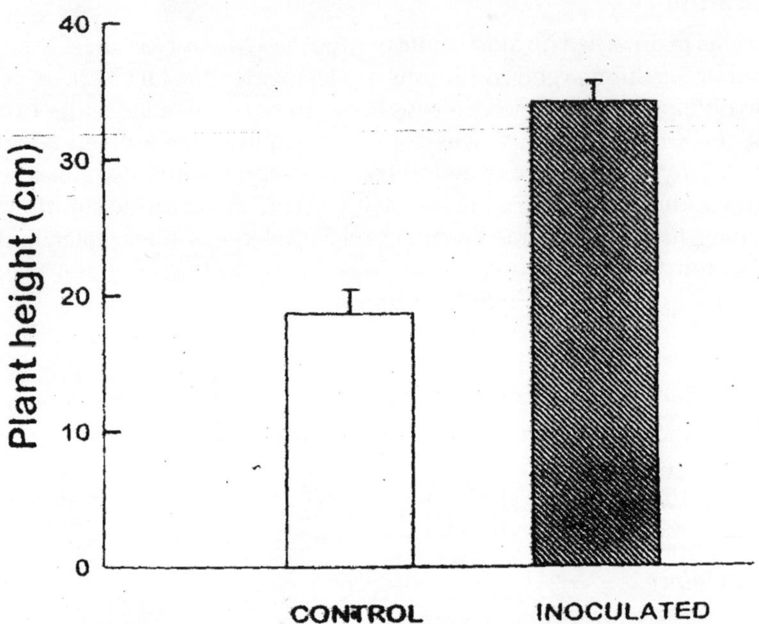

Figure 29.1: Effect on Inoculation of *Laccaria laccata* on *Acacia nilotica* in the Sterile Soil-vermiculite Mixture where Control Seedlings did not Contain Inoculum while Inoculated Seedlings Contained Inoculum of *Laccaria laccata* at the Rate of Ten Beads/Seedlings

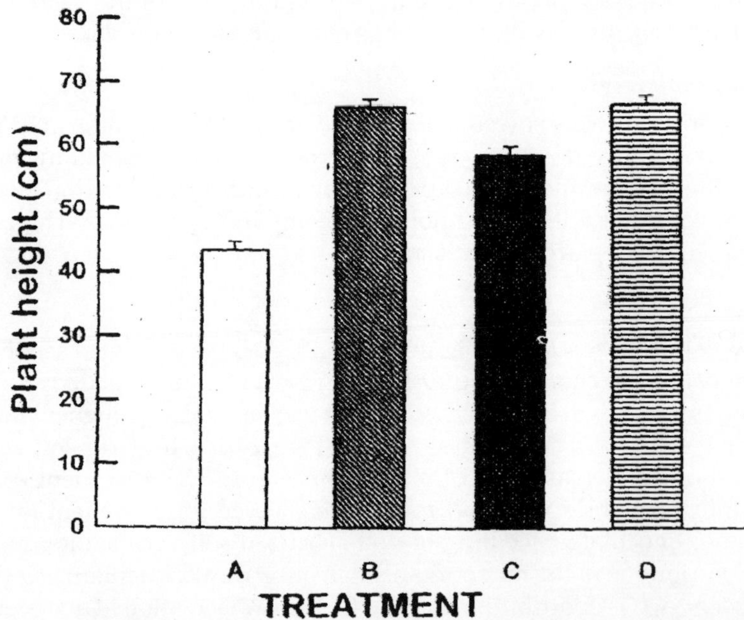

Figure 29.2: Effect on Inoculation of *Laccaria laccata* on *Eucalyptus* Hybrid where various Treatments are Sterile Soil-vermiculite Mixture without Inoculum (A) and with Inoculum (B) and Non-sterile Soil-vermiculite Mixture without Inoculum (C) and with Inoculum (D)

Discussion

The alginate entrapped inoculum of *Laccaria laccata* for inoculation of forest tree seedling is very handy and robust in comparison to its counterpart peat-vermiculite based inoculum. We have produced the inoculum based on alginate matrix. In this matrix the inoculum is physiologically consistent and remaining stable for six months. *Laccaria laccata* is an early stage ectomycorrhizal fungus which has been used for inoculation of various tree seedlings. Inoculation of *Laccaria laccata* increased the growth of *Acacia nilotica* and *Eucalyptus* hybrid seedlings in comparison to uninoculated seedlings in sterilised and non-sterilised soil. Endomycorrhiza is very common in *Acacia* species, but very few reports show the possibility of ectomycorrhizal synthesis in this genus. Effect of inoculation on growth of *Eucalyptus* agrees with the previous studies. With this work author conclude that the granulated alginate entrapped *Laccaria laccata* can be used for the inoculation of *Acacia nilotica* and *eucalyptus* hybrid.

Chapter 30
Tree Legumes Seedlings

In general the tree crops establishment in problem soil will be severely interfered by the non-availability of basic nutrient and possible inhibition by the soil borne pathogen and other parasitic attack. For the tree crop establishment none of the inoculants are given and hence the tree crops grown in virgin soil or the waste land are very much affected by the environmental stresses and certain biotic factors. By the involvement of certain microbial interaction might be possible to alleviate the circumstances to enable them grow and establish in further proliferation. The present investigation is aimed to ascertain the role of certain nitrogen fixing root nodulating bacterial inoculation with P mobilizing VAM fungi as co-inoculant in alfisol, since no attempt was made earlier in acid soil. The possibility of these microbes not only created better environment but also slowly makes the soil to be suited for other crops to be introduced at an easiest way.

Material and Methods

A nursery study was conducted to assess the significance of nitrogen fixing *Rhizobium* specific for *Acacia* spp (fast grower) along with and without soil incorporation of three VAM (*Glomus fasciculatum, Glomus mosseae* and *Gigaspora calospora*) *Rhizobium* was given as seed bacterization at 600 g/20 kg of Acacia seeds (with a cell population of 10^7/g of peat based carrier). VAM fungi was inoculated at the rate of 2 g/plant seed in the polythene pouch (with 75 per cent infective propagate). These microbes were treated on five Acacias: (*Acacia nlangium, A. crassicarpa, A. concianna, A. holosericea* and *A. hispida-exatic* from Australia). To compare the efficacy of these bioinoculants, N and P controls were also maintained. Each treatment was replicated twenty five times to obtain the effective information and conducted in a Randomized Block Design.

At 20 days after sowing the seedling germinated on the treated and untreated pouches were assessed for germinability and vigour index. During 40 days after sowing the *Acacia* seedlings were ascertained for various growth attributes *viz.,* seedling biomass, nodulation and its biomass, shoot and root growth and percentage of VAM infection by employing the method of Haymen. The data were analyzed and nodulated (Tables 30.1 to 30.5).

Table 30.1: Influence of Endomycorrhizae and *Rhizobium* on Growth and Establishment of *Acacia mangium*

Treatment	*20 DAS (Initial)*		*40 DAS (days after sowing)*				
	Vigour Index	*Plant Biomass (g/Pl)*	*Nodule No./Pl*	*Nodule Biomass (mg/Pl)*	*Root Growth (cm/Pl)*	*Shoot Growth (cm/Pl)*	*Percent of VAM Infection*
Control	640	1.96	1.0	4	7.8	13.2	10
N-control	820	3.21	6.3	19	13.9	21.9	40
P-control	710	2.62	5.0	17	11.6	18.6	60
Rhizobium	790	2.91	14.0	53	12.5	19.6	45
VAM-1	770	2.73	10.0	28	11.9	22.0	90
VAM-2	740	2.61	6.0	19	11.7	16.9	80
VAM-3	700	2.53	5.6	17	11.3	18.7	70
R + VAM-1	870	3.42	16.6	64	14.5	24.9	90
R + VAM-2	830	3.22	12.0	49	12.7	22.3	85
R + VAM-3	810	3.01	10.0	41	12.2	19.7	80

VAM-1: *Glomus fasciculatum;* VAM-2: *Glomus mosseae;* VAM-3: *Gigaspora calospora.*

CD (P = 0.05) NS 0.4 2.9 0.7 1.2

Table 30.2: Influence of Endomycorrhizae and *Rhizobium* on Growth and Establishment of *Acacia crassicarpa*

Treatment	*20 DAS (Initial)*		*40 DAS (days after sowing)*				
	Vigour Index	*Plant Biomass (g/Pl)*	*Nodule No./Pl*	*Nodule Biomass (mg/Pl)*	*Root Growth (cm/Pl)*	*Shoot Growth (cm/Pl)*	*Percent of VAM Infection*
Control	600	2.10	4	9	9.8	15.2	15
N-control	790	5.97	10	30	16.7	26.1	30
P-control	700	3.85	8	26	12.5	21.2	45
Rhizobium	780	4.31	17	42	14.3	22.3	45
VAM-1	760	4.11	11	24	15.3	19.8	80
VAM-2	740	3.41	9	18	14.3	19.0	80
VAM-3	755	3.61	8	15	16.3	21.9	60
R + VAM-1	950	6.32	23	78	18.3	29.1	90
R + VAM-2	890	5.11	18	47	14.2	23.8	80
R + VAM-3	870	4.95	16	39	13.7	21.7	76

VAM-1: *Glomus fasciculatum;* VAM-2: *Glomus mosseae;* VAM-3: *Gigaspora calospora.*

CD (P = 0.05) 0.2 0.5 12 1.5 2.3 NS

Table 30.3: Influence of Endomycorrhizae and *Rhizobium* on Growth and Establishment of *Acacia holoserisea*

Treatment	20 DAS (Initial)		40 DAS (days after sowing)				
	Vigour Index	Plant Biomass (g/Pl)	Nodule No./Pl	Nodule Biomass (mg/Pl)	Root Growth (cm/Pl)	Shoot Growth (cm/Pl)	Percent of VAM Infection
Control	710	2.06	2	4	6.6	13.9	15
No-control	960	4.22	5	13	9.8	24.6	40
P-control	840	3.27	4	11	7.8	22.1	45
Rhizobium	920	3.78	13	45	8.9	23.1	35
VAM-1	860	3.54	9	29	8.4	21.7	80
VAM-2	830	3.22	8	21	7.9	19.7	70
VAM-3	845	3.11	7	24	7.9	16.5	60
R + VAM-1	995	4.55	18	76	10.6	27.4	90
R + VAM-2	940	4.66	13	54	9.4	25.5	80
R + VAM-3	900	4.21	10	46	9.8	23.7	78

VAM-1: *Glomus fasciculatum;* VAM-2: *Glomus mosseae;* VAM-3: *Gigaspora calospora.*

CD (P = 0.05) 45 0.12 0.6 1.4 1.8 0.99 NS

Table 30.4: Influence of Endomycorrhizae and *Rhizobium* on Growth and Establishment of *Acacia conciaana*

Treatment	20 DAS (Initial)		40 DAS (days after sowing)				
	Vigour Index	Plant Biomass (g/Pl)	Nodule No./Pl	Nodule Biomass (mg/Pl)	Root Growth (cm/Pl)	Shoot Growth (cm/Pl)	Percent of VAM Infection
Control	580	1.80	1.3	3	7.3	16.7	15
N-control	765	3.23	4	11	16.4	29.3	35
P-control	660	2.33	3	7	13.2	22.7	50
Rhizobium	730	3.70	12	32	14.6	26.9	40
VAM-1	655	2.99	10	24	16.4	24.5	80
VAM-2	610	2.66	9	21	15.2	21.1	75
VAM-3	650	2.61	8	19	14.7	23.4	75
R + VAM-1	810	3.99	15	69	18.1	32.0	90
R + VAM-2	796	3.76	11	32	16.3	29.1	80
R + VAM-3	770	3.22	10	31	15.2	26.7	75

VAM-1: *Glomus fasciculatum;* VAM-2: *Glomus mosseae;* VAM-3: *Gigaspora calospora.*

CD (P = 0.05) 0.6 0.2 0.8 1.7 2.1 15

Table 30.5: Influence of Endomycorrhizae and *Rhizobium* on Growth and Establishment of *Acacia hispida*

Treatment	20 DAS (Initial)		40 DAS (days after sowing)				
	Vigour Index	Plant Biomass (g/Pl)	Nodule No./Pl	Nodule Biomass (mg/Pl)	Root Growth (cm/Pl)	Shoot Growth (cm/Pl)	Percent of VAM Infection
Control	560	0.99	1	3	5.2	14.2	10
N-control	740	2.24	3	11	10.5	19.3	40
P-control	660	1.75	2	9	9.9	16.9	50
Rhizobium	730	1.94	11	37	10.1	18.5	45
VAM-1	690	1.99	9	29	11.2	15.2	80
VAM-2	670	1.54	8	25	10.3	16.2	80
VAM-3	650	1.77	7	22	9.8	16.1	75
R + VAM-1	795	2.38	14	44	13.9	23.5	90
R + VAM-2	740	2.10	12	34	12.1	21.3	80
R + VAM-3	720	2.16	10	31	12.4	19.7	75

VAM-1: *Glomus fasciculatum;* VAM-2: *Glomus mosseae;* VAM-3: *Gigaspora calospora.*

CD (P = 0.05) 4.5 0.4 1.2 1.6 2.4 20

Results and Discussion

The results on the influence of *Rhizobium* along with and without bioinoculation of VAM on the five Acacias have revealed that the biogrowth, nodulation shoot and root growth were significantly augmented in the inoculated treatment over their controls. This information found correlated with the report of Murugesan in which they have reported the response of ten species of Acacias to Rhizobial interference. Similar to present investigation the several reports had also indicated the role of VAM and nitrogen fixer on crops. According to the report of Sekar the inoculation of biofertilizer tends to enhance the seedling growth, root colonization and uptake on P nutrient in three trees in Shola forest ecosystem. In the present investigation the dual inoculation of *Rhizobium* with *Glomus fasciculatum* registered better symbiosis and field establishment and nodule formation followed by the combination of *Rhizobium* with either of *Glomus mosseae* by *Gigaspora calospora* in all the five tree legumes in alfisol.

Natarajan indicated effective growth of *Acacia nilotica* by the *in vitro* inoculation of mycorrhizae. Thomas indicated the possible increase in root growth by phosphate mobilizing fungi. The role of VAM on uptake of micronutrient with special reference to Zn was reported by Burgres.

Dual inoculation of favoring the biomass production and yield aspect on various crop along with nutritive mobility were also documented observed with *A. holoserecia* and in *A. crassicarpa* under alfisol and able to put forth better wood lot as compared to other Acacias.

The possible influence on the enhancement of symbiosis and plant establishment in alfisol might be due to the following reasons:

1. Inherent capacity of nitrogen fixation of *Rhizobium* and mobilization of fixed P by VAM might have registered better effects over their individual inoculation.

2. The possible approach on arresting the invading soil borne pathogen by VAM proliferation can able to increases the germinability and vigour of the crop, and
3. Under triangular symbiosis plant obtains N and P from these in endosymbionts, whereas *Rhizobium* provides N for VAM and plant and it obtain energy from plant and P from VAM. In turn VAM mobilise fixed P (iron and Alumina P) to polyphosphate form and able to supply for host species and able to get energy from it.

Chapter 31

Infectivity on Growth of *Cajanus Cajan*

Cement dust is a common air and soil pollutant around cement factories and construction sites. It is reported that the particular matter (dust) from kiln exhaust falling on the leaves affect the plant growth by clogging stomata due to formation of crust on leaves, cause foliar injuries, bring changes in photosynthesis and transpiration, turn stigma surfaces to alkalinity and there by reducing the pollen germination and primary, fertilisation, all leading to reduce agricultural production. As a soil pollutant it is known to decrease water holding capacity of soils and affect the uptake of minerals from soils. Emanuelssan has observed, that cement kiln dust, in mild doses inducing lateral root formation at an early stage and reduction of shoot length, number of branches, leaves, inflorescence and fruits. Arul have observed poor rhizobial nodulation in crop legumes in kiln exhaust dust polluted soil.

Gulbarga District alone has 4 major and 5 mini cement industries. The impact of cement dust pollution, and of effluents from these cement factories, mining and quarrying activities for the last two decades in the district has lead to land dereliction (due to pollution and erosion) and considerable damage to vegetation.

Vesicular arbuscular mycorrhizae (VAM) formation are now considered to be very important in establishment and growth of plants, particularly in an inhospitable sites. Literature survey indicated that the effect of cement pollution on VA mycorrhizal association is very meagre and the preliminary survey on the occurrence of VA mycotrophy in plants growing around cement factories has revealed that, the plants from cement polluted soils are more intensely arbusculous than plants from non polluted areas. Hence an experiment with pigeon pea (*Cajanus cajan*), a major crop of this district, to study the impact of cement on the infectivity and efficacy of a VAM fungus.

Material and Methods

Growth Response of *Cajanus cajan* to *Glomus aggregatum* with Cement Dust Amendments

A pot experiment was set-up by using a local soil (loamy sand with low fertility). Twelve earthenware pots each with 5 kg of autoclaved soil were taken. Three of the pots were treated as control without adding cement dust and in the other pots cement dust was added at the rate of 1 g/kg soil (treatment-1), 2 g/kg soil (treatment-2) and 4 g/kg soil (treatment-3) each in three replicates. All the twelve pots were adged with 50 g of pure *Glomus aggregatum* inoculum (having the infective propagules level approximately 0.2×10^4 propagules/g soil) as an uniform layer. Ten surface sterilized seeds of *Cajanus cajan* (cv PT 221) were sown in each pot by pushing down to 1 cm depth. All the pots were regularly watered and maintained in green house beds.

After 75 days of growth, the plants were harvested and assessed for percent mycorrizhal association and for the determination of dry weight.

Assessment of Per cent Mycorrhizal Association

The percent mycorrhizal association was done by using the simplified slide technique of Giovannetti 40-50 root segments (1 cm) were selected for each treatment and stained by following the root cleaning and staining technique at random and the root segments were mounted on slides at the rate of 10 segments per slide and observed under the microscope to record the absence or presence and to count VAM structures (vesicles and arbuscules) in each root segment and expressed the results as percent VAM association/colonisation.

$$\%\ \text{VAM colonisation} = \frac{\text{Number of root bits positive for colonisation}}{\text{Total number of root bits observed}} \times 100$$

Estimation of Dry Weight

The plants after 75 days of the growth from all the sets (control and cement dust treated) were uprooted taking care not to damage the roots. The roots were washed in running water and dipped in water several times till the adhering soil particles were completely removed, then the roots and shoots were separated and oven dried at 70°C for 72 hours. The dry weights of root and shoot were separately recorded.

Results

Infectivity of *G. aggregatum* and its efficacy as judged by the growth response of pigeon pea in 3 different concentrations of cement dust amendment in soils is given in Tables 31.1 and 31.2.

Table 31.1: Effect of Cement Dust Amendment in Soils Infested with *Glomus aggregatum* on the % VAM Association and Associated Fungal Structures in *Cajanas cajan* (L.) Millsp. (cv. PT-221) Plants at the Age Level of 75 Days

Treatment*	pH	% VAM Association	* Vesicles	% Arbuscules
Glomus aggragatum (control)	8.5	65	65	65
Glomus aggragatum + 1 g cement/kg soil	8.7	80	45	80
Glomus aggragatum + 2 g cement/kg soil	9.2	90	45	90
Glomus aggragatum + 4 g cement/kg soil	9.5	90	35	85

*F = 10.6 ($P < 0.01$).

Infectivity

It is clearly evident from the present investigation that the per cent VAM association in pigeon pea roots increased with the increase of cement content in the soil (Table 31.1). The VAMF structures, both arbuscules and vesicles were also found to vary quantitatively. The formation of vesicles inside the host root in cement treated soils was found gradually declined depending upon the concentration as compared to control. On the contrary the intense arbuscule formation was observed inside the roots of plants growing in cement dust polluted soils and this arbusculous nature was found to be very dense as the concentration of cement dust increased. The pH of the soil was also found increased due to the addition of cement in soil.

Efficacy

The effectiveness of *G. tiggregatum* under the influence of cement dust in a sterilized 'p' deficient soil when assessed by studying the growth response of pigeon pea after 75 days growth showed that the cement dust amendment in soils could affect the growth of the plants and vis-à-vis the effectiveness of mycobiont (Table 31.2). There is a significant (at 5 per cent level) decrease in shoot length and dry weight of both root and shoot systems due to cement addition in soil, and it was found to be in relation with the cement content in soil.

Table 31.2: Effect of Cement Dust Amendment in Soils Infested with *Glomus aggregatum* on the Growth of *Cajanus cajan* (L.) Millsp. (cv. PT-221) Plants at the Age Level of 75 Days

*Treatments***	*Shoot Length (cm)**	*Dry Weight**	
		Root	*Shoot*
Glomus aggregatum (Control)	31.31±6.63	4.66±0.16	2.30±0.57
Glomus aggregatum + 1 g cement/kg soil	26.44±2.83	4.92±0.19	2.51±0.71
Glomus aggregatum + 2 g cement/kg soil	24.42±1.87	3.20±0.96	1.70±0.57
Glomus aggregatum + 4 g cement/kg soil	21.00±2.03	2.96±0.98	1.32±0.43

* Average value of 3 replicates with 10 plants in each replicate.

** $P < 0.05$.

Similarly, the proliferation of roots system in cement dust treated plants was also high as compared to control. However, the dry weight of the root system of cement treated plants decreased when compared to control plants.

Discussion

The results obtained on the effect of cement dust amendment in soils in the infectivity and effectiveness of VAM fungus (*G. aggregatum*) clearly indicated that the cement dust pollution enhanced the degree of VAM fungal colonisation in pigeon pea as compared to control (Table 31.1). It has been reported by Arul that the kiln exhaust induced alkaline soil did not affect the existence of mycorrhizal spores and their colonization of a few popularly grown legumes in Tamil Nadu. Phosphorus deficiency in an alkaline pH and high levels of Ca associated (calcarious) soils may practically explain the higher levels of VAM colonisations as reported by Lesica. However, the associated structures, both arbuscules and vesicles inside the root cortex differed quantitatively. The intensity of formation of arbuscules increased significantly (significant at 1 per cent level) with the addition of cement dust and equally there is a precipitous decline in the formation of vesicles (Table 30.1). This indicated that there is a tilt in the degree of dependency and it is an indicative of change in the degree of benefit that a plant species receives when grown at various soil conditions.

The growth stimulation in plant due to VAM fungal association is well documented and the growth promoting ability of *G. aggregatum* in pigeon pea is known. Cement dust pollution in soils was found to affect the growth of pigeon pea and there is a significant (significant at 5 per cent level) decrease in shoot length and dry weight of both shoot and root systems (Table 30.1). Similar effect on rhizobial nodulation was observed by Arul in crop legumes in cement kiln exhaust dust polluted soil.

Host plant growing normally in soils with relatively low nutrient status are predisposed to effective endophyte fungal invasion. In the present study similar situation was observed in the loamy sand soil with low fertility where extensive root proliferation and endophyte invasion is evident. Similar observation of intensive VAM development in *Cassia occidentalis, C. sericea* plants was observed in loamy sand soil by Narayana Reddy and Ramchander Goud. The importance of soil type and the pollutants like heavy metals or sewage sludge in affecting VAM formation and function was discussed earlier.

Chapter 32
Saline Soil Tolerance

Saline, saline-alkali soils commonly referred as salt affected soils are quite extensive in India and occupy over 12 million hectare. Salt affected soils contains various cations and anions which interact with Na^+ and Cl^- and influence the effect on plant response. The plant species mostly crop plants differ in their response to salinity level. Salt affected soils are widely distributed in India and occupy an area of 7 million hectares. In Andhra Pradesh it comprises an area of 2.40 lakh hectares.

Endomycorrhizae (Vesicular Arbuscular Mycorrhizae) from an intimate association with plant roots. Majority of plant species depend upon mycorrhizal associates for adequate nutrient uptake, those lacking mycorrhizae can be severely stunted with low growth. In addition to greatly enhanced uptake of nutrients they confer other benefits to their host. They are known to increase drought resistance of young seedlings detoxify certain soil toxins, enable seedlings to with stand high soil temperature extreme acidity and alkalinity. Thus these fungi play an important role in difficult soils and has a potential application in afforestation and reforestation programmes.

Sapindus emarginatus is an economically important tropical deciduous species, usually confined to dry deciduous forests. A moderate hand some tree, it is frequently cultivated for ornamental purpose or for the sake of its fruits, the pulp of which is used as a substitute for soap and also have medicinal value. The fruit is emetic, tonic, astringent and anthelmintic. It is used in the treatment of asthma, colic due to indigestion, diarrhoea, cholera and paralysis of the limbs and lumbago. Fruit powder is taken with honey in the treatment of tonsils. Roots and bark are employed as mild expectorant and demulcent.

The knowledge on response of *Sapindus emarginatus* to salinity is scanty. In the present study efforts had been made to study the response of *Sapindus emarginatus* to various levels of saline alkaline soils in a pot experiment at nursery level after establishing infection with *Glomus fasciculatum*.

Material and Methods

Fruits of *Sapindus emarginatus* were obtained from the Forest Department of Tirupati Division, India. The seeds were separated carefully without affecting the seed coat. The seeds were surface sterilized by 0.1 per cent H_2O_2 for 15 minutes, rinsed with tap water followed by sterile distilled water. Half strength Murashige and Skoog medium was prepared as described by Narayanaswamy. Single surface sterilized seed was placed in each slant containing MS medium under aseptic condition. The tubes were placed in seed incubator for germination.

Culture of *Glomus fasciculatum* (Thaxt) Gerd. and Trappe (obtained from ICFRI, Dehra Dun, India) was established in 4-L pots on Sorghum (Co26) in sand, soil, red earth (1 : 1) rooting medium for several 21 days cycle to increase the mycorrhizal inoculum potential of the medium. The spores of *Glomus fasciculatum* were isolated from the rooting medium by sieving and decanting technique described by Gerdman. Each individual spore is carefully picked up by observing under dissecting microscope placed in a inoculation chamber under aseptic conditions to the emerged radicle of *Sapindus emerginatus* on MS medium. The tubes were incubated for establishing infection with *Glomus fasciculatum* to the young roots for one month.

Saline alkaline lands were selected from three areas *viz.*, Akasaganga, Jeevakona and Thumburavanam in Rayalaseema region of Andhra Pradesh. Soil samples from different horizons *i.e.*, 0 cm, 30 cm, 60 cm and 90 cm were collected from each area separately. The soil from different horizons was mixed individually, dried, processed and passed through 2 mm sieve for further analysis. Analysis of soil samples was done for particle size distribution, cation exchange capacity, organic matter, exchangeable cations and composition of saturation extract. The soil from each area was autoclaved at 121°C for 1 h for three consecutive days and filled earthen pots of 25 cm and 13 cm depth. Thirty earthen pots were used for experiment and 10 were used as control with fertile autoclaved soil. The seedlings grown on MS medium were uprooted and washed with sterile distilled water to remove the media attached to the roots. Two seedlings were transplanted to each pot filled separately with the saline-alkaline soil collected from three different areas and also to the fertile soil collected from agricultural field. All the forty pots were placed in green house for four months and watered with 700 ml sterilized tap water on every alternate day. The percentage of survival in control and experimental pots was recorded.

The seedlings were carefully uprooted to avoid the possible loss of root system and were thoroughly washed first with tap water than in 0.1 N HCl followed by distilled water. The root system of four plants per each type of soil was separated and stored in preserving and fixing solution of 37 per cent formaldehyde solution, glacial acetic acid and 95 per cent ethanol with a 0.05/0.05/0.9/v/v/v ratio until mycorrhizal colorization could be assessed. Spore density in the pots with three different saline alkaline soils were determined by the procedure described by Gerdman and Nicolson. For VA mycorrhizal fungal colorization assessment. The designated root samples were cleared and stained for fungal structures as described by Brundrett *et al.* and stored in glycerine. The VA mycorrhizal fungal assessment method is described by McGonigle. Approximately 100 intersects of each species were assessed for the population of the root colorized. The shoot height and root height of root height of the seedlings were measured. The leaves were separated and leaf area was measured using Skye leaf area meter. The fresh weight of the seedlings were determined. The plant material was dried at 80°C and dry matter yield was recorded. The data obtained was subjected to ANOVA.

Results and Discussion

The saline alkaline soils collected from three different areas were characterized by high pH 8.9 to 9.6, ESP 32-46, ECE 6.1-11.6 dSm–1, water soluble salts are CO_3^{-2}, 43.8–62.5 and HCO_3^- 43.8–62.5 Cl^- 9.11–14.93 and SO_4^{-2} 8.14–10.63 (Table 32.1). The amount of nitrogen and organic carbon was found significantly more ($p < 0.05$) in the soil collected from Jeevakona. The amount of sodium was significantly high ($P < 0.05$) in the soil collected from Akashaganga and Thumburavanam than the soil from Jeevakona. The other cations Ca^{+2}, Mg^{+2}, K^{+} were also found more in the soils collected from other than Jeevakona area.

Table 32.1: Composition of Saturation Extract, pH, $CaCO_3$ of Saline Alkaline Soils Collected from Different Areas

Location	ECe DSm^{-1}	Cations (me/L)				Anions (me/L)				pH	ESP	$CaCO_3$ %
		Ca^{2+}	Mg^{2+}	Na^{+}	K^{+}	CO_3^{2-}	HCO_3^{-}	Cl^{-}	SO_4^{2-}			
Akasaganga	11.6	0.19	0.09	88.0	0.06	81.36	62.5	14.93	10.63	9.6	46.0	4.5
Jeevakona	6.1	0.11	0.05	69.0	0.02	59.40	43.8	9.11	8.14	8.9	32.0	3.6
Thumburavanam	8.4	0.14	0.07	73.0	0.04	67.89	59.0	11.63	9.76	9.2	40.0	8.0

Table 32.2: Physical and Chemical Properties of Saline Alkaline Soils Collected from Three Different Areas

Location	Particle Size Distribution			Organic Matter %	Total Nitrogen %	Exchangeable Cations			
	Sand %	Silt %	Clay %			Ca^{2+}	Mg^{2+}	Na^{+} [mol (P+)] kg^{-1}	K^{+}
Akasaganga	63.81	22.24	13.95	0.31	0.028	6.92	4.6	14.37	0.92
Jeevakona	58.90	24.98	16.12	0.46	0.048	4.68	2.4	10.63	0.58
Thumburavanam	60.13	21.63	18.24	0.26	0.032	5.79	3.2	13.60	0.89

Table 32.3: Percentage of Survival, Shoot Height, Leaf Area, Fresh and Dry Matter Yield of *Sapindus emarginatus* Seedlings

Area of Soil Collection	% of Survival	Shoot Height (cm)	Root Height (cm)	Leaf Area (cm)	Fresh Matter Yield (g)	Dry Matter Yield (g)	% of VAM Infection	No. of Spores g^{-1} Soil
Akasaganga	58	11.6	5.0	116	9.2	4.3	52	24
Jeevakona	91	19.4	7.6	184	14.3	8.0	76	38
Thumburavanam	76	14.8	5.9	138	10.6	5.2	59	29
Fertile soil	94	19.8	7.9	193	15.1	7.9	78	41

Maximum percentage of survival (94 per cent) of *Sapindus emerginatus* seedlings were found in fertile soil (pH 6.8) followed by soil collected from Jeevakona area (pH 8.9). However the percent of germination between these two soils is not significant at 5 per cent level. The minimum percentage of germination was found in the soil from Akashaganga (pH 9.6). Seventy six percent of germination was found in the soil of Thumubravanam area (pH 9.2). There was no significant difference of shoot height and root height between the fertile soil and soil collected from Jeevakona. Whereas the shoot and root height was found to be inhibited significantly ($P < 0.05$) in the two soils other than Jeevakona. Maximum leaf area (193 cm) was found in the seedlings grown on fertile soil, followed by soil from Jeevakona, Thumburavanam and Akashaganga. The fresh and dry matter yield of the seedlings were found significantly lower ($P < 0.05$) in the soils from Akashaganga and Thumburavanum. Similarly there was not significant different in per cent of VAM infection and number of spores between soil from Jeevakona and fertile soil. The percent of VAM infection and occurrence of number of spores inhibited in the soils of Thumburavanam and Akashaganga. The difference was significant at 5 per cent level. A positive relationship was observed between the per cent infection, fresh and dry matter yield of the seedlings.

In the present experimental conditions VAM treatment to the *Sapindus emarginatus* could able to increase the tolerance of the seedlings to pH 8.9 and high slat concentration. Establishment of VAM infection to the seedlings is therefore essential to undertake national development programmes of afforestation in saline alkaline soils to maintain ecological balance and environmental stability.

Chapter 33

Importance of VAM Mycorrhizae

Vesicular arbuscular mycorrhizal (VAM) fungi are known to improve growth and yield of many crop plants mainly through phosphorus nutrition. However application of this technology in commercial crop production is minimal as VAM fungi are obligate symbionts and are not cultured under laboratory conditions. The obligate symbiotic nature of VAM fungi presently dictates that all VAM inoculum be grown on roots of an appropriate host plant. The best way to utilize VAM fungi for crop production may be to concentrate on transplanted crops like chilli and tomato which are normally raised in nursery beds. They can be easily inoculated to nursery beds at the time of sowing and precolonized seedlings can be transplanted to main field to harness the benefits of mycorrhization. Some researchers have observed wide variations among and with in different species of VAM fungi in their ability to promote plant growth. Recently some scientists have noticed host preference for VAM endophytes. Looking into these findings it seems advantageous to select VAM fungus suitable for a particular host-soil-climate combination. Hence the present investigations were carried out to select an efficient VAM fungus each for chilli and tomato, to determine their optimum dares of inoculation and to assess their performance on plant growth, and yield at different P levels in versitol.

Material and Methods

Glomus fasciculatum, G. macrocarpum, Gigaspora margarita, Acaulospora laevis, and *Sclerocystis dussii* were tried for the selection of efficient VAM fungi for chilli var. Byadagi and tomato var L-15 based on their inoculum potential. Earthen pots of 30 cm diameter were filled with unsterile vertisol which was P-deficient (28 kg P_2O_5 per hectare). VAM inocula were inoculated to pots with uniform number of infective propagules. Comparable uninoculated control pots were maintained. Fertilizer NPK were given at the recommended dose (Chilli 150 : 75 : 75 and Tomato 115 : 100 : 60 kg NPK/ha). Bold healthy of chilli var. Byadagi or tomato var L-15 were sown to pots. All agronomic practices were followed. After crop harvest, shoot dry weight and weight of fruits were recorded. Per cent mycorrhizal root colonization and mycorrhizal spore counts were estimated. Shoot P concentration was determined by vanadomolybdate phosphoric yellow method.

After analysing the results of this pot trial, nursery beds of chilli and tomato were raised. The efficient VAM inoculum for each crop was tried at different levels (0, 0.5, 1.0, 1.5, 2.0 and 2.5 kg per bed) to know the optimum does of inoculum. The size of each nursery bed was one square meter. Comparable uninoculated control beds were maintained precolonized seedlings were transplanted to microplots of size 5.4 × 5.4 square meters for chilli and 4.5 × 3.75 square meters for tomato. Fertilizer NPK were given at recommended level (as mentioned above) for each crop. The plant growth, yield and mycorrhizal parameters were estimated by following standard procedures (as mentioned above).

After determining the optimum level of efficient inoculum for each crop, another field trial with field levels of P (0, 20, 40, 60, 80 and 100 per cent of recommended dose for each crop) was conducted to workout the P-savings with inoculation of efficient VAM fungi. All agronomic practices and parameters determined were similar as in the previous trial.

Results and Discussion

Chilli and tomato responded well to the inoculation of VAM fungi. Among different VAM fungi tried, *Glomus macrocarpum* and *G. fasciculatum* caused significantly maximum root colonization, spore number, shoot dry weight, shoot P concentration and fruit yield in chilli and tomato respectively (Tables 33.1 and 33.2) as compared to other VAM fungi. Inoculation of respective efficient VAM fungi to the nursery beds resulted in precolonized seedlings. Per cent root colonization and spore numbers increased with increase in inoculum levels in both the crops. Both these parameters did not differ significantly between 2.0 and 2.5 kg (Tables 33.3 and 33.4). A matching trend was observed in plant dry weight fruit yield and plant P concentration also (Tables 33.3 and 33.4) in both the crops. Thus inoculation of 2.0 kg of respective efficient VAM inoculum was found be optimum.

Table 33.1: Effect of Different VAM Fungi on Mycorrhizal and Plant Parameters in Chilli var. Byadagi

VAM Fungi	*Per cent Root Colonization*	*Spore Count per 50 g Soil*	*Shoot P Conc. (%)*	*Shoot Dry Weight (g/plant)*	*Weight of Green Fruits (g/pot)*
Gtomus fasciculatum	88	446	0.19	19	106
Gigasora margarita	77	396	0.16	15	89
Acaulospora lacvis	70	361	0.14	11	73
Sclerocystis dussii	71	369	0.14	11	76
Glomus macrocarpum	90	478	0.24	22	135
Uninoculated Control	41	98	0.06	6	48
CD at P = 0.05	4.5	101.8	0.02	1.06	4.89

Table 33.2: Influence of Different VAM Fungi on Mycorrhizal and Plant Parameters in Toamto var. L-5

VAM Fungi	*Per cent Root Colonization*	*Spore Count per 50 g Soil*	*Shoot P Conc. (%)*	*Shoot Dry Weight (g/plant)*	*Weight of Green Fruits (g/pot)*
Gtomus fasciculatum	93	476	0.35	23	1.10
Gigasora margarita	29	455	0.27	20	0.98
Acaulospora lacvis	86	439	024	19	0.98
Sclerocystis dussii	82	438	0.22	19	0.85
Glomus macrocarpum	90	470	0.33	22	1.00
Uninoculated Control	38	86	0.17	15	0.60
CD at P = 0.05	4.1	9.8	0.02	3.0	0.26

In the last trial, per cent root colonization and spore counts were found to increase with increase in P-level up to 80 per cent of recommended dose beyond which these parameters decreased in both the crops (Tables 33.5 and 33.6). The plant dry weight, plant P concentration and fruit yield increased with increase in P level. However these parameters did not differ significantly between 80 and 100 per cent of recommended P in both the crops (Tables 33.5 and 33.6).

Table 33.3: Effect of Different Levels of Efficient VAM Fungus, *Glomus macrocarpum* on Mycorrhizal and Plant Parameters Under Field Conditions in Chilli var. Byadagi

Inoculum Level (kg/bed)	*Per cent Root Colonization*	*Spore Count 50 g Soil*	*Plant P Conc. (%)*	*Plant Dry Biomass (g/plant)*	*Fruit (g/plant)*	*Yield (Q/ha)*
0	26	70	0.18	49	132	20.2
0.5	47	103	0.23	63	137	20.5
1.0	61	141	0.30	72	151	21.5
1.5	71	183	0.40	80	166	24.0
2.0	79	202	0.46	87	182	26.2
2.5	79	202	0.47	89	185	26.6
SE m (±)	1.13	2.19	0.01	0.97	2.03	0.43
CD at P = 0.05	3.42	6.61	0.02	2.94	6.10	1.31

Note: Size of Nursery bed 1 m × 1 m.

Table 33.4: Influence of Different Levels of Efficient VAM Fungus, *G. fascilulatum* on Mycorrhizal and Plant Parameters in Tomato var. L-15 Under Field Conditions

Inoculum Level (kg/bed)	*Per cent Root Colonization*	*Spore Count 50 g Soil*	*Plant P Conc. (%)*	*Plant Dry Biomass (g/plant)*	*Fruit (g/plant)*	*Yield (Q/ha)*
0	27	72	0.18	46	1950	40.1
0.5	49	125	0.26	54	1980	41.0
1.0	65	164	0.32	67	2096	43.4
1.5	73	195	0.38	75	2232	46.2
2.0	79	224	0.48	83	2318	48.06
2.5	79	225	0.48	83	2328	48.27
SE m (±)	1.36	2.09	0.01	0.96	13.6	0.74
CD at P = 0.05	4.10	6.30	0.012	2.90	48.9	2.22

Glomus macrocarpum and *G. fasciculatum* exhibited host preference in chilli and tomato respectively perhaps because of optimum inflow rates of P. Two kilogrammes of respective efficient VAM inoculum seems to contain optimum number of infective propagules to cause maximum plant growth and yield in both chilli and tomato. Thus it was found to be an economical and optimum dose. Inhibition of per cent root colonization and spore counts at higher P level is well documented. This may be ascribed to decrease in membrane mediated root exudated. The benefits of mycorrhization in terms of plant growth and yield can be obtained at 80 percent of recommended P in both the crops suggesting the possibility of savings in P to an extent of 20 per cent of recommended dose.

Table 33.5: Performance of Efficient VAM Fungus, *G. marcrocarpum* at Different P Levels in Chilli var. Byadagi Under Field Conditions

P-level (% Rec. dose)		Per cent Root Colonization	Spore Count 50 g Soil	Plant P Conc. (%)	Plant Dry Biomass (g/plant)	Fruit (g/plant)	Yield (Q/ha)
Inoculated	0	49	70	0.09	32	95	15.5
	20	51	71	0.14	37	103	18.3
	40	65	92	0.22	62	151	26.5
	60	73	139	0.31	81	196	34.5
	80	82	189	0.40	93	254	44.2
	100	69	144	0.41	96	258	45.0
Uninoculated	100	44	54	0.36	91	230	39.0
Sem ±		1.71	5.9	0.01	1.5	6.2	1.3
CD at P: 0.05		5.30	18.3	0.04	4.6	19.0	3.9

Table 33.6: Performance of Efficient VAM Fungus, *G. fasciculatum* at Different P Levels in Tomato var. L-15 Under Field Conditions

P-level (% Rec. dose)		Per cent Root Colonization	Spore Count 50 g Soil	Plant P Conc. (%)	Plant Dry Biomass (g/plant)	Fruit (g/plant)	Yield (Q/ha)
Inoculated	0	53	79	0.08	26.4	403	15.1
	20	59	112	0.15	38.5	619	24.6
	40	70	152	0.27	54.7	1091	34.3
	60	84	196	0.34	65.7	1790	49.9
	80	69	135	0.42	71.4	2395	56.7
	100	68	114	0.45	78.0	2513	58.2
Uninoculated	100	44	35	0.39	69.8	2334	53.7
Sem ±		1.8	6.4	0.01	2.4	55.2	0.7
CD at P = 0.05		5.6	19.8	0.04	7.5	170.3	2.3

Thus these trials clearly revealed *Glomus macrocarpum* and *G. fasciculatum* to be efficient VAM fungi for chilli and tomato, two kg of respective efficient VAM inoculum to be optimum dose to harness the benefits of inoculation in terms of P savings to an extent of 20 per cent in both the crops.

Chapter 34
Biochemical and Genetic Characterization of Mineral Phosphate

Phosphate solubilizing microorganisms (PSMs) are considered to play a significant role in making available soil phosphates for the growth of plants. PSMs solubilize mineral phosphates by bringing about a drop in the pH of the surrounding either by proton extrusion or by the secretion of mono, di and tricarboxylic acids.

We have earlier reported that two PSMs, *Citrobacter koseri* and *Bacillus coagulans,* could not solubilize mineral phosphates present in alkaline vertisol soil when carbon and nitrogen sources were provided. They also did not solubilize rock phosphate when the medium was buffered. These PSMs secreted low levels of organic acids whereas 20–50 fold higher concentration of organic acids are necessary for solubilizing soil phosphates. This factor may therefore be responsible for the variations in the efficacy of PSMs in plant-PSMs inoculation experiments.

Since alkaline soils have a high buffering capacity, we screened rhizobacteria which could grow and solubilize rock phosphate under buffered media conditions. We could isolate three isolates from pigeon pea (*Cajanus cajan*) rhizosphere which secrete approximately 50mM gluconic acid. These bacteria have been identified as *Enterobacter asburiae* and could solubilize soil phosphates when they were grown in the presence of glucose as C source and ammonia or nitrate as the N source.

It is known that pyrroloquinoline quinone (PQQ)-dependent glucose dehydrogenase (GDH) (EC 1.1.99.17) is responsible for gluconic acid secretion in *Erwinia herbicola, Pseudomonas cepacia* and *Acinetobacter calcoaceticus. A. c.alcoaceticus* has two GDH isozymes. GDH-A is a monomer of 82Kd and is localized in the plasma membrane whereas GDH-B is a homodimer of 50 Kd and is present on the outer side of periplasmic membrane. While both the enzymes can use glucose as the substrate, GDH-B can also act on disaccharides like maltose and lactose whereas deoxyglucose is a specific substrate of GDH-A. The GDH from *E. coli* and *Pseudomonas* is like GDH-A. GDH from *P. aeruginosa* is 20 per cent active with galactose as substrate as compared to GDH from *P. cepacia* which is as active with galactose as with glucose.

In this chapter, we show that the activity of GDH of P-solubilizing *E. asburiae* is increased by about 5-fold under P starvation conditions and protein synthesis is required for this increase. The GDH is similar to the B type enzyme found in *A. calcoaceticus.*

Material and Methods

Culturing Procedures

The *E. asburiae* strains were cultured on media containing 100 mM glucose, 25 mM $MgSO_4$, 10 mM NH_4Cl and the following micronutrients (mg/L) $FeSO_4$. $7H_2O$ (3.5), Zn SO_4 $7H_2O$ (0.16), $CuSO_4.5H_2O$ (0.08), H_3BO_3 (0.5), $CaCl_2$ $2H_2O$ (0.03) and $MnSO_4$ $4H_2O$ (0.4). Phosphate source was 0.1 mM KH_2PO_4. The medium was buffered with 100 mM Tris-Cl pH 8.0. These cells were used to inoculate to fresh medium as described below.

Induction of *GDH* and Alkaline Phosphatase (*AP*) of Soil Isolates

The isolates were inoculated in minimal medium containing 100 mM glucose, 10 mM NH_4Cl, 10 mM K_2HPO_4 and the micronutrients. The medium was buffered with 100 mM Tris-HCl pH 8.0. The cells were harvested after growth to 0.4 O.D. and washed twice with normal saline. The cells were then resuspended in same medium with 100 µM P, To monitor the effect of protein synthesis inhibitor on the induction of GDH, chloramphenicol was added at 170 µg/ml. The cells were harvested at different time intervals, washed thrice with saline and resuspended in 50 mM Tris-HCl pH 8.0 and were used for GDH and alkaline phosphatase assay. To study the levels of P concentration that was sufficient for GDH repression, different concentrations of K_2HPO_4 were added. The cells were harvested after 6 hours washed and used for GDH assay. Glucose was replaced with other carbon sources at 100 mM concentration for determining their effect on GDH. The cells were harvested by centrifugation in a table top centrifuge, washed twice with 0.9 per cent saline and resuspended in the above described media either with 10 mM P as P sufficient or with 10 mM P as P deficient conditions.

Enzyme Assays

The assay mixture for GDH consisted of 1 ml of 50 mM Tris-HCl buffer pH 8.75, 0.1 ml of 6.7 mM 2.6 Di-chlocophenolindophenol (DCIP), 0.1 ml 20 mM phenazine methosulphate (PMS) and 0.1 ml of cells in a final volume of 3 ml. GDH activity was determined by measuring the decrease in absorbence at 600 nm of DCIP mediated with PMS using glucose to begin the reaction. For alkaline phosphatase assay 0.1 ml of the cells were added in the assay system consisting of 1 mg/ml P-Nitrophenylphosphate (pNPP) in 30 mM Tris-HCl pH 9.0. The activity was determined by estimating P-Nitrophenol liberated from pNPP and comparing it to absorbence of standard P-nitrophenol at 420 nm. Protein estimations were done by the dye binding method.

Results

Induction of *GDH* and *AP*

The specific activity of GDH was found to increase upon P starvation. There was basal level of GDH activity which was increased to 5 fold in 6 hours. By P starvation in all the isolates (Figure 34.1). The increase in GDH activity correlated with the decrease of media pH. GDH of these isolates was induced only upon P starvation and was independent of the carbon source used for the growth of the cells (Table 34.1). The increase in activity upon P starvation required *de novo* protein synthesis as addition of chloramphenicol abolished the induction of GDH (Table 34.2).

Table 34.1: GDH Activity Soil Isolates with Various C Sources

C Source	GDH Activity (Units/mg total protein)	
	+ P	− p
Glucose		
Isolate 1	262±4.1	1390±149
Isolate 2	265±7.5	1737±125
Isolate 3	285±8.5	1500±55
Glycecol		
Isolate 1	242±7.5	1303±100
Isolate 2	274±10	1600±60
Isolate 3	236±5.1	1418±125
Mannitol		
Isolate 1	225±6.0	1193±130
Isolate 2	250±6.5	1395±75
Isolate 3	240±5.0	1158±100
Gluconate		
Isolate 1	200±10.0	1000±100
Isolate 2	220±5.5	900±65
Isolate 3	250±5.0	1200±120

Values are expressed as mean ± SD of three independent experiments.

Table 34.2: Effect of Chloramphenicol of Induction of GDH

	GDH Activity (Units/mg total protein)	
	Chl +	Chl −
Isolate 1	300±55	1000±100
Isolate 2	250±40	1200±125
Isolate 3	200±50	900±75

Values are expressed as mean ± SD of three independent experiments.

Substrate Specificity of *GDH* of Soil Isolates

In order to determine whether the GDH of these isolates was of GDH-A or GDH-B type the GDH activity was assayed with different substrates. The whole cells of all the three soil isolates could show activity with maltose and galactose as substrates but could not use deoxyglucose as a substrate. The activity with these substrates also increased upon P starvation similar to the increase in GDH (Table 34.3).

Discussion

The P-solubilizing *E. asburiae* isolates under study, secrete high concentration of gluconic acid and showed the induction of glucose dehydrogenase activity upon P starvation. The induction required *de novo* protein synthesis and different concentration of free P was required to repress the GDH

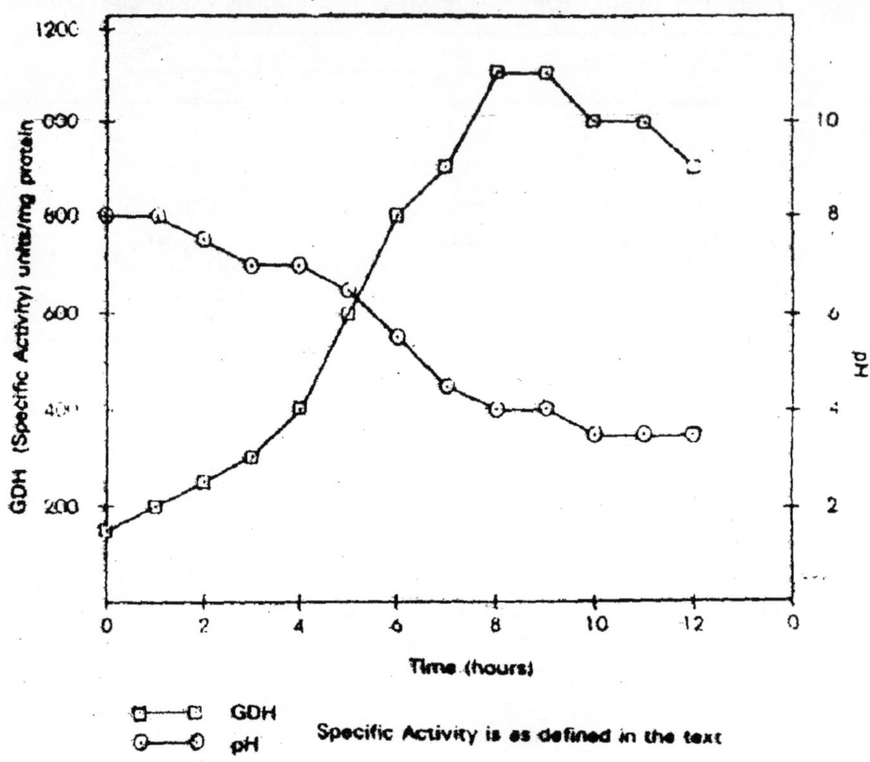

Figure 34.1: Induction of GDH of Soil Isolate 3

activity with 5 mM P repressing the activity to about 50 per cent for isolates 1 and 2 and more than 75 per cent for isolate 3. This indicates that the GDH of these isolates is regulated by the P concentration in the media and could be part of phosphate starvation inducible mechanism of these bacteria. That the soil isolates expressed genes induced for P starvation under conditions allowing maximum GDH activity is seen from the induction of alkaline phosphatase under the same conditions. It is known that various strains of *Rhizobium* are P limited at 100–150 μm concentration of P as determined by the depression of AP activity.

Table 34.3: Substrate Specifity of GDH of High-potency PS Bacteria

Substrate	*GDH Activity (Units/mg total protein)*					
	Isolate1		*Isolate 2*		*Isolate 3*	
	+P	*–P*	*+P*	*–P*	*+P*	*–P*
Maltose	100	600	150	700	150	750
Galactose	50	200	50	150	60	250
Lactose	UD	UD	UD	UD	UD	UD

Values are expressed as mean of three independent experiments.

In *E. herbicola* the P solubilizing property was shown to be inducible by P starvation. 20mM P was required to completely abolish HAP solubilization by *E. herbicola.* The mineral phosphate solubilizing gene involved in PQQ biosynthesis cloned from *E. herbicola* in *E. coli* also show P induction/repression phenotype indicating it to be part of phosphate starvation system. Similarly the DCP solubilization by *E. coli* is repressed by P concentration. However, the Ed pathway enzymes are not induced by P starvation in *E. coli.* In Pseudomonas GDR is induced by gluconate or 2-ketogluconate. The GDH of the soil isolates reported here was induced only by P starvation as the activity was similar with all the C sources used for the growth of the bacterial strains.

The GDH of these isolates was active with maltose as substrate and could not use deoxyglucose indicating it to be like GDR-B of *A. calcoaceticus* which is a periplasmic protein. The whole cells also showed activity with galactose although it was substantially less as compared to glucose.

It would be of interest to characterize further the glucogenic acid secretion by these *E. asburiae* strains and also clone the gene (s) involved in the overproduction of gluconic acid from these soil isolates. The cloned gene (s) could then be transferred to PGPR like *Rhizobium* and *Pseudomonas* allowing them to provide P to the plants they colonize.

Chapter 35
Effect of Phosphobacterium on Growth

Swordbean, otherwise called Jackbean is a dual crop and contains a protein content of around 24.5 per cent. It can playa vital role in filling up the protein shortage. It is nontraditional pulse and relatively an unexplored legume with higher productivity and wider adaptability. The tender pods are delicious and become fibrous on maturity. Seeds are large and contain an alkaloid canavalin. This crop can also be grown as an ornamental hedge or border plant or as a forage legume.

Material and Methods

The present study on swordbean, variety SBS-1 (Swordbean Selection-1), was designed to study the effect of phosphobacterium (*Bacillus megaterium*) under graded levels of phosphorus at the farm of Gandhigram Rural Institute (Deemed University), Tamil Nadu, during the year 1995–96. The experiment was carried out in randomised block design with three replications. There ware three methods of phosphobacterium application *viz.*, seed treatment (500 g of inoculum for 120 kg of seed per hectare), soil application (2000 g of inoculum per hectare mixed with dried cowdung manure) both basal along with an untreated control. Three levels of phosphorus were tried *viz.*, 25, 50 75 kg P_2O_5 per hectare with a common dose of 25 kg N as basal. Sowing was done in plots of size 4.5 × 3.0 m at a spacing of 45 × 30 cm under ridges and furrow system. Observations on 25 randomly selected plants per plot were recorded for plant height (cm), number of flowers and fruits per plant, fruit setting percentage, number of seeds per pod and seed yield (kg per hectare). One hundred seed weight (g) and germination percentage were assessed from each replication. Seeds are sown in germination trays in incubator at 25°C and 95 per cent RH. Seedling count was taken after seven days and germination was expressed in percentage.

Results and Discussion

Results obtained indicated that there was a significant effect of seed inoculation of phosphobacterium on swordbean in increasing the fruit setting percentage (49.27 per cent), number of seeds per pod (10.52) and seed yield (1410 kg per ha) as compared to soil application and untreated control (Table 35.1). This may be attributed to the presence of seed borne inoculum within rhizosphere where root hairs absorb the nutrients. Different levels of phosphorus application showed no significant difference except in influencing the number of fruits and fruit setting percentage. They were found to be more with either 50 or 75 kg P_2O_5 application as compared to 25 kg P_2O_5 per hectare. One hundred seed weight and germination percentage of seeds were not influenced by different treatments.

Table 35.1: Effect of Phosphobacterium on Growth and Seed Yield of Swordbean

Treatment	*Plant Height at Harvest (cm)*	*No. of Flowers per Plant (75th day)*	*No. of Fruits per Plant (75th day)*	*Fruit Setting Percentage*	*No. of Seeds per Fruit*	*One Hundred Seeds Weight (g)*	*Seed Yield (kg/ha)*	*Germination (%)*
Phosphobacterium								
P_0–Untreated control	111.2	36.18	14.73	40.28	8.67	132.60	1050	98.7
P_1–Seed inoculation	113.1	39.44	18.33	49.27	10.52	136.70	1410	99.1
P_2–Soil application	112.1	36.98	15.20	40.64	9.83	136.20	1250	99.6
CD (P = 0.05)	NS	0.58	0.89	1.64	0.04	0.51	140	NS
Phosphorus levels								
F_1–25 kg/ha	111.7	37.69	15.47	41.91	9.63	132.60	1200	99.1
F_2–50 kg/ha	111.2	37.20	16.33	43.51	9.70	133.80	1250	99.6
F_3–75 kg/ha	111.6	37.71	16.47	44.74	9.66	133.80	1260	99.7
CD (P = 0.05)	NS	NS	0.89	1.64	NS	0.51	NS	NS

NS: Not significant.

Application of pliosphobacterium seemed to be environment friendly and the study revealed the efficiency of seed inoculation of phosphobacterium in reducing the phosphorus loss due to insolubility thereby increasing the seed yield of swordbean.

Chapter 36

Effect of Phosphomicrobes

Groundnut requires large amount of phosphorus for growth and high phosphorus availability to the crops in the form of different phosphatic fertilizer is of great economic importance, especially in country like ours, where the soils are poor in available phosphorus content and super phosphate in expensive. Insoluble inorganic compounds of phosphorus are largely unavailable to the plants. Phosphomicrobes have a capacity in bringing the insoluble phosphate into solution. Apart from increasing the availability in soil, they are known to produce growth promoting substances which in turn enhance growth and yield. The fertility status of the soils is likely to go down unless adequate quantities of plant nutrients are added to the soil. The information on the use of phosphomicrobes on the growth and yield of crops in vertisols of Karnataka is scanty and inconsistent. Therefore long term field trial investigation was undertaken to know the performance of phosphomicrobes on yield and nutrient uptake in groundnut.

Material and Methods

A field experiment conducted at Main Research Station, University of Agricultural Sciences, Dharwad for three consecutive years of Kharif 1992, 1993 and 1994 on medium black soils. The soil having a pH of 8 organic carbon 0.54 per cent, available N, P_2O_5 and K_2O were 254, 35 and 335 kg/ha respectively. The treatments consisted of phosphorus application (50 kg P_2O_5/ha) through two sources *viz.*, Mussoorie rock phosphate (MRP) and single super phosphate (SSP) with 0, 25, 50, 75 and 100 per cent of the recommended dose. The phosphomicrobes includes *Aspergillum awamori* and *Pseudomonas striata.* Total of 18 treatments were replicated thrice in a Randomised Block Design. The fertilizers N and K_2O were given as per the recommended dose (each 25 kg/ha). After harvest of the crop yield and yield components recorded. The nodule number and weight was taken at 45 DAS. The nutrient uptake studies were made at harvest. The nitrogen in the plant sample was estimated by modified Kjeldahl's method and the phosphorus by vanadomolybdo phosphoric yellow colour method using spectrophotometer at 470 nm as described by Jackson.

Results and Discussion

Yield and Yield Components

The pod yield recorded with inoculation of *A. awamori* (3399 kg/ha) of *P. striata* (3416 kg/ha) alone were 12 and 13 per cent higher respectively when compared to control (3026 kg/ha). The beneficial effect of P-solubilizers on grain yield of rice have been reported by Gaur. Inoculation of *P. striata* along with the application of 75 per cent MRP + 26 per cent SSP recorded (4648 kg/ha) and differed significantly over uninoculated treatments (Table 36.1). Similar studies were corroborated by Manjaiah. The number of pods per plant and pod weight per plant followed the similar trends as that of pod yield per hectare.

Table 36.1: Number of Pods per Plants, Pod Weight (g) per Plant, Pod Yield (kg/ha) and Haulm Yield of Groundnut as Influenced by Phosphomicrobes (Average of three years)

Sl.No.	Treatments	Pod Numbers Per Plant	Pod Weight (g) per Plant	Pod Yield (kg/ha)	Haulm Yield (kg/ha)
1.	Control	12.86	14.23	3026	3137
2.	MRP 0 : SSP 100	17.93	19.16	3543	3743
3.	MRP 25 : SSP 75	17.20	20.10	3492	3627
4.	MRP 50 : SSP 50	18.10	20.10	3413	3637
5.	MRP 75 : SSP 25	16.60	20.63	3368	3605
6.	MRP 100 : SSP 0	16.63	17.90	3336	3445
7.	*Aspergillus awamori* (A)	17.70	19.80	3399	3360
8.	MRP 0 : SSP 100 + A	18.63	22.03	3613	3871
9.	MRP 25 : SSP 75 + A	17.96	21.76	4100	3504
10.	MRP 50 : SSP 50 + A	17.73	20.70	3600	3761
11.	MRP 75 : SSP 25 + A	20.70	20.36	3869	3173
12.	MRP 100 : SSP 0 + A	17.36	20.43	3518	3187
13.	*Pseudomonas striata* (P)	18.43	21.23	3416	3623
14.	MRP 0 : SSP 100 + P	20.40	22.53	4445	3853
15.	MRP 25 : SSP 75 + P	18.66	21.70	3842	3731
16.	MRP 50 : SSP 50 + P	19.30	21.43	3865	3531
17.	MRP 75 : SSP 25 + P	22.40	23.90	4648	3900
18.	MRP 100 : SSP 0 + P	10.06	20.50	3571	3464
	S.Em. ±	1.40	0.47	139	300
	C.D. at 5%	4.10	1.35	401	NS

MRP: Mussoorie rock phosphate; SSP: Single sugar phosphate; NS: Not significant.

The haulm yield of groundnut was unaffected due to various treatments. But inoculation of phosphomicrobes significantly influenced the nodule number and nodule dry weight at 45 DAS. Inoculation of *P. striata* with the application of 75 per cent MRP + 25 per cent SSP recorded significantly higher nodule number per plant (110.20) when compared to uninoculated treatments. Nodule dry weight was highest in *Pseudomonas striata* with 100 per cent SSP (119 mg/plant). Similarly Manjaiah reported that phosphorus application increased the nodule number and nodule dry weight and SSP was more effective than MRP.

Uptake of Nitrogen and Phosphorus

Inoculation of *P. striata* with 75 per cent MRP + 26 per cent SSP recorded significantly highest nitrogen uptake (83.29 kg/ha) over control (46.80 kg/ha) indicating 1.7 fold increase in N uptake by groundnut (Table 36.2). The increased in N-uptake by groundnut crop was ascribed due to the application of phosphorus and their synergistic interaction.

Table 36.2: Nodule Number, Nodule Dry Weight (mg/plant), Nitrogen and Phosphorus Uptake as Influenced by Phosphomicrobes (average of three years)

Sl.no.	*Treatments*	*Nodule Number/ Plant at 45 DAS*	*Nodule Dry Weight (mg/plant at 45 DAS)*	*N-uptake (kg/ha)*	*P-uptake (kg/ha)*
1.	Control	49.73	46	46.80	5.85
2.	MRP 0 : SSP100	78.73	81	66.20	9.00
3.	MRP 25 : SSP 75	64.66	75	63.00	8.50
4.	MRP 50 : SSP 50	80.83	65	60.25	8.22
5.	MRP 75 : SSP 25	90.73	50	55.40	6.90
6.	MRP 100 : SSP 0	75.50	83	59.50	7.26
7.	*Aspergillus awamori* (A)	66.00	75	61.25	7.00
8.	MRP 0 : SSP 100 + A	98.80	104	71.00	10.50
9.	MRP 25 : SSP 75 + A	77.20	68	68.25	9.25
10.	MRP 50 : SSP 50 + A	90.86	70	60.00	9.90
11.	MRP 75 : SSP + A	93.93	99	63.28	10.25
12.	MRP 100 : SSP + A	81.83	76	57.94	10.00
13.	*Pseudomonas striata* (P)	73.06	84	64.00	8.30
14.	MRP 0 : SSP 100 + P	115.46	119	80.25	11.31
15.	MRP 25 : SSP 75 + P	90.60	69	70.20	10.20
16.	MRP 50 : SSP 50 + P	102.60	80	69.60	10.70
17.	MRP 75 : SSP 25 + P	110.20	89	83.29	13.00
18.	MRP 100 : SSP 0 + P	88.06	93	69.40	9.65
	S.Em.±	4.07	2.50	0.84	0.40
	C.D. at 5%	11.72	7.45	2.50	1.20

MRP: Mussoorie rock phosphate; SSP: Single sugar phosphate; NS: Not significant.

Inoculation of *A. awamori* or *P. striata* alone recorded 20 (7 kg/ha) and 30 (8.3 kg/ha) per cent higher total P-uptake by groundnut plant than control (5.85 kg/ha). The higher P-uptake may be due to the increased availability of phosphorus in the soil due to solubilization of native phosphorus by phosphomicrobes. Ahmed and Jha reported increased P-uptake by wheat and gram with the inoculation of phosphobacterin. Among all the treatments, inoculation of *P. striata* with 75 per cent MRP + 25 per cent SSP recorded highest P-uptake (13.00 kg/ha).

The results clearly indicates the use of phosphomicrobes to augment the yield and nutrient uptake in groundnut crop. The Indian farmers succumbs to a greater loss due to the high cost incurred by application of phosphorus through SSP. Hence application of 75 per cent of the recommended dose of P through MRP and 25 per cent through SSP along with the inoculation of *P. striata* is more lucrative than applying 100 per cent of phosphorus through SSP alone.

Chapter 37
Recommendations

1. It is recognised that bulky organic manures, in particular, play their role in increasing crop yields through (*a*) supply of plant nutrients; and (*b*) improvement in several factors of soil fertility like the physical, chemical and biological properties of soils. Variable and often non-significant responses that are obtained through the use of these manures are now understood to be owing mainly to the low availability of nitrogen in them and consequently their inability to make good the primary deficiency of nitrogen in the dose applied. It is, therefore, recommended that effective use of such manures should be made by (*a*) use in adequate quantity, if available, on the basis of their content of plant utilisable nitrogen, (*b*) use in combination with fertilizer nitrogen in quantity to provide adequate nitrogen for the requirements of crops, and (*c*) use after enrichment with nitrogen as described earlier. Since the quantity of manure available with farmers is much short of their requirements, the latter two alternatives provide a solution that would invariably give an assured response.
2. It must be realized that high crop yields induced by adequate fertilization with NPK salts are bound to project sooner or later the need for adjustment of other elements, from time to time to maintain the productive capacity of soils. It is in this role of supplying various secondary and trace elements that complex organic manures of this type can be relied upon to make good known and unknown deficiencies and rectify any adverse residual effects of fertilizers, particularly in their long-term use.
3. The urine component of cattle and buffalo excreta contains 50 to 70 per cent of total nitrogen and 90 to 95 per cent of the total potash excretion. The separate collection and utilization of urine absorbed and dehydrated in waste materials offers great scope for exploiting a potential source of highly effective manure available with every farmer. As an alternative to the recommended practice of adding urine-soaked litters to the manure heap, this approach would enable the limitation of fermented manures to be overcome and a high efficiency obtained from the liquid and solid excreta.
4. The saving of about 29 per cent of dung that is presently being burnt away as fuel in villages and its used as manure is a consideration of great practical importance in the context of the new strategy for increasing crop production. There is, no doubt, that in the present economy,

the practice of burning cattle-dung as fuel to a limited extent is inescapable in rural areas and will continue at least for some more time. The question has, however, received attention and various solutions have been suggested, *viz.*, (*a*) the programme of installing and popularizing gas plants in the rural areas should be intensified; (*b*) since the use of soft coke in rural areas is not going to be cheaper in view of the tremendous cost of transport and other factors involved, the popularization of soft coke in urban and semi-urban areas may release wood to some extent for use as fuel in rural areas, thereby releasing cattle-dung for manurial purposes; (*c*) to augment the wood supply for fuel purposes in rural areas, stress must be laid on farm-forestry programme; and (*d*) the quality of manure should be improved by adoption of scientific methods of composting and minimizing losses of plant nutrients. The programme of setting up biogas plants involves high cost of installation with KVIC plants whereas the Janata biogas plant is cheaper by 40 to 50 per cent. Hence, the biogas recovery in rural farm-sheds through the Janata biogas plant programme should be intensified.

5. The technology on the method of composting available in the country is sound. However, the methods followed by the farmers are substandard. Composting of surplus and available farm wastes such as straw, leaf fall, weeds, water hyacinth and animal dung should be practised by improved methods of composting available in the country for augmenting/ supplementing the nutrients supply and humus substances for higher crop yields.
6. Preparation of rock phosphate digested compost is recommended. Low-grade rock phosphate @ 10 kg/tonne of organic material is recommended to apply while preparing the compost by pit or heap method. It has been found that rock phosphate application hastens the process of composting and improves the quality of compost.
7. The use of mesophilic cellulolytic (*Aspergillus* spp., *Penicillium* spp., *Trichurus spiralis*) inoculants is recommended for rapid composting of dry and high C/N ratio farm wastes such as straw of different types, leaf fall, etc. It has been observed that by inoculating the compost heaps by an efficient and suitable culture, the composting period can be reduced with a net gain in nitrogen.
8. Mechanical composting has made good progress in the country and semi-mechanised plants are more suitable to Indian conditions because of the advantage of low cost. The success of mechanical composting plants should not be judged purely on economic considerations because public health and checking of environmental pollution by systematic disposal of solid and liquid wastes is an important factor.
9. Concentrated organic manures like oil-cakes, blood, meat, fish and bone-meals are sold in our country at abnormal premium as compared with chemical fertilizers of equal plant nutrient capacity. This does not appear to be justified on the basis of responses obtained with them which are at best equal to or slightly better than fertilizers. Considering also that the quantity of organic matter that gets applied by the use of such organic manures is rather small and of little practical consequence, the prices paid for organic nitrogen from such sources are not commensurate with the returns obtained in crop yields, unless their prices are reduced. It should be realized that organic forms of nitrogen are first converted into mineral forms of ammonia and nitrate, before this element is assimilated by plants so that there is no particular superiority in effect that can be expected from this. Actually, organic manures do not nitrify completely and a part of their nitrogen always remains untransformed and, therefore, unutilizable. Low-grade oil-cakes can be ammoniated of enrichcd with fertilizer for fuller and more effective use.

10. Sewage water besides irrigation, adds plant nutrients to the soil and increases crop yields from 30 per cent to 50 per cent in food crops and 100 per cent in fodder crops as compared with ordinary irrigation water. The sewage water must, therefore, be fully exploited for increased crop production.
11. Organic materials rich in cellulose and lignin such as crop wastes, straw @ 5 to 10 tonnes/ha can be recycled in the first instance as surface mulch and subsequently after one season partially degraded straw should be incorporated in the soil. This is a cheap method of utilisation of surplus crop residues instead of burning it. The results have shown that organic mulch is helpful in increasing the yield of crops significantly as well as the second crop when the left-over organic matter is ploughed in the soil.
12. Another promising method for direct disposal of crop wastes such as paddy or wheat straw can be followed by incorporating the organic material @ 5 tonnes/ha and allowed it to decompose for a week followed by growing of a legume crop to advantage. The results have shown that the yield of legume crops can be increased by ploughing in straw.
13. It is recommended that adequate quantity of organic manure should be used in alkali soils in combination with gypsum as combined application interacts significantly in increasing the crop yields and also produces favourable effect on soil physical properties.
14. Further experimental evidence is needed on the comparative response and economics of various organic manures in relation to fertilizer and the effect of their combined application. Experiments are also necessary to study the effect of smaller applications or bulky organic manures, fortified with nitrogen and non-symbiotic nitrogen fixing bacteria (*Azotobacter*) and phosphorus solubilising bacteria (*Pseudomonas striata, Bacillus polymyxa*).

Experience of users of organic manures has indicated that the quality of crops, particularly fruits and vegetables is superior to the quality obtained with chemical fertilizers. It would be useful to obtain some conclusive experimental evidence on this aspect relating to determinable quality factors like nutritive value, storage properties, colour, mineral and organic composition, etc.

It is often argued that even if dung is not at all used as fuel and all available organic wastes with manurial value are carefully conserved and returned to the land, the supplies of plant nutrients cannot be adequate to improve and maintain optimum soil fertility. This may be true but in view of the various advantages of organic manures, it is not at all convincing that the improvement that can be brought about by the development of local manurial resources is not worth the trouble. The manurial policy of the farmers should, therefore, be to take every possible step to ensure the return to the land of all organic waste materials and supplementing such materials with chemical fertilizers adequately.

Bibliography

Chandra S., and Ali, M. 1986. Recent achievements in pulses production. Tech Bull. No. 1. Directorate of Pulses Research, Kanpur, pp. 44.

Dargan, K.S. *et al.* 1975. In alkali soils–green manuring for more paddy. Indian Fmg., 25(3): 13–15.

FAI. 1990. Fertilizer Statistics for 1989–90. New Delhi. P. 1–199.

FAO. 1988. Bio- and organic fertiliser: Prospects and progress in Asia. FAO/RAPA publication 1988/10, Bangkok, pp. 78.

Gaur, A.C. *et al.* 1984. Organic manures. ICAR, New Delhi, pp. 159.

Goel, B.B.P.S. *et al.* 1973. Availability and disposal of dung in India. Indian J. Animal Sci., 43: 671–676.

Grewal, J.S., Trehan, S.P. 1990. Micronutrients for potato. Tech. Bull. 23. CPRI, Shimla, pp. 24.

Gupta, R. and Sandhar, N.S. 1990. Biogas wet slurry can be an effective fertiliser. Indian Fmg., 40: 13–14.

Hazra, C.R. 1992. Fertilizer use in forages, pastures and grasslands. In: Non-traditional Sector for Fertilizer Use (Ed. HLS Tandon). FDCO, New Delhi (in press).

Hegde, B.R. *et al.* 1988. In Natn. Symp. Recent Advances in Dryland Agriculture. CPIDA, Hyderabad.

IARI. 1978. Algal technology for rice. Res. Bull. No. 9, New Delhi, pp. 9.

IARI. 1980. Soil fertility maps or India. Res., Bull. No. 22, New Delhi, pp. 16+maps.

ICAR. 1987. Biological N-fixation in groundnut. Technologies for Better Crops No. 37. ICAR, New Delhi.

Juwarkar, A.S. *et al.* 1991. Biological and industrial wastes as a source of plant nutrients. In Fertilisers, Organic Manures, Recyclable Wastes and Biofertilizer (ed. HLS Tandon). FDCO, New Delhi.

Katyal, J.C. 1985. Research achievements of All-India Coordinated Scheme on Micronutrients in Soils and Plants. Fert. News, 30(4): 67–81.

Leelavalhi, C.R. *et al.* 1986. Revised yardsticks of additional production of rice due to improved measures. IASIU, New Delhi, pp. 108.

Mahapatra, B.S. and Shamra, G.L. 1988. Effect of *azospirillum*, cyanobacteria and *azolla* biofertilizer on productivity or lowland rice. Indian J. Agron, 33: 368–371.

Mahapatra, I.C. *et al.* 1981. Fertilizer use in rice-rice cropping systems. Fert. News, 26(9): 3–15.

Meelu, O.P. and Morris, R.A. Green manure management in rice-based cropping systems. In: Proc. Symp. Sustainable Agriculture–the role of green manure crops in rice farming systems. IRRI, Los Banos, Philippines, p. 208-222.

Mishra, M.M. and Tauro, P.M. 1983 Use of organic vis-à-vis mineral fertiliser. In: Recycling Organic Matter in Asia for Fertiliser. APO, Tokyo, p. 29–39.

Nambiar, K.K.M. and Abrol, I.P. 1989. Long-term fertilizer experiments in India: An overview. Fert. News, 34(4): 11–20, 26.

Nambiar KKM, Ghosh, A.B. 1984. Highlights or research or a long-term fertiliser experiment in India (1971-82). IARI, New Delhi, pp. 100.

Palaniappan, S.P. 1991. Green manures and their management. In: Fertilizers, Organic Manure, Recyclable Wastes and Biofertilizers (ed. HLS Tandon). FDCO, New Delhi.

Patil, A.J. and Kulkarni, S.D. 1988. Effect or organic recycling of Subabul (*L. leucocephala*) on yield, nutrient uptake and moisture utilisation by Rabi sorghum. In: Natn. Symp. Recent Advances in Dryland Agriculture. CRIDA, Hyderabad, Abstracts, p. 30–31.

Pillai, K.G.K. *et al.* 1980. Biofertilizers in rice culture-problems and prospects for large scale adoption. Fert. News. 25(12): 40–45.

Pillai, K.G.K. *et al.* 1985. Crop responses to fertilizers and fertilizer use efficiency in different soil and agro-climatic regions. Bull. No. 2. AICARP, Bangalore, pp. 50.

Prihar, S.S. 1986. Fertilizer and water use efficiency through the aid of mulching. Proc. FAI-NR Seminar Fertilizer Use Efficiency Through Agronomic Management, p. 31–44.

Rai, P. *et al.* 1979. Effect of various sources and levels of N on the botanical composition, dry matter production and forage quality of Sehima-Heteropogon grassland. Forage Res. 5: 37–48.

Randhawa, N.S. *et al.* 1985. Current status of yardstick of crop response to fertilizer. Proc. FAI Group Discussion on Measures to Increase Crop Response to Fertiliser, p. 25–46.

Randhawa, N.S. and Tandon, H.L.S. 1982. Advances in soil fertility and fertilizer use research in India. Fert. News, 27(2): 11–26.

Roy, R.N. and Braun, H. 1987. Development of integrated plant nutrition systems. Proc. FAI Seminar Fertilizer Industry Challenges and Strategies, New Delhi, S-IV 12/1–12.

Rupela, O.P. and Saxena, M.C. 1987. Nodulation and nitrogen fixation in chickpea. In: The Chickpea (Eds.) M.C. Saxena and K.B. Singh. CAB International, UK, p. 191–206.

Sarkar, A.N. 1990. Using biogas slurry with chemical fertilisers is beneficial. Indian Fmg, 40(4): 23–27.

Singh, N.T. 1983. Green manuring in cropping systems in recycling organic matter in Asia for Fertiliser. APO. Tokyo, 11–28.

Singh, P.K. 1982. *Azolla* as an organic nitrogen fertilizer for medium and lowland rice. In: Review of Soil Research in India. Part I. ISSS, New Delhi, p. 236–242.

Singh, P.K. 1991. Biofertilizers for flooded rice ecosystem. In: Fertilisers, Organic Manures, Recyclable Wastes and Biofertilizers. (ed.) H.L.S. Tandon. FDCO, New Delhi.

Swaminathan, M.S. 1982. Biotechnology research and third world agriculture. Science, 218: 967–972.

Tandon, H.L.S. 1980. Soil fertility and fertiliser use research on wheat in India: A review. Fert. News, 25(10): 45–78.

Tandon, H.L.S. 1983. Fertilizer use efficiency systems: components and their status. Fert. News, 28(12): 46–56.

Tandon, H.L.S. 1987. Phosphorus Research and Agricultural Production in India. FDCO, New Delhi, p. 160+viii.

Tandon, H.L.S. 1989. Long-term fertiliser experiments in India: Lessons and practical expectations. Fert. News, 34(4): 21–26.

Tandon, H.L.S. 1991a. Sulphur Research and Agricultural Production in India, 3rd edition. TSI, Washington D.C., U.SA., pp. 140+viii.

Tandon, H.L.S. 1991b. Secondary and Micronutrients in Agriculture. FDCO, New Delhi, pp. 122+viii.

Tandon, H.L.S. and Pratap Narayaran. 1990. Fertilizers in Indian Agriculture: Past, present and future (1950–2000). FDCO, New Delhi, pp. 160+viii.

Tandon, H.L.S. and Sekhon, G.S. 1988. Potassium Research and Agricultural Production in India. FDCO, New Delhi, pp. 144+viii.

Tauro, P. and Khurana, A.L. 1986. Problems and prospects of growth, extension and promotion of biofertilizers. Proc. FAI Seminar Growth and Modernisation of Fertilizer Industry. PS1/1/1–4.

TNDA. 1989. Crop Production Guide 1989–90. Tamil Nadu Directorate or Agriculture, Madras, pp. 299.

Verma, L.N. and Bhattacharya, P. 1991. Production, promotion and distribution of biofertilizers. In: Fertilizers, Organic Manures, Recyclable Wastes, and Biofertilizers (ed.) H.L.S. Tandon. FDCO, New Delhi.

Wani, S.P. and Lee, K.K. 1991. Biofertilizers for upland crop production. In: Fertilizers, Organic Manures, Recyclable Wastes and Biofertilizers. (ed.) H.L.S. Tandon. FDCO, New Delhi.

Index

O

P

R